AF167748

Nitric Oxide in Biology and Medicine

Nitric oxide (NO) is one of the most important molecules in biological systems, and there has been a tremendous growth in research on NO and its properties over the last decade. Although one of the simplest biologically-active molecules in nature, NO has found its way into nearly every area of biology and medicine. Nitric oxide plays a critical role in many aspects of human physiology ranging from memory, diabetes, cancer, and drug addiction. Nitric oxide is also the key neurotransmitter mediating erectile function, a major pathophysiological mediator of inflammation and host defense, and the principal endogenous regulator of blood flow and thrombosis. These major discoveries have stimulated intense and extensive research into a widespread array of fields including chemistry, molecular biology, and gene therapy. The newly discovered role of NO in pathophysiology and disease has resulted in the development of new therapeutic strategies.

This series will contain books that review particular areas of NO research and will provide useful summaries of our current state of knowledge into the role of NO in biology and medicine. Researchers and graduate students in physiology, pharmacology, and cell and molecular biology who are interested in the role of NO in living systems will find the books in this series to be the essential guide to the subject.

Series Editors

Jack R. Lancaster
Thomas D. Giles
Louisiana State University Medical Center
New Orleans, LA 70112

Editorial Board

Peter Barnes
London, England

Philip J. Kadowitz
New Orleans, Louisiana

Rudi Busse
Graz, Austria

Larry Keefer
Frederick, Virginia

Kenneth Clark
Cincinnati, Ohio

Betty Sue Masters
University of Texas at Dallas

O.W. Griffith
University of Wisconsin

Dennis B. McNamara
New Orleans, Louisiana

J.R. Hibbs, Jr.
Salt Lake City, Utah

S. Saito
Osako, Japan

Louis J. Ignarro
Los Angeles, California

Forthcoming Books in the Series

Nitric Oxide and Free Radicals in Peripheral Neurotransmission
Stanley Kalsner, Editor

Nitric Oxide and the Regulation of the Peripheral Circulation

Philip J. Kadowitz
Dennis B. McNamara
Editors

Springer Science+Business Media, LLC

Philip J. Kadowitz, Ph.D.
Dennis B. McNamara, Ph.D.
Department of Pharmacology
Tulane University Medical Center
New Orleans, LA 70112-2699
USA

Library of Congress Cataloging-in-Publication Data
Nitric oxide and the regulation of the peripheral circulation /
    editors, P.J. Kadowitz, D.B. McNamara.
        p.   cm.—(Nitric oxide in biology and medicine)
    Includes bibliographical references and index.
    ISBN 978-0-8176-4046-0   ISBN 978-1-4612-1326-0 (eBook)
    DOI 10.1007/978-1-4612-1326-0
    1. Peripheral circulation—Regulation.   2. Nitric oxide—
Physiological effect.   I. Kadowitz, Philip J., 1941–
II. McNamara, D.B.   III. Series.
QP108.5.P45N56   2000
612.1'3—dc21   99-27053
                                                        CIP

Printed on acid-free paper.
© 2000 Springer Science+Business Media New York   ®
Originally published by Birkhäuser Boston in 2000

ISBN 978-0-8176-4046-0   SPIN 19901571

Typeset by Impressions Book and Journal Services, Inc., Madison, WI.

9 8 7 6 5 4 3 2 1

Contents

CONTRIBUTORS vii

1. Nitric Oxide in the Regulation of Blood Flow: A Historical Overview   1
  Louis J. Ignarro

PART I: SELECTED MECHANISMS OF ACTION 13

2. Regulation of Intracellular Ca^{2+} by Cyclic GMP-Dependent Protein
  Kinase in Vascular Smooth Muscle Cells 15
  Padmini Komalavilas and Thomas M. Lincoln
3. Oxidants and Vascular Nitric Oxide Signaling 33
  *Michael S. Wolin, Sachin A. Gupte, Takafumi Iesaki, and
  Kamal M. Mohazzab-H.*
4. Role of Na^+-K^+-ATPase in Nitric Oxide-Induced Relaxation 49
  Sandeep Gupta
5. Purinergic Receptors, Nitric Oxide, and Regional Blood Flow 65
  Vera Ralevic and Geoffrey Burnstock
6. Nitric Oxide Derived from Perivascular Nerves and Endothelium ... 85
  Tomio Okamura and Noboru Toda
7. Nitric Oxide and Hypertension 99
  Roberto Zatz and Christine Baylis

PART II: VASCULAR BED 111

A. CEREBRAL 111
8. Nitric Oxide in Control of the Cerebral Circulation 113
  Charles W. Leffler

B. PULMONARY 129
9. Intravenous Anesthetics in the Pulmonary Circulation 131
  Alan D. Kaye, Bobby D. Nossaman, and Philip J. Kadowitz
10. Inhaled Nitric Oxide Therapy for Acute Respiratory Failure 148
  William E. Hurford, Wolfgang Steudel, and Warren M. Zapol

11. Clinical Applications of Inhaled Nitric Oxide 166
    Elie Haddad, Lesley J. Millatt, and Roger A. Johns
12. Nitric Oxide and the Perinatal Pulmonary Circulation 185
    Sidney Cassin
13. Nitric Oxide and the Pulmonary Circulation in the Adult 197
    Bobby D. Nossaman, Alan D. Kaye, and Philip J. Kadowitz

C. SKELETAL 225
14. Role of Nitric Oxide in the Skeletal Muscle Vascular Bed 227
    Hunter C. Champion and Philip J. Kadowitz

D. GI/GU 241
15. Nitric Oxide and the Hepatic Circulation 243
    W. Wayne Lautt and M. Paula Macedo
16. Nitric Oxide and the Gastrointestinal Circulation 259
    Philippe Bauer, Zsuzsanna Rozsa, and D. Neil Granger
17. Uterine Effects of Nitric Oxide 270
    Kenneth E. Clark and Leslie Myatt
18. NO Effect on Penile Blood Flow and Lower Genitourinary
    Tract Function 286
    *Trinity J. Bivalacqua, Hunter C. Champion, Philip J. Kadowitz,
    and Wayne J.G. Hellstrom*

E. RENAL 303
19. Role of Nitric Oxide in the Regulation of Renal Blood Flow 305
    Dewan S.A. Majid and L. Gabriel Navar
20. Role of Nitric Oxide in the Regulation of Renal Function in
    Conscious Animals 320
    R. Davis Manning, Jr., Lufei Hu, and Dunyong Y. Tan
21. Modulation of Renal Microvascular Responsiveness by
    Nitric Oxide 337
    Edward W. Inscho

INDEX 347

Contributors

Philippe Bauer
Research Fellow
Department of Molecular and Cellular
  Physiology
Louisiana State University Medical
  Center
1501 Kings Highway
Shreveport, LA 71130-3932
USA

Christine Baylis
Department of Physiology
West Virginia University
Morgantown, WV 26506
USA

Trinity J. Bivalacqua
Department of Urology
Tulane University School of Medicine
1430 Tulane Ave.
New Orleans, LA 70112
USA

Geoffrey Burnstock
Autonomic Neuroscience Institute
Royal Free Hospital School of Medicine
Rowland Hill Street
London NW3 2PF
UK

Sidney Cassin
Departments of Physiology and
  Pediatrics
College of Medicine
University of Florida
Gainesville, FL 32610
USA

Hunter C. Champion
Department of Pharmacology
Tulane University School of
  Medicine
1430 Tulane Ave.
New Orleans, LA 70112
USA

Kenneth E. Clark
Division of Maternal-Fetal Medicine
Department of Obstetrics and
  Gynecology
University of Cincinnati College of
  Medicine
PO Box 670526
Cincinnati, OH 45267
USA

D. Neil Granger
Boyd Professor & Head
Department of Molecular and Cellular
  Physiology
Louisiana State University Medical
  Center
1501 Kings Highway
Shreveport, LA 71130-3932
USA

Sandeep Gupta
Principal Scientist
Department of Pharmacology and
  Toxicology
Forest Laboratories, Inc.
909 Third Avenue
New York, NY 10022-4731
USA

Sachin A. Gupte
Department of Physiology
New York Medical College
Basic Science Building
Valhalla, NY 10595
USA

Elie Haddad
University of Virginia
Department of Anesthesiology
Charlottesville, VA 22906
USA

Wayne J. G. Hellstrom
Department of Urology
Tulane University School of Medicine
1430 Tulane Ave.
New Orleans, LA 70112
USA

Lufei Hu
Department of Physiology and
  Biophysics
University of Mississippi Medical
  Center
2500 North State St.
Jackson, MS 39216
USA

William E. Hurford
Department of Anesthesia and Critical
  Care
Massachusetts General Hospital
55 Fruit Street
Boston, MA 02114
USA

Takafumi Iesaki
Department of Physiology
New York Medical College
Basic Science Building
Valhalla, NY 10595
USA

Louis J. Ignarro
Department of Molecular and Medical
  Pharmacology
UCLA School of Medicine
Center for Health Sciences
23-120CHS, Box 951735
Los Angeles, CA 90095
USA

Edward W. Inscho
Department of Physiology
Tulane University School of Medicine
1430 Tulane Avenue
New Orleans, LA 70112
USA

Roger A. Johns
Professor and Chairman
Anesthesiology and Critical Care
  Medicine
Johns Hopkins University
School of Medicine
600 North Wolfe St.
Baltimore, MD 21287-4965
USA

Alan D. Kaye
Departments of Anesthesiology and
  Pharmacology
Tulane University Medical School
1430 Tulane Avenue
New Orleans, LA 70112-2699
USA

Philip J. Kadowitz
Departments of Anesthesiology and
  Pharmacology
Tulane University Medical School
1430 Tulane Ave.

New Orleans, LA 70112-2699
USA

Padmini Komalavilas
Department of Pathology
Division of Molecular and Cellular
  Pathology
The University of Alabama at
  Birmingham
1670 University Blvd.
Birmingham, AL 35294-0019
USA

W. Wayne Lautt
Department of Pharmacology and
  Therapeutics
Faculty of Medicine
University of Manitoba
Winnipeg, Manitoba R3E OW3
Canada

Charles W. Leffler
Departments of Physiology and
  Pediatrics
Laboratory for Research in Neonatal
  Physiology
University of Tennessee
Memphis, TN 38163
USA

Thomas M. Lincoln
The Department of Pathology
Division of Molecular and Cellular
  Pathology
The University of Alabama at
  Birmingham
1670 University Blvd.
Birmingham, AL 35294-0019
USA

M. Paula Macedo
Institute of Health Sciences
Qunita da Granja-Trav. da Granja
Monte da Caparica 2825
Portugal

Dewan S. A. Majid
Department of Physiology SL39
Tulane University School of Medicine
1430 Tulane Avenue
New Orleans, LA 70112
USA

R. Davis Manning, Jr.
Department of Physiology and
  Biophysics
University of Mississippi Medical
  Center
2500 North State Street
Jackson, MS 39216
USA

Lesley J. Millatt
University of Virginia
Department of Anesthesiology
Charlottesville, VA 22906
USA

Kamal M. Mohazzab-H.
Department of Physiology
New York Medical College
Basic Science Building
Valhalla, NY 10595
USA

Leslie Myatt
Division of Maternal-Fetal Medicine
Department of Obstetrics and
  Gynecology
University of Cincinnati College of
  Medicine
PO Box 670526
Cincinnati, OH 45267
USA

L. Gabriel Navar
Department of Physiology SL39
Tulane University School of Medicine
1430 Tulane Avenue
New Orleans, LA 70112
USA

Bobby D. Nossaman
Department of Anesthesiology
Tulane University Medical School
1430 Tulane Avenue
New Orleans, LA 70112-2699
USA

Tomio Okamura
Department of Pharmacology
Shiga University of Medical Science
Seta, Ohtsu 520-2192
Japan

Vera Ralevic
School of Biomedical Sciences
Queen's Medical Centre
University of Nottingham
Nottingham NG7 2UH
and
Autonomic Neuroscience Institute
Royal Free Hospital School of
  Medicine
Rowland Hill Street
London NW3 3PF
UK

Zsuzsanna Rozsa
Visiting Professor
Department of Molecular and Cellular
  Physiology
Louisiana State University Medical
  Center
1501 Kings Highway
Shreveport, LA 71130-3932
USA

Wolfgang Steudel
Department of Anesthesia
  and Critical Care
Massachusetts General Hospital
55 Fruit Street
Boston, MA 02114
USA

Dunyong Y. Tan
Department of Physiology and
  Biophysics
University of Mississippi Medical
  Center
2500 North State St.
Jackson, MS 39216
USA

Noboru Toda
Department of Pharmacology
Shiga University of Medical Science
Seta, Ohtsu 520-2192
and
Nippon Shinyaku Co., Ltd.
14, Kisshoin Nishinosho-Monguchi-Cho
Minami-ku, Kyoto 601-8550
Japan

Michael S. Wolin
Professor
Department of Physiology
New York Medical College
Basic Science Building
Valhalla, NY 10595
USA

Warren M. Zapol
Department of Anesthesia and Critical
  Care
Massachusetts General Hospital
55 Fruit Street
Boston, MA 02114
USA

Roberto Zatz
Renal Division
Department of Clinical Medicine
Faculty of Medicine
Av. Dr. Arnaldo, 455, 3-s/3342
University of São Paulo
São Paulo, 01246-903
Brazil

1
Nitric Oxide in the Regulation of Blood Flow: A Historical Overview

Louis J. Ignarro

Introduction

Nearly 20 years have gone by since we discovered the vascular smooth muscle relaxant properties of nitric oxide (NO). The original observation was made by delivering a gaseous mixture of NO in nitrogen by means of a gas-tight microliter syringe into a tissue bath containing isolated precontracted strips of bovine coronary artery (Gruetter et al. 1979). The pharmacological response was a marked but transient relaxation that was blocked by the prior addition of hemoproteins or methylene blue. Moreover, NO activated soluble guanylate cyclase isolated from bovine coronary artery. The closely similar pharmacological profile of action of NO and nitroglycerin led us to suspect that nitroglycerin causes vascular smooth muscle relaxation by mechanisms involving NO, perhaps by generating or donating NO. Although the terms "NO donor drugs" or "nitrovasodilators" were not used in 1979, we were accumulating solid experimental evidence that nitroglycerin, other organic nitrate esters, some organic nitrite esters, and nitroprusside all caused vasodilation by acting as NO donor agents on contact with tissues in aqueous solution. Struck by the extraordinarily high potency of nitroglycerin as a vascular smooth muscle relaxant both in vivo and in vitro, we suspected, as pharmacologists, that tissue receptors for nitroglycerin probably exist, because there must be an endogenous nitroglycerin or similar NO donor or NO itself in mammalian tissues. Of course, we did not appreciate at this time that the vascular endothelium generated and delivered the NO to the underlying smooth muscle cells. In fact, our original observations on the vasorelaxant properties of NO (Gruetter et al. 1979) were published 1 year before the discovery of endothelium-dependent vasorelaxation and endothelium-derived relaxing factor (Furchgott and Zawadzki 1980). The answer to the puzzle did not become known for an additional seven years.

While conducting experiments dealing with the vasorelaxant properties of NO, we discovered that NO also potently inhibits platelet aggregation (Mellion et al. 1981). Indeed, the mechanism by which certain nitrovasodilators inhibit platelet aggregation was found to be identical to the mechanism by which these agents cause vascular smooth muscle relaxation, namely, through the action of NO (Mellion et al.

1983). A series of S-nitrosothiols were synthesized and found to be potent inhibitors of platelet aggregation and NO-activated platelet-derived soluble guanylate cyclase.

The objective of this chapter is not to review all the evidence for the role of endogenous NO in the control of blood flow, but rather to highlight the early research that led to our current understanding of the physiological and pathophysiological roles of NO in regulating vascular smooth muscle tone and blood flow.

Mechanism of Action of Nitroglycerin

Starting in 1979, efforts in this laboratory were focused on elucidating the mechanism of vasodilator action of nitroglycerin and other nitrovasodilators. Evidence accumulated that these NO-containing chemical agents either spontaneously released NO in aqueous solution or reacted with tissue thiols to generate chemically unstable intermediates, S-nitrosothiols, that subsequently decomposed with the liberation of NO (Ohlstein et al. 1979; Gruetter et al. 1980a,b, 1981; Ignarro and Gruetter 1980; Ignarro et al. 1980a,b, 1981). Based on these observations, the hypothesis was formulated that lipophilic vasodilators such as nitroglycerin, other organic nitrate esters, and organic nitrite esters (isoamyl nitrite) permeate vascular smooth muscle cells and react with tissue thiols to form S-nitrosothiols, which then liberate NO, the common active vasodilator species (Ignarro et al. 1981). Several years earlier, NO was suspected to be the common species responsible for activation of soluble guanylate cyclase by nitroglycerin and related agents (Katsuki et al. 1977).

The discovery that S-nitrosothiols were intermediates that served as NO donor agents in expressing the vasorelaxant action of many nitrovasodilators originated from experiments on the activation of soluble guanylate cyclase (Ohlstein et al. 1979; Gruetter et al. 1980a; Ignarro and Gruetter 1980; Ignarro et al. 1980a,b, 1981). Thiols such as cysteine and glutathione markedly enhanced the activation of soluble guanylate cyclase by nitrite and isoamyl nitrite. Cysteine was required for enzyme activation by nitroglycerin and other organic nitrate esters. Thiols were found to decrease the chemical stability of nitrites and nitrosoguanidines by liberating NO gas. The liberation of NO from these nitrovasodilators involved the formation of intermediate S-nitrosothiols, which turned out to be excellent NO donor agents. Like the direct vasorelaxant effect of NO, vasorelaxation elicited by S-nitrosothiols was accompanied by the accumulation of tissue cyclic GMP, and both relaxation and cyclic GMP formation were inhibited by hemoglobin and methylene blue. In vivo experiments in anesthetized cats revealed that the profile of hemodynamic effects of the S-nitrosothiols was virtually identical to that of nitroglycerin and nitroprusside (Ignarro et al. 1981). On the basis of these observations, we proposed that nitrovasodilators elicit vasorelaxation by undergoing metabolic conversion to NO in vascular smooth muscle cells. Accordingly, nitrovasodilators act as parent drugs that must be metabolized to the active species, NO.

The elucidation of the mechanism of action of nitroglycerin led to a better understanding of the mechanism by which tolerance can develop to the vasodilator action of organic nitrate esters. An earlier hypothesis stated that tissue sulfhydryl groups (-SH) were required for expression of the vasodilator action of nitroglycerin and other organic nitrate esters (Needleman and Johnson 1973; Needleman et al. 1973). The view was that repeated administration of relatively large doses of nitroglycerin led to the depletion or oxidation of tissue thiols, thereby leading to the gradual diminution of the action of nitroglycerin. Our studies on the mechanism of action of nitroglycerin are consistent with this earlier hypothesis on tolerance. That is, tissue thiols are required for the vasorelaxation effect of nitroglycerin, because these thiols are required for activation of soluble guanylate cyclase by nitroglycerin. Since the vasorelaxant effect of nitroglycerin, like that of NO, is cyclic GMP-dependent, thiols are necessary for nitroglycerin to relax vascular smooth muscle. The likely explanation for the mechanism of development of tolerance is that tissue thiols are required for chemical reaction with nitroglycerin to liberate NO from the intermediate S-nitrosothiols that are formed. Administration of sulfhydryl compounds often reverses or prevents tolerance to nitroglycerin in laboratory animals and in patients, just as sulfhydryl compounds facilitate the activation of guanylate cyclase by nitroglycerin (Axelsson et al. 1982; Keith et al. 1982; Axelsson and Andersson 1983; Horowitz et al. 1983; Winniford et al. 1986; May et al. 1987; Packer et al. 1987).

Mechanism of Action of NO

Activation of soluble guanylate cyclase and increased synthesis of cyclic GMP in tissues were demonstrated before the vasorelaxant effect of NO was appreciated (DeRubertis and Craven 1976). The mechanism by which NO activates soluble guanylate cyclase was first suggested by Craven and DeRubertis (Ignarro and Kadowitz 1985). The hypothesis was that soluble guanylate cyclase contained heme iron, which was required for the binding and interaction with NO to cause enzyme activation. In equipment using purified guanylate cyclase, heme was found to be bound to the enzyme as a prosthetic group (Gerzer et al. 1981, 1982; Ignarro et al. 1982a,b, 1984a,b; Ohlstein et al. 1982; Wolin et al. 1982), and studies from this laboratory revealed the precise mechanism by which NO activates guanylate cyclase (Ignarro et al. 1982a; Wolin et al. 1982). In experiments designed to ascertain whether the iron in heme was required for binding of the protoporphyrin ring structure to guanylate cyclase, enzyme that was rendered free of bound heme was reacted with protoporphyrin IX (heme without iron), passed through a gel filtration column to remove free unbound porphyrin, and examined spectrophotometrically. Two important observations were made. Guanylate cyclase readily bound stoichiometric quantities of protoporphyrin IX, and the porphyrin caused maximal enzyme activation. Heme-free guanylate cyclase could not be activated by NO but was maximally activated by nanomolar concentrations of protoporphyrin IX. Heme-containing or heme-reconstituted guanylate cyclase was activated by both NO and protoporphyrin IX. In the latter case, the porphyrin displaced heme from the common binding site on guanylate cyclase to cause enzyme activation.

On the basis of information available for hemoglobin and myoglobin, we formulated the hypothesis that NO binds to the heme iron to form the nitrosylheme adduct of guanylate cyclase. It was envisioned that heme binds to the enzyme in the same manner as it does to the protein hemoglobin, namely, as a five-coordinate complex with the fifth bond being the axial ligand between heme iron and histidine in the enzyme protein. On interaction of the heme iron with NO, the resulting NO–heme complex was believed to remain as a five-coordinate complex, which means that the axial ligand must undergo cleavage, leading to the projection of heme iron away from the enzyme protein and out of the plane of the porphyrin ring configuration (Ignarro et al. 1984c). This conformational change at the porphyrin binding site of guanylate cyclase was thought to modify the nearby catalytic site as well, perhaps by exposing the catalytic site to the surface where GTP substrate and magnesium must bind. This hypothesis would explain why NO causes a 100-fold increase in the V_{max} (maximal velocity) as well as a threefold decrease in the Michaelis constant (K_m) for GTP (Wolin et al. 1982).

Soluble guanylate cyclase was also reported to contain bound copper, although the reason for this was unknown (Gerzer et al. 1981). Based on our studies with S-nitrosothiols and recent reports that copper can potently liberate NO from S-nitrosothiols (Dicks and Williams 1996; Gorren et al. 1996; Williams 1996), we can formulate a hypothesis on the role of copper in guanylate cyclase. S-nitrosothiols are well known to activate soluble guanylate cyclase by heme-dependent mechanisms involving the action of NO (Wolin et al. 1982; Ignarro and Kadowitz 1985). The activation of guanylate cyclase by S-nitrosothiols, like that of NO, is extremely rapid and may not be readily explainable on the basis of the slow release of NO from most S-nitrosothiols. Therefore, one role for copper may be to facilitate the release of NO from S-nitrosothiols, thereby facilitating the activation of guanylate cyclase. This may be an important mechanism of activation of guanylate cyclase by endogenous S-nitrosothiols, ranging from S-nitroso-amino acids to S-nitroso-proteins.

Endogenous NO

The realization that mammalian cells do indeed possess endogenous nitroglycerin or NO came a little more than a decade ago when the elusive endothelium-derived relaxing factor (EDRF) was identified as NO (Ignarro et al. 1987a,b; Palmer et al. 1987). Before that, experimental evidence was mounting in several laboratories that EDRF possessed pharmacological, biochemical, and chemical properties that were similar to those of NO. For example, studies from this laboratory published in 1986 revealed that EDRF from artery and vein could activate soluble guanylate cyclase (Ignarro et al. 1986a). Additional experiments revealed that the mechanism of activation of guanylate cyclase by EDRF was closely similar or identical to that for NO, thereby allowing us to first propose that EDRF is NO or a closely related nitroso compound (Ignarro et al. 1986b). These observations explained why hemoglobin and myoglobin antagonize endothelium-dependent relaxation as well as the direct relaxant effect of EDRF. Subsequent studies showed that the activation of soluble guanylate cyclase by EDRF, like its activation by NO, was heme-dependent (Ignarro et al. 1987b). Other observations were also consistent with the view that EDRF was NO. For example, the short half-life of NO could be prolonged by the addition of superoxide dismutase to tissue baths or bioassay cascade superfusion systems, whereas superoxide anion-generating agents decreased the half-life of NO (Gryglewski et al. 1986; Rubanyi and Vanhoutte 1986). We now understand the mechanism of these effects, namely, the rapid reaction between NO and superoxide anion to generate peroxynitrite (Beckman et al. 1990), which is less potent than NO as a vasorelaxant by orders of magnitude. The finding in 1986 that EDRF inhibits platelet aggregation (Azuma et al. 1986) was also consistent with the developing hypothesis that EDRF could be NO. Although these earlier studies had suggested that EDRF might be NO or a closely related nitroso species, the definitive studies on the chemical and biochemical characterization and identification of EDRF as NO came in 1987 (Ignarro et al. 1987a,b; Palmer et al. 1987).

After the discovery that EDRF is NO, many laboratories jumped into this field of cardiovascular research to study the physiological and pathophysiological roles of endothelium-dependent vasodilation and endogenous NO and cyclic GMP in the regulation of systemic blood pressure, organ blood flow, hemostasis, and cell proliferation. Since most of the pharmacological properties of nitroglycerin and other nitrovasodilators in laboratory animals and in humans were already well appreciated, a systematic approach could be taken to better understand or appreciate the actions of endogenous NO.

Regulation of Vascular Smooth Muscle Tone by NO

We now appreciate that EDRF or endothelium-derived NO is continuously generated from vascular endothelial cells in the absence of added endothelium-dependent vasodilators. One early clue for continuous or basal generation of NO came from studies showing that vascular tissue cyclic GMP levels were higher in endothelium-intact than in endothelium-denuded vascular preparations (Ignarro 1989). Studies with hemoglobin and cyclic GMP phosphodiesterase inhibitors, both of which cause endothelium-dependent vasoconstriction, provided additional indirect evidence for basal NO generation. Direct evidence for basal generation of NO came from bioassay cascade experiments that revealed the continuous formation and release of NO in effluents collected from perfused or superfused vascular preparations (Ignarro 1989). The basal formation of NO varied considerably from one vessel type to another, and among parts of the same vessel segment with different diameters. For example, in bovine pulmonary arteries and veins, the basal formation of NO was high in vessels of smaller diameter (Ignarro et al. 1987c). The endothelium-intact rings of smaller diameter contained higher resting levels of smooth muscle cyclic GMP and were considerably less likely than rings of larger diameter to maintain a steady level of tone when mounted and equilibrated under optimal resting tensions. Normal contractile responses to phenylephrine were restored by endothelium denudation or by addition of hemoglobin or methylene blue to tissue baths. In contrast, when cyclic GMP phosphodiesterase inhibitors were added to endothelium-denuded rings, which otherwise maintain constant tone, the rings rapidly lost tone and in this way resembled endothelium-intact rings. On the basis of these original observations, we proposed the hypothesis that the basal or continuous formation of arterial and venous NO may be important for the continuous modulation not only of vascular smooth muscle tone but also of circulating platelet adhesion and aggregation (Ignarro et al. 1987c). Later studies revealed that interference with the continuous production of endothelium-derived NO in animals by administration of NO synthase (NOS) inhibitors caused a prompt and sustained increase in systemic blood pressure (Aisaka et al. 1989; Rees et al. 1989; Vargas et al. 1991).

Although the intrinsic stimulus for basal generation of NO was not appreciated in the early 1980s, later studies revealed that the shear stress or tangential shear force generated by flowing blood against the endothelial cell surface triggered the generation of NO in the endothelial cells. Several important early studies contributed to the development of this concept that flow-dependent vasodilation is endothelium-dependent (Holtz et al. 1983, 1984; Hintze and Vatner 1984; Tesfamariam et al. 1985; Hull et al. 1986; Kaiser et al. 1986; Rubanyi et al. 1986; Bevan and Joyce 1988; Buga et al. 1991). It now appears that shear forces trigger the opening of calcium channels on endothelial cells, thereby leading to the calcium-dependent activation of endothelial NOS and increased local production of NO (Ayajiki et al. 1996; Fleming et al. 1997). However, both calcium-dependent and calcium-independent activation of endothelial NOS can occur and lead to NO-mediated vasorelaxation in response to shear stress (Ayajiki et al. 1996; Fleming et al. 1997). In the case of calcium-dependent NOS activation, the increase in intracellular calcium concentration may be the result of tyrosine phosphorylation and activation of phospholipase C as well as protein phosphatases. Calcium-independent activation of NOS may involve tyrosine phosphorylation of endothelial NOS or the action of another regulatory protein.

The chemical and biological properties of NO endow this potent endogenous mediator with the capacity to act as a local modulator of blood flow and hemostasis. The vascular cell origin of NO is ideal for the local and immediate delivery of this lipophilic and labile vasodilator directly to the underlying smooth muscle as well as to the endothelial cell surface for interaction with nearby circulating platelets. The small size and lipophilic nature of NO are conducive to the rapid diffusion of NO through cell membranes to reach its target cells. The chemically labile property of NO allows for a truly local action, as does the high binding affinity of erythrocytes for NO. All of these properties of NO endow this biological mediator with the unique capacity to engage in cell-to-cell communication. In this manner, NO can recruit the functions of various cell types to elicit a concerted physiological or pathophysiological response, such as improved local blood flow to an injured tissue. Damage to the endothelial cell surface can interfere with the normal functions of NO in regulating local blood flow, as occurs in numerous pathophysiological disorders ranging from atherosclerosis to angioplasty techniques used to surgically treat patients with atherosclerosis. There may be several physiological control mechanisms that diminish NO production by inhibiting NO synthase or by down-regulating the level of NOS protein. We first demonstrated that NO itself can act as a negative feedback modulator of NOS catalytic activity (Rogers and Ignarro 1992). This action could also be demonstrated in vitro using isolated vascular preparations or superfused vessels in a bioassay cascade (Buga et al. 1993) as well as in vivo in anesthetized rabbits (Cohen et al. 1996). The mechanism of this negative feedback effect was elucidated and determined to be an interaction between NO and the heme iron in NOS (Abu-Soud et al. 1995).

Physiological Role of NO in Erectile Function

In 1990 we first reported that the relaxation of isolated rabbit corpus cavernosum smooth muscle by electrical stimulation was accompanied by the production of NO and cyclic GMP and was prevented by treatment of tissues with NOS inhibitors, hemoglobin, and methylene blue but not by indomethacin (Ignarro et al. 1990). Relaxation of corpus cavernosum was mediated by nonadrenergic, noncholinergic (NANC) neurons and attributed to the generation and release of NO as the primary neurotransmitter. Based on these observations, we proposed the hypothesis that mammalian penile erection is mediated by NO released from NANC neurons and that cyclic GMP serves as the signal transduction mechanism for smooth muscle relaxation. Additional studies from this laboratory found that exactly the same physiological mechanism for penile erection exists in human (Rajfer et al. 1992) and canine (Trigo-Rocha et al. 1993) corpus cavernosum and that electrical stimulation results in calcium-dependent activation of neuronal NOS present in corporal smooth muscle (Ignarro et al. 1990; Bush et al. 1992). These original observations provided a rational basis for investigating the etiology and therapy of impotence. At least one form of impotence may be attributed to a lesion or defect in the arginine–NO–cyclic GMP pathway that results in diminished relaxation of corpus cavernosum smooth muscle in response to stimulation of the NANC nerves.

Summary

In retrospect, basic research in the fields of NO and cyclic GMP during the past two decades appears to have followed a logical course, beginning with the findings that NO and cyclic GMP are vascular smooth muscle relaxants, that nitroglycerin relaxes smooth muscle by metabolism to NO, progressing to the discovery that mammalian cells synthesize NO, and finally the revelation that NO is a neu-rotransmitter mediating vasodilation in specialized vascular beds. A great deal of basic and clinical research on the physiological and pathophysiological roles of NO in cardiovascular function has been conducted since the discovery that EDRF is NO. Many aspects of the regulation of the peripheral circulation by NO are ad-dressed in the chapters that follow in this text.

References

Abu-Soud HM, Wang J, Rousseau DL, Fukuto JM, Ignarro, LJ, & Stuehr DJ. 1995. Neu-ronal nitric oxide synthase self-inactivates by forming a ferrous-nitrosyl complex during aerobic catalysis. *J Biol Chem 270:*22997–23006.

Aisaka K, Gross SS, Griffith OW, & Levi R. 1989. N^G-Methylarginine, an inhibitor of endothelium-derived nitric oxide synthesis, is a potent pressor agent in the guinea pig: does nitric oxide regulate blood pressure in vivo? *Biochem Biophys Res Commun 160:*881–886.

Axelsson KL, Anderson RGG, & Wikberg JES. 1982. Vascular smooth muscle relaxation by nitro compounds: reduced relaxation and cGMP elevation in tolerant vessels and re-versal of tolerance by dithiothreitol. *Acta Pharmacol Toxicol 50:*350–357.

Axelsson KL, & Andersson RG. 1983. Tolerance towards nitroglycerin, induced in vivo, is correlated to a reduced cGMP response and an alteration in cGMP turnover. *Eur J Phar-macol 88:*71–79.

Ayajiki K, Kindermann M, Hecker M, Fleming I, & Busse R. 1996. Intracellular pH and tyrosine phosphorylation but not calcium determine shear stress-induced nitric oxide production in native endothelial cells. *Circ Res 78:*750–758.

Azuma H, Ishikawa M, & Sekizaki S. 1986. Endothelium-dependent inhibition of platelet aggregation. *Br J Pharmacol 88:*411–415.

Beckman JS, Beckman TW, Chen J, Marshall PA, & Freeman BA. 1990. Apparent hy-droxyl radical production by peroxynitrite: implications for endothelial injury from ni-tric oxide and superoxide. *Proc Natl Acad Sci USA 87:*1620–1624.

Bevan JA, & Joyce EH. 1988. Flow-dependent dilation in myographmounted resistance ar-tery segments. *Blood Vessels 25:*101–104.

Buga GM, Gold ME, Fukuto JM, & Ignarro LJ. 1991. Shear stress-induced release of ni-tric oxide from endothelial cells grown on beads. *Hypertension 17:*187–193.

Buga GM, Griscavage JM, Rogers NE, & Ignarro LJ. 1993. Negative feedback regulation of endothelial cell function by nitric oxide. *Circ Res 73:*808–812.

Bush PA, Gonzalez NE, & Ignarro LJ. 1992. Biosynthesis of nitric oxide and citrulline from L-arginine by constitutive nitric oxide synthase present in rabbit corpus caver-nosum. *Biochem Biophys Res Commun 186:*308–314.

Cohen GA, Hobbs AJ, Fitch RM, Zinner MJ, Chaudhuri G, & Ignarro LJ. 1996. Nitric oxide regulates endothelium-dependent vasodilator responses in rabbit hindquarters vascular bed in vivo. *Am J Physiol 271:*H133–H139.

DeRubertis FR, & Craven PA. 1976. Calcium-independent modulation of cyclic GMP and activation of guanylate cyclase by nitrosamines. *Science 193:*897–899.

Dicks AP, & Williams DL. 1996. Generation of nitric oxide from *S*-nitrosothiols using protein-bound Cu^{2+} sources. *Chem Biol 3:*655–659.

Fleming I, Bauersachs J, & Busse R. 1997. Calcium-dependent and calcium-independent activation of the endothelial NO synthase. *J Vasc Res 34:*165–174.

Furchgott RF, & Zawadzki JV. 1980. The obligatory role of endothelial cells in the relaxation of arterial smooth muscle by acetylcholine. *Nature 288:*373–376.

Gerzer R, Hofmann F, & Schultz G. 1981. Purification of a soluble, sodium-nitroprusside-stimulated guanylate cyclase from bovine lung. *Eur J Biochem 116:*479–486.

Gerzer R, Radany EW, & Garbers DL. 1982. The separation of the heme and apoheme forms of soluble guanylate cyclase. *Biochem Biophys Res Commun 108:*678–686.

Gorren AC, Schrammel A, Schmidt K, & Mayer B. 1996. Decomposition of *S*-nitrosoglutathione in the presence of copper ions and glutathione. *Arch Biochem Biophys 330:*219–228.

Gruetter CA, Barry BK, McNamara DB, Gruetter DY, Kadowitz, PJ, & Ignarro LJ. 1979. Relaxation of bovine coronary artery and activation of coronary arterial guanylate cyclase by nitric oxide, nitroprusside and a carcinogenic nitrosoamine. *J Cyclic Nucleotide Protein Phosphor Res 5:*211–224.

Gruetter CA, Barry BK, McNamara DB, Kadowitz PJ, & Ignarro LJ. 1980a. Coronary arterial relaxation and guanylate cyclase activation by cigarette smoke, *N'*-nitrosonornicotine and nitric oxide. *J Pharmacol Exp Ther 214:*9–15.

Gruetter DY, Gruetter CA, Barry BK, Baricos WH, Hyman AL, Kadowitz PJ, & Ignarro LJ. 1980b. Activation of coronary arterial guanylate cyclase by nitric oxide, nitroprusside, and nitrosoguanidine: inhibition by calcium, lanthanum, and other cations, enhancement by thiols. *Biochem Pharmacol 29:*2943–2950.

Gruetter CA, Gruetter DY, Lyon JE, Kadowitz PJ, & Ignarro LJ. 1981. Relationship between cyclic guanosine 3',5'-monophosphate formation and relaxation of coronary arterial smooth muscle by glyceryl trinitrate, nitroprusside, nitrite and nitric oxide: effects of methylene blue and methemoglobin. *J Pharmacol Exp Ther 219:*181–186.

Gryglewski RJ, Palmer RMJ, & Moncada S. 1986. Superoxide anion is involved in the breakdown of endothelium-derived vascular relaxing factor. *Nature 320:*454–456.

Hintze TH, & Vatner SF. 1984. Reactive dilation of large coronary arteries in conscious dogs. *Circ Res 54:*50–57.

Holtz J, Giesler M, & Bassenge E. 1983. Two dilatory mechanisms of antianginal drugs on epicardial coronary arteries in vivo: indirect, flow-dependent, endothelium-mediated dilation and direct smooth muscle relaxation. *Z Kardiol 72:*98–106.

Holtz J, Förstermann U, Pohl U, Giesler M, & Bassenge E. 1984. Flow-dependent, endothelium-mediated dilation of epicardial coronary arteries in conscious dogs: effects of cyclooxygenase inhibition. *J Cardiovasc Pharmacol 6:*1161–1169.

Horowitz JD, Antman EM, Lorell BH, Barry WH, & Smith, TW. 1983. Potentiation of the cardiovascular effects of nitroglycerin by *N*-acetylcysteine. *Circulation 68:*1247–1253.

Hull SS, Kaiser L, Jaffe MD, & Sparks HV. 1986. Endothelium-dependent flow-induced dilation of canine femoral and saphenous arteries. *Blood Vessels 23:*183–198.

Ignarro LJ, & Gruetter CA. 1980. Requirement of thiols for activation of coronary arterial guanylate cyclase by glyceryl trinitrate and sodium nitrite: possible involvement of *S*-nitrosothiols. *Biochim Biophys Acta 631:*221–231.

Ignarro LJ, Edwards JC, Gruetter DY, Barry BK, & Gruetter CA. 1980a. Possible involvement of *S*-nitrosothiols in the activation of guanylate cyclase by nitroso compounds. *FEBS Lett. 110:*275–278.

Ignarro LJ, Barry BK, Gruetter DY, Edwards JC, Ohlstein EH, Gruetter CA, & Baricos WH. 1980b. Guanylate cyclase activation of nitroprusside and nitrosoguanidine is related to formation of *S*-nitrosothiol intermediates. *Biochem Biophys Res Commun 94:*93–100.

Ignarro LJ, Lippton H, Edwards JC, Baricos WH, Hyman AL, Kadowitz PJ, & Gruetter CA. 1981. Mechanism of vascular smooth muscle relaxation by organic nitrates, nitrites, nitroprusside and nitric oxide: evidence for the involvement of *S*-nitrosothiols as active intermediates. *J Pharmacol Exp Ther 218:*739–749.

Ignarro LJ, Wood KS, & Wolin MS. 1982a. Activation of purified soluble guanylate cyclase by protoporphyrin IX. *Proc Natl Acad Sci USA 79:*2870–2873.

Ignarro LJ, Degnan JN, Baricos WH, Kadowitz PJ, & Wolin MS. 1982b. Activation of purified guanylate cyclase by nitric oxide requires heme: comparison of heme-deficient, heme-reconstituted and heme-containing forms of soluble enzyme from bovine lung. *Biochim Biophys Acta 718:*49–59.

Ignarro LJ, Wood KS, Ballot B, & Wolin MS. 1984a. Guanylate cyclase from bovine lung: evidence that enzyme activation by phenylhydrazine is mediated by iron-phenyl hemoprotein complexes. *J Biol Chem 259:*5923–5931.

Ignarro LJ, Ballot B, & Wood KS. 1984b. Regulation of soluble guanylate cyclase activity by porphyrins and metalloporphyrins. *J Biol Chem 259:*6201–6207.

Ignarro LJ, Wood KS, & Wolin MS. 1984c. Regulation of purified soluble guanylate cyclase by porphyrins and metalloporphyrins: a unifying concept. *Adv Cyclic Nucleotide Protein Phosphor Res 17:*267–274.

Ignarro LJ, & Kadowitz PJ. 1985. The pharmacological and physiological role of cyclic GMP in vascular smooth muscle relaxation. *Annu Rev Pharmacol Toxicol 25:*171–191.

Ignarro LJ, Harbison RG, Wood KS, & Kadowitz PJ. 1986a. Activation of purified soluble guanylate cyclase by endothelium-derived relaxing factor from intrapulmonary artery and vein: stimulation by acetylcholine, bradykinin and arachidonic acid. *J. Pharmacol. Exp. Ther. 237:*893–900.

Ignarro LJ, Wood KS, and Byrns RE. 1986b. Pharmacological and biochemical properties of endothelium-derived relaxing factor (EDRF): evidence that EDRF is closely related to nitric oxide (NO) radical (abstract). *Circulation 74:*II–287.

Ignarro LJ, Buga GM, Wood KS, Byrns RE, and Chaudhuri G. 1987a. Endothelium-derived relaxing factor produced and released from artery and vein is nitric oxide. *Proc. Natl. Acad. Sci. USA 84:*9265–9269.

Ignarro LJ, Byrns RE, Buga GM, and Wood KS. 1987b. Endothelium-derived relaxing factor from pulmonary artery and vein possesses pharmacologic and chemical properties identical to those of nitric oxide radical. *Circ. Res. 61:*866–879.

Ignarro LJ, Byrns RE, and Wood KS. 1987c. Endothelium-dependent modulation of cGMP levels and intrinsic smooth muscle tone in isolated bovine intrapulmonary artery and vein. *Circ. Res. 60:*82–92.

Ignarro LJ. 1989. Biological actions and properties of endothelium-derived nitric oxide formed and released from artery and vein. *Circ. Res. 65:*1–21.

Ignarro LJ, Bush PA, Buga GM, Wood KS, Fukuto JM, and Rajfer J. 1990. Nitric oxide and cyclic GMP formation upon electrical field stimulation cause relaxation of corpus cavernosum smooth muscle. *Biochem. Biophys. Res. Commun. 170:*843–850.

Kaiser L, Hull SS Jr, & Sparks HV Jr. 1986. Methylene blue and ETYA block flow-dependent dilation in canine femoral artery. *Am. J. Physiol. 250:*H974–H981.

Katsuki S, Arnold W, Mittal C, and Murad F. 1977. Stimulation of guanylate cyclase by sodium nitroprusside, nitroglycerin and nitric oxide in various tissue preparations and comparison to the effects of sodium azide and hydroxylamine. *J. Cyclic. Nucleotide Res. 3:*23–35.

Keith RA, Burkman AM, Sokoloski TD, and Fertel RH. 1982. Vascular tolerance to nitroglycerin and cyclic GMP generation in rat aortic smooth muscle. *J. Pharmacol. Exp. Ther. 221:*525–531.

May DC, Popma JJ, Black WH, Schaefer S, Lee HR, Levine BD, and Hillis LD. 1987. In vivo induction and reversal of nitroglycerin tolerance in human coronary arteries. *N. Engl. J. Med. 317:*805–809.

Mellion BT, Ignarro LJ, Ohlstein EH, Pontecorvo EG, Hyman AL and Kadowitz PJ. 1981. Evidence for the inhibitory role of guanosine 3',5'-monophosphate in ADP-induced human platelet aggregation in the presence of nitric oxide and related vasodilators. *Blood. 57:*946–955.

Mellion BT, Ignarro LJ, Myers CB, Ohlstein EH, Ballot BA, Hyman AL, and Kadowitz PJ. 1983. Inhibition of human platelet aggregation by *S*-nitrosothiols. Heme-dependent activation of soluble guanylate cyclase and stimulation of cyclic GMP accumulation. *Mol. Pharmacol. 23:*653–664.

Needleman P, and Johnson EM. 1973. Mechanism of tolerance development to organic nitrates. *J. Pharmacol. Exp. Ther. 184:*709–715.

Needleman P, Jakschik B, and Johnson EM. 1973. Sulfhydryl requirement for relaxation of vascular smooth muscle. *J. Pharmacol. Exp. Ther. 187:*324–331.

Ohlstein EH, Barry BK, Gruetter DY, and Ignarro LJ. 1979. Methemoglobin blockade of coronary arterial soluble guanylate cyclase activation by nitroso compounds, and its reversal with dithiothreitol. *FEBS Lett. 102:*316–320.

Ohlstein EH, Wood KS, and Ignarro LJ. 1982. Purification and properties of heme-deficient hepatic soluble guanylate cyclase: effects of heme and other factors on enzyme activation by NO, NO-heme, and protoporphyrin IX, *Arch. Biochem. Biophys. 218:*187–198.

Packer M, Lee WH, Kessler PD, Gottlieb SS, Medina N, and Yushak M. 1987. Prevention and reversal of nitrate tolerance in patients with congestive heart failure. *N. Engl. J. Med. 317:*799–804.

Palmer RM, Ferrige AG, and Moncada S. 1987. Nitric oxide release accounts for the biological activity of endothelium-derived relaxing factor. *Nature 327:*524–526.

Rajfer J, Aronson WJ, Bush PA, Dorey FJ, and Ignarro LJ. 1992. Nitric oxide as a mediator of relaxation of the corpus cavernosum in response to nonadrenergic, noncholinergic neurotransmission. *N. Engl. J. Med. 326:*90–94.

Rees DD, Palmer RM, and Moncada S. 1989. Role of endothelium-derived nitric oxide in the regulation of blood pressure. *Proc. Natl. Acad. Sci. USA 86:*3375–3378.

Rogers NE, and Ignarro LJ. 1992. Constitutive nitric oxide synthase from cerebellum is reversibly inhibited by nitric oxide formed from L-arginine. *Biochem. Biophys. Res. Commun. 189:*242–249.

Rubanyi GM, and Vanhoutte PM. 1986. Superoxide anions and hyperoxia inactivate endothelium-derived relaxing factor. *Am. J. Physiol. 250:*H822–H827.

Rubanyi GM, Romero JC, and Vanhoutte PM. 1986. Flow-induced release of endothelium-derived relaxing factor. *Am. J. Physiol. 250:*H1145–H1149.

Tesfamariam B, Halpern W, and Osol G. 1985. Effects of perfusion and endothelium on the reactivity of isolated resistance arteries. *Blood Vessels 22:*301–305.

Trigo-Rocha F, Aronson WJ, Hohenfellner M, Ignarro LJ, Rajfer J, and Lue TF. 1993. Nitric oxide and cGMP: mediators of pelvic nerve-stimulated erection in dogs. *Am. J. Physiol 264:*H419–H422.

Vargas HM, Cuevas JM, Ignarro LJ, and Chaudhuri G. 1991. Comparison of the inhibitory potencies of $N^{(G)}$-methyl-, $N^{(G)}$-nitro- and $N^{(G)}$-amino-L-arginine on EDRF function in the rat: evidence for continuous basal EDRF release. *J. Pharmacol. Exp. Ther. 257:*1208–1215.

Williams DL. 1996. *S*-Nitrosothiols and role of metal ions in decomposition to nitric oxide. *Methods Enzymol. 268:*299–308.

Winniford MD, Kennedy PL, Wells PJ, and Hillis LD. 1986. Potentiation of nitroglycerin-induced coronary dilatation by *N*-acetylcysteine. *Circulation 73:*138–142.

Wolin MS, Wood KS, and Ignarro LJ. 1982. Guanylate cyclase from bovine lung: a kinetic analysis of the regulation of the purified soluble enzyme by protoporphyrin IX, heme, and nitrosyl-heme. *J. Biol. Chem. 257:*13312–13320.

Part I
Selected Mechanisms of Action

2
Regulation of Intracellular Ca^{2+} by Cyclic GMP-Dependent Protein Kinase in Vascular Smooth Muscle Cells

PADMINI KOMALAVILAS AND THOMAS M. LINCOLN

Introduction

Although nitric oxide (NO)-mediated relaxation of vascular smooth muscle was described several years ago, the downstream effects of NO and the mechanism of relaxation of vascular smooth muscle are still not well understood. NO produced endogenously or from nitrovasodilators such as nitroglycerin or sodium nitroprusside generates cGMP (Arnold et al. 1977; DeRubertis and Craven 1977; Gruetter et al. 1980) which in turn produces vascular smooth muscle relaxation. There have been several reviews of the roles of NO and cGMP in cellular regulation (Walter 1989; Nathan 1992; Marletta 1993; Lincoln 1994; Lincoln et al. 1996). This article is focused on the effects of cGMP-dependent protein kinase (cGMP kinase), the major receptor protein for cGMP in vascular smooth muscle, and our current understanding on how this enzyme mediates relaxation by lowering intracellular Ca^{2+} ($[Ca^{2+}]_i$).

Nitric Oxide and cGMP Signaling in Vascular Smooth Muscle

Furchgott and Zawadzki (1980) and Furchgott et al. (1981) identified the endothelium-derived relaxing factor (EDRF) released by endothelial cells that was found to control the tone of the blood vessels. Studies by several investigators further demonstrated that EDRF indeed is NO, identical to the NO produced from nitrovasodilators such as nitroglycerin and nitroprusside (Furchgott 1988; Ignarro et al. 1987; Palmer et al. 1987). The classical pathway of endogenous NO production leading to smooth muscle relaxation is the generation of inositol 1,4,5-Triphosphate (IP_3) in the endothelial cells in response to shear stress and hormonal stimuli (e.g., acetylcholine, bradykinin, histamine), resulting in the release of calcium. The increase in cytosolic Ca^{2+} leads to Ca^{2+}–calmodulin complex formation which activates the enzyme NO synthase (NOS) (reviewed by (Marletta 1983; Nathan 1992; Moncada and Higgs 1993). Nitric oxide synthase catalyzes the conversion of L-arginine to NO and citrulline (Palmer and Ashton 1988;

Schmidt et al. 1988). Nitric oxide, being a diatomic free radical gas, diffuses readily to the adjacent smooth muscle cells, causing relaxation of the muscle. Nitric oxide produces relaxation of smooth muscle at a concentration in the nanomolar range. Higher concentrations of NO, which are generated pathophysiologically [e.g., from stimulation of inducible NOS (iNOS) by cytokines] or by the use of higher amounts (micromolar to millimolar) of nitrovasodilators can, however, produce other effects from its interactions with ferrous-sulfhydryl groups and heme-containing enzymes, the production of peroxynitrite, and the ADP-ribosylation of proteins (Beckman et al. 1990; Lancaster and Hibbs 1990; Kwon et al. 1991; Ischiropoulos et al. 1992; Salvemini et al. 1993). The muscle-relaxing effects of low concentrations of NO generated by endothelial cells or NO generated from low concentrations of nitrovasodilators in the smooth muscle cells are only described.

The well-studied action of NO in mediating relaxation is the activation of soluble guanylate cyclase, elevating cGMP levels in vascular smooth muscle cells (Arnold et al. 1977; Gruetter et al. 1980). Vascular smooth muscle cells have at least two types of guanylate cylases that generate cGMP from GTP. The majority of the activity is cytosolic and exists as a heterodimer consisting of two different subunits: an α subunit and a β subunit. The cDNAs encoding the subunits of the enzyme have been cloned (Koestling et al. 1988; Nakane et al. 1988, 1990). Soluble guanylate cyclase is a heme-containing enzyme (Waldman and Murad 1987) and has very high affinity for NO. Studies by Ignarro et al. (1984) have suggested that the mechanism of activation of guanylate cyclase by NO may be due to the binding of NO to the iron atom of the heme, causing a conformational change of the structure of the enzyme. Thus, activation of soluble guanylate cyclase by NO can increase cGMP levels in smooth muscle cells severalfold, producing relaxation. However the levels of cGMP, once elevated, are regulated by phosphodiesterases, which are a class of enzymes that catalyze the hydrolysis of cyclic nucleotides to their inactive 5' nucleotides (Weishaar 1987; Beavo and Reifsnyder 1990; Charbonneau 1990). The particulate form of guanylate cyclase is found to be activated by biologically active peptides such as atrial natriuretic peptides, and the properties of these enzymes have been reviewed elsewhere (Schultz et al. 1989).

Cyclic GMP and Vascular Smooth Muscle Relaxation

Though the use of nitroglycerin for relieving angina is more than 100 years old, its mechanism of action was elucidated in the 1970s by several laboratories (Katsuki et al. 1977; Schultz et al. 1977; Gruetter et al. 1979). These investigations demonstrated that nitrovasodilators such as nitroglycerin and sodium nitroprusside evoke vascular smooth muscle relaxation by generating cGMP. Further research identified NO as the free radical derived from these compounds that activated soluble guanylate cyclase to produce cGMP (Arnold et al. 1977; Gruetter et al. 1980). Early studies demonstrated that cGMP relaxes smooth muscle contracted with agonists or with depolarizing concentrations of potassium, which led to further study on the mechanism of cGMP-mediated relaxation. Agents that increase the level of cGMP, such as phosphodiesterase (PDE) inhibitors, also increase smooth muscle relaxation, whereas agents that decrease cGMP levels, such as guanylate cyclase inhibitors, also reduce smooth muscle relaxation, giving further evidence of the role of cGMP in vascular smooth muscle relaxation. Analogues of cGMP, such as 8-bromo-cGMP, also evoke vascular smooth muscle relaxation, confirming the results obtained with nitrovasodilators.

Cyclic GMP-Dependent Protein Kinase

The effects of cGMP are mediated through the receptor proteins for cGMP (Lincoln and Cornwell 1993; Lincoln 1994). There are three major types of receptor proteins for cGMP: the cGMP-binding PDEs, cGMP-dependent protein kinases (cGMP kinases), and cGMP-regulated ion channels. These receptor proteins are expressed in a cell type-specific manner, and each cell type may have one or more of these receptor proteins mediating the effects of cGMP. Since cGMP kinases are the major receptor proteins expressed in vascular smooth muscle cells (Lincoln and Corbin 1983), this discussion is mainly focused on their role in mediating the effects of cGMP in vascular smooth muscle relaxation. However, contributions from PDEs that regulate the levels of cGMP and ion channels that regulate the ionic concentrations in the cell should not be ignored in investigating the mechanism of relaxation.

Cyclic GMP kinases are serine threonine kinases that have higher affinity for activation by cGMP than by cAMP. Two types of cGMP kinases are found in mammalian cells, type I and type II. Type I is expressed as two isoforms, cGMP kinase I α (~78 kDa) and cGMP kinase I β (~80 kDa). Analyses of partial cDNA sequences of bovine cGMP kinase 1α, and 1β (Wernet et al. 1989) and the gene encoding human cGMP kinase 1β (Orstavik et al. 1992) provide additional evidence that these two forms arise by alternate mRNA splicing (Francis et al. 1989). Two classes of cDNAs coding for cGMP kinase were identified from mouse brain (Uhler 1993). Although cGMP kinase I is widely distributed in mammalian cells it is most abundant in smooth muscle cells (both α and β) (Francis et al. 1988), lung (Kuo and Greengard 1975), platelets (Walter 1988), and Purkinje cells (Lohmann and Walter 1984). The presence of high levels of cGMP kinase in the vasculature may account for the small amounts detected in organs such as the liver, kidney, and heart (Lincoln 1994). Low levels are present in neutrophils and endothelial cells (Pryzwansky et al. 1990; MacMillan-Crow et al. 1994). In vascular smooth muscle cells, cGMP kinase has been demonstrated to be localized partially to the sarcoplasmic reticulum, where some of the substrates for cGMP kinase regulating cytosolic Ca^{2+} concentration are localized (Cornwell et al. 1991).

Earlier studies (Lincoln 1983; Johnson and Lincoln 1985; Lincoln et al. 1985) using intact vascular strips led to the proposal that reduction of cytosolic Ca^{2+} is an important mechanism by which cGMP evokes relaxation, since contraction of smooth muscle is initiated by a rise in $[Ca^{2+}]_i$ and removal of Ca^{2+} will lead to an inhibition of contraction causing relaxation. In arterial smooth muscle cells treated with either angiotensin II or depolarizing concentrations of KCl, 8-bromo-cGMP reduced cytosolic Ca^{2+} (Rashatwar et al. 1987), confirming the Ca^{2+} reduction hypothesis. This study also suggested a potential role for cGMP kinase in the reduction of $[Ca^{2+}]_i$. By introducing a catalytically active fragment of cGMP kinase to isolated tracheal smooth muscle cells, Felbel and co-workers (1988) demonstrated that cGMP kinase lowered $[Ca^{2+}]_i$, mediating smooth muscle relaxation. Further studies using isolated smooth muscle cells from rat aorta confirmed the role of cGMP kinase in mediating the reduction of $[Ca^{2+}]_i$ evoked by cGMP (Cornwell and Lincoln 1989). It was clear from early on that cGMP was able to reduce Ca^{2+} in cells stimulated by G protein-coupled agonists, such as vasopressin and angiotensin II, that increase $[Ca^{2+}]_i$ by activation of phopholipase Cγ and generation of IP_3 as well as by depolarizing agents such as KCl, which elevate $[Ca^{2+}]_i$ through opening of voltage-dependent channels. Therefore, to understand the mechanism of reduction of $[Ca^{2+}]_i$ by cGMP, one must look at multiple pathways that control Ca^{2+} levels in cells. This also suggests that cGMP kinase, the major mediator of cGMP effects, may have as substrates several proteins that are involved in Ca^{2+} homeostasis. Thus, several mechanisms have been proposed for the reduction of Ca^{2+} by cGMP kinase. These are described below.

Mechanisms of cGMP Kinase-Mediated Reduction of $[Ca^{2+}]_i$

Since Ca^{2+} is an important mediator of all physiological functions in cells, its concentration is regulated very judiciously. Since there are two pathways by which Ca^{2+} is elevated in the cell during contraction, uptake from extracellular fluids through plasma membrane Ca^{2+} channels and release from the sarcoplasmic reticulum, the lowering of Ca^{2+} by cGMP kinase during relaxation involves several mechanisms that affect both these pathways. The proposed mechanisms include activation of Ca^{2+}-ATPases to increase Ca^{2+} uptake or extrusion from the cytoplasm, inhibition of Ca^{2+} release by the sarcoplasmic reticulum, inhibition of phospholipase C activation and IP_3 formation, inhibition of G protein coupling to phospholipase C, activation of Ca^{2+}-activated K^+ channels, and regulation of contractile protein sensitivity.

Activation of Ca^{2+}-ATPase Activity (Increased Uptake of Ca^{2+} by Sarcoplasmic Reticulum)

Since cGMP has been demonstrated to decrease $[Ca^{2+}]_i$ levels in K^+-treated arterial smooth muscle cells, it was proposed that cGMP leads to activation of Ca^{2+}-ATPase activity (Felbel et al. 1988). Other studies suggested that cGMP lowers intracellular Ca^{2+} by a mechanism that involves sequestration or removal of Ca^{2+} from the cytoplasm (Karaki et al. 1988; Twort and van Breemen 1988; Hassid and

Yu 1989; Magliola and Jones 1990), along with direct demonstration of the activation of Ca^{2+}-ATPase (Rashatwar et al. 1987; Twort and van Breemen 1988; Vrolix et al. 1988; Yoshida et al. 1991). One potential mechanism is activation of the sarcoplasmic reticulum Ca^{2+}-ATPase through the phosphorylation of the Ca^{2+}-ATPase regulatory protein phospholamban which is phosphorylated by cGMP kinase in vitro and in intact smooth muscle cells (Raeymaekers et al. 1988; Huggins et al. 1989; Sarcevic et al. 1989; Cornwell et al. 1991; Karczewski et al. 1992). Phosphorylation of phospholamban is correlated with increased stimulation of Ca^{2+} sequestration as well as Ca^{2+}-ATPase activation (Raeymaekers et al. 1988; Cornwell et al. 1991). Selective inhibitors of the sarcoplasmic reticulum Ca^{2+}-ATPase also inhibited cGMP-mediated relaxation of rabbit aorta (Luo et al. 1993). Reconstitution studies using purified cardiac sarcoplasmic reticulum Ca^{2+}-ATPase and phospholamban in phospholipid vesicles demonstrated that phospholamban is an inhibitor of Ca^{2+}-ATPase and that phosphorylation of phospholamban by cAMP kinase reverses the inhibitory effect on the Ca^{2+} pump (Kim et al. 1989; Sasaki et al. 1992). Both the cytoplasmic and the transmembrane domains of phospholamban are required for phosphorylation-dependent modulation of the Ca^{2+} pump activity (Jones and Field 1993). Though phospholamban is also a substrate for cAMP-dependent protein kinase (cAMP kinase), only cGMP kinase was found to be colocalized with phospholamban in the sarcoplasmic reticulum of rat aortic smooth muscle cells, which is an important element of cGMP-dependent phosphorylation of this protein during relaxation (Cornwell et al. 1991). It is also evident that lowering of Ca^{2+} by cGMP involves more than removal of Ca^{2+} from the cytoplasm, since agonist-contracted smooth muscle is relaxed to a greater extent by cGMP than is depolarized smooth muscle. Also, phospholamban is not expressed uniformly in all smooth muscle tissues. Therefore, activation of sarcoplasmic reticulum Ca^{2+}-ATPase through phosphorylation of phospholamban is only one of the mechanisms of lowering of Ca^{2+} by cGMP in vascular smooth muscle cells.

Regulation of IP$_3$ Production

There are several kinds of evidence that suggest that cGMP may also mediate cytosolic Ca^{2+} reduction by regulating the levels of IP_3. This could be a mechanism of relaxation in the excitable cells but not in the depolarized cells. Studies by Rapoport (1986) showed that elevations in cGMP inhibited the accumulation of inositol phosphate in aortic smooth muscle strips, which is similar to the results obtained by Takai and co-workers (1981) in platelets. In rabbit aorta, EDRF has been shown to inhibit IP_3 formation (Lang and Lewis 1989). Hirata and co-workers (1990) proposed a mechanism for inhibition of inositol phosphate formation by cGMP, involving inhibition of activation of a guanine nucleotide regulatory (G) protein and interaction of G protein with phospholipase C. Ruth and co-workers (1993) demonstrated that in Chinese hamster ovary (CHO) cells over-expressing cGMP kinase, thrombin-evoked IP_3 formation and the increase in $[Ca^{2+}]_i$ were inhibited by cGMP analogue, whereas control cells not expressing cGMP kinase were insensitive to the effects of cGMP analogue. These effects were later shown to be due to the phosphorylation of the α subunits of G_i protein by cGMP kinase (Pfeifer et al. 1995). However, the mechanism is still not understood. G proteins (e.g., G_i, G_o, and G_z) that have the potential to stimulate phospholipase C when treated with agonists were not phosphorylated by cGMP kinase (Lincoln 1991). None of the phospholipase C isoforms identified so far have been demonstrated to be phosphorylated by cGMP kinase. Besides, cGMP relaxes depolarized vascular smooth muscle, which does not involve the formation of IP_3, and therefore further studies are needed to establish the mechanism of cGMP-mediated inhibition of IP_3 formation.

Inhibition of Ca^{2+} Release by Sarcoplasmic Reticulum

Early studies demonstrated that cGMP-elevating agents inhibited agonist-mediated Ca^{2+} release and contraction in rabbit aortic strips, possibly by inhibiting the action of IP_3, on the sarcoplasmic reticulum (Collins et al. 1986; Meisheri et al. 1986). More recently, in gastric smooth muscle cells, IP_3 generation and IP_3-dependent Ca^{2+} mobilization were inhibited by cGMP as well as cAMP, whereas cGMP also stimulated Ca^{2+} uptake in these cells, supporting the hypothesis that cGMP acts on multiple sites to reduce cytosolic Ca^{2+} (Jin et al. 1993; Murthy et al. 1993; Murthy and Makhlouf 1995). Since during agonist-mediated contraction, the majority of cytosolic Ca^{2+} is released from the intracellular stores by the action of IP_3 on its specific receptor (IP_3 receptor), which is the major Ca^{2+} release channel of the sarcoplasmic reticulum, cGMP kinase action on the regulation of the IP_3 receptor was proposed. The purified IP_3 receptor (M_r ~240 kDa) is an excellent substrate for cGMP kinase, phosphorylated on the serine residue 1755 (Koga et al. 1994; Komalavilas and Lincoln 1994). The same site is phosphorylated by cAMP kinase also (Ferris et al. 1991). Our laboratory has also demonstrated that the IP_3 receptor is phosphorylated by cGMP kinase in response to elevation of cGMP as well as cAMP during relaxation of intact rat aorta (Komalavilas and Lincoln 1996), suggesting that cGMP kinase preferentially phosphorylates the IP_3 receptor in vivo, even though both kinases can phosphorylate the receptor in vitro. As mentioned earlier, one explanation for this effect may be the localization of the cGMP kinase and not the cAMP kinase to the sarcoplasmic reticulum (Cornwell et al. 1991), where the IP_3 receptor is also located, and the proximity of the kinase to the substrate may enable the phosphorylation to occur efficiently. Cyclic GMP kinase has been demonstrated to bind reversibly to the intermediate filament protein vimentin (MacMillan-Crow et al. 1994), which may also help to anchor the cGMP kinase to enable rapid phosphorylation of substrates. Moreover, "cross activation" of cGMP kinase by cAMP during relaxation of vascular smooth muscle was proposed earlier (Francis et al. 1988; Cavallini et al. 1996) and has been confirmed in intact smooth muscle tissue and cells using different approaches. In swine coronary arteries, cGMP kinase has been shown to be activated with forskolin (Jiang et al. 1992). In rat aortic smooth muscle cells, the growth-inhibitory actions of NO have been attributed at least partly to the activation of cAMP kinase by cGMP (Cornwell et al. 1994), once again demonstrating that "cross activation" of the respective kinases by the nucleotide other than the one for which it is specific does occur in the cell.

Although there are no reports to date describing the effect of cGMP kinase phosphorylation of the IP_3 receptor, the reduction of $[Ca^{2+}]_i$ release elicited by cGMP in agonist-stimulated cells demonstrated by several investigators may be explained by the inhibition of IP_3 receptor activity by cGMP kinase phosphorylation of the receptor. This reduction in $[Ca^{2+}]_i$ by inhibition of IP_3 receptor activity may contribute to the greater sensitivity of cGMP-evoked relaxation in agonist-contracted muscle tissue compared with depolarized tissue. However, there are studies that describe the effect of cAMP kinase phosphorylation of the IP_3 receptor, and the results are rather controversial. Supattapone and co-workers (1988) first reported that cAMP kinase-mediated phosphorylation of the IP_3 receptor decreases the potency of IP_3 in releasing Ca^{2+} from cerebellar microsomes. In platelet membranes, cAMP kinase-mediated phosphorylation of the IP_3 receptor also resulted in diminished potency of IP_3 in releasing Ca^{2+} (Quinton and Dean 1992). Nitroprusside and prostacyclin were shown to diminish IP_3-mediated Ca^{2+} release in intact platelets, suggesting a cGMP- and cAMP-mediated IP_3 receptor desensitization (Cavallini et al. 1996). However, there are studies demonstrating that cAMP kinase-mediated phosphorylation of the IP_3 receptor increases the potency of IP_3 in releasing Ca^{2+} in platelets (Knouf et al. 1987), hepatocytes (Burgess et al. 1991; Bird et al. 1993; & Hajnoczky et al. 1993), and reconstituted lipid vesicles with the purified IP_3 receptor (Nakade et al. 1994). This variation in results may be due to the tissue-specific-expression of different IP_3 receptor proteins, which are regulated differently by cAMP kinase phosphorylation, or the capacity of cAMP kinase to phosphorylate an additional site (serine 1584) besides the serine 1755 on the type I receptor. The functional consequences of the cGMP kinase-mediated phosphorylation of the IP_3 receptor remain to be established. Further studies using the purified receptor in reconstituted systems as well as vascular smooth muscle microsomes may help to elucidate the functions of cGMP kinase-mediated phosphorylation of the IP_3 receptor. Because of the complex nature of the IP_3 receptor and the fact that it is phosphorylated by different kinases, the regulation by phosphorylation in the intact cells may be different from the regulation of purified receptor in reconstituted systems, as the in vitro system may also be devoid of any additional factors that may be required for the regulation.

Activation of Ca^{2+}-Activated K^+ Channels

Several investigators have reported NO-mediated hyperpolarization and relaxation in vascular smooth muscle cells (Tare et al. 1990; Fujino et al. 1991; Thornbury et al. 1991; Chen and Rembold 1992; Krippeit-Drews et al. 1992), but there are also reports in which NO or cGMP analogues were not found to elicit hyperpolarization in some vascular smooth muscle cells (Meisheri et al. 1991; Chen and Cheung 1992; Garland and McPherson 1992). Hyperpolarization inhibits voltage-dependent Ca^{2+} channels and elicits relaxation. Studies using gastrointestinal smooth muscle (Thornbury et al. 1991; Ward et al. 1992) and vascular smooth muscle (Tare et al. 1990; Chen and Rembold 1992; Krippeit-Drews et al. 1992) have demonstrated that analogues of cGMP increase Ca^{2+}-activated K^+-channel activity, resulting in hyperpolarization and relaxation. The Ca^{2+}-activated K^+-channel inhibitor charybdotoxin has been shown to decrease NO- and cGMP-mediated relaxation in several muscle preparations, such as coronary arterial strips and tracheal smooth muscle strips, but not in contracted rat aortic smooth muscle strips (Hamaguchi et al. 1991). Robertson and co-workers (1993) demonstrated that cGMP kinase-activated the Ca^{2+}-activated K^+ channel when added to patches of depolarized smooth muscle cell membranes of cerebral artery. Cyclic GMP and cGMP kinase were also demonstrated to stimulate the Ca^{2+}-activated K^+-channel from pituitary tumor cells, leading to hyperpolarization of the membrane and inhibition of hormone secretion. This effect was blocked by the phosphatase inhibitor okadaic acid, suggesting the activation of a protein phosphatase leading to the dephosphorylation and activation of the K^+ channel (White et al. 1993). Activation of the Ca^{2+}-activated K^+ channel by cGMP kinase has been demonstrated to require protein phosphatase 2A activity in tracheal smooth muscle (Zhou et al. 1996). Also, cGMP kinase activation has been demonstrated to stimulate the Ca^{2+}-activated K^+ channel in rat pulmonary arterial smooth muscle (Archer et al. 1994; Hampl et al. 1995) and in human mesangial cells (Stockand and Sansom 1996). The mechanism of activation of protein phosphatase 2A by cGMP kinase is not well understood at this time. Direct activation of the K^+ channel derived from trachael smooth muscle reconstituted into lipid bilayers by cGMP kinase-dependent phosphorylation of the α subunit of the channel has also been demonstrated (Alioua et al. 1995). However, NO itself can stimulate the Ca^{2+}-activated K^+ channel in the presence of guanylate cyclase inhibitors in rabbit aorta, suggesting a cGMP-independent regulation of the K^+ channel in this cell type (Lei et al. 1992; Bolotina et al. 1994). These results do not support the notion that activation of the K^+ channel is a universal mechanism of cGMP-evoked relaxation. Rather, it is a cell type-specific regulation and contributes to reduction of cytosolic Ca^{2+} by cGMP in certain types of smooth muscle cells.

Inhibition of Ca^{2+} Influx Through Ion Channels

Since drugs that inhibit the L-type Ca^{2+} channels, such as verapamil and nifedipine, are also potent vascular smooth muscle relaxants and antihypertensive agents, several studies have focused on the effect of cGMP on Ca^{2+} channel activity. Besides, regulation of L-type Ca^{2+} channels by cGMP is important in cardiac function (Lindemann and Watanabe 1988). In isolated cardiac myocytes, cGMP inhibits the inward Ca^{2+} (Tohse and Sperelakis 1991; Quignard et al. 1997), and high concentrations of cGMP kinase can mimic the effects of cGMP (Mery et al. 1991). In vascular smooth muscle, Ca^{2+} channels have been demonstrated to be regulated by NO (Blatter and Wier 1994), and cGMP (Ishikawa et al. 1993). The effects of cAMP in this case are mediated by "cross activation" of cGMP kinase and inhibition of the Ca^{2+} channels, instead of by activation of Ca^{2+} channels by cAMP, which is the usual mechanism observed. However, the mechanism by which cGMP kinase regulates Ca^{2+} channels is still not understood. Phosphorylation of purified Ca^{2+} channel by cGMP kinase does not inhibit the Ca^{2+} current, suggesting that the effects of cGMP kinase are indirect (Welling et al. 1992). As in the case of K^+ channels, the cGMP kinase-mediated regulation of the Ca^{2+} channels appears to be complex and may play a role in the lowering of Ca^{2+} by cGMP kinase, but it does not account for the total effect of cGMP, since cGMP analogue inhibits norepinephrine-elicited contraction of smooth muscle, even if the tissue is incubated in Ca^{2+}-free media.

Regulation of Contractile Protein Sensitivity

Cyclic GMP and cGMP kinase have been demonstrated to reduce contractile protein force generation in "skinned" muscle fibers (Pfitzer et al. 1982; Nishimura and van Breeman 1989). Cyclic GMP kinase has been demonstrated to mediate a reduction in Ca^{2+} sensitivity induced by both cGMP and cAMP in smooth muscle obtained from skinned rat mesenteric artery (Kawada et al. 1997). The mechanisms by which cGMP inhibits contractile protein function are unrelated to the regulation of cytosolic Ca^{2+}. The decreased sensitivity would inhibit contractile activity, even in the presence of elevated cytosolic Ca^{2+}. Studies by McDaniel and co-workers (1992) have suggested that relaxation of intact swine carotid artery by nitrovasodilators involves a reduction in intracellular Ca^{2+} levels that may be due to activation of Ca^{2+}-activated K^+ channels and the uncoupling of Ca^{2+} from force generation. Since cGMP kinase has not been demonstrated to phosphorylate any of the regulatory contractile proteins, the mechanism of decreased sensitivity of contractile proteins may be indirect; nevertheless, decreased sensitivity of the contractile proteins could be a mechanism of cGMP kinase-mediated relaxation.

Summary

From the above discussion, it is clear that cGMP kinase mediates the reduction of intracellular Ca^{2+} and relaxation of vascular smooth muscle by several mechanism, which are shown in Fig. 2.1. This is not surprising in view of the importance of Ca^{2+} as a second messenger involved in the regulation of several physiological functions in cells. Since smooth muscle cells function against a high Ca^{2+} gradient, with an intracellular concentration of $0.1\ \mu M$ and an extracellular concentration of 1 mM, inadequate regulation of Ca^{2+} would result in damage to the muscle tissue. Deficient Ca^{2+} regulation would result in vascular hypercontractility and spasm, vessel occlusion, decreased tissue perfusion, and eventual tissue death. Thus, as reviewed earlier, the pleiotropic role of cGMP kinase in mediating vascular smooth muscle relaxation is very important in maintaining vascular tone

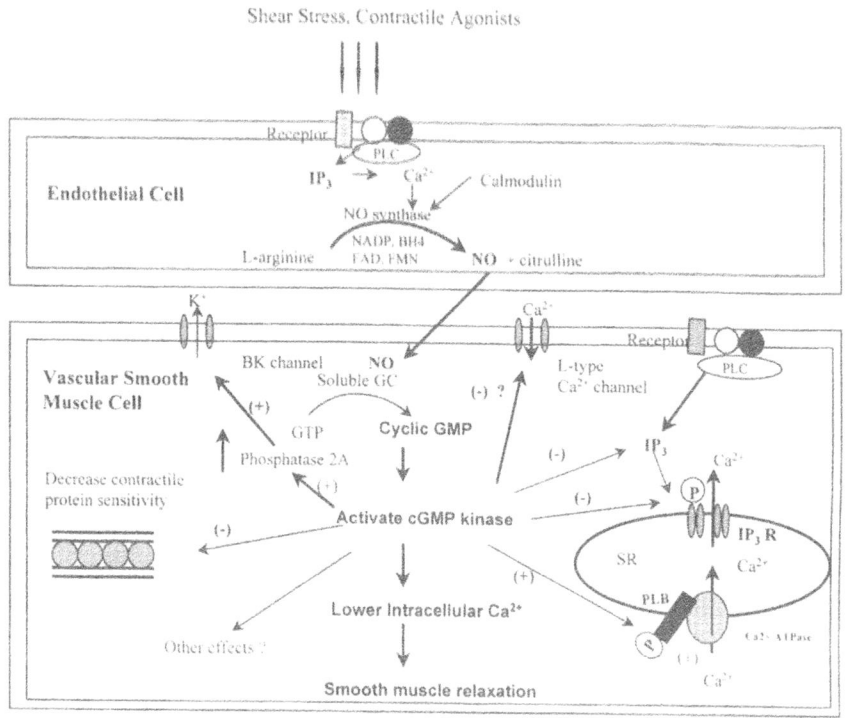

FIGURE 2.1. Schematic diagram showing the mechanism of cGMP kinase-mediated relaxation of vascular smooth muscle. NO generated by shear stress or by the action of contractile agonists in endothelial cells diffuses to the adjacent smooth muscle cells and activates soluble guanylate cyclase (GC), which in turn generates cGMP. Cyclic GMP activates cGMP kinase, which is localized to the sarcoplasmic reticulum (SR), phosphorylates proteins in the SR such as the IP_3 receptor (IP_3 R) and phospholamban (PLB), resulting in the lowering of $[Ca^{2+}]_i$ by the decreased release of Ca^{2+} from the SR, and increased uptake of Ca^{2+} by the SR, respectively. Cyclic GMP kinase activation of protein phosphatase 2A may lead to activation of the Ca^{2+}-activated K^+ channels (BK, channel), which leads to hyperpolarization and subsequent closing of the Ca^{2+} channels, contributing to lowering of $[Ca^{2+}]_i$. Cyclic GMP kinase can inhibit the formation of IP_3 through inactivation of phospholipase C (PLC) or G protein coupling. Inhibition of the L-type Ca^{2+} channels by cGMP kinase also contributes toward the reduction of $[Ca^{2+}]_i$ along with a reduction of the contractile protein sensitivity.

(Lincoln et al. 1994). For efficient regulation, muscle cells from different vascular beds may choose one mechanism over another or a combination of mechanisms. In rat aortic smooth muscle cells, phosphorylation of phospholamban and the IP_3 receptor may be the major pathways used for regulating Ca^{2+} fluxes from the sarcoplasmic reticulum, since cGMP kinase is localized to the sarcoplasmic reticulum, where phospholamban and the IP_3 receptor are located. In resistance vessels, cytosolic Ca^{2+} may be controlled mainly through the regulation of Ca^{2+}-activated K^+ channels and voltage-dependent Ca^{2+} channels.

References

Alioua A, Huggins JP, & Rousseau E. 1995. PKG-I alpha phosphorylates the alpha-subunit and upregulates reconstituted GKCa channels from tracheal smooth muscle. *Am J Physiol 268*:L1057–L1063.

Archer SL, Huang JMC, Hampl V, Nelson DP, Shultz PJ, et al. 1994. Nitric oxide and cGMP cause vasorelaxation by cGMP-kinase-dependent activation of a charybdotoxin-sensitive K channel. *Proc Natl Acad Sci USA. 91*:7583–7587.

Arnold WP, Mittal CK, Katsuki S, A Murad F. 1977. Nitric oxide activates guanylate cyclase and increases guanosine 3':5'-cyclic monophosphate levels in various tissue preparations. *Proc Natl Acad Sci USA. 74*:3203–3207.

Beavo JA, & Reifsnyder DH. 1990. Primary sequence of cyclic nucleotide phosphodiesterase isozymes and the design of selective inhibitors. *TIBS 11*:150–155.

Beckman JS, Beckman TW, Chen J, Marshall PA, & Freeman BA. 1990. Apparent hydroxyl radical production by peroxynitrite: implications for endothelial injury from nitric oxide and superoxide. *Proc Natl Acad Sci USA 87*:1620–1624.

Bird GSJ, Burgess JM, & Putney JW. 1993. Sulfhydryl reagents and cAMP-dependent protein kinase increases the sensitivity of the inositol 1,4,5-triphosphate receptor in hepatocytes. *J Biol Chem 268*:17917–17923.

Blatter LA, & Wier WG. 1994. Nitric oxide decreases $[Ca^{++}]_i$ in vascular smooth muscle by inhibition of calcium current. *Cell Calcium 15*:122–131.

Bolotina VM, Najibi S, Palacino JJ, Pagano PJ, & Cohen RA. 1994. Nitric oxide directly activates potassium channels in vascular smooth muscle. *Nature 368*:850–853.

Burgess GM, Bird GSJ, Obie JF, & Putney JWJ. 1991. The mechanism of synergism between phospholipase C- and adenylylcyclase-linked hormones in the liver. Cyclic AMP-dependent kinase augments inositol triphosphate-mediated Ca^{2+} mobilization without increasing the cellular levels of inositol polyphosphates. *J Biol Chem 266*:4772–4781.

Cavallini L, Coassin M, Borean A, & Alexandre A. 1996. Prostacyclin and sodium nitroprusside inhibit the activity of the platelet inositol 1,4,5-triphosphate receptor and promote its phosphorylation. *J Biol Chem 271*:5545–5551.

Charbonneau H. 1990. Structure-function relationships among cyclic nucleotide phosphodiesterases. In: Beavo JA, and Housley MD, eds. *Cyclic Nucleotide Phosphodiesterases:Structure, Regulation, and Drug Action.* West Sussex, England: John Wiley & Sons, p. 267–296.

Chen GF, & Cheung DW. 1992. Characterization of acetylcholine-induced membrane hyperpolarization in endothelial cells. *Circ Res 70*:257–263.

Chen X-L, & Rembold CM. 1992. Cyclic nucleotide-dependent regulation of Mn^{2+} influx, $[Ca^{2+}]_i$, and arterial smooth muscle relaxation. *Am J Physiol 263*:C468–C473.

Collins P, Griffith TM, Henderson AH, & Lewis MJ. 1986. Endothelium-derived relaxing factor alters calcium fluxes in rabbit aorta: a cyclic guanosine monophosphate-mediated effect. *J Physiol (London) 381*:427–437.

Cornwell TL, & Lincoln TM. 1989. Regulation of intracellular Ca^{2+} levels in cultured vascular smooth muscle cells: reduction of Ca^{2+} by atriopeptin and 8-bromo-cyclic GMP is mediated by cGMP-dependent protein kinase. *J Biol Chem 264*:1146–1155.

Cornwell TL, Pryzwansky KB, Wyatt TA, & Lincoln TM. 1991. Regulation of sarcoplasmic reticulum phosphorylation by localized cyclic GMP-dependent protein kinase in vascular smooth muscle cells. *Mol Pharmacol 40:*923–931.

Cornwell TL, Arnold E, Boerth NJ, & Lincoln TM. 1994. Inhibition of smooth muscle cell growth by nitric oxide and activation of cAMP-dependent protein kinase by cGMP. *Am J Physiol 267:*C1405–C1413.

DeRubertis FR, & Craven PA. 1977. Activation of the renal cortical and hepatic guanylate cyclase-guanosine 3',5'-monophosphate systems by nitrosoureas. Divalent cation requirements and relationship to thiol reactivity. *Biochim Biophys Acta 499:*337–351.

Enouf J, Giraud F, Bredoux R, Bourdeau N, & Levy-Toledano S. 1987. Possible role of a cAMP-dependent phosphorylation in the calcium release mediated by inositol 1,4,5-triphosphate in human platelets membrane vesicles. *Biochem Biophys Acta 928:*76–82.

Felbel J, Trockur B, Ecker T, Landgraf W, & Hofmann F. 1988. Regulation of cytosolic calcium by cAMP and cGMP in freshly isolated smooth muscle cells from bovine trachea. *J Biol Chem 263:*16764–16771.

Ferris CD, Cameron AM, Bredt DS, Huganir RL, & Snyder SH. 1991. Inositol 1,4,5-triphosphate receptor is phosphorylated by cyclic AMP-dependent protein kinase at serines 1755 and 1589. *Biochem Biophys Res Commun 175:*192–198.

Francis SH, Noblett BD, Todd BW, Wells JN, & Corbin JD. 1988. Relaxation of vascular and tracheal smooth muscle by cyclic nucleotide analogs that preferentially activate purified cGMP-dependent protein kinase. *Mol Pharmacol 34:*506–517.

Francis SH, Woodford TA, Wolfe L, & Corbin JD. 1989. Types Iα and Iβ isozymes of cGMP-dependent protein kinase: alternative mRNA splicing may produce different inhibitory domains. *Second Messengers Phosphoproteins 12:*301–310.

Fujino K, Nakaya S, Wakatsuki T, Miyoshi Y, Hakaya Y, et al. 1991. Effects of nitroglycerin on ATP-induced Ca^{2+} mobilization, Ca^{2+} activated K^+ channels and contraction of cultured smooth muscle cells of porcine coronary artery. *J Pharmacol Exp Ther 256:*371–377.

Furchgott RF. 1988. Studies on relaxation of rabbit aorta by sodium nitrate: the basis for the proposal that the acid-activatable inhibitory factor from bovine retractor penis is inorganic nitrate and the endothelium derived relaxing factor is nitric oxide. In: Vanhoutte P, ed. *Mechanisms of Vasodilation.* New York: Raven Press, pp 401–414.

Furchgott RF, & Zawadzki JV. 1980. The obligatory role of endothelial cells in the relaxation of arterial smooth muscle by acetylcholine. *Nature (London) 288:*373–376.

Furchgott RF, Zawadzki JV, & Cherry PD. 1981. Role of endothelium in the vasodilator response to acetylcholine. In: Leusen P, Vanhoutte PM, eds. *Vasodilation,* New York: pp 49–66.

Garland CJ, & McPherson GA. 1992. Evidence that nitric oxide does not mediate the hyperpolarization and relaxation to acetylcholine in the rat mesenteric artery. *Br J Pharmacol 105:*429–435.

Gruetter CA, Barry BK, McNamara DB, Gruetter DY, Kadowitz PJ, et al. 1979. Relaxation of bovine coronary artery and activation of coronary arterial guanylate cyclase by nitric oxide, nitroprusside and a carcinogenic nitrosoamine. *J Cyclic Nucl Res 5:*211–224.

Gruetter CA, Barry BK, McNamara DB, Kadowitz PJ & Ignarro LJ. 1980. Coronary arterial relaxation and guanylate cyclase activation by cigarette smoke, *N'*-nitrosonornicotine and nitric oxide. *J Pharmacol Exp Ther 214:*9–15.

Hajnoczky G, Gao E, Nomura T, Hoek JB, & Thomas AP. 1993. Multiple mechanisms by which protein kinase A potentiates inositol 1,4,5-triphosphate-induced Ca^{2+} mobilization in permabilized hepatocytes. *Biochem J 293:*413–422.

Hamaguchi M, Ishibashi T, & Imai S. 1991. Involvement of charybdotoxin-sensitive K$^+$ channel in the relaxation of bovine tracheal smooth muscle by glyceryl trinitrate and sodium nitroprusside. *J Pharmacol Exp Ther 262:*263–270.

Hampl V, Huang JM, Weir EK, & Archer SL. 1995. Activation of the cGMP-dependent protein kinase mimics the stimulatory effect of nitric oxide and cGMP on calcium-gated potassium channels. *Physiol Res 44:*39–44.

Hassid A, & Yu Y. 1989. Mechanism of atriopeptin-induced decrease of cytosolic free Ca in rat vascular smooth muscle cells: evidence for an intracellular locus of action. *J Cardiovasc Pharmacol 14:*S34–S38.

Hirata M, Kohse KP, Chang C, Ikebe T, & Murad F. 1990. Mechanism of cyclic GMP inhibition of inositol phosphate formation in rat aorta segments and cultured bovine aortic smooth muscle cells. *J Biol Chem 265:*1268–1273.

Huggins JP, Cook EA, Piggott JR, Mattinsley TJ, & England PJ. 1989. Phospholamban is a good substrate for cyclic GMP-dependent protein kinase *in vitro,* but not in intact cardiac or smooth muscle. *Biochem J 260:*829–835.

Ignarro LJ, Wood KS, & Wolin MS. 1984. Regulation of purified soluble guanylate cyclase by porphyrins and metalloporphyrins: a unifying concept. *Adv Cyclic Nucl Prot Phosphorylation Res 17:*267–274.

Ignarro LJ, Byrns RE, Buga GM, & Wood KS. 1987. Endothelium-derived relaxing factor from pulmonary artery and vein possesses pharmacologic and chemical properties identical to those of nitric oxide radical. *Circ Res 61:*866–879.

Ischiropoulos H, Zhu L, Chen J, Tsai M, Martin JC, et al. 1992. Peroxynitrite-mediated tyrosine nitration catalyzed by superoxide dismutase. *Arch Biochem Biophys 298:*431–437.

Ishikawa T, Hume JR, & Keef KD. 1993. Regulation of Ca^{2+} channels by cAMP and cGMP in vascular smooth muscle cells. *Circ Res 73:*1128–1137.

Jiang H, Colbran JL, Francis SH, & Coribin JD. 1992. Direct evidence for cross-activation of cGMP-dependent protein kinase by cAMP in pig coronary arteries. *J Biol Chem 267:*1015–1019.

Jin JG, Murthy KS, Grider JR, & Makhlouf GM. 1993. Activation of distinct cAMP- and cGMP-dependent pathways by relaxant agents in isolated gastric muscle cells. *Am J Physiol 264:*G470–G477.

Johnson RM, & Lincoln TM. 1985. Effects of nitroprusside, glyceryl trinitrate and 8-bromo cyclic GMP on phosphorylase *a* formation and myosin light chain phosphorylation in rat aorta. *Mol Pharmacol 27:*333–342.

Jones LR, & Field LJ. 1993. Residues 2-25 of phospholamban are insufficient to inhibit Ca^{2+} transport ATPase of cardiac sarcoplasmic reticulum. *J Biol Chem 268:*11486–11488.

Karaki H, Sato K, Ozaki H, & Murakami K. 1988. Effects of sodium nitroprusside on cytosolic calcium level in vascular smooth muscle. *Eur J Pharmacol 156:*259–266.

Karczewski P, Kelm M, Hartmann M, & Schrader J. 1992. Role of phospholamban in NO/EDRF-induced relaxation in rat aorta. *Life Sci 51:*1205–1210.

Katsuki S, Arnold WP, & Murad F. 1977. Effects of sodium nitroprusside, nitroglycerin, and sodium azide on levels of cyclic nucleotides and mechanical activity of various tissues. *J Cyclic Nucl Res 3:*239–247.

Kawada T, Toyosato A, Islam MO, Yoshida Y, & Imai S. 1997. cGMP-kinase mediates cGMP- and cAMP-induced Ca^{2+} desensitization of skinned rat artery. *Eur J Pharmacol 323*:75–82.

Kim HW, Steenaart NAE, Ferguson DG, & Kranias EG. 1989. Functional reconstitution of the cardiac sarcoplasmic reticulum Ca^{2+}-ATPase with phospholamban in phospholipid vesicles. *J Biol Chem 265*:1702–1709.

Koesling D, Herz J, Gausepohl HH, Niroomand F, Hinsch KD, et al. 1988. The primary structure of the 70 KDa subunit of bovine soluble guanylate cyclase. *FEBS Lett 239*:29–34.

Koga T, Yoshida Y, Cai JQ, Islam MO, & Imai S. 1994. Purification and characterization of a 240-kDa cGMP-dependent protein kinase substrate of vascular smooth muscle: close resemblance to inositol 1,4,5-triphosphate receptor. *J Biol Chem 269*:11640–11647.

Komalavilas P, & Lincoln TM. 1994. Phosphorylation of the inositol 1,4,5-trisphosphate receptor by cyclic GMP-dependent protein kinase. *J Biol Chem 269*:8701–8707.

Komalavilas P, & Lincoln TM. 1996. Phosphorylation of the inositol 1,4,5-trisphosphate receptor: cyclic GMP-dependent protein kinase mediates cAMP and cGMP dependent phosphorylation in the intact rat aorta. *J Biol Chem 271*:21933–21938.

Krippeit-Drews P, Norel N, & Godfraind T. 1992. Effect of nitric oxide on membrane potential and contraction of rat aorta. *J Cardiovasc Pharmacol 20*:S72–S75.

Kuo JF, & Greengard P. 1975. Cyclic nucleotide-dependent protein kinase VI. Isolation and partial purification of a protein kinase activated by guanosine 3′,5′-monophosphate. *J Biol Chem 245*:2493–2498.

Kwon NS, Stuehr DJ, & Nathan CF. 1991. Inhibition of tumor cell ribonucleotide reductase by macrophage-derived nitric oxide. *J Exp Med 174*:761–767.

Lancaster JR, & Hibbs JB. 1990. EPR demonstration of iron-nitrosyl complex formation by cytotoxic activated macrophages. *Proc Natl Acad Sci USA 87*:1223–1227.

Lang D, & Lewis MJ. 1989. Endothelial-derived relaxing factor inhibits the formation of inositol trisphosphate by rabbit aorta. *J Physiol (London) 441*:45–52.

Lei SZ, Pan ZH, Aggarwal SK, Chen HS, Hartman J, et al. 1992. Effect of nitric oxide production on the redox modulatory site of the NMDA receptor-channel complex. *Neuron 8*:1087–1089.

Lincoln TM. 1983. Effects of nitroprusside and 8-bromo-cyclic GMP on the contractile activity of the rat aorta. *J Pharmacol Exp Ther 224*:100–107.

Lincoln TM. 1991. Pertussis toxin-sensitive and insensitive guanine nucleotide binding proteins (G-proteins) are not phosphorylated by cyclic GMP-dependent protein kinase. *Second Messengers Phosphoproteins 13*:99–109.

Lincoln TM. 1994. *Cyclic GMP: Biochemistry, Physiology, and Pathophysiology.* Austin, Texas: Landes.

Lincoln TM, & Corbin JD. 1983. Characterization and biological role of the cGMP-dependent protein kinase. *Adv Cyclic Nucl Res 15*:139–192.

Lincoln TM, & Cornwell TL. 1993. Intracellular cyclic GMP receptor proteins. *FASEB J 7*:328–338.

Lincoln TM, Laks JA, & Johnson RM. 1985. Ultraviolet radiation-induced decreases in tension and phosphorylase *a* formation in rat aorta. *J Cyclic Nucl Prot Phosphorylation Res 10*:525–533.

Lincoln TM, Cornwell TL, Komalavilas P, MacMillan-Crow, LA, & Boerth, N. 1996. *The Nitric Oxide-Cyclic GMP Signaling system. Biochemistry of Smooth Muscle Contraction.* New York: Academic Press.

Lincoln TM, Cornwell TL, & Taylor AE. 1990. cGMP-dependent protein kinase mediates the reduction of Ca^{2+} by cAMP in vascular smooth muscle cells. *Am J Physiol* 258:C399–C407.

Lincoln TM, Komalavilas P, & Cornwell TL. 1994. Pleiotropic regulation of vascular smooth muscle tone by cyclic GMP-dependent protein kinase. *Hypertension* 23:1141–1147.

Lindemann JP, & Watanabe AM. 1988. Mechanisms of Adrenergic and Cholinergic regulation of myocardial contractility. In: Sperelakis, N, ed. *Physiology and Pathophysiology of the Heart* Norwell, Mass. Kluwer Academic Publishers, pp 423–452.

Lohmann SM, & Walter U. 1984. Regulation of cellular and subcellular concentrations and distribution of cyclic nucleotide-dependent protein kinases. *Adv Cyclic Nucl Prot Phosphorylation Res* 18:63–117.

Luo DL, Nakazawa M, Ishibashi T, Kato K, & Imai S. 1993. Putative, selective inhibitors of sarcoplasmic reticulum Ca^{++} pump ATPase inhibit relaxation by nitroglycerin and atrial natriuretic factor of the rabbit aorta contracted by phenylephrine. *J Pharmacol Exp Therap* 265:1187–1192.

MacMillan-Crow LA, Murphy-Ullrich JE, & Lincoln TM. 1994. Identification and possible localization of cGMP-dependent protein kinase in bovine aortic endothelial cells. *Biochem Biophys Res Commun* 201:531–537.

Magliola L, & Jones AW. 1990. Sodium nitroprusside alters Ca^{2+} flux components and Ca^{2+} dependent fluxes of K^+ and Cl^- in rat aorta. *J Physiol (London)* 421:411–424.

Marletta M. 1993. Nitric oxide synthase structure and mechanism. *J Biol Chem* 268:12231–12234.

McDaniel NL, Chen XL, Singer HA, Murphy RA, & Rembold CM. 1992. Nitrovasodilators relax arterial smooth muscle by decreasing $[Ca^{2+}]_i$ and uncoupling stress from myosin phosphorylation. *Am J Physiol* 263:C461–C467.

Meisheri KD, Taylor CJ, & Saneii H. 1986. Synthetic atrial peptide inhibits intracellular calcium release in smooth muscle. *Am J Physiol* 250:C171–C174.

Meisheri KD, Cipkus-Dubray L, Hosner JM, & Khan SA. 1991. Nicorandil-induced vasorelaxation: functional evidence for K^+ channel-dependent and cyclic GMP-dependent components in a single vascular preparation. *J Cardiovasc Pharmacol* 17:903–912.

Mery P, Lohmann SM, Walter U, & Fischmeister R. 1991. Ca^{2+} current is regulated by cyclic GMP-dependent protein kinase in mammalian cardiac myocytes. *Proc Natl Acad Sci USA* 88:1197–1201.

Moncada S, & Higgs A. 1993. The L-arginine-nitric oxide pathway. *N Engl J Med* 329:2002–2012.

Murthy KS, & Makhlouf GM. 1995. Interaction of cA-kinase and cG-kinase in mediating relaxation of dispersed smooth muscle cells. *Am J Physiol* 268:C171–C180.

Murthy KS, Severi C, Grider JR, & Makhlouf GM. 1993. Inhibition of IP3 and IP3-dependent Ca^{2+} mobilization by cyclic nucleotides in isolated gastric muscle cells. *Am J Physiol* 264:G967–G974.

Nakade S, Rhee SK, Hamanaka H, & Mikoshiba K. 1994. Cyclic AMP-dependent phosphorylation of an immunoaffinity-purified homotetrameric inositol 1,4,5-trisphosphate receptor (Type I) increases Ca^{2+} flux in reconstituted lipid vesicles. *J Biol Chem* 269:6735–6742.

Nakane M, Saheki S, Kuno T, Ishii K, & Murad F. 1988. Molecular cloning of a cDNA coding for 70 kilodalton subunit of soluble guanylate cyclase from rat lung. *Biochem Biophys Res Commun* 157:1139–1147.

Nakane M, Arai K, Saheki S, Kuno T, Buechler W, et al. 1990. Molecular cloning and expression of cDNAs coding for soluble guanylate cyclase from rat lung. *J Biol Chem* 265:16841–16845.

Nathan C. 1992. Nitric oxide as a secretory product of mammalian cells. *FASEB J* 6:3051–3064.

Nishimura J, & van Breemen C. 1989. Direct regulation of smooth muscle contractile elements by second messengers. *Biochem Biophys Res Commun 163:*929–935.

Orstavik S, Sandberg M, Berube D, Natarajan V, Simard J, et al. 1992. Localization of the human gene for the type I cyclic GMP-dependent protein kinase to chromosome 10. *Cytogenet Cell Genet 59:*270–273.

Palmer RMJ, Ferrige AG, & Moncada S. 1987. Nitric oxide release accounts for the biological activity of endothelium-derived relaxing factor. *Nature 327:*524–526.

Palmer RMJ, & Ashton DS, Moncada S. 1988. Vascular endothelial cells synthesize nitric oxide from L-arginine. *Nature 333:*664–666.

Pfeifer A, Nurnberg B, Kamm S, Uhde M, & Schultz G. 1995. Cyclic GMP-dependent protein kinase blocks pertussis toxin-sensitive hormone receptor signaling pathways in Chinese hamster ovary cells. *J Biol Chem 270:*9052–9059.

Pfitzer G, Ruegg JC, Flockerzi V, & Hofmann F. 1982. cGMP-dependent protein kinase decreases calcium sensitivity of skinned cardiac fibers. *FEBS Lett 149:*171–175.

Pryzwansky KB, Wyatt TA, Nichols H, & Lincoln TM. 1990. Compartmentalization of cyclic GMP-dependent protein kinase in formyl-peptide stimulated neutrophils. *Blood 76:*612–618.

Quignard JF, Frapier JM, Harricane MC, Albat B, Nargeot J, et al. 1997. Voltage-gated calcium channel currents in human-coronary myocytes. *J Clin Invest 99:*185–193.

Quinton TM, & Dean WL. 1992. Cyclic AMP-dependent phosphorylation of the inositol-1,4,5-triphosphate receptor inhibits Ca^{2+} release from platelet membranes. *Biochem Biophys Res Commun 184:*893–899.

Raeymaekers L, Hofmann F, & Casteels R. 1988. Cyclic GMP-dependent protein kinase phosphorylates phospholamban in isolated sarcoplasmic reticulum from cardiac and smooth muscle. *Biochem J 252:*269–273.

Rapoport RM. 1986. Cyclic guanosine monophosphate inhibition of contraction may be mediated through inhibition of phosphatidylinositol hydrolysis in rat aorta. *Circ Res 58:*407–410.

Rashatwar SS, Cornwell TL, & Lincoln TM. 1987. Effects of 8-bromo-cGMP on Ca^{2+} levels in vascular smooth muscle cells: possible regulation of Ca^{2+}-ATPase by cGMP-dependent protein kinase. *Proc Natl Acad Sci USA 84:*5685–5689.

Robertson BE, Schubert R, Hescheler J, & Nelson MT. 1993. cGMP-dependent protein kinase activates Ca-activated K channels in cerebral artery smooth muscle cells. *Am J Physiol 265:*C299–C303.

Ruth P, Wang G-X, Boekhoff I, May B, Pfeifer A, et al. 1993. Transfected cGMP-dependent protein kinase suppresses calcium transient by inhibition of inositol 1,4,5-trisphosphate production. *Proc Natl Acad Sci USA 90:*2623–2627.

Salvemini D, Misko TP, Masferrer JL, Seibert K, Currie MG, et al. 1993. Nitric oxide activates cyclooxygenase enzymes. *Proc Natl Acad Sci USA 90:*7240–7244.

Sarcevic B, Brookes V, Martin TJ, Kemp BE, & Robinson PJ. 1989. Atrial natriuretic peptide-dependent phosphorylation of smooth muscle cell particulate fraction proteins is mediated by cGMP-dependent protein kinase. *J Biol Chem 264:*20648–20654.

Sasaki T, Inui M, Kimura Y, Kuzuya T, & Tada M. 1992. Molecular mechanisms of regulation of Ca^{2+} pump ATPase by phospholamban in cardiac sarcoplasmic reticulum. Effects of synthetic phospholamban peptides on Ca^{2+} pump ATPase. *J Biol Chem* 267:1674–1679.

Schmidt HW, Klein MM, Niroomand F, & Bohme E. 1988. Is arginine a physiological precursor for endothelium-derived nitric oxide? *Eur J Pharmacol 148:*293–295.

Schultz KD, Schultz K, & Schultz G. 1977. Sodium nitroprusside and other smooth muscle relaxants increase cyclic GMP levels in rat ductus deferens. *Nature 265:*750–751.

Schultz S, Chinkers M, & Garbers DL. 1989. The guanylate cyclase/receptor family of proteins. *FASEB J 3:*2026–2035.

Stockand JD, & Sansom SC. 1996. Mechanism of activation by cGMP-dependent protein kinase of large Ca$^{(2+)}$-activated K$^+$ channels in mesangial cells. *Am J Physiol 271:*C1669–C1677.

Supattapone S, Danoff SK, Thiebert A, Joseph SK, Steiner J, et al. 1988. Cyclic AMP-dependent phosphorylation of a brain inositol trisphosphate receptor decreases its release of calcium. *Proc. Natl. Acad. Sci. USA* 85:8747–8750.

Takai Y, Kaibuchi K, Matsubara T, & Nishizuka Y. 1981. Inhibitory action of guanosine 3′, 5′-monophosphate on thrombin-induced phosphatidylinositol turnover and protein phosphorylation in human platelets. *Biochem Biophys Res Commun 101:*61–67.

Tare M, Parkington HC, Coleman HA, Neild TO, & Dusting GJ. 1990. Hyperpolarization and relaxation of arterial smooth muscle caused by nitric oxide derived from the endothelium. *Nature 346:*69–71.

Thornbury KD, Ward SM, Dalziel HH, Carl A, Westfall DP, et al. 1991. Nitric oxide and nitrosocysteine mimic nonadrenergic noncholinergic hyperpolarization in canine proximal colon. *Am J Physiol 261:*G553–G557.

Tohse N, & Sperelakis N. 1991. cGMP inhibits the activity of single calcium channels in embryonic chick heart cells. *Circ Res 69:*325–331.

Twort CHC, & van Breemen C. 1988. Cyclic guanosine monophosphate-enhanced sequestration of Ca^{2+} by sarcoplasmic reticulum in vascular smooth muscle. *Circ Res* 62:961–964.

Uhler M. 1993. Cloning and expression of a novel cyclic GMP-dependent protein kinase from mouse brain. *J. Biol. Chem. 268:*13586–13591.

Vrolix M, Raeymaekers L, Wuytack F, Hofmann F, & Casteels R. 1988. Cyclic GMP-dependent protein kinase stimulates the plasmalemmal Ca^{2+} pump of smooth muscle via phosphorylation of phosphatidylinositol. *Biochem J 255:*855–863.

Waldman SA, & Murad F. 1987. Cyclic GMP synthesis and function. *Pharmacol Rev 39:*163–196.

Walter U. 1988. Distribution of cyclic GMP-dependent protein kinase in various rat tissues and cell lines determined by a sensitive and specific radioimmunoassay. *Eur J Biochem 118:*339–346.

Walter U. 1989. Physiological role of cGMP and cGMP-dependent protein kinase in the cardiovascular system. *Rev Physiol Biochem Pharmacol 113:*41–88.

Ward SM, Dalziel HH, Bradley ME, Buxton IL, Keef K, et al. 1992. Involvement of cyclic GMP in non-adrenergic, non-cholinergic inhibitory neurotransmission in dog proximal colon. *Br J Pharmacol 107:*1075–1082.

Weishaar RE. 1987. Multiple forms of phosphodiesterase: an overview. *J Cyclic Nucl Prot Phosphorylation Res 11:*463–472.

Welling A, Felbel J, Peper K, & Hofmann F. 1992. Hormonal regulation of calcium current in freshly isolated airway smooth muscle cells. *Am J Physiol 262:*L351–L359.

Wernet W, Flockerzi V, & Hofmann F. 1989. The cDNA of the two isoforms of bovine cGMP-dependent protein kinase. *FEBS Lett 251:*191–196.

White RE, Lee AB, Shcherbatko AD, Lincoln TM, Schonbrunn A, et al. 1993. Potassium channel stimulation by natriuretic peptides through cGMP-dependent dephosphorylation. *Nature 361*:263–266.

Yoshida Y, Sun HT, Cai JQ, & Imai S. 1991. Cyclic GMP-dependent protein kinase stimulates the plasma membrane Ca^{2+} pump ATPase of vascular smooth muscle via phosphorylation of a 240 kDa protein. *J Biol Chem 266*:19819–19825.

Zhou XB, Ruth P, Schlossmann J, Hofmann F, & Korth M. 1996. Protein phosphatase 2A is essential for the activation of Ca^{2+} activated currents by cGMP-dependent protein kinase in tracheal smooth muscle and Chinese hamster ovary cells. *J Biol Chem 271*:19760–19767.

3
Oxidants and Vascular Nitric Oxide Signaling

Michael S. Wolin, Sachin A. Gupte, Takafumi Iesaki, and Kamal M. Mohazzab-H.

Introduction

Oxidant species and related redox processes appear to have the potential for multiple interactions with vascular nitric oxide (NO) signaling mechanisms. These interactions can affect the production, stability, and mechanisms of action of NO. Oxidants that affect the production of NO are likely to function through signaling systems that alter the activity or expression of the different forms of nitric oxide synthase (NOS) present in the vessel wall. The stability of NO and signaling mechanisms influenced by NO–oxidant interactions are more likely to be controlled by the levels of the species present and by the function of antioxidant and metabolizing systems that interact with each active species. This chapter will focus on processes shown in Figure 3.1, which are likely to contribute to the control of vascular tone and organ blood flow in the peripheral circulation.

Influence of Oxidant Processes on the Production of NO

Recent studies have provided evidence that peroxides stimulate NO-associated endothelium-dependent relaxation (Furchgott 1991; Zembowicz et al. 1993). A potential explanation for the origin of this action of NO is suggested by the observation that peroxides can increase endothelial cell levels of calcium (Elliott and Koliwad 1995; Shimizu et al. 1997), an activator of the endothelial NOS (eNOS) present in these cells. This action of peroxides does not seem to be expressed in all vascular preparations, since endothelium-derived NO-associated responses have been demonstrated to be very sensitive to inhibition by low micromolar levels of hydrogen peroxide (H_2O_2) in the cerebral microcirculation (Wei and Kontos 1990). It appears that NOS catalyzes the production of superoxide anion ($O_2^{\cdot-}$) even under conditions that optimize the production of NO (Mayer et al. 1998). In addition, when cellular levels of its substrate L-arginine and/or cofactor tetrahydrobiopterin are decreased, an increased production of $O_2^{\cdot-}$ by NOS seems to be observed (Cosentino and Katusic 1995; Pritchard et al. 1995). Thus, NOS appears to generate $O_2^{\cdot-}$ in quantities that significantly influence the amounts

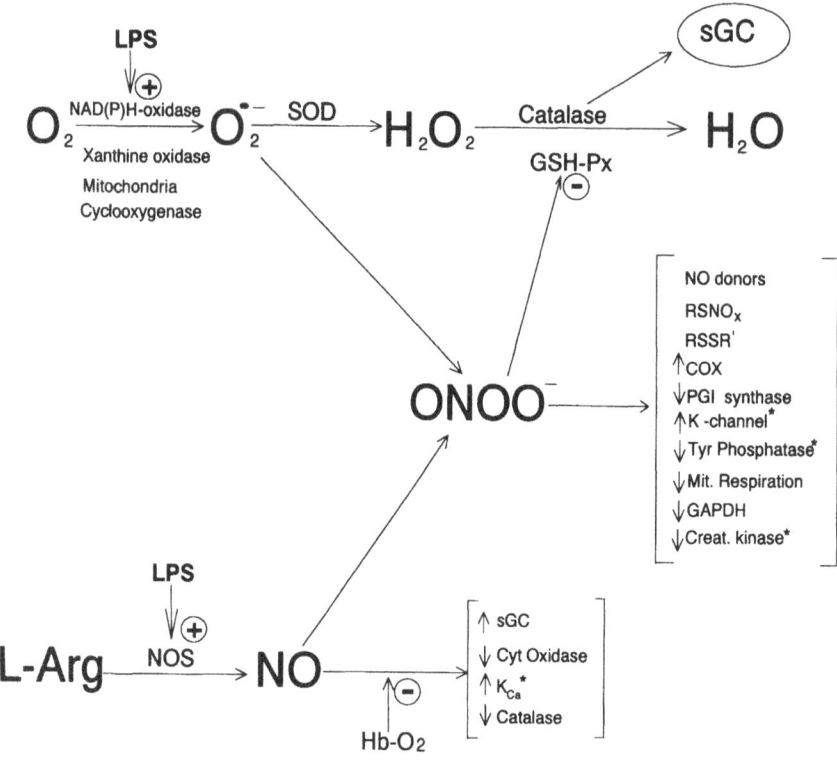

FIGURE 3.1. Model of potential mechanisms discussed in this chapter through which oxidants influence NO-mediated signaling. * Hypothesized interaction

of NO produced by the NOS reaction (Mayer et al. 1998). Under these conditions, species that form from the simultaneous presence of NO and $O_2{}^{-}$, such as peroxynitrite ($ONOO^-$), are also likely to participate in NOS-derived signaling. The activity of NOS is inhibited by NO, and an additional effect of the scavenging of NO by $O_2{}^{-}$ in vascular tissue could be the suppression of this action of NO. It appears that the production of $ONOO^-$ by vascular endothelium is detectable even in the absence of stimulation of NO biosynthesis (Wolin and Mohazzab-H 1997). Thus, reactive O_2 species may have significant stimulatory and inhibitory effects on NOS activity, which are likely to be influenced by the reactive species present and the levels of scavenging activities for each species. In addition, the production of $O_2{}^{-}$ by NOS may contribute significantly to the levels of oxidant species present.

Recent evidence suggests that the expression of the NOS activity could be influenced by oxidants and redox processes. The expression of eNOS was shown to be increased by 13-hydroperoxyoctadecadienoic acid, a biologically active component of oxidized low-density lipoprotein, (Ramasamy et al. 1998). The promoter region of the bovine and human eNOS gene has binding sites for the redox-regulated transcription factor activating protein 1 (AP-1) (Venema et al. 1994) and

3. Oxidants and Vascular Nitric Oxide Signaling   35

hydroperoxyeicosatetraenoic acids have been demonstrated to increase AP-1 in vascular smooth muscle cells (Rao et al. 1996). Thus, the expression of vascular NOS activity is also likely to be regulated by oxidant processes.

Influence of Oxidant Processes on the Stability of NO

Oxidant processes appear to have a spectrum of mechanisms through which they can influence the stability of NO. The direct reaction of NO with $O_2^{\cdot-}$ seems to be the most important interaction of an oxidant species with NO. This reaction has a rate constant of 6.7×10^9 $M^{-1}s^{-1}$, which is threefold faster than the reaction of superoxide dismutase (SOD) with $O_2^{\cdot-}$ (Huie and Padmaja 1993). Thus, as NO levels increase through the nanomolar concentration range, NO will compete with SOD in the intracellular environment for the scavenging of $O_2^{\cdot-}$. As a consequence of this reaction, $ONOO^-$ is likely to be generated in vascular tissue, and the formation of this species could result in the activation of signaling mechanisms that are sensitive to low levels of $ONOO^-$. These mechanisms are discussed in the next section of this chapter. Nitric oxide also has an extremely efficient reaction with oxyhemoglobin (and perhaps oxymyoglobin), which results in an oxidative inactivation of NO associated with the production of nitrate (Doyle and Hoekstra 1981). It appears that this may be the most important mechanism for the removal of NO in vivo (Wennmalm et al. 1993). Cytochrome oxidase in the mitochondria has been reported to reduce NO to nitrous oxide (Zhao et al. 1995; Borutaite and Brown 1996). However, the contribution of this process to the metabolism of NO is not currently known. Thus, our present understanding of the metabolism of NO suggests that its consumption by hemoproteins appears to be the major pathway of its inactivation in vivo, but conditions are likely to exist where localized increased levels of $O_2^{\cdot-}$ react with NO in a manner that activates additional signaling mechanisms.

The interaction of NO with $O_2^{\cdot-}$ has a major role in the influence and expression of NO signaling mechanisms in vascular tissue. Studies on the impact of the inhibition of Cu, Zn-SOD with the copper chelator diethyldithiocarbamate (DETCA) have resulted in the realization that SOD functions to prevent $O_2^{\cdot-}$ produced by endothelium from inactivating NO prior to its release from this key cellular source of vascular NO generation (Mugge et al. 1991; Omar et al. 1991). Additional studies with this inhibitor also demonstrated the important role Cu, Zn-SOD has in protecting NO from inactivation by $O_2^{\cdot-}$ produced within vascular smooth muscle cells before it activates relaxing mechanisms, such as those elicited through the stimulation of the soluble or cytosolic form of guanylate cyclase (sGC) (Cherry et al. 1990; Omar et al. 1991). The data in Figure 3.2 show how a 30-minutes pretreatment of endothelium-removed bovine coronary arteries with an inhibitor of SOD (10 mM DETCA) results in an attenuation of relaxation to the NO donor S-nitroso-N-acetylpenicillamine (SNAP) and how this effect of inhibition of SOD is diminished by the intracellular scavenger of $O_2^{\cdot-}$, 10 mM Tiron. These data show how the scavenging of endogenous $O_2^{\cdot-}$ by SOD protects

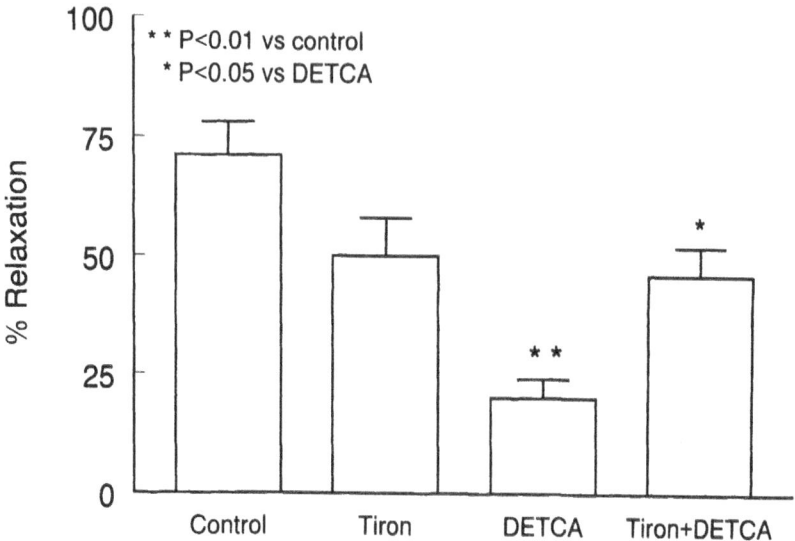

FIGURE 3.2. Relaxation of endothelium-removed bovine coronary arteries precontracted with 30 mM KCl to a NO donor (1 μM SNAP, control) is attenuated by a 30-minute pretreatment with an inhibitor of SOD (10 mM DETCA), and this effect is prevented by a scavenger of intracellular $O_2^{\cdot-}$ (10 mM Tiron) (N = 12).

NO from being inactivated by this oxygen species. Our previous studies have suggested that it requires 50 nM NO to produce a ~50% reduction of the lucigenin-detectable source of $O_2^{\cdot-}$ in the bovine coronary and pulmonary arteries (Davidson et al. 1997a,b). Under these conditions, a prominent increase in ONOO⁻ formation is detected by luminol chemiluminescence during the period of exposure to 50 nM NO, and the amount of ONOO⁻ produced appears to promote a prolonged depression of force generation. Data shown in Figure 3.3 indicate that the intracellular source of $O_2^{\cdot-}$ detected by lucigenin chemiluminescence in endothelium-removed bovine coronary arteries is only modestly depressed in the presence of 10 μM SNAP, a dose of this NO-releasing agent that typically causes maximal relaxation of vascular preparations. The data in Figure 3.3 employing luminol for the detection of ONOO⁻ also demonstrate that its formation is only modestly increased by the 10 μM dose of SNAP. Thus, vasodilator doses of NO scavenge only a small amount of the lucigenin-detectable source of intracellular $O_2^{\cdot-}$, and the amounts of ONOO⁻ produced under these conditions are probably below the levels that affect vascular force generation. There is also an extracellular form of SOD (EC-SOD) present in the vascular wall. This form of SOD appears to function to protect NO from inactivation during its diffusion from the endothelium to the vascular smooth muscle. (Abrahamsson et al. 1992). Thus, SOD enzymes in the vessel wall seem to have a critical role in the expression of NO-dependent vascular regulation.

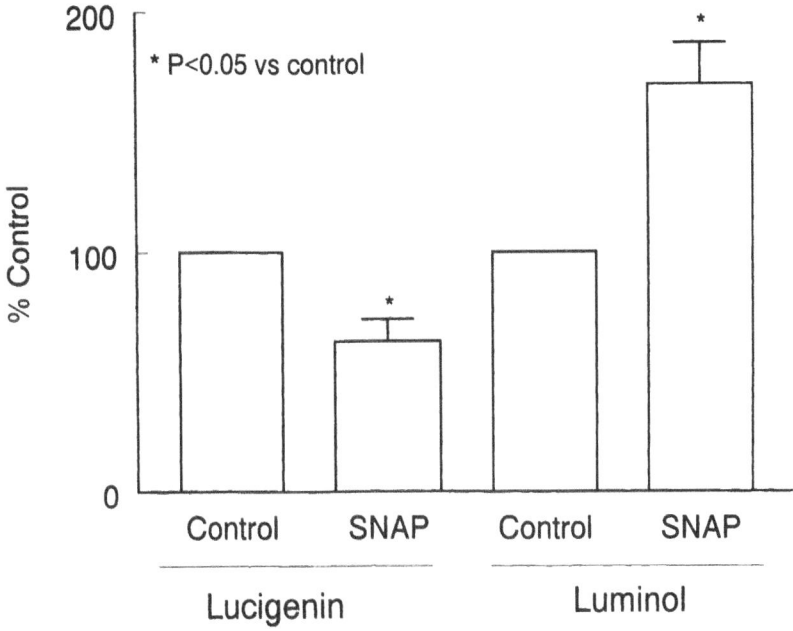

FIGURE 3.3. Exposure of endothelium-removed bovine coronary arteries (control) to a maximal relaxant dose of a NO donor (10 μM SNAP, control) causes a decrease in endogenous $O_2{}^-$ detected by lucigenin chemiluminescence ($N = 6$) and an increase in $ONOO^-$, detected by luminol chemiluminescence ($N = 5$).

Influence of Oxidant Processes on the Mechanisms of Action of NO

Cellular levels of NO and species derived from the oxidative metabolism of NO have a major impact on the expression of signaling mechanisms that control vascular reactivity. Low nanomolar levels of NO are likely to produce relaxation through increased cGMP production as a result of the stimulation of (sGC) activity. As shown by the data in Figure 3.3, relaxant concentrations of NO are likely to have only minimal scavenging effects on basal $O_2{}^-$ levels. As $O_2{}^-$ levels increase as a result of enhanced production or reduced scavenging, the major signaling effect observed is likely to be an attenuation of NO-elicited relaxation. Conditions that result in an increased interaction of NO with $O_2{}^-$ potentially activate additional signaling mechanisms that affect vascular reactivity. As mentioned previously, it has been observed in bovine coronary and pulmonary arterial smooth muscle that a 2-minute exposure to ~50 nM NO results in the expression of a prolonged relaxation or depressed reactivity to contractile agents (Davidson et al. 1997a,b). The mechanism responsible for this observed response seems to involve $ONOO^-$ formation and a glutathione-dependent process that contributes to the trapping and subsequent regeneration of NO. Micromolar levels of $ONOO^-$ were observed to elicit a similar prolonged relaxation response, which appeared to be mediated by NO (Liu et al. 1994; Wu et al. 1994). Studies on the effects of ONOO⁻ on vascular sGC activity have also detected evidence for an involvement of what appears to be a thiol-dependent mechanism of stimulation of cGMP production (Mayer et al. 1995; Tarpey et al. 1995). Although species including nitrogen dioxide (NO_2), N_2O_3, nitrosated and nitrated thiols, nitrated alcohols (e.g., glucose), hydroxyl radical, and perhaps iron–thiol complexes of NO [$(RS)_2FeNO$] are being considered as participants in responses that originate from ONOO⁻ or interactions of NO with $O_2^{\cdot-}$ (Wu et al. 1994; Mayer et al. 1995; Moro et al. 1995; Davidson et al. 1996, 1997a,b; Wei et al. 1996), the actual species involved remain to be elucidated.

Several studies have detected evidence that NO-dependent mechanisms have roles in regulating prostaglandin (PG) production by vascular tissue (e.g., Davidge et al. 1995). For example, cyclooxygenase (COX; prostaglandin H synthase) may be directly activated by ONOO⁻ (Ladino et al. 1996) and not by NO itself (Tsai et al. 1994). In addition, ONOO⁻ has been reported to inhibit the conversion of PGH_2 to PGI_2 by PGI_2 synthase (Zou and Ullrich 1996). In isolated saline-perfused rat hearts, oxidant stress caused by a 20-minute perfusion with an $O_2^{\cdot-}$-generating system composed of 0.03 U/ml xanthine oxidase, 2.5 mM purine, and 300 U/ml catalase resulted in the expression of a thromboxane $A_2/PGF_{2\alpha}$-mediated contractile response upon infusion of nitroglycerin, and a PGI_2-mediated relaxant response was elicited by infusion of N^G-nitro-L-arginine (L-NA), an inhibitor of endogenous NO biosynthesis (Gupte et al. 1996). It is likely that oxidized metabolites of NO may have a major impact on these observed prostaglandin-mediated responses in the isolated rat heart and in many other systems where prostaglandin-associated vascular pathophysiology has been observed.

Nitric oxide, which originates from sources including the vascular endothelium, appears to be an important physiological regulator of tissue respiration (Shen et al. 1994, 1995; Wolin et al. 1997). The interaction of $O_2^{\cdot-}$ with NO converts this reversible inhibitory effect of NO on mitochondrial oxygen metabolism to what appears to be a markedly more prolonged and perhaps irreversible event (Xie and Wolin 1996). Mitochondrial respiration in vascular tissue is important for energy metabolism that controls vascular force generation (Wingard et al. 1997) and is inhibited by NO (Geng et al. 1992). Thus, the control of respiration by NO may also have important roles in controlling blood flow both through direct effects on energy metabolism in vascular tissue and through its influence on parenchymal tissue-derived mechanisms of metabolic regulation of organ blood flow.

Other signaling mechanisms that control vascular reactivity may also be modulated by NO and its derived species. It has been reported that NO can activate the opening of calcium-regulated potassium channels through what appears to be a cGMP-independent and thiol modification-associated process (Bolotina et al. 1994). A mechanism of this type needs to be further investigated to determine if oxidized species of NO are important in the activation of this response. Vascular catalase activity seems to be very sensitive to inhibition by NO, and this process appears to attenuate the expression of cGMP-associated relaxation to H_2O_2 (Mohazzab-H et al. 1996b). The inhibition of this mechanism of relaxation to H_2O_2 by NO could participate in alterations in some of the vascular responses to lactate and changes in oxygen tension. Peroxynitrite may also cause alterations in the expression of certain receptor-coupling mechanisms (Lipton et al. 1993) and inhibition of tyrosine phosphatases (Caselli et al. 1995), glutathione peroxidase (Asahi et al. 1995), and other enzymes involved in energy metabolism, such as creatine kinase (Gross et al. 1996) and glyceraldehyde-3-phosphate dehydrogenase (GAPDH) (Mohr et al. 1994). Thus, processes originating from the interaction of $O_2^{\cdot-}$ with NO have the potential to activate additional signaling mechanisms involved in the control of vascular tone and organ blood flow.

Sources of Oxidants That Influence Vascular NO Signaling

Several sources of production of reactive oxygen species appear to be able to influence NO-related signaling mechanisms. Inflammatory cells possess an NADPH oxidase that is activated by cellular stimulation (Thelen et al. 1993). However, it has been observed that cardiac ischemia–reperfusion results in a loss of endothelium-derived NO-associated coronary arterial relaxation from a source of increased $O_2^{\cdot-}$ production that is activated prior to the accumulation of inflammatory cells adhering to the vessel wall (Lefer and Lefer 1991). A NADH oxidase has been observed to be the major source of lucigenin chemiluminescence-detectable $O_2^{\cdot-}$-synthesizing activity in endothelium and vascular smooth muscle (Mohazzab-H and Wolin 1994; Mohazzab-H et al. 1994, 1996a). The activity of this system is present in unstimulated tissue, and its rate of $O_2^{\cdot-}$ production appears to be dependent on the availability of cytosolic NADH (Omar et al. 1993; Wolin 1994; Mohazzab-H et al. 1994; Wolin et al. 1996; Gupte et al. 1997). It has recently been observed that increasing cytosolic NADH through the lactate dehydrogenase reaction potentiates the attenuation of NO-elicited vascular smooth muscle relaxation caused by inhibition of SOD with DETCA pretreatment (Gupte et al. 1997). Angiotensin II and tumor necrosis factor α (TNFα) have been observed to increase the expression of $O_2^{\cdot-}$ production by NADH oxidase (Griendling et al. 1994; De Keulenaer et al. 1998). Increased NADH oxidase activity has been associated with a loss of NO-associated endothelium-derived relaxation in an elevated angiotensin II model of hypertension (Rajagopalan et al. 1996). Although exposure of vascular tissue to endotoxin lipopolysaccharide (LPS) has been considered to be a method of increasing the expression of NO production by the inducible form of NOS, there is evidence that it may also increase the production of reactive O_2 species (Brigham et al. 1987). Treatment of isolated endothelium-removed bovine pulmonary arteries with 2 µg/ml endotoxin for 4 h causes an increase in the detection of $O_2^{\cdot-}$ (Figure 3.4). This increase in $O_2^{\cdot-}$ in LPS-treated pulmonary arteries appears to originate from NADH oxidase, because the observed level of $O_2^{\cdot-}$ is potentiated by the presence of lactate. Since L-NA did not alter the detection of $O_2^{\cdot-}$ (Figure 3.4) or the generation of force (not shown) in LPS-treated tissue, a significant production of NO was not detectable.

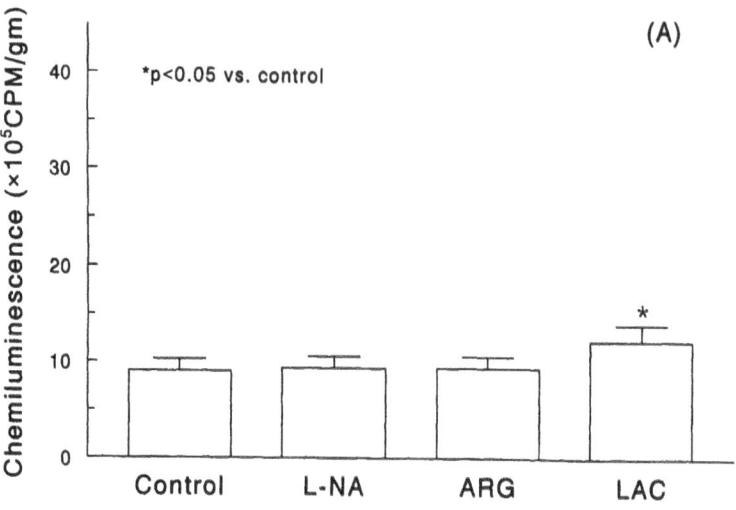

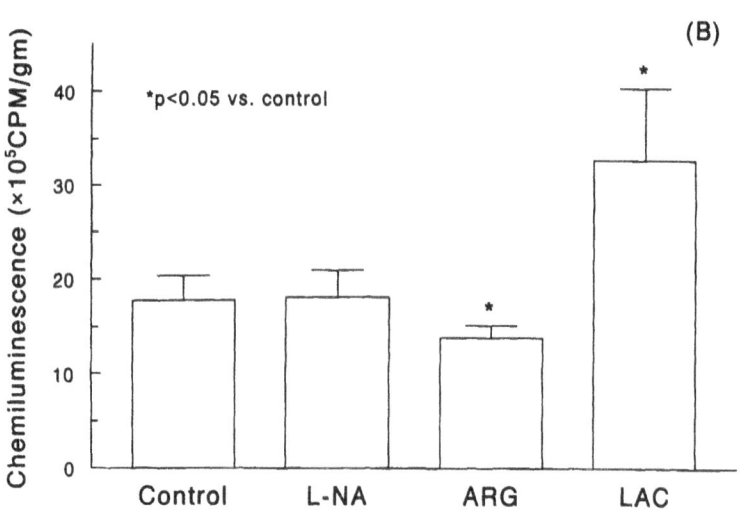

FIGURE 3.4. Exposure of endothelium-removed bovine pulmonary arteries (A, untreated) to 2 μg/ml of endotoxin for 4 hours (B, LPS treated) causes an increase ($P < 0.05$) in endogenous $O_2^{\cdot-}$ that is detected by lucigenin chemiluminescence. This increase in $O_2^{\cdot-}$ is attenuated by a substrate for NO biosynthesis (1 mM Arg) and not altered by an inhibitor of NOS (0.1 mM L-NA), suggesting that LPS caused an induction of NO biosynthesis that required Arg to observe its expression. Lactate (10 mM, LAC) potentiated the LPS-elicited rise in $O_2^{\cdot-}$, suggesting that NADH oxidase activity was increased ($N = 18-22$).

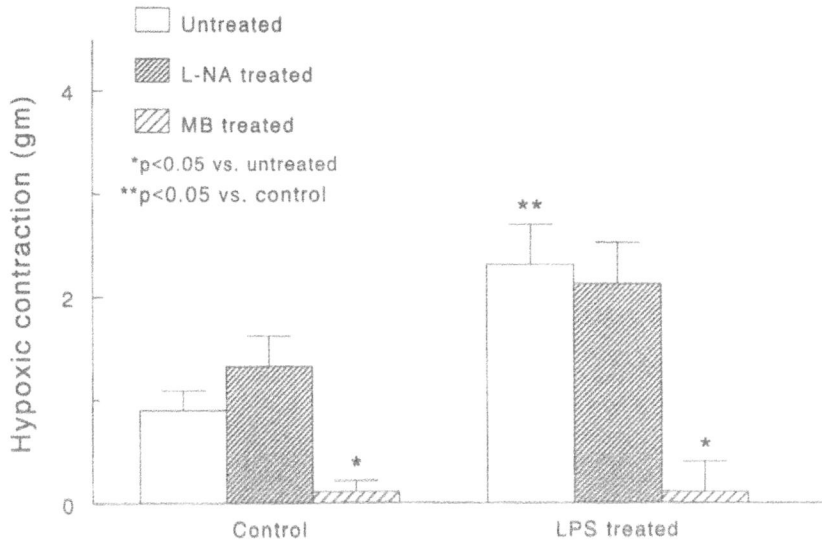

FIGURE 3.5. Exposure of endothelium-removed bovine pulmonary arteries precontracted with a thromboxane A_2 receptor agonist (5 nM U46619) under normoxia ($PO_2 = 150$ torr) to hypoxia ($PO_2 = 8$-10 torr) causes a hypoxic contraction (control), which is eliminated by the presence of methylene blue (10 μM MB) and not altered by inhibition of NO biosynthesis (0.1 mM L-NA). This contractile response to hypoxia is increased in arteries pretreated for 4 hours with endotoxin (2 μg/ml LPS) ($N = 7$-14).

However, in the presence of L-arginine, a reduction in $O_2^{\cdot -}$ was seen in the LPS-treated tissue, suggesting that L-arginine was needed for the anticipated increase in NOS activity in the LPS-treated tissue. It has been suggested that the contractile response of endothelium-removed bovine pulmonary arteries to hypoxia is mediated by a decrease in H_2O_2-elicited stimulation of sGC (Burke-Wolin and Wolin 1989). Methylene blue (MB), an agent that inhibits the stimulation of sGC by H_2O_2, mimics the effects of hypoxia in these pulmonary vessels. The data in Figure 3.5 are consistent with enhancement of the contractile response to hypoxia by LPS through increasing the tonic relaxation under normoxia mediated by H_2O_2 and not NO, because the contraction to hypoxia was eliminated in the presence of 10 μM MB, but not L-NA. Lipopolysaccharide was also observed to depress force generation under normoxia in a manner that was not altered by L-NA, whereas MB increased force generation to the level observed under hypoxia, and MB eliminated all of the effects of hypoxia and LPS (not shown). These data also suggest that the LPS-elicited impairment of relaxation to the NO donor sodium nitroprusside (SNP) shown in Figure 3.6 is likely to be a result of increased $O_2^{\cdot -}$ generation by NADH oxidase, because H_2O_2 levels in excess of 100 μM are required to inhibit SNP-mediated relaxation of pulmonary arteries (not shown). Endothelium contains xanthine dehydrogenase activity, which is converted to xanthine oxidase activity through oxidant- and proteolysis-associated mechanisms

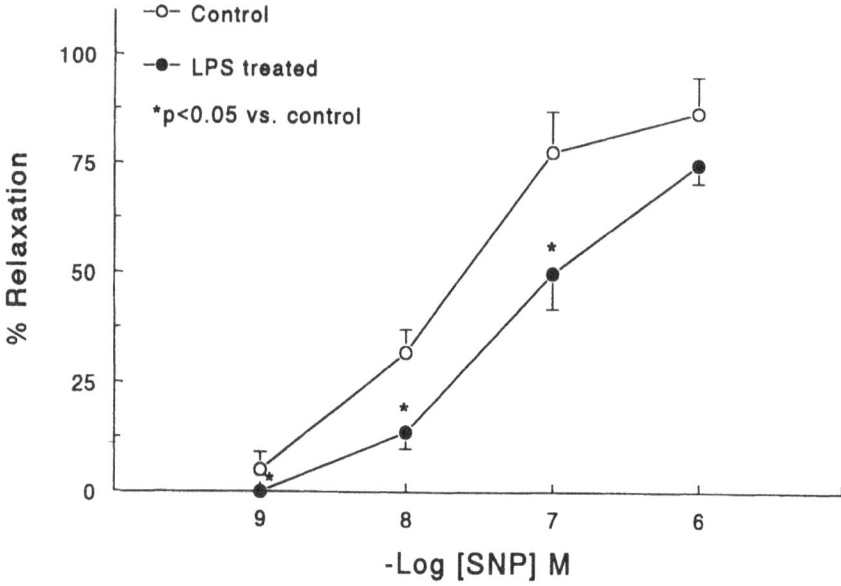

FIGURE 3.6. Exposure of endothelium-removed bovine pulmonary arteries precontracted with 30 mM KCl to endotoxin (2 μg/ml LPS for 4 hours) causes an attenuation of the concentration-dependent relaxation to sodium nitroprusside (SNP, $N = 7–8$).

(Granger 1988). A hypercholesterolemia model of atherosclerosis has been observed to express diminished NO-associated endothelium-dependent relaxation originating from elevated $O_2^{\cdot-}$ production by endothelial xanthine oxidase activity (Ohara et al. 1993). Although the effect of $O_2^{\cdot-}$ originating from the NADPH oxidase activity of NOS on endothelium-associated NO signaling is difficult to experimentally define in intact tissue, it has been observed that a deficiency of its cofactor tetrahydrobiopterin can cause NOS to become a source of vascular relaxant levels of H_2O_2 (Cosentino and Katusic 1995). Cyclooxygenase can be a significant source of $O_2^{\cdot-}$ generation as a result of a one-electron oxidation of cosubstrates such as NAD(P)H to intermediates that autooxidize (Kukreja et al. 1986). This system has been observed to be a major source of oxidants that appear to contribute to altered endothelial function in the cerebral microcirculation (Marshall and Kontos 1991). Although the mitochondrial electron transport chain is thought to produce $O_2^{\cdot-}$ from sites in the region of NADH dehydrogenase and coenzyme Q when these proteins are in their reduced forms (Forman and Boveris 1982), the influence of mitochondrial-derived oxidants on endothelial NO-associated signaling is currently not well defined. It appears that impairment of mitochondrial function in vascular smooth muscle may increase the level of interaction of NO with NADH oxidase-derived $O_2^{\cdot-}$ as a result of increasing cytosolic NADH (Wolin et al. 1996). Thus, the vessel wall has multiple sources of production of oxidants that can participate in the alteration of NO-associated signaling.

Evidence for the Importance of Interactions of Oxidants and NO in Vascular Pathophysiology

Many of the major diseases that affect vascular regulation appear to show evidence of increased interactions between oxidants and NO. The evidence for an increased level of $O_2^{\cdot-}$ production in hypertension is quite strong (Suzuki et al. 1995; Rajagopalan et al. 1996). For example, an acute increase in pressure has been demonstrated to cause elevated $O_2^{\cdot-}$ production in the cerebral microcirculation of cats (Wei et al. 1985), and a more prolonged exposure to hypertension observed in the previously mentioned angiotensin II infusion rat model resulted in impairment of endothelium-dependent relaxation by elevated NADH oxidase-derived $O_2^{\cdot-}$ production (Rajagopalan et al. 1996). There is substantial evidence that diabetes is associated with an impairment of NO-associated endothelium-dependent relaxation and an increased level of oxidant stress in the vessel wall (Pieper et al. 1992; Tesfamariam 1994). Human aortic endothelium has been observed to increase its expression of its endothelium (eNOS) activity and generation of $O_2^{\cdot-}$ during prolonged exposure to elevated levels of glucose (Cosentino et al. 1997). Although it is likely that the vessel wall is exposed to an increased level of $ONOO^-$ in diabetes, the consequences of this chronic exposure are not yet known. There is much evidence that the vessel wall is chronically exposed to elevated levels of NO and $O_2^{\cdot-}$ for a prolonged period of time in regions of atherosclerosis, and it is likely that these processes markedly contribute to the evolution of this disease (Ohara et al. 1993; White et al. 1994). It has recently been demonstrated that humans with atherosclerosis-related coronary artery disease (Levine et al. 1996), hypercholesterolemia (Ting et al. 1997), diabetes (Ting et al. 1996), and chronic heart failure (Hornig et al. 1998) show improved NO-associated endothelium-dependent relaxation with acute treatment with high levels of ascorbate. The xanthine oxidase inhibitor oxypurinol has been observed (Cardillo et al. 1997) to improve endothelial function in patients with hypercholesterolemia but not to improve similar responses in hypertensive patients, an observation consistent with the results of previously discussed studies of these diseases in animal models. These are only a few of many additional pieces of evidence suggesting that increased interactions between NO and oxidants are an important component in the expression of some of the key cardiovascular diseases. Certain of the systems that produce oxidants and NO and the enzymes that scavenge reactive oxidant species appear to readily undergo changes in the expression of their activities, and this phenomenon may have a major influence on the processes of adaptation that appears to occur during the chronic expression of oxidant-associated vascular diseases.

Acknowledgements. We wish to thank members of our laboratory and collaborators quoted in the References for their essential contributions to developing an understanding of signaling mechanisms reported in the papers quoted. Studies from the authors' laboratory have been funded by grants from the American Lung Association, the American Heart Association, and the National Institutes of Health (HL31069 and HL43023).

References

Abrahamsson T, Brandt U, Marklund SK, Sjoquist PO. 1992. Vascular bound recombinant extracellular superoxide dismutase type C protects against the detrimental effects of superoxide radicals on endothelium-dependent relaxation. *Circ Res 70:*264–271.

Asahi M, Fujii J, Suzuki K, Seo HG, Kuzuya T, Hori M, Tada M, Fuji S, & Taniguchi N. 1995. Inactivation of glutathione peroxidase by nitric oxide. *J Biol Chem 270:*21035–21039.

Bolotina VM, Najibi S, Palacino JJ, Pagano PJ, & Cohen RA. 1994. Nitric oxide directly activates calcium-dependent potassium channels in vascular smooth muscle. *Nature 368:*850–853.

Borutaite V, & Brown GC. 1996. Rapid reduction of nitric oxide by mitochondria, and reversible inhibition of mitochondrial respiration by nitric oxide. *Biochem J 315:*295–299.

Brigham KL, Meyrick B, Berry LC, & Repine JE. 1987. Antioxidants protect cultured bovine lung endothelial cells from injury by endotoxin. *J Appl Physiol 63:*840–850.

Burke-Wolin TM, & Wolin MS. 1989. H_2O_2 and GMP may function as an O_2 sensor in the pulmonary artery. *J Appl Physiol 66:*167–170.

Cardillo C, Kilcoyne CM, Cannon RO, Ouyyumi AA, & Panza JA. 1997. Xanthine oxidase inhibition with oxypurinol improves endothelial vasodilator function in hypercholesterolemic but not hypertensive patients. *Hypertension 30:*57–63.

Caselli A, Chiarugi P, Camici G, Manao G, & Ramponi G. 1995. In vivo activation of phosphotyrosine protein phosphatases by nitric oxide. *FEBS Lett 374:*249–252.

Cherry PD, Omar HA, Farrell KA, Stuart JS, & Wolin MS. 1990. Superoxide anion inhibits cGMP-associated bovine pulmonary arterial relaxation. *Am J Physiol 259:*H1056–H1062.

Cosentino F, & Katusic ZS. 1995. Tetrahydrobiopterin and dysfunction of endothelial nitric oxide synthase in coronary arteries. *Circulation 91:*139–144.

Cosentino F, Hishikawa K, Katusic ZS, & Luscher TF. 1997. High glucose increases nitric oxide synthase expression and superoxide anion generation in human aortic endothelial cells. *Circulation 96:*25–28.

Davidge ST, Baker PN, McLaughlin MK, & Roberts JM. 1995. Nitric oxide produced by endothelial cells increases production of eicosanoids through activation of prostaglandin H synthase. *Circ Res 77:*274–283.

Davidson CA, Kaminski PM, Wu M, & Wolin MS. 1996. Nitrogen dioxide causes pulmonary arterial relaxation via thiol nitrosation and NO formation. *Am J Physiol 270:*H1038–H1043.

Davidson CA, Kaminski PM, & Wolin MS. 1997a. Nitric oxide elicits prolonged relaxation of bovine pulmonary arteries via endogenous peroxynitrite generation. *Am J Physiol 273:*L437–L444.

Davidson CA, Kaminski PM, & Wolin MS. 1997b. Endogenous peroxynitrite generation causes a subsequent suppression of coronary arterial contraction to serotonin. *Nitric Oxide Biol Med 1:*244–253.

De Keulenaer GW, Alexander RW, Ushio-Fukai M, Ishizaka N, & Griendling KK. 1998. Tumor necrosis factor α activates a p22[phox]-based NADH oxidase in vascular smooth muscle. *Biochem J 329:*653–657.

Doyle MP, & Hoekstra JW. 1981. Oxidation of nitrogen oxides by bound dioxygen in hemoproteins. *J Inorg Biochem 14:*351–358.

Elliott S, & Koliwad SK. 1995. Oxidant stress and endothelial membrane transport. *Free Radical Biol Med 19:*649–658.

Forman HJ, & Boveris A. 1982. Superoxide radical and hydrogen peroxide in mitochondria. In: Pryor WA, ed. *Free Radicals in Biology,* Vol 5. New York: Academic Press, pp 65–90.

Furchgott RF. 1991. Interactions of H_2O_2 and NO in modifying tone in vascular smooth muscle: the SOD paradox. In: Mulvany MJ, ed. *Resistance Arteries, Structure and Function.* New York: Elsevier Science Publishers, pp 216–220.

Geng Y-J, Hansson GK, & Holme E. 1992. Interferon-τ and tumor necrosis factor synergize to induce nitric oxide production and inhibit mitochondrial respiration in vascular smooth muscle cells. *Circ Res 71:*1268–1276.

Granger DN. 1988. Role of xanthine oxidase and granulocytes in ischemia-reperfusion injury. *Am J Physiol 255:*H1269–H1275.

Griendling KK, Minieri CA, Ollerenshaw JD, & Alexander RW. 1994. Angiotensin stimulates NADH and NADPH oxidase activity in cultured vascular smooth muscle cells. *Circ Res 74:*1141–1148.

Gross WL, Bak MI, Ingwall JS, Arstall MA, Smith TW, Balligand J-L, & Kelley RA. 1996. Nitric oxide inhibits creatine kinase and regulates rat heart contractile reserve, *Proc Natl Acad Sci USA 93:*5604–5609.

Gupte SA, Okada T, & Ochi R. 1996. Superoxide and nitroglycerin stimulate release of $PGF_{2\alpha}$ and TxA_2 in isolated rat heart. *Am J Physiol 271:*H2447–H2453.

Gupte SA, Rupawalla TA, Mohazzab-H KM, & Wolin MS. 1997. Role of NADH oxidase and SOD in the modulation of nitric oxide-elicited pulmonary artery relaxation and guanylate cyclase activation. *Circulation 96:*I–44.

Hornig B, Arakawa N, Kohler C, & Drexler H. 1998. Vitamin C improves endothelial function of conduit arteries in patients with chronic heart failure. *Circulation 97:*363–368.

Huie RE, & Padmaja S. 1993. The reaction of NO with superoxide. *Free Radic Res Commun 18:*195–199.

Kukreja RC, Kontos HA, Hess ML, & Ellis EF. 1986. PGH synthetase and lipoxygenase generate superoxide in the presence of NADH or NADPH. *Circ Res 59:*612–619.

Landino LM, Crews BC, Timmons MD, Morrow JD, & Marnett LJ. 1996. Peroxynitrite, the coupling product of nitric oxide and superoxide, activates prostaglandin biosynthesis. *Proc Natl Acad Sci USA 93:*15069–15074.

Lefer AM, & Lefer DJ. 1991. Endothelial dysfunction in myocardial ischemia and reperfusion: role of oxygen-derived free radicals. In: Drexler H, Zeiher AM, Bassenge E, & Just H, eds. *Endothelial Mechanisms of Vasomotor Control.* New York: Springer-Verlag, pp 111–116.

Levine GN, Frei B, Koulouris SN, Gerhard MD, Keaney JF, & Vita JA. 1996. Ascorbic acid reverses endothelial vasomotor dysfunction in patients with coronary artery disease. *Circulation 93:*1107–1113.

Lipton SA, Choi Y-B, Pan Z-H, Lei SZ, Chen H-SV, Scher NJ, Loscalzo J, Singel DJ, & Stamler JS. 1993. A redox-based mechanism for the neuroprotective and neurodestructive effects of nitric oxide and nitroso-compounds. *Nature 364:*626–632.

Liu S, Beckman JS, & Ku DD. 1994. Peroxynitrite, a product of superoxide and nitric oxide, produces coronary vasorelaxation in dogs. *J Pharmacol Exp Ther 268:*1114–1121.

Marshall JJ, & Kontos HA. 1991. Endothelium and cerebral vascular diseases. In: Rubanyi GM, ed. *Cardiovascular Significance of Endothelium-Derived Vasoactive Factors.* Mount Kisco, NY: Futura Publishing Co., pp 125–145.

Mayer B, Pfeiffer S, Schrammel A, Koesling D, Schmidt K, & Brunner F. 1998. A new pathway of nitric oxide/cyclic GMP signalling involving S-nitrosoglutathione, *J Biol Chem 273:*3264–3270.

Mayer B, Schrammel A, Klatt P, Koesling D, Schmidt K. 1995. Peroxynitrite-induced accumulation of cyclic GMP in endothelial cells and stimulation of purified guanylate cyclase. *J Biol Chem 270:*17355–17360.

Mohazzab-H KM, Wolin MS. 1994. Sites of superoxide anion production detected by lucigenin in calf pulmonary artery smooth muscle. *Am J Physiol 267:*L815–L822.

Mohazzab-H KM, Kaminski PM, Wolin MS. 1994. NADH-oxidoreductase is a major source of superoxide anion in bovine coronary endothelium. *Am J Physiol 266:*H2568–H2572.

Mohazzab-H KM, Fayngersh RP, Kaminski PM, Wolin MS. 1996a. Oxygen-elicited responses in calf coronary arteries: Role of H_2O_2 production via NADH-derived superoxide. *Am J Physiol 270:*H1044–H1053.

Mohazzab-H KM, Fayngersh RP, Wolin MS. 1996b. Nitric oxide inhibits pulmonary artery catalase and H_2O_2-associated relaxation. *Am J Physiol 271:*H1900–H1906.

Mohr S, Stamler JS, Brune B. 1994. Mechanism of covalent modification of glyceraldehyde-3-phosphate dehydrogenase at its active site thiol by nitric oxide, peroxynitrite and related nitrosating agents. *FEBS Lett 348:*223–227.

Moro MA, Darley-Usmar VM, Lizasoain I, Su Y, Knowles RG, Radomski MW, Moncada S. 1995. The formation of nitric oxide donors from peroxynitrite. *Br J Pharmacol 116:*1999–2004.

Mugge A, Elwell JH, Peterson TE, Harrison DG. 1991. Release of intact endothelium-derived relaxing factor depends on endothelial superoxide dismutase activity. *Am J Physiol 260:*C219–C225.

Ohara Y, Peterson TE, Harrison DG. 1993. Hypercholesterolemia increases endothelial superoxide anion production. *J Clin Invest 91:*2546–2551.

Omar HA, Cherry PD, Mortelliti MP, Burke-Wolin T, & Wolin MS. 1991. Inhibition of coronary artery superoxide dismutase attenuates endothelium-dependent and independent nitrovasodilator relaxation. *Circ Res 69:*601–608.

Omar HA, Mohazzab-H KM, Mortelliti MP, & Wolin MS. 1993. O_2 dependent modulation of calf pulmonary artery tone by lactate: role of H_2O_2 and cGMP. *Am J Physiol 264:*L141–L145.

Pieper G, Mei DA, Langenstroer P, & O'Rourke. 1992. Bioassay of endothelium-derived relaxing factor in diabetic rat aorta. *Am J Physiol 263:*H676–H680.

Rajagopalan S, Kurz S, Munzel T, Tarpey M, Freeman B, Griendling KK, & Harrison DG. 1996. Angiotensin mediated hypertension in the rat increases vascular superoxide production via membrane NADH/NADPH oxidase activation: contributions to altered vasomotor tone. *J Clin Invest 97:*1916–1923.

Ramasamy S, Parthasarathy S, & Harrison DG. 1998. Regulation of endothelial nitric oxide synthase gene expression by oxidized linoleic acid. *J Lipid Res 39:*268–276.

Rao GN, Glasgow WC, Eling TE, & Runge MS. 1996. Role of hydroperoxyeicosatetraenoic acids in oxidative stress-induced activating protein (AP-1) activity. *J Biol Chem 271:*27760–27764.

Shen W, Xu X, Ochoa M, Zhao G, Wolin MS, & Hintze TH. 1994. Role of nitric oxide in the regulation of oxygen consumption in conscious dogs. *Circ Res 75:*1086–1095.

Shen W, Hintze TH, & Wolin MS. 1995. Nitric oxide: an important signalling mechanism between the vascular endothelium and parenchymal cells in the regulation of oxygen consumption. *Circulation 92:*3505–3512.

Shimizu S, Ishii M, Yamamoto T, & Momose K. 1997. Mechanism of nitric oxide production induced by H_2O_2 in cultured endothelial cells. *Res Commun Mol Pathol Pharmacol 95:*227–239.

Pritchard KA, Groszek L, Smalley DM, Sessa WC, Wu M, Villalon P, Wolin MS, Stemerman MB. 1995. Native low density lipoprotein enhances endothelial cell nitric oxide synthase generation of superoxide anion. *Circ Res 77:*510–518.

Suzuki H, Swei A, Zweifach BW, & Schmidt-Schonbein GW. 1995. In vivo evidence for microvascular oxidative stress in spontaneously hypertensive rats. Hydroethidine microfluorography. *Circ Res 25:*1083–1089.

Tarpey MM, Beckman JS, Ischripoulos H, Gore JZ, & Brock TA. 1995. Peroxynitrite stimulates vascular smooth muscle cell cGMP synthesis. *FEBS Lett 364:*314–318.

Tesfamariam B. 1994. Free radicals in diabetic endothelial cell dysfunction. *Free Radical Biol Med 16:*383–391.

Thelen M, Dewald B, & Baggiolini M. 1993. Neutrophil signal transduction activation and the respiratory burst. *Physiol Rev 73:*797–821.

Ting HH, Timimi FK, Boles KS, Creager SJ, Ganz P, & Creager MA. 1996. Vitamin C improves endothelium-dependent vasodilation in patients with non-insulin dependent diabetes mellitus. *J Clin Invest 97:*22–28.

Ting HH, Timimi FK, Haley EA, Roddy M-A, Ganz P, & Creager MA. 1997. Vitamin C improves endothelium-dependent vasodilation in forearm resistance vessels of humans with hypercholesterolemia. *Circulation 95:*2617–2622.

Tsai A-L, Wei C, & Kulmacz R. 1994. Interactions between nitric oxide and prostaglandin H synthase. *Arch Biochem Biophys 313:*367–372.

Venema RC, Nishida K, Alexander RW, Harrison DJ, & Murphy TJ. 1994. Organization of the bovine gene encoding the endothelial nitric oxide synthase. *Biochim Biophys Acta 1218:*1–8.

Wei EP, & Kontos HA. 1990. H_2O_2 and endothelium-dependent cerebral arteriolar dilation. Implications for the identity of endothelium-derived factor generated by acetylcholine. *Hypertension 16:*162–169.

Wei EP, Kontos HA, & Beckman JS. 1996. Mechanisms of cerebral vasodilation by superoxide, hydrogen peroxide, and peroxynitrite. *Am J Physiol 271:*H1262–H1266.

Wei EP, Kontos HA, Christman CW, DeWitt DS, & Povlishock JT. 1985. Superoxide generation and reversal of acetylcholine-induced cerebral arteriolar dilation after acute hypertension. *Circ Res 57:*781–787.

Wennmalm A, Benthin G, Edlund A, Jungersten L, Kieler-Jensen N, Lundin S, Westfelt UN, Petersson A-S, & Waagstein F. 1993. Metabolism and excretion of nitric oxide in humans. An experimental and clinical study. *Circ Res 73:*1121–1127.

White CR, Brock TA, Chang LY, Crapo J, Briscoe P, Ku D, Bradley WA, Gianturco SH, Gore J, & Freeman BA. 1994. Superoxide and peroxynitrite in atherosclerosis. *Proc Natl Acad Sci USA 91:*1044–1048.

Wingard CJ, Paul RJ, & Murphy RA. 1997. Energetic cost of activation processes during contraction of swine arterial smooth muscle. *J Physiol 501:*213–223.

Wolin MS, & Mohazzab-H KM. 1997. Mediation of signal transduction by oxidants. In: Scandalios JG, ed. *Oxidative Stress and the Molecular Biology of Antioxidant Defenses.* Cold Spring Harbor, NY: Cold Spring Harbor Laboratory Press, pp 21–48.

Wolin MS, Burke-Wolin TM, Kaminski PM, & Mohazzab-H KM. 1996. Reactive oxygen species and vascular oxygen sensors. In: Weir EK, Archer SL, Reeves JT, eds. *Nitric Oxide and Radicals in the Pulmonary Vasculature*. Armonk, NY: Futura Publishing Co., pp 245–263.

Wolin MS, Hintze TH, Shen W, Mohazzab-H KM, & Xie Y-W. 1997. Involvement of reactive O_2 and N_2 species in signalling mechanisms that control tissue respiration in muscle. *Biochem Soc Trans 25:*929–934.

Wu M, Pritchard KA, Kaminski PM, Fayngersh RP, Hintze TH, & Wolin MS. 1994. Involvement of nitric oxide and nitrosothiols in relaxation of pulmonary arteries to peroxynitrite. *Am J Physiol 266:*H2108–H2113.

Xie Y-W, & Wolin MS. 1996. Role of nitric oxide and its interaction with superoxide in the suppression of cardiac muscle mitochondrial respiration: involvement in response to hypoxia/reoxygenation. *Circulation 94:*2580–2586.

Zembowicz A, Hatchett RJ, Jakubowski AM, & Gryglewski RJ. 1993. Involvement of nitric oxide in the endothelium-dependent relaxation induced by hydrogen peroxide in the rabbit aorta. *Br J Pharmacol 110:*151–158.

Zhao XJ, Sampath V, & Caughey WS. 1995. Cytochrome *c* oxidase catalysis of the reduction of nitric oxide to nitrous oxide. *Biochem Biophys Res Commun 212:*1054–1060.

Zou M-H, & Ullrich V. 1996. Peroxynitrite formed by simultaneous generation of nitric oxide and superoxide selectively inhibits bovine aortic prostacyclin synthase. *FEBS Lett 382:*101–104.

4
Role of Na^+-K^+-ATPase in Nitric Oxide-Induced Relaxation

SANDEEP GUPTA

Introduction

Relaxation of vascular smooth muscle by nitric oxide (NO) has been attributed mainly to its ability to activate soluble guanylate cyclase and increase intracellular cGMP concentration (Craven and DeRubertis 1978; Furchgott and Vanhoutte 1989). The resultant activation of cGMP-dependent protein kinase (PKG) has been shown to increase the activity of plasma membrane Ca^{2+}-ATPase and Ca^{2+}-activated potassium channels, and consequently lowering of free cytoplasmic Ca^{2+} and relaxation of vascular smooth muscle (Lincoln and Cornwell 1993). Nitric oxide also causes relaxation of vascular smooth muscle by increasing the activity of Ca^{2+}-activated K^+ channels by a cGMP-independent mechanism (Bolotina et al., 1994).

Na^+-K^+-ATPase is a ubiquitously expressed enzyme that utilizes the energy derived from ATP hydrolysis to transport Na^+ and K^+ across the plasma membrane. Na^+-K^+-ATPase is electrogenic, since it extrudes 3 Na^+ while bringing 2 K^+ into the cell, and by virtue of this, it plays a role in the regulation of membrane potential and the tone of vascular and nonvascular smooth muscle cells and cardiac myocytes (Sweadner 1989; Skou 1992). Stimulation of Na^+-K^+-ATPase activity, similar to activation of K^+ channels, results in hyperpolarization leading to inhibition of Ca^{2+} influx via voltage-operated channels, and secondarily to relaxation of smooth muscle. In contrast, inhibition of Na^+-K^+-ATPase activity leads to depolarization of smooth muscle cells and contraction (Blaustein 1977, 1993). This chapter summarizes studies demonstrating that NO stimulates Na^+-K^+-ATPase activity independently of its ability to elevate the intracellular cGMP concentration, and that this is one of the mechanisms responsible for NO-induced relaxation of vascular and penile corpus cavernosum smooth muscle. The possible relationship of NO-induced stimulation of Na^+-K^+-ATPase activity and the pathogenesis of diabetic vascular complications and erectile dysfunction is also discussed.

Role of Endothelium in Regulation of Na$^+$ Pump Activity in Blood Vessels

Vascular endothelium plays a critical role in the regulation of underlying smooth muscle tone and local hemodynamics. Endothelial cells also contribute to the maintenance of resting Na$^+$-K$^+$-ATPase activity in the aorta, since removal of endothelium, which lowers intracellular cGMP concentration, diminished ouabain-sensitive uptake of ^{86}Rb$^+$ (a measure of Na$^+$-K$^+$-ATPase activity) (Figure 4.1) (Gupta et al. 1992). Likewise, endothelium removal has been shown to lower Na$^+$-K$^+$-ATPase activity in the carotid artery (Rodriguez-Manas et al. 1992). The diminution in

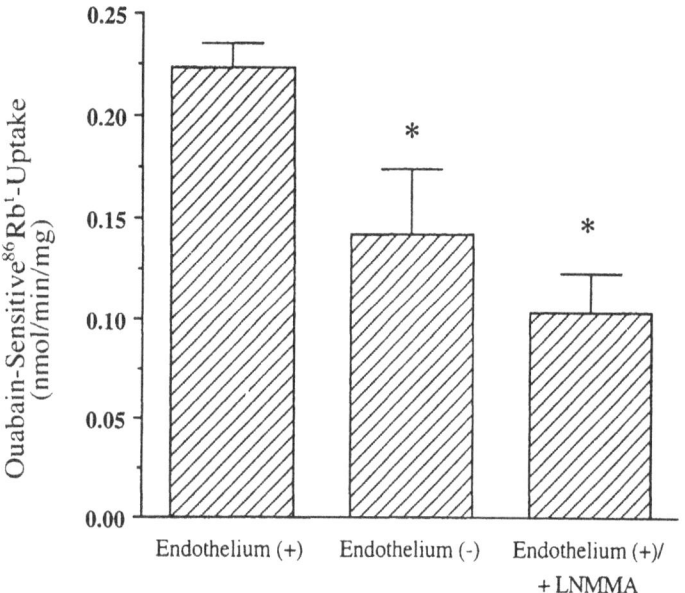

FIGURE 4.1. Effects of endothelial denudation and inhibition of the NOS inhibitor I-NMMA, on ouabain-sensitive ^{86}Rb$^+$ uptake. Aortic rings with or without endothelium were incubated for 3 hours in physiological salt solution containing (mM): 118 NaCl, 4.5 KCl, 0.54 MgCl$_2$, 2.5 CaCl$_2$, 1.2 Na$_2$HPO$_4$, 25 NaHCO$_3$, and 5.5 D-glucose. At the end of this time, ^{86}RbCl (2 μCi/ml) was added to the vial for 10 minutes. The tissues were next washed in cold (4°C) unlabeled physiological salt solution and dried. The ^{86}Rb$^+$ content of tissue was determined by gamma counting. Ouabain-sensitive ^{86}Rb uptake was calculated by subtracting ^{86}Rb uptake in the presence of a maximally effective concentration of ouabain (0.2 mM present during the final 20 minutes) from total ^{86}Rb uptake. L-NMMA (300 μM) was present for the final 30 minutes. Results are means ± SE. * Values significantly different from that of aorta with endothelium, P<0.05. From Gutpa et al. (1992.) Copyright 1992 by the American Society for Clinical Investigation. Adapted with permission.

Na$^+$-K$^+$-ATPase activity in endothelium-denuded rings was not related to uptake by endothelial cells, since Na$^+$-K$^+$-ATPase activity in rings from which endothelium was removed after the completion of ^{86}Rb$^+$ uptake, was not different from that of rings with an intact endothelium (Gupta et al. 1992). Furthermore, inhibition of NO synthase (NOS) activity in endothelium-intact rings, but not endothelium-denuded rings, decreased tissue cGMP levels and Na$^+$-K$^+$-ATPase activity, suggesting that NO released from vascular endothelial cells may be responsible for endothelium-dependent Na$^+$-K$^+$-ATPase activity in blood vessels (Gupta et al. 1994b).

Nitric Oxide Stimulates Na$^+$ Pump Activity in Vascular Smooth Muscle

The involvement of Na$^+$-K$^+$-ATPase activity in endothelium-dependent relaxation of vascular smooth muscle was first suggested by DeMey and Vanhoutte (1980), who showed that the inhibitory effects of acetylcholine on blood vessel tone were prevented by ouabain. This was later confirmed by a number of investigators (Foley 1984; Rappaport 1985a, 1985b). Although it was known for some time that acetylcholine causes endothelium-dependent relaxation of vascular smooth muscle by releasing NO, the relationship between NO and Na$^+$-K$^+$-ATPase activity was not evident until recently (Gupta et al. 1992). In aortic smooth muscle, NO gas dissolved in solution as well as NO donor, sodium nitroprusside (SNP), increased Na$^+$-K$^+$-ATPase activity with concomitant increases in intracellular cGMP concentration (Figure 4.2). The effect of SNP on Na$^+$-K$^+$-ATPase activity in endothelium-denuded rings was concentration dependent. Half-maximal stimulation occurred at ~0.3 μM SNP (Gupta et al. 1994b). These observations, taken together with studies showing inhibition of NO relaxation responses by ouabain, indicate a role for Na$^+$-K$^+$-ATPase activity in NO-induced relaxation of vascular smooth muscle. As mentioned earlier, stimulation of Na$^+$-K$^+$-ATPase activity by NO would result in hyperpolarization and secondarily relaxation of vascular smooth muscle. In further support of this, acetylcholine- and NO-induced hyperpolarization of arterial smooth muscle has been shown to be sensitive to ouabain and membrane-depolarizing agents (Feletou and Vanhoutle et al. 1988; Tare et al. 1990; Rand and Garland 1992).

cGMP-Independent Stimulation of Na$^+$-K$^+$-ATPase Activity by NO

Stimulation of Na$^+$-K$^+$-ATPase activity has been proposed as one of the mechanisms underlying cGMP-induced vasorelaxation, since inhibitors of Na$^+$-K$^+$-ATPase activity prevent relaxation induced by the guanylate cyclase activators as well as cGMP (Rappaport et al. 1985a, b). However, there is no direct evidence to date in

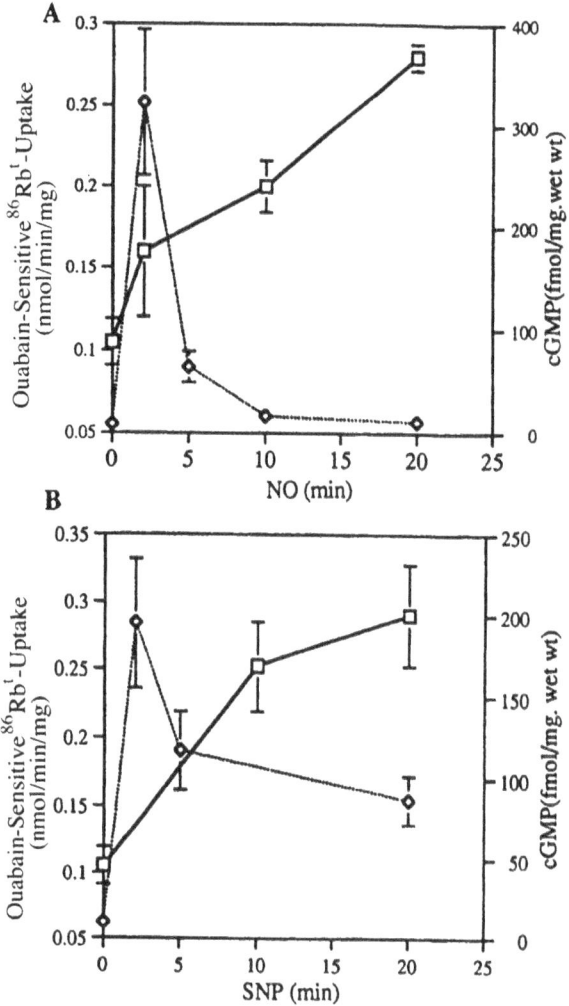

FIGURE 4.2. Time-dependent stimulation of ouabain-sensitive ^{86}Rb$^+$ uptake and cGMP formation by NO and SNP in endothelium-denuded aortic rings. (A) Nitric oxide (NO, 1 μM) and (B) sodium nitroprusside (SNP, 10 μM) were added for the indicated periods at the end of the preincubation period. For ^{86}Rb$^+$ uptake studies, ^{86}RbCl (2 μCi/ml) was added to incubation media along with NO and SNP. Results are means ± SE of three to five experiments. (From Gupta et al. 1994b.) Copyright 1994 by the American Physiological Society. Reprinted with permission.

support of this hypothesis. On the contrary, NO-induced stimulation of Na$^+$-K$^+$-ATPase activity was independent of cGMP formation, as the guanylate cyclase inhibitor/cGMP depletor LY83583 lowered cGMP levels by approximately 80% but had no effect on Na$^+$-K$^+$-ATPase activity in endothelium-intact aortic rings. Similarly, the phosphodiesterase inhibitor isobutylmethylxanthine (IBMX) increased cGMP levels without affecting Na$^+$-K$^+$-ATPase activity (Gupta et al. 1994b). Furthermore, permeable cGMP analogues that caused relaxation of precontracted vas-

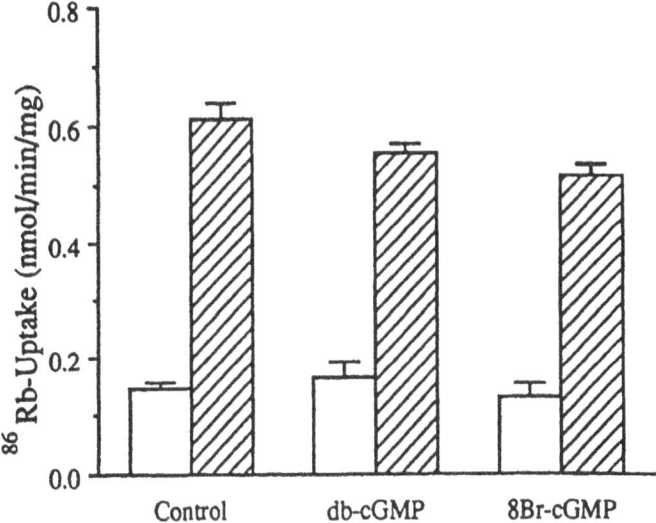

FIGURE 4.3. Lack of effect of cGMP analogues on ouabain-sensitive ^{86}Rb$^+$ uptake in endothelium-denuded rings. 8-Bromo- or dibutyryl-cGMP (100 μM) was added to the incubation medium during the final 20 minutes of incubation. Results are means ± SE of three to five experiments. (From Gupta et al. 1994b.) Copyright 1994 by the American Physiological Society. Reprinted with permission.

cular smooth muscle did not stimulate Na$^+$-K$^+$-ATPase activity in the aorta (Figure 4.3). These results suggest that NO stimulates vascular smooth muscle Na$^+$-K$^+$-ATPase activity independently of its ability to increase cGMP concentration.

NO Stimulates Na$^+$-K$^+$-ATPase Activity by Interacting with Sulfhydryl Groups of the Enzyme

Nitric oxide has been shown to cause ADP ribosylation of actin (Clancy et al. 1991) and regulate activity of N-methyl-D-aspartate (NMDA) receptors via S-nitrosylation of free sulfhydryl groups (Lipton et al. 1993). Furthermore, activation of Ca^{2+}-activated K$^+$ channels in vascular smooth muscle, which was sensitive to the sulfhydryl group alkylating agent N-ethylmaleimide, has been shown to be independent of intracellular cGMP concentration (Bolotina et al. 1994). Na$^+$-K$^+$-ATPase is a cysteine-rich protein with multiple sulfhydryl (-SH) groups, and its activity is modulated by agents that modify sulfhydryl groups (Sweadner 1989). Thus, NO-induced stimulation of Na$^+$-K$^+$-ATPase may also involve nitrosylation of sulfhydryl groups of the enzyme. As shown in Figure 4.4, NO caused a concentration-dependent increase in the activity of a partially purified preparation of Na$^+$-K$^+$-ATPase. The ability of NO to directly stimulate Na$^+$-K$^+$-ATPase activity was prevented upon pretreatment of the enzyme with N-ethylmaleimide. Furthermore, the sulfhydryl oxidizing agent

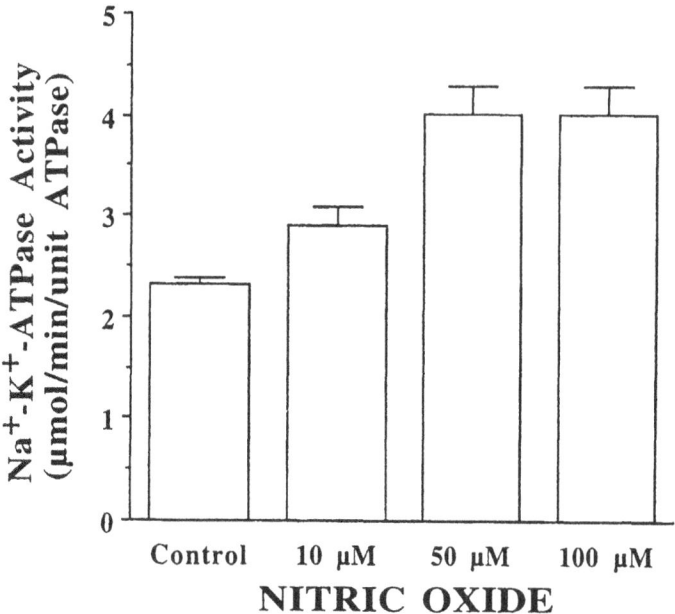

FIGURE 4.4. Nitric oxide stimulates Na^+-K^+-ATPase activity in vitro. The direct effects of NO on Na^+-K^+-ATPase activity were studied in a partially purified enzyme preparation. Enzyme activity was measured at 37°C by coupling ADP production to oxidation of NADH and recording the absorbance change versus time at 340 nm in the absence and presence of 0.1 mM ouabain. The assay mixture contained 0.3 mM NADH, 1 mM phosphoenolpyruvate, 5 mM $MgCl_2$, 120 mM NaCl, 25 mM KCl, 0.1 mM EGTA, 2.5 mM Na-ATP, 4 U/ml of pyruvate kinase, 20 U/ml of lactic dehydrogenase, and 50 mM triethanolamine, pH 7.4. The reaction was started with the addition of ~20 µg/ml Na^+-K^+-ATPase. In some experiments, the enzyme was pretreated with 5 mM N-ethylmaleimide for 30 minutes. (Unpublished data; Daley, J. and Gupta, S.)

2,2-dithiodipyridine mimicked the effect of NO on Na^+-K^+-ATPase activity in endothelium-denuded aortic rings (unpublished data).

Role of NO in Stimulation of Na^+-K^+-ATPase Activity by Agonists

Vasoactive agonists and hormones (norepinehrine, endothelin-1) modulate Na^+-K^+-ATPase activity by altering the intracellular Na^+ concentration and/or by protein kinase A (PKA)- and protein kinase C (PKC)-mediated phosphorylation of the catalytic α-subunit of the enzyme (Brock et al. 1982; Navran et al. 1988; Sweadner 1988; Bertorello et al. 1991; Gupta et al. 1991). Endothelin-1 stimulates Na^+-K^+-ATPase activity in vascular smooth muscle in a concentration-dependent

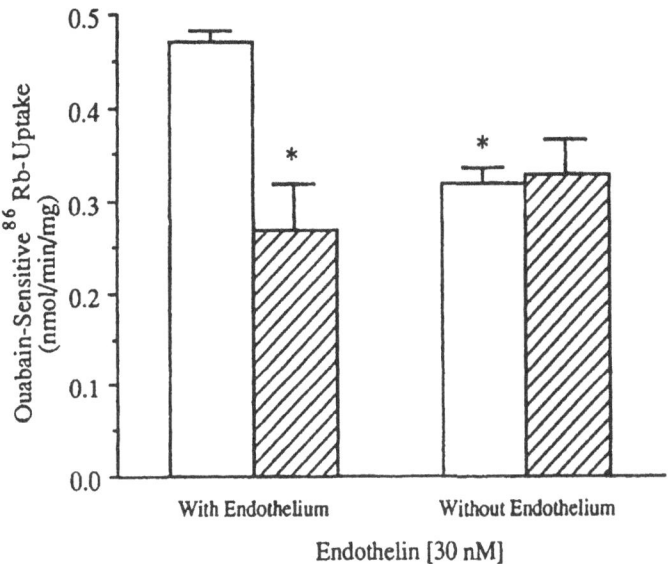

FIGURE 4.5. Inhibition of endothelin-stimulated, ouabain-sensitive ^{86}Rb$^+$ uptake by L-NMMA. Aortic rings with or without endothelium were incubated in PSS for 3 hours. L-NMMA (0.3 mM) and endothelin (30 nM) were added to the incubation medium during the final 30 and 20 minutes, respectively. Results are means ± SE of five experiments.* Values significantly different from those in the endothelin-stimulated group (with endothelium) in the absence of LNMMA. (From Gupta et al. 1994a.) Copyright 1994 by the American Physiological Society. Reprinted with permission.

manner (Gupta et al. 1991, 1994a). Endothelin-1 has also been shown to modulate NO release in blood vessels by interacting with specific receptors on endothelial cells. In this context, endothelin lowers blood pressure in rats in vivo (Whittle et al. 1989) and causes endothelium-dependent relaxation of blood vessels via generation of NO (Suzuki et al. 1991). In addition, PKC-activating phorbol esters have been suggested to modulate NO release (Sakata and Karaki 1990; Nakane et al. 1991). In rabbit aorta, the ability of endothelin to stimulate Na$^+$-K$^+$-ATPase activity was partly dependent on its ability to release NO from endothelium. In support of this, removal of endothelium significantly diminished endothelin-induced Na$^+$-K$^+$-ATPase activity (Gupta et al. 1994a). In addition, the NOS inhibitor N^G-monomethyl-L-arginine (L-NMMA) diminished endothelin-stimulated Na$^+$-K$^+$-ATPase activity in endothelium-intact rings, but not in endothelium-denuded rings (Figure 4.5). Furthermore, endothelin caused a significant increase in intracellular cGMP concentration in endothelium-intact but not in endothelium-denuded aortic rings. In contrast to endothelin, increases in Na$^+$-K$^+$-ATPase activity caused by the α_1-adrenergic receptor agonist phenylephrine and by phorbol 12–13 dibutyrate (PDBu) were independent of endothelium-derived NO, since removal of endothelium did not diminish PDBu-stimulated Na$^+$-K$^+$-ATPase activity and it enhanced stimulation by phenylephrine (Gupta et al. 1994a).

Stimulation of Na⁺-K⁺-ATPase Activity by Insulin

Insulin stimulates the activity of Na⁺-K⁺-ATPase in a number of tissues, including brain, adipocytes, skeletal muscle, and vascular smooth muscle (McGill 1991; McGill and Guidotti 1991; Tirupattur et al. 1993; Sampson et al. 1994; Gupta et al. 1996). Recent evidence suggests that insulin acts in vascular smooth muscle by stimulating PKC and/or increasing NO synthesis (Cooper et al. 1993; Baron 1994; Wu et al. 1994). However, endothelium-derived NO does not seem to play a role in the stimulation of Na⁺-K⁺-ATPase activity by physiological concentrations of insulin in vascular smooth muscle. In the rabbit aorta, insulin caused a greater increase in Na⁺-K⁺-ATPase activity in endothelium-denuded rings than in endothelium-intact rings (Figure 4.6). Insulin did not increase intracellular cGMP concentration in endothelium-intact aortic rings, and its ability to increase Na⁺-K⁺-ATPase activity was only marginally affected by L-NMMA. The effects of insulin and the NO donor SNP on Na⁺-K⁺-ATPase activity were additive. In contrast, the effects of insulin and PDBu on Na⁺-K⁺-ATPase activity were not additive. Pretreatment with PKC inhibitors totally inhibited the ability of insulin to increase Na⁺-K⁺-ATPase activity in both endothelium-intact and endothelium-denuded aorta, suggesting that insulin-induced increase in Na⁺-K⁺-ATPase activity may be related to its ability to activate PKC (Gupta et al. 1996).

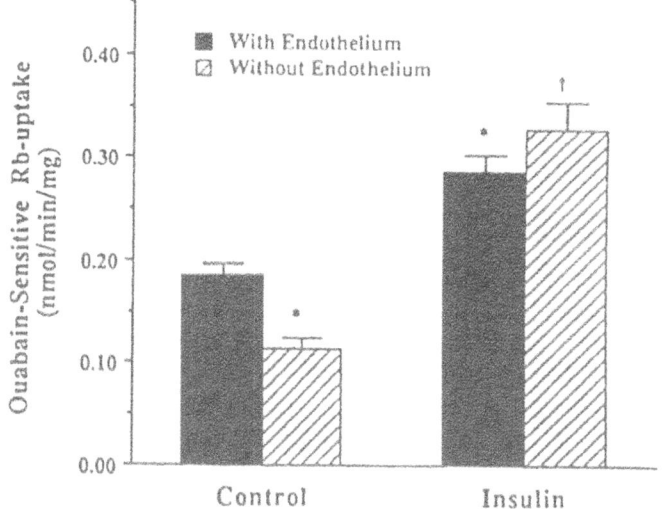

FIGURE 4.6. Effect of endothelium removal on insulin-stimulated, ouabain-sensitive ⁸⁶Rb⁺ uptake. Aortic rings with or without endothelium were incubated in PSS glucose for 3 hours. Insulin (1 mU/ml) was added during the final 20 minutes of incubation. Results are means ± SE of six experiments. *Significantly different from values in endothelium-intact and denuded controls. (†Significantly different from values) in endothelium denuded controls. (From Gupta et al. 1996.) Copyright 1996 by the American Physiological Society. Reprinted with permission.

Nitric Oxide, Na$^+$-K$^+$-ATPase Activity, and Penile Erection

It is now well recognized that NO is a key mediator of penile erection. Nitric oxide is synthesized by both the endothelium lining the lacunar spaces and the autonomic dilator nerves of the corpora cavernosa, and causes relaxation of penile arterial and trabecular smooth muscle (Ignarro et al. 1990; Kim et al. 1991; Rajfer et al. 1992). Furthermore, inhibitors of NOS activity block penile erection induced by electrical stimulation of pelvic nerves in vivo (Burnette et al. 1992). As in vascular smooth muscle, Na$^+$-K$^+$-ATPase activity plays an important role in NO-induced relaxation of corpus cavernosum smooth muscle (Gupta et al. 1995). In support of this, the Na$^+$-K$^+$-ATPase inhibitor ouabain inhibited relaxation of human and rabbit corpus cavernosum smooth muscle induced by endothelium-derived NO as well as NO donors (Figure 4.7). Nitric oxide also stimulated Na$^+$-K$^+$-ATPase activity in cultured human corpus cavernosum smooth muscle cells (HCCSMC). The ability of NO to stimulate Na$^+$-K$^+$-ATPase activity was not affected by the guanylate cyclase inhibitor methylene blue, which prevents NO-induced increases in intracellular cGMP concentration in HCCSMC. Furthermore, as in vascular smooth muscle, permeable cGMP analogues did not increase Na$^+$-K$^+$-ATPase activity in HCCSMC, indicating that stimulation of Na$^+$-K$^+$-ATPase activity by NO in HCCSMC was independent of increases in the intracellular cGMP concentration (Gupta et al. 1995).

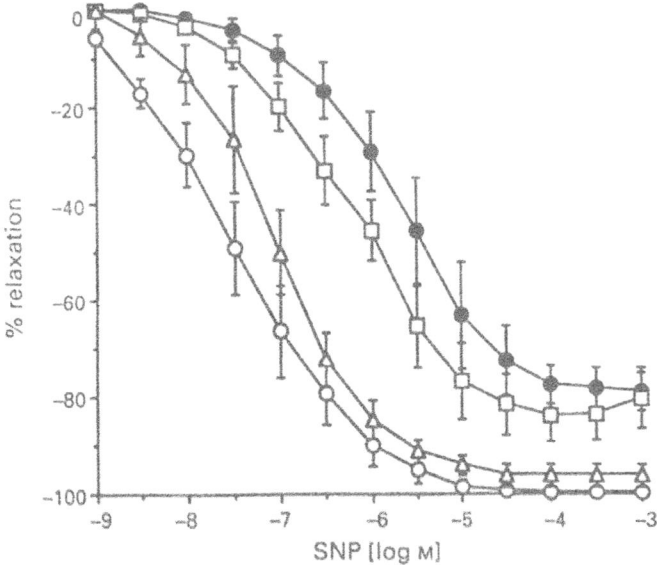

FIGURE 4.7. Inhibition of SNP-induced relaxation by ouabain. Human corpus cavernosum strips were mounted on force transducers in the organ chambers for isometric tension measurements. Human corpus cavernosum smooth muscle strips were incubated in the absence or presence of 0.1, 1.0, and 10 μM ouabain. Results are means ± SE of four experiments. (From Gupta et al. 1995.) Copyright 1995 by Stockton Press, MacMillan Publishers Ltd. Reprinted with permission.

Nitric Oxide–Na^+-K^+-ATPase System in Digoxin-Associated Male Erectile Dysfunction

Erectile dysfunction, defined as the consistent inability to attain and maintain an erection sufficient to permit satisfactory sexual performance, has been reported with drug use, particularly for cardiovascular disorders (Brock and Lue 1993; Feldman et al. 1994). The cardiac glycoside digoxin is one of the most commonly prescribed drugs for the treatment of heart failure. Digoxin is also among the 10 most commonly dispensed medications whose use has been associated with male sexual dysfunction (Hubbard et al. 1991; Feldman et al. 1994). Since digoxin has a cyclopentanoperhydrophenanthrene nucleus similar to that of steroid hormones, some investigators have suggested that it affects sexual function by depressing libido as a result of an altered sex hormone milieu (Stoffer et al. 1973; Neri et al. 1980, 1987). However, other investigators have failed to confirm these findings (Kley et al. 1982, 1984).

Na^+-K^+-ATPase is the physiological receptor for digoxin, and the proposed mechanism for the positive inotropic action of digoxin in cardiac tissue involves inhibition of Na^+-K^+-ATPase activity (Schwartz et al. 1969; Blaustein 1993). The results of recent in vitro and in vivo studies indicate that the direct effects of digoxin on the NO–Na^+-K^+-ATPase system contribute to erectile dysfunction associated with its use in humans. In support of this, physiological concentrations of digoxin inhibited $^{86}Rb^+$ uptake (half-maximum effect at 0.01 μM) and caused contraction of corporal smooth muscle (half-maximum effect at 0.8 μM). Therapeutic concentrations of digoxin (2 nM) also inhibited relaxation induced by NO released from corpus cavernosum endothelial cells as well as nonadrenergic, noncholinergic nerves (Figure 4.8). A prospective, double-blind, placebo-controlled, crossover, four-period investigation in healthy male volunteers revealed that digoxin diminishes penile rigidity during visual sexual stimulation and nocturnal penile tumescence testing without influencing libido or serum levels of testosterone, estrogen, or luteinizing hormone (Gupta et al. 1998).

Use of the Na^+ Pump Inhibitor Digoxin for the Treatment of Idiopathic Recurrent Venocclusive Priapism

Idiopathic recurrent venoocclusive priapism (IRVOP) is a rare medical condition of persistent erection without sexual stimulation secondary to the inability to regulate venous outflow. If left untreated, it may lead to permanent erectile dysfunction (Bertram et al. 1985; Levine et al. 1991). The underlying pathophysiologic mechanism for IRVOP is unknown, although it is hypothesized to be due to a local imbalance in the relaxation/contraction mechanisms regulating corporal smooth muscle contractility. Therefore, nonsurgical strategies for the management of IRVOP have been directed toward increasing smooth muscle tone. An ideal pharmacological agent to treat IRVOP would not only promote contractility of smooth muscle but also inhibit its ability to relax. Recently, digoxin, due to its inhibitory

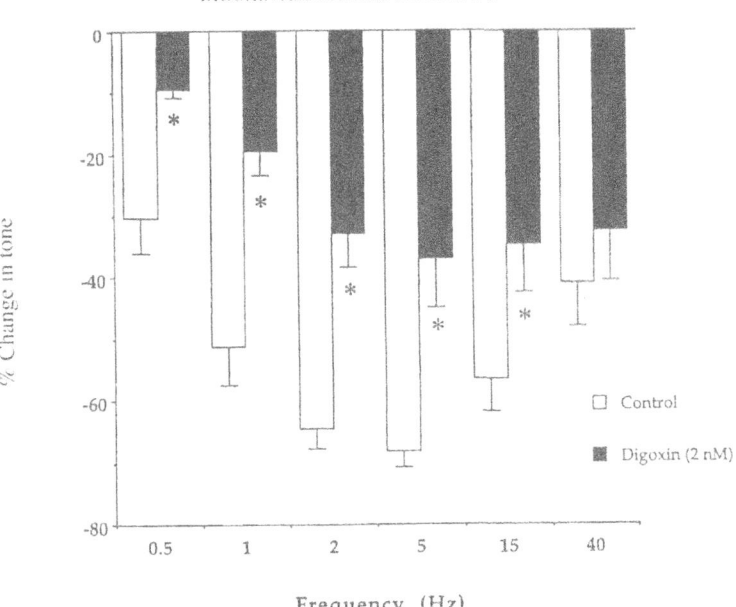

FIGURE 4.8. Effect of digoxin on relaxation induced by electrical field stimulation. Human corpus cavernosum strips were mounted on force transducers in the organ chambers. Results are means ± SE of six experiments in strips from different patients. Addition of vehicle dimethyl sulfoxide (final concentration, 0.02%) did not affect electrical field stimulation-induced relaxation. * Values significantly different from control (no digoxin) group as determined by unpaired t-test ($P < 0.05$). (From Gupta et al. 1998.) Copyright 1998 by Lippincott Williams & Wilkens. Reprinted with permission.

effects on corpus cavernosum Na⁺-K⁺-ATPase activity leading to contraction of corpus cavernosum smooth muscle and as inhibition of NO-induced relaxation, has recently been utilized for this purpose. In a 6-month follow-up study, digoxin treatment significantly reduced the frequency as well as the duration of IRVOP episodes in affected humans (Hossein, Gupta et al. 1997).

Role of the NO–Na⁺-K⁺-ATPase System in the Development of Diabetic Complications

Diabetes mellitus induces a decrease in Na⁺-K⁺-ATPase activity in humans and animals, and this decrease plays a role in the development of diabetic neuropathy as well as in the development of diabetic vascular and other complications (Greene et al. 1987). Diabetes and associated hyperglycemia have been shown to

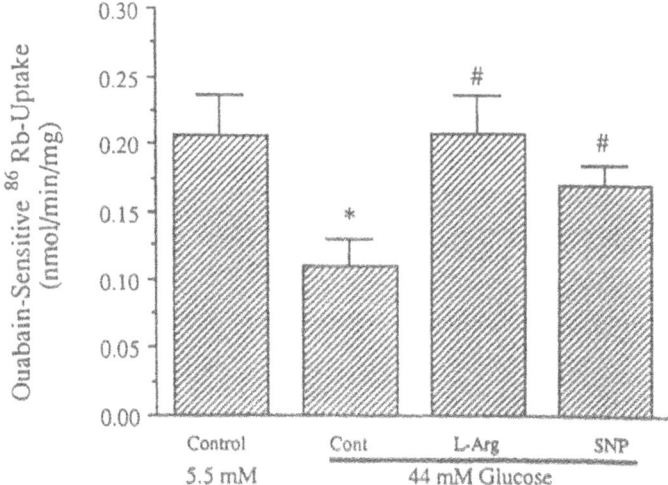

FIGURE 4.9. Reversal of hyperglycemia-induced inhibition of ouabain-sensitive ^{86}Rb$^+$ uptake by L-arginine and SNP in endothelium-intact aorta. L-Arginine (0.3 mM) and SNP (10 μM) were added to the incubation medium during the final 30 and 10 minutes, respectively, of the 3-hour incubation. When L-arginine (0.3 mM) was present in the medium during the entire 3-hour incubation, hyperglycemia failed to decrease ouabain-sensitive ^{86}Rb$^+$ uptake. Results are means ± SE of five experiments. Values significantly different from that of aorta incubated in *5.5 or #44 mM glucose, $P < 0.05$. (From Gupta et al. 1992.) Copyright 1992 by the American Society for Clinical Investigation. Reprinted with permission.

inhibit endothelium-dependent responses in blood vessels (Gupta et al. 1992, 1994a; Tesfamariam et al. 1991; Johnstone et al. 1993). The mechanisms underlying diabetes and hyperglycemia-induced alteration of endothelium-dependent function are not clearly understood. Nonetheless, decreases in NOS and Na$^+$ pump activities have been implicated (Gupta et al. 1992; Williamson et al. 1993; Pieper and Peltier 1995). Diabetes and hyperglycemia-induced lessening of Na$^+$-K$^+$-ATPase activity in the aorta has been shown to be related to endothelial dysfunction, most probably associated with diminished synthesis or availability of NO. In support of this, the effects of hyperglycemia on Na$^+$ pump activity were mimicked by NOS inhibitors and reversed by L-arginine and exogenously added NO (Figure 4.9). Similar diabetes-associated defects in the NO–Na$^+$-K$^+$-ATPase system have also been implicated in slowing nerve conduction (Stevens et al. 1994). Along these lines, the NO–Na$^+$-K$^+$-ATPase system also seems to play a crucial role in the pathogenesis of erectile dysfunction in diabetes. In support of this, corpus cavernosum tissues from impotent men with diabetes mellitus have impaired NO-mediated relaxation of the trabecular smooth muscle (Saenz De Tejada et al. 1989), probably due to either reduced synthesis or reduced availability of NO. Furthermore, corpus cavernosum Na$^+$-K$^+$-ATPase activity is depressed in diabetes.

Conclusions

In conclusion, NO stimulates Na^+-K^+-ATPase activity in vascular and corpus cavernosum smooth muscle. This effect of NO is independent of its ability to increase intracellular cGMP concentration. The stimulation of Na^+-K^+-ATPase activity, which would cause hyperpolarization, is another mechanism in addition to increasing intracellular cGMP concentration by which NO causes relaxation of vascular smooth muscle. An impairment in this NO–Na^+-K^+-ATPase system contributes to diabetes-associated vascular and erectile dysfunction and underlies digoxin-associated erectile dysfunction.

References

Baron AD. 1994. Hemodynamic actions of insulin. *Am. J. Physiol. 267:*E187–202.

Bertorello AM, Aperia A, Walaas SI, & Nairni AC. 1991. Phosphorylation of the catalytic subunit of Na^+K^+ATPase inhibits the activity of the enzyme. *Proc. Natl. Acad. Sci. USA 88:*11359–11362.

Bertram RA, Carson CC III, & Webster GD. 1985. Implantation of penile prostheses in patients impotent after priapism. *Urology 26:*325–327.

Blaustein MP. 1977. Sodium ions, calcium ions, blood pressure regulation and hypertension: a reassessment and a hypothesis. *Am. J. Physiol. 245:*H604–H609.

Blaustein MP. 1993. Physiological effects of endogenous ouabain: control of intracellular Ca^{2+} stores and cell-responsiveness. *Am. J. Physiol. 232:*C165–C173.

Bolotina VM, Najibi S, Palacino JJ, Pagano PJ, & Cohen RA. 1994. Nitric oxide directly activates calcium-dependent potassium channels in vascular smooth muscle. *Nature 368:*850–853.

Brock GB, & Lue TF. 1993. Drug-induced male sexual dysfunction. An update. *Drug Safety 8:*414–426.

Brock TA, Lewis LJ, & Smith JB. 1982. Angiotensin increases Na entry and Na/K pump activity in cultures of smooth muscle from rat aorta. *Proc. Natl. Acad. Sci. USA 79:*1438–1442.

Burnette AL, Lowenstein CJ, Bredt DS, Chang TSK, & Snyder SH. 1992. Nitric oxide: a physiologic mediator of penile erection. *Science 257:*401–403.

Clancy RM, Leszcynska-Piziak J, & Abramson SB. 1991. Nitric oxide stimulates the ADP-ribosylation of actin in human neutrophils. *Biochem. Biophys. Res. Commun. 191:*847–852.

Cooper DR, Khalakdina A, & Watson JE. 1993. Chronic effects of glucose on insulin signaling in A-10 vascular smooth muscle cells. *Arch. Biochem. Biophys. 302:*490–498.

Craven PA, & DeRubertis FR. 1978. Restoration of responsiveness of purified guanylate cyclase to nitrosoguanidine, nitric oxide and related activators by heme and heme proteins: evidence for the involvement of the papamagnetic nitrosyl-heme complex in enzyme activation. *J. Biol. Chem. 253:*8433–8438.

DeMey JG, & Vanhoutte PM. 1980. Interaction between Na^+, K^+ exchanges and the direct inhibitory effect of acetylcholine on canine femoral arteries. *Circ. Res. 46:*826–835.

Feldman H, Goldstein I, Hatzichristou D, Krane R, & McKinlay J. 1994. Impotence and its medical and psychosocial correlates: results of the Massachusetts male aging study. *J. Urol. 151:*54–61.

Feletou M, & Vanhoutte PM. 1988. Endothelium-dependent hyperpolarization of canine coronary smooth muscle. *Br. J. Pharmacol. 93:*515–524.

Foley DH. 1984. Diminished arterial smooth muscle response to sodium nitroprusside during Na-K pump inhibition. *Pharmacology 28:*95–103.

Furchgott RF, & Vanhoutte PM. 1989. Endothelium-derived relaxing and contractor factors. *FASEB J. 3:*2007–2018.

Greene DA, Lattimer SA, & Sima AAF. 1987. Sorbitol, phosphoinositides and sodium-potassium-ATPase in the pathogenesis of diabetic complications. *N Engl J Med 316:*599–606.

Gupta S, Ruderman NB, Cragoe EJ Jr., & Sussman I. 1991. Endothelin stimulates Na^+/K^+ ATPase activity by a protein kinase C-dependent pathway in rabbit aorta. *Am. J. Physiol. 261:*H38–H45.

Gutpa S, Sussman I, McArthur CS, Tornheim K, Cohen RA, & Ruderman NB. 1992. Endothelium-dependent inhibition of Na^+-K^+ ATPase activity in rabbit aorta by hyperglycemia. Possible role of endothelium-derived nitric oxide. *J. Clin. Invest. 90:*727–732.

Gupta S, McArthur C, Grady C, & Ruderman NB. 1994a. Role of endothelium-derived nitric oxide in the stimulation of Na^+-K^+ ATPase activity by endothelin in rabbit aorta. *Am. J. Physiol. 266:*H577–H582.

Gupta S, McArthur C, Grady C, & Ruderman NB. 1994b. Stimulation of Na^+K^+ ATPase activity by nitric oxide: a cGMP-independent effect. *Am. J. Physiol. 266:*H2146–2151.

Gupta S, Moreland RB, Munarriz R, Daley J, Goldstein I, & Saenz de Tejada I. 1995. Possible role of Na^+-K^+-ATPase in the regulation of human corpus cavernosum smooth muscle contractility by nitric oxide. *Br. J. Pharmacol. 116:*2201–2206.

Gupta S, Tieken K, & Ruderman NB. 1996. Differential stimulation of vascular Na^+-K^+-ATPase activity by insulin and nitric oxide. *Am. J. Physiol. 270:*H1287–H1293.

Gupta S, Salimpour P, Saenz de Tejada I, Daley J, Gholami S, Krane R, Traish A, & Goldstein I. 1998. A possible mechanism for alteration of human erectile function by digoxin: inhibition of corpus cavernosum smooth muscle Na^+-K^+-ATPase activity. *J. Urol. 159:*1529–1536.

Hossein SN, Gupta S, Salimpour P, Krane RJ, & Goldstein I. 1997. Treatment of recurrent veno-occlusive priapism by digoxin. *J. Urol. 157:*202(787).

Hubbard JR, Levenson JL, & Patrick G. 1991. Psychiatric side effects associated with the ten most commonly dispensed prescription drugs: a review. *J. Fam. Pract. 33:*177–186.

Ignarro LJ, Bush PA, Buga GM, Wood KS, Fukuto JM, & Rajfer J. 1990. Nitric oxide and cyclic GMP formation upon electrical stimulation cause relaxation of corpus cavernosum smooth muscle. *Biochem Biophys Res Commun 170:*843–850.

Johnstone MT, Creager SJ, Scales KM, Cusco JA, Lee BK, & Creager MA. 1993. Impaired endothelium-dependent vasodilation in patients with insulin-dependent diabetes mellitus. *Circulation 88:*2510–2516.

Kim N, Azadzoi KM, Goldstein I, & Saenz De Tejada I. 1991. A nitric oxide-like factor mediates nonadrenergic-noncholinergic neurogenic relaxation of penile corpus cavernosum smooth muscle. *J. Clin. Invest. 88:*112–118.

Kley HK, Muller A, Pecrenboom H, & Kruskemper HL. 1982. Digoxin does not alter plasma steroid levels in healthy men. *Clin. Pharmacol. Ther. 32:*12–17.

Kley HK, Abendroth H, Hehrmann R, Muller A, Keck E, & Schneitler H. 1984. No effect of digitalis on sex and adrenal hormones in healthy subjects and in patients with congestive heart failure. *Klin. Wochenschr. 62:*65–73.

Levine JF, Saenz de Tejada I, Payton, TR, & Goldstein I. 1991. Recurrent prolonged erections and priapism as a sequela of priapism: pathophysiology and management. *J Urol* *145*:764–767.

Lincoln TM, & Cornwell TL. 1993. Intracellular cyclic GMP receptor proteins. *FASEB J.* *7*:328–338.

Lipton SA, Choi Y-B, Pan Z-H, Lei SZ, Chen H-SV, Sucher NJ, Loscalzo J, Singel DJ, & Stamler JS. 1993. A redox-based mechanism for the neuroprotective and neurodestructive effect of nitric oxide and related nitroso-compounds. *Nature 364*:626–632.

McGill DL. 1991. Characterization of the adipocyte ghost (Na$^+$, K$^+$) pump: insights into the insulin regulation of the adipocyte (Na$^+$, K$^+$) pump. *J. Biol. Chem. 266*:15817–15823.

McGill DL, & Guidotti G. 1991. Insulin stimulates both the alpha 1 and alpha 2 isoforms of the rat adipocyte (Na$^+$, K$^+$) ATPase: two mechanisms of stimulation. *J. Biol. Chem. 266*:15824–15831.

Nakane M, Mitchelle J, Forstermann U, & Murad F. 1991. Phosphorylation by calcium calmodulin-dependent protein kinase II and protein kinase C modulates the activity of nitric oxide synthase. *Biochem. Biophys. Res. Commun. 180*:1396–1402.

Navran SS, Adair SE, Jemelka SK, Seidel CL, & Allen JC. 1988. Sodium pump stimulation by activation of two alpha adrenergic receptor subtypes in canine blood vessels. *J. Pharmacol. Exp Ther. 245*:608–613.

Neri A, Aygen M, Zukerman Z, & Bahary C. 1980. Subjective assessment of sexual dysfunction of patients on long-term administration of digoxin. *Arch. Sex Behav. 9*:319–326.

Neri A, Zukerman Z, Aygen M, Lidor Y, & Kaufman H. 1987. The effect of long-term administration of digoxin on plasma androgens and sexual dysfunction. *J. Sex. Mar. Ther. 13*:58–63.

Pieper G, & Peltier B. 1995. Amelioration by L-arginine of a dysfunctional arginine/nitric oxide pathway in diabetic endothelium. *J. Cardiovasc. Pharmacol. 25*:397–403.

Rajfer J, Aronson WJ, Bush PA, Dorey FJ, & Ignarro, L. 1992. Nitric oxide as a mediator of relaxation of corpus cavernosum from rabbit and man. *N. Engl. J. Med. 326*:90–94.

Rand VE, & Garland CJ. 1992. Endothelium-dependent relaxation to acetylcholine in the rabbit basilar artery: importance of membrane hyperpolarization. *Br. J. Pharmacol. 106*:143–150.

Rappaport RM, Schwartz K, & Murad F. 1985a. Effects of sodium-potassium pump inhibitors and membrane depolarizing agents on sodium nitroprusside-induced relaxation and cyclic guanosine monophosphate accumulation in rat aorta. *Circ. Res. 57*:164–170.

Rappaport RM, Waldan SA, Schwartz K, Winquist RJ, & Murad F. 1985b. Effects of atrial natriuretic factor, sodium nitroprusside, and acetylcholine on cyclic GMP levels and relaxation in rat aorta. *Eur. J. Pharmacol. 115*:219–229.

Rodriguez-Manas L, Pareja A, Sanchez-Ferrer CF, Casado MA, Salaices M, & Marin J. 1992. Endothelial role in ouabain-induced contractions in guinea pig carotid arteries. *Hypertension 20*:674–681.

Saenz de Tejada I, Goldstein I, Azadzoi K, Krane RJ, & Cohen RA. 1989. Impaired neurogenic and endothelium-mediated relaxation of penile smooth muscle from diabetic men with impotence. *N. Engl. J. Med. 320*:1025–1030.

Sakata K, & Karaki H. 1990. Phorbol ester-induced release of endothelium-derived relaxing factor. *Eur. J. Pharmacol. 179*:207–210.

Sampson SR, Brodie C, & Alboim SV. 1994. Role of protein kinase C in insulin activation of the Na-K-pump in cultured skeletal muscle. *Am. J. Physiol. 266 (Cell Physiol 35)*:C751–C758.

Schwartz A, Allen JC, & Harigaya S. 1969. Possible involvement of cardiac Na-K-adenosine triphosphatase in the mechanism of action of cardiac glycosides. *J. Pharmacol. Exp. Ther 168:*31–41.

Skou JC, & Esmann M. The Na, K-ATPase. 1992. *J. Bioenerget. Biomembr. 24:*249–261.

Stevens MJ, Danaberg J, Feldman EL, Lattimer SA, Kamijo M, Thomas TP, Shindo H, Sima AAF, & Green D. 1994. The linked roles of nitric oxide, aldose reductase and Na$^+$-K$^+$-ATPase in the slowing of nerve conduction in the steptozotocin diabetic rat. *J. Clin. Invest. 94:*853–859.

Stoffer SS, Hynes KM, Jiang N-S, & Ryan RJ. 1973. Digoxin and abnormal serum hormone levels. *JAMA 225:*1643–1644.

Suzuki S, Kajukuri J, Suzuki A, & Itoh T. 1991. Effect of endothelin-1 on endothelial cells in the porcine coronary artery. *Circ. Res. 69:*1361–1368.

Sweadner KJ. 1989. Isozymes of the Na$^+$/K$^+$-ATPase. *Biochim. Biophys. Acta. 988:*185–220. 1989

Tare M, Parkington HC, Coleman HA, Nerld TO, & Dusting GJ. 1990. Hyperpolarization and relaxation of arterial smooth muscle caused by nitric oxide derived from the endothelium. *Natur. 346:*69–71.

Tesfamariam B, Brown ML, & Cohen RA. 1991. Elevated glucose impairs endothelium-dependent relaxation by activating protein kinase C. *J. Clin. Invest. 87:*1643–1648.

Tirupattur PR, Ram JL, Standley PR, & Sowers JR. 1993. Regulation of Na, K-ATPase gene expression by insulin in vascular smooth muscle cells. *Am. J. Hypertens 6:*626–629.

Whittle BJR, Lopez-Belmonte J, & Rees DD. 1989. Modulation of the vasodepressor actions of acetylcholine, bradykinin, substance P and endothelin in the rat by a specific inhibitor of nitric oxide formation. *Br. J. Pharmacol. 98:*646–652.

Williamson JR, Kilo C, & Tilton RG. 1993. Hyperglycemic psuedohypoxia and diabetic complications. *Diabetes 42:*801–813.

Wu H-Y, Young YJ, Yue C-J, Chyu K-Y, Hsuch WA, & Chan TM. 1994. Endothelial-cell dependent vascular effects of insulin and insulin-like growth factor I in the Perfused rat mesenteric artery and aortic ring. *Diabetes 43:*1027–1032.

5
Purinergic Receptors, Nitric Oxide, and Regional Blood Flow

Vera Ralevic and Geoffrey Burnstock

Summary

Extracellular purines and pyrimidines regulate blood flow by pronounced effects on vascular tone mediated via cell surface purinergic receptors. There are two superfamilies of purinergic receptors: P1 receptors, recognizing adenosine, and P2 receptors, recognizing primarily ATP, ADP, UTP, and UDP. P1 receptors are further subdivided into four subtypes, A_1, A_{2A}, A_{2B}, and A_3, all of which couple to G proteins. P2 receptors are divided according to whether they are ligand-gated ion channels (P2X receptors) or are coupled to G proteins (P2Y receptors). Seven P2X and seven P2Y receptor proteins have been cloned to date. Members of both the P1 and P2 receptor families are expressed on blood vessels, smooth muscle, endothelium, and perivascular nerves, according to distinct patterns of distribution, where they are associated with specific effects on vascular tone. Nitric oxide (NO) is released from the endothelium and from perivascular nerves in some vessels, to act as a mediator and modulator of responses elicited by purines and pyrimidines. This chapter examines the relationship between vascular purinergic receptors and NO in the control of blood flow.

Introduction

The concept of extracellular purines as modulators of cardiovascular tone was introduced by Drury and Szent-Györgyi in 1929 in a report showing that adenosine and AMP have pronounced biological effects, including heart block, arterial dilatation, and lowering of blood pressure. Subsequently, extracellular ATP and UTP were also shown to have important effects on the heart and vasculature, and there is now a substantial body of evidence concerning the actions of purines and pyrimidines in the modulation of cardiovascular tone (Su 1985; Burnstock and Kennedy 1986; Burnstock 1987; Olsson and Pearson 1990; Ralevic and Burnstock 1991b, 1998). Adenosine acts on P1 receptors, of which four subtypes, A_1, A_{2A}, A_{2B}, and A_3, all coupling to G proteins, have been identified. ATP, ADP, UTP, and UDP act on P2 receptors, which have been divided into two superfamilies of ligand-gated cation channels and G protein-coupled receptors, termed P2X and P2Y receptors, respectively (Abbracchio and Burnstock 1994; Burnstock and King 1996; Fredholm et al. 1994, 1996). Further subdivision of P2 receptors is based on structurally distinct cloned receptors, and a new system of nomenclature (P2X$_{1-7}$ and P2Y$_{1-7}$) has arisen from this and will replace the earlier subtype nomenclature (including P$_{2X}$, P$_{2Y}$, P$_{2U}$, P$_{2T}$, and P$_{2Z}$ receptors), as correlations between cloned and endogenous receptors are established. It is important to recognize that whereas cloned P2Y proteins represent single protein receptors, cloned P2X proteins are receptor subunits, not actual receptors; P2X receptors are formed from combinations of two or more, probably four, subunits, implying that there may be a large number of endogenous heteromeric P2X receptors, not adequately represented by the seven homomeric P2X receptors cloned to date.

The recognition that endothelial cells can be stimulated to release a mediator of vasodilatation, endothelium-derived relaxing factor (EDRF) (Furchgott and Zawadzki 1980) or nitric oxide (NO) (Ignarro et al. 1987; Palmer et al. 1987), is now known to be relevant to the vasodilator actions of adenosine and ATP via endothelial P1 (subtype A$_2$) and P2 (P2Y family) receptors, respectively, in most blood vessels. Prostacyclin (prostaglandin I$_2$, PGI$_2$) and endothelium-derived hyperpolarizing factor (EDHF) may also contribute to purinergic endothelium-dependent vasodilatation in some blood vessels, whereas in others, purine receptors mediating vasodilatation are present on the vascular smooth muscle. It is now clear that in addition to its actions as a mediator of endothelium-dependent vasodilatation to purines, pyrimidines, and other agonists, NO can modulate contractile responses mediated by purinergic receptors on vascular smooth muscle, modulate purinergic neurotransmitter release, and modulate platelet aggregation and, thus, the release of purines from platelets.

The present chapter considers the relationship between NO and purinergic receptors in the control of blood flow. General concepts for local regulation of blood flow by NO and purinergic receptors are discussed, which are relevant to most blood vessels and vascular beds. In addition to its role as a mediator and modulator of responses effected at purinergic receptors, the effect of regulation by NO of purine and pyrimidine release is considered. Highlighted is the fact that in pathophysiological conditions, disruption of the endothelium and a deficit in the production of NO can shift the balance from purinergic endothelium-dependent vasodilatation to contractile responses mediated by purinergic receptors on the underlying vascular smooth muscle, resulting in local vasospasm.

Sources and Release of Purines and Pyrimidines in Blood Vessels

Purines and pyrimidines are released from a number of different sources relevant to the activation of specific purinergic receptors throughout the blood vessel wall. The vascular endothelium is an important source of ATP and UTP, which are released by stimuli such as hypoxia and shear stress (Milner et al. 1990a,b; Bodin et al. 1992; Ralevic et al. 1992; Saiag et al. 1995). ATP and UTP act on P2 receptors on endothelial cells adjacent to or downstream from the site of nucleotide release to elicit vasodilatation (via NO and possibly other EDRFs) and an increase in blood flow. In addition, they may release further ATP and, presumably, UTP in an autocatalytic manner, which would serve to amplify the vasodilator response (Yang et al. 1994). ATP and UTP are rapidly broken down by ectonucleotidases to ADP and UDP, respectively, and ADP is further degraded to adenosine, implying that multiple endothelial P1 and P2 receptors may be involved in this process. NO is also released from the endothelium by mechanical and physical factors such as shear stress, blood viscosity, and blood pressure (Holtz et al. 1984; Hull et al. 1986; Kaiser et al. 1986; Rubanyi et al. 1986), raising interesting parallels between NO and purines and pyrimidines in the local control of blood flow.

Erythrocytes are important sources of ATP (and presumably ADP and adenosine following degradation of ATP) released during shear stress or hypoxia (Forrester 1990; Ellsworth et al. 1995). Release of purines from erythrocytes into the lumen of the blood vessel implies that their principal actions are at endothelial cell purinergic receptors that mediate vasodilatation and an increase in local blood flow.

ATP, ADP, and UTP are stored in high concentrations in platelets and are released from activated platelets during aggregation (Goetz et al. 1971; Born and Kratzer 1984). This is a pathophysiological process, occurring when there is damage to the function or integrity of the endothelium, and is associated with activation of P2 purinergic receptors on smooth muscle cells underlying the endothelium, and severe vasoconstriction or vasospasm and consequent ischemia.

ATP is released as a contransmitter in various combinations with other neurotransmitters from perivascular nerves. In most blood vessels, ATP is released as a contransmitter with noradrenaline (NA) and neuropeptide Y (NPY) from sympathetic perivascular nerves and acts on smooth muscle P2X receptors to mediate rapid membrane depolarization (or excitatory junction potentials, EJPs), which may summate, resulting in further membrane depolarization, calcium influx via voltage-dependent Ca^{2+} channels, and vasoconstriction (Burnstock 1990; Burnstock and Ralevic 1996). Evidence that ATP acting at P2X receptors mediates the EJPs and purinergic smooth muscle vasoconstriction comes from pharmacological studies showing that both of these responses are blocked by P2X receptor desensitization with the selective P2X receptor agonist and desensitizing agent α,β-methylene ATP (α,β-meATP), as well as by the P2 receptor antagonists suramin and pyridoxalphosphate-6-azophenyl-2′,4′-disulfonic acid (PPADS) (Sneddon and Burnstock 1984; Burnstock and Warland 1987; Angus et al. 1988; Ziganshin et al. 1994) (Fig. 5.1).

Considerable variation exists in the relative proportions of NA and ATP acting as cotransmitters in sympathetic nerves. For example, in guinea pig submucosal arterioles, both vasoconstriction and EJPs evoked in response to electrical stimulation of sympathetic nerves are mediated exclusively by ATP, with NA assuming the role of a neuromodulator by acting on prejunctional α_2-adrenoceptors to depress transmitter release (Evans and Surprenant 1992). At the other extreme, the

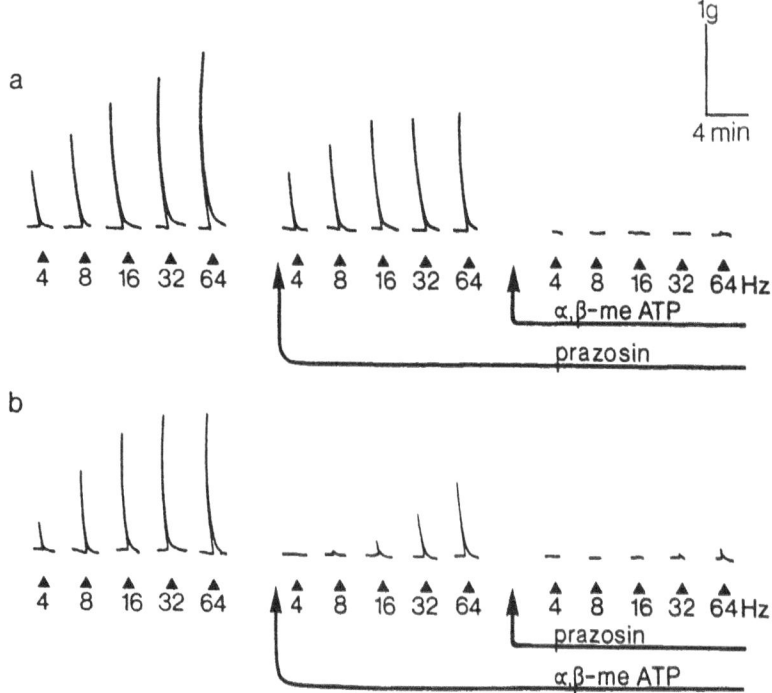

FIGURE 5.1. Contractions produced in the isolated saphenous artery of the rabbit on neurogenic transmural stimulation (0.08 – 0.1 ms; supramaximal voltage) for 1 s at the frequencies (Hz) indicated (▲). Nerve stimulations were repeated in the presence of 10 μM prazosin added before (a) or after (b) desensitization of the P2 purinoceptor with 10 μM α,β-methylene ATP (α,β-me ATP), as indicated on the figure by the arrows. The horizontal bar signifies 4 minutes and the vertical bar 1 g. (From Burnstock and Warland 1987.) Copyright 1987 by Stockton Press, MacMillan Publishers Ltd. Reprinted with permission.

purinergic component of sympathetic neurotransmission in rat mesenteric arteries is relatively small (Sjöblom-Widfeldt 1990; Ralevic and Burnstock 1991a). The release of ATP from sympathetic nerve terminals in the adventitia, the outermost layer of the blood vessels, is most relevant to the activation of purinergic receptors situated on adjacent smooth muscle cells. Although this is typically far from the innermost endothelial cell layer of the blood vessel (except in some small blood vessels), as discussed later in this chapter, NO is a readily diffusible molecule, and both endothelial and nonendothelial NO can modulate the sympathetic response by effects on the release and postjunctional actions of ATP and NA.

ATP has been shown to be released from perivascular sensorimotor nerves in the rabbit ear artery (Holton and Holton 1953, 1954; Holton 1959) but has not yet been demonstrated in other vessels. Neurogenic nonadrenergic, noncholinergic (NANC) relaxation mediated by ATP has been shown in guinea pig pulmonary arteries (Liu et al. 1992), rabbit portal vein (where ATP appears to be coreleased with NO) (Brizzolara et al. 1993) (Fig. 5.2), and lamb small coronary arteries (Simonsen et al. 1997). By analogy with ATP released as a neurotransmitter from sympathetic perivascular nerves, it is possible that ATP release from NANC nerves and its subsequent postjunctional actions may be modulated by NO (see Nitric Oxide Mediation and Modulation of Purinergic Responses, page 75).

Purinergic Receptors on Vascular Endothelium

Although the A_2 subtype of P1 receptors in the vasculature are located primarily on smooth muscle, both A_{2A} and A_{2B} receptors are expressed by endothelial cells, in various proportions, in a number of blood vessels. A_{2A} receptors have been demonstrated on endothelial cells cultured from human umbilical vein (Sobrevia et al. 1997) and guinea pig coronary arteries (Schiele and Schwabe 1994), and both A_{2A} and A_{2B} receptors have been demonstrated on human aortic endothelial cells in culture (Iwamoto et al. 1994). An endothelial component of relaxation mediated by A_{2A} receptors has been demonstrated in a number of vessels, including rat aorta (Conti et al. 1993; Lewis et al. 1994; Monopoli et al. 1994; Prentice and Hourani 1996), porcine coronary artery (Balwierczak et al. 1991; Abebe et al. 1994; Monopoli et al. 1994), guinea pig coronary artery and bed (Martin et al. 1993; Vials and Burnstock 1993), and human mammary artery (Makujina et al. 1992). Endothelially mediated relaxation via A_{2B} receptors has been shown in rabbit corpus cavernosum (Chiang et al. 1994) and rat renal artery (Martin and Potts 1994).

In most of the isolated blood vessels in which this has been studied, the endothelial component of A_2 receptor-mediated vasodilatation has been identified by removal of the endothelium. Since selective antagonism of endothelially derived mediators has generally not been studied, it is not clear whether NO, EDHF, or PGI_2 is involved in this response. However, a role for NO in A_{2A} receptor-mediated vasodilatation has clearly been shown in the guinea pig coronary bed, where inhibitors of NO synthase (NOS) attenuate vasodilatation mediated by A_{2A} receptor agonists (Vials and Burnstock 1993). The mechanism by which A_2 receptors stimulate NOS and subsequent NO formation is not clear, but an increase in intracellular Ca^{2+} seems obligatory. The biological significance of A_2 receptor-mediated endothelium-dependent vasodilatation remains to be determined. Endothelial P1 (A_2) receptors are generally less sensitive to activation by agonist than are endothelial P2 (P2Y) receptors, suggesting that one possible role of adenosine is to supplement vasodilatation during conditions leading to high local concentrations of ATP (and thus to biologically active concentrations of the breakdown product adenosine).

The principal P2 receptors expressed by vascular endothelial cells are $P2Y_1$ and $P2Y_2$ receptors. These mediate vasodilatation to purine and pyrimidine nucleotides via generation primarily of NO, with a possible contribution of EDHF or PGI_2. The $P2Y_1$ receptor is selective for the ligands ADP and ATP and is not activated by UTP

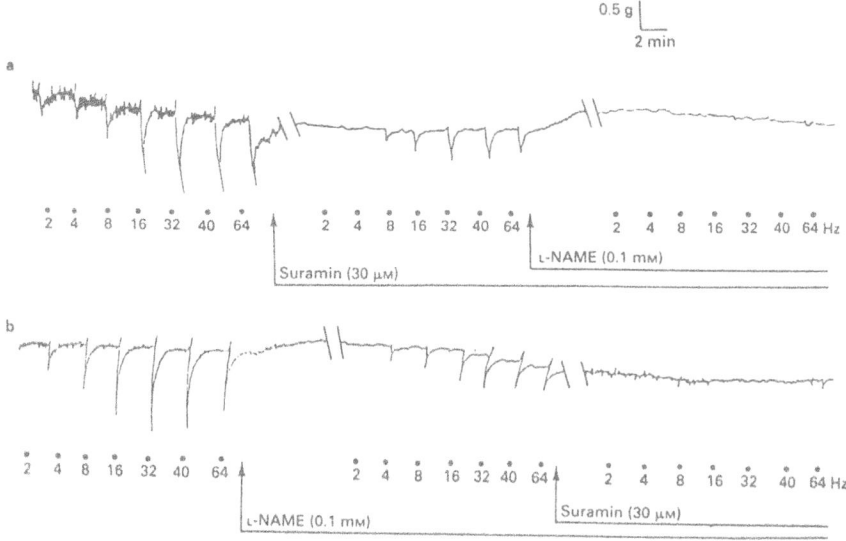

FIGURE 5.2. Relaxations of the rabbit portal vein to neurogenic transmural stimulation for 10 s (2–64 Hz, 0.7 ms, 100 V) at 5-minute intervals. Guanethidine (3.4 μM) and atropine (0.114 μM) were present throughout to block adrenergic and cholinergic neurotransmission, respectively. Tone was induced with ergotamine (8.6 μM). Panel (a) shows that preincubation with suramin (30 μM) for 20 minutes reduced the nerve-mediated relaxations compared with controls, and that the suramin-resistant neurogenic relaxations were abolished 20 minutes after the addition of N_G-nitro-L-arginine methylester (L-NAME, 0.1 mM). Panel (b) shows that neurogenic relaxations remaining after 20 minutes of pretreatment of the tissue with L-NAME (0.1 mM) were abolished after the addition of suramin (30 μM).

and UDP. It is more sensitive to adenine nucleotide diphosphates than to triphosphates. The P2Y$_2$ receptor is distinct from other P2Y receptors in that it is activated by ATP and UTP with approximately equal potency. The main signal transduction pathway for both P2Y$_1$ and P2Y$_2$ receptors is activation of phospholipase C (PLC) via coupling with G$_a$ and G$_{io}$ proteins, respectively. Activation of PLC results in the formation of inositol 1,4,5 trisphosphate (IP$_3$) and diacylglycerol (DAG) and the mobilization of Ca^{2+}, which in turn are responsible for the activation of a variety of signaling pathways. Elevation of intracellular Ca^{2+} may stimulate NOS and NO formation, whereas the main physiological target of DAG is activation of protein kinase C (PKC), which may subsequently cause tyrosine phosphorylation of mitogen-activated protein kinases (MAPKs), leading to generation of PGI$_2$ (Bowden et al. 1995; Patel et al. 1996). The P2Y$_1$ receptor pathway is more sensitive to manipulations of PKC such that stimulation of PKC inhibits, and inhibition of PKC activates, the P2Y$_1$ receptor, whereas the P2Y$_2$ response is less affected or is unaffected (Purkiss et al. 1994; Communi et al. 1995; Gallinaro et al. 1995; Chen et al. 1996). P2Y$_1$ receptor-mediated inhibition of adenylate cyclase has been described in a clonal population of rat brain capillary endothelial

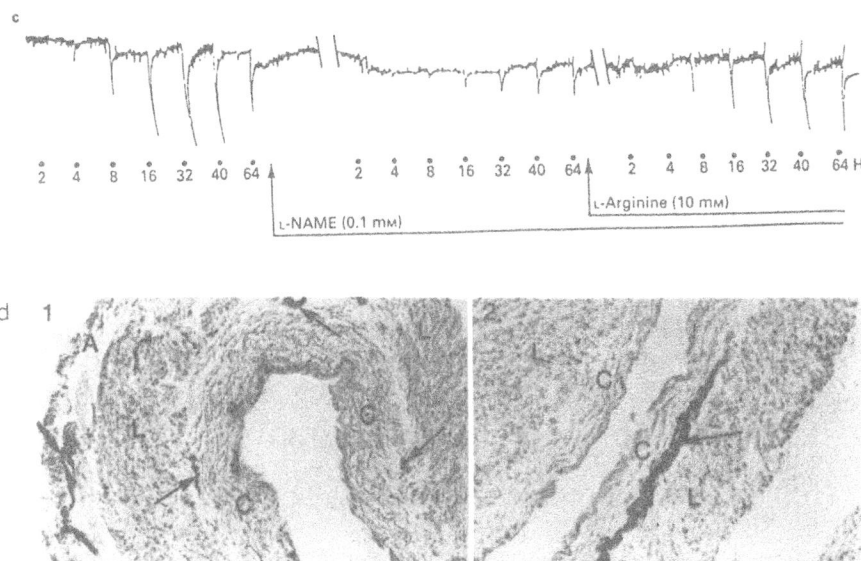

FIGURE 5.2. (*Continued*) Panel (c) shows the effect of adding L-NAME (0.1 mM) to the tissue; there was an additional rise in tone and inhibition of the response to nerve stimulation after a 20-minute incubation period. The subsequent treatment of tissues with L-arginine (10 mM) for 20 minutes reversed this effect. Panel (d) shows the histochemical localization of reduced nicotinamide adenine dinucleotide phosphate (NADPH)-diaphorase in the rabbit portal vein. (d1) NADPH-diaphorase reaction seen in nerve fibers between the inner circular (C) and the outer longitudinal muscle coats (thin arrows). Note also the NADPH-diaphorase positive nerves in adventitia (A) (thick arrows). Tangential section. (d2) NADPH-diaphorase reaction seen in a dense nerve plexus between the inner circular (C) and outer longitudinal muscle coats of the portal vein (thin black arrow). Transverse section. Calibration bar for d1 and d2 is 30 μm. Each of the traces in (a), (b), and (c) is representative of similar results in six separate experiments. (From Brizzolara et al. 1993.) Copyright 1993 by Stockton Press, MacMillan Publishers, Ltd. Reprinted with permission.

cells (B10 cells); this alternative signaling mechanism seems to be due to different G protein coupling of the $P2Y_1$ receptor (Webb et al. 1996).

The location of P2Y ($P2Y_1$ and $P2Y_2$) receptors on endothelial cells forming the innermost surface of blood vessels, implies that these receptors are important as sensors and effectors of responses to local changes in purine concentrations in blood. Thus, endothelium-dependent vasodilatation via P2Y receptors is likely to be linked to the release of purines at the intimal surface from erythrocytes and from endothelial cells themselves in response to physiological stimuli such as shear stress and hypoxia, and possibly from platelets during aggregation. In most blood vessels, $P2Y_1$ and $P2Y_2$ receptors coexist on endothelial cells, albeit in variable proportions. For example, pharmacological studies suggest that hamster mesenteric arteries express predominantly $P2Y_2$, receptors and few $P2Y_1$ receptors (Ralevic and Burnstock 1996b), whereas the converse seems to be true for piglet aorta (Martin et al. 1985) and lamb small coronary arteries (Simonsen et al. 1997), where UTP is a very weak agonist. A uridine nucleotide-specific receptor activated by UTP, but not by ATP or ADP, coexisting with $P2Y_1$ and $P2Y_2$ receptors, has been identified on guinea pig cardiac endothelial cells (Yang et al. 1996). The reason for this coexistence is currently unclear, although the different agonist selectivities of the receptors, namely for ADP by the $P2Y_1$ receptor and for ATP and UTP by the $P2Y_2$ receptor, imply differential release of these nucleotides from different sources, or from the same source but in response to different stimuli. Interestingly, the $P2Y_1$ receptor has been implicated in the autocatalytic release of ATP (ATP-induced release of ATP) from guinea pig coronary artery endothelial cells (Yang et al. 1994) which suggests that a component of the basal release of NO from endothelial cells may be because of a priori release of ATP.

Purinergic Receptors on Vascular Smooth Muscle

Smooth muscle A_2 (A_{2A} and A_{2B}) receptors mediating relaxation have been described in a large number of different vascular smooth muscle cells, presumably through activation of cAMP (see Ralevic and Burnstock 1998). Contractile effects of adenosine at smooth muscle receptors in the vasculature are rare, although A_1 receptors have been shown to mediate contraction of porcine coronary artery (Merkel et al. 1992), guinea pig aorta (Stogall and Shaw 1990) and coronary artery (Szentmiklósi et al. 1995) and dog coronary vasculature (Akatsuka et al. 1994).

The most significant P2 purinergic receptors in vascular smooth muscle are P2X receptors. These have been shown to mediate vasoconstriction to ATP released as a cotransmitter with NA (and often NPY) from sympathetic nerves (Burnstock 1990; Burnstock and Ralevic 1996). P2X receptors mediate the rapid (onset within 10 ms) nonselective passage of cations across the cell membrane, resulting in an increase in intracellular Ca^{2+} and depolarization (Bean 1992; Dubyak and El-Moatassim 1993). Summation of EJPs leads to further depolarization and the activation of voltage-dependent Ca^{2+} channels, which results in further Ca^{2+} influx and leads to smooth muscle contraction (Burnstock and Ralevic 1996). Thus, a component of the sympathetic contractile response can be blocked by purine receptor antagonists but not by adrenoceptor antagonists (Fig. 5.1).

The principal vascular smooth muscle P2X receptor appears to be the $P2X_1$ receptor. This is implied by the similar pharmacological response profiles of native vascular smooth muscle P2X receptors and recombinant smooth muscle $P2X_1$ receptors cloned from vas deferens and bladder (P2X receptors have not yet been cloned from vascular smooth muscle); both the P2X receptors on isolated smooth muscle cells and the recombinant $P2X_1$ receptors are activated potently by α,β-meATP and rapidly desensitize (Evans and Kennedy 1994; Valera et al. 1994, 1995). Antibody staining has provided direct evidence for the expression of $P2X_1$ receptors in vascular smooth muscle. However, smooth muscle $P2X_2$ and $P2X_4$ proteins have also been identified, and it remains to be determined whether the predominant P2X receptor in the vasculature is indeed a homomeric $P2X_1$ receptor, or whether it is a heteromer that includes the P2X$_1$ receptor protein as one of its subunits (giving the smooth muscle P2X receptor its characteristic sensitivity to α,β-meATP).

Vascular smooth muscle P2 receptors that mediate vasoconstriction and are activated equipotently by ATP and UTP, but not by α,β-meATP or 2-methylthio ATP (2MeSATP), have been described in a number of isolated blood vessels (Eltze and Ullrich 1996; Miyagi et al. 1996; Rubino and Burnstock 1996; Ishizaki et al. 1997; Qasabian et al. 1997). According to the classical system of receptor nomenclature, this agonist potency profile identifies these receptors as P$_{2U}$ receptors. Current nomenclature based on molecular structure suggests that these receptors should be renamed as P2Y$_2$ receptors. However, this new terminology implies G protein coupling, which, although verified for P$_{2U}$ receptors on human coronary artery smooth muscle cells (Strøbæk et al. 1996), has not yet been confirmed for the other smooth muscle P$_{2U}$ receptors.

Uridine nucleotide-specific receptors mediating contraction to UTP (but not to ATP), which are resistant to desensitization by α,β-meATP and do not show cross-tachyphylaxis with responses to ATP, or are unaffected by P2 receptor antagonists, including PPADS and suramin, have been described on vascular smooth muscle of a variety of vessels and species (von Kügelgen et al. 1987, 1990; Saiag et al. 1990, 1992; Ralevic and Burnstock 1991a; Juul et al. 1992; Windscheif et al. 1994; Lagaud et al. 1996). The physiological role of these-receptors is currently unclear, reflecting a general lack of information about sources of and stimuli for the release of UTP.

In some mammalian blood vessels, P2Y receptors are present on the smooth muscle and mediate vasodilatation to ATP and 2MeSATP. The fact that these receptors are most potently activated by 2MeSATP and ADP suggests that these are most likely to be P2Y$_1$ receptors, although this classification awaits confirmation from structural data when the smooth muscle P2Y receptor is cloned. Relaxation mediated by smooth muscle P2Y (P2Y$_1$-like) receptors has been shown in rabbit hepatic artery (Brizzolara and Burnstock 1991), rabbit mesenteric artery (Mathieson and Burnstock 1985; Burnstock and Warland 1987), rabbit portal vein (Kennedy and Burnstock 1985), rabbit pumonary artery (Qasabian et al. 1997), and human small pulmonary arteries (Liu et al. 1989). In guinea pig and rabbit large coronary arteries (Keefe et al. 1992; Corr and Burnstock 1994) and in lamb small coronary arteries (Simonsen et al. 1997), vasodilatation is mediated by both P2Y (P2Y$_1$-like) receptors on smooth muscle and their counterparts, endothelial P2Y$_1$ receptors. Both P2Y (P2Y$_1$-like) and P$_{2U}$ (P2Y$_2$-like) receptors have been described in human coronary smooth muscle cells in culture (Strøbæk et al. 1996). The mechanism of relaxation via smooth muscle P2Y$_1$ receptors is unknown, but it may involve activation of K$^+$ channels. The physiological significance of smooth muscle P2Y (P2Y$_1$-like) receptors has been established for rabbit portal vein (where ATP appears to be coreleased with NO) (Kennedy and Burnstock 1985; Brizzolara et al. 1993) (Fig. 5.2), guinea pig pulmonary artery (Liu et al. 1992), and lamb small coronary arteries (Simonsen et al. 1997), where neurogenic NANC relaxation is mediated by ATP.

Purinergic Receptors on Perivascular Nerves

Activation of A_1 adenosine receptors located prejunctionally on the terminals of perivascular sympathetic and sensorimotor nerves mediates inhibition of neurotransmitter release (see Burnstock 1990). One source of the adenosine is that formed from ATP released from perivascular nerves, identifying a role for A_1 receptors in feedback autoregulation of ATP neurotransmission. A_{2A} receptors have been described on sympathetic nerves in rat tail artery, where they augment electrically evoked NA release (Gonçalves and Queiroz 1993).

P2Y (P2Y$_1$-like) receptors have been described on postganglionic sympathetic neurons and mediate inhibition of neurotransmission, presumably being involved in feedback inhibition of coreleased NA and NPY as well as of ATP. Prejunctional P2Y (P2Y$_1$-like) receptors have been shown to inhibit NA release in postganglionic sympathetic neurons in a variety of tissues, including the vas deferens of the mouse and rat and the iris, atria, and tail artery of the rat (von Kügelgen et al. 1989, 1993, 1994a,b, 1995; Fuder and Muth 1993; Kurz et al. 1993; Gonçalves and Queiroz 1993).

The expression of P2X receptors on peripheral sensory nerves (Pelleg and Hurt 1996; Cook et al. 1997) is consistent with a role for P2X receptors in nociception and raises the possibility of a widespread distribution of P2X receptors on sensory nerve terminals throughout the cardiovascular system. At the recombinant P2X$_3$ receptor cloned from rat dorsal root ganglion, currents evoked by ATP and α,β-meATP rapidly desensitize, whereas ATP-gated currents at endogenous P2X receptors in adult rat nodose and dorsal root ganglion neurons do not desensitize (Lewis et al. 1995). When P2X$_3$ receptors are coexpressed in HEK293 cells with P2X$_2$ receptors (but not with other subtypes), a nondesensitizing response to ATP is observed, which mimicks that seen in adult rat sensory neurons and cannot be explained by additive effects of the two homomeric channels (Lewis et al. 1995). It seems that a new heteromeric receptor, P2X$_{2/3}$ is formed from the P2X$_2$ and P2X$_3$ subunits. Thus, one type of P2X receptor on the peripheral terminals of sensory neurons may be a heteromeric P2X$_{2/3}$ receptor, although other heteromeric combinations are also possible.

Purinergic Receptors on Platelets

Platelets express a receptor for ADP, currently termed the P2Y$_{ADP}$ or P$_{2T}$ receptor (this receptor has not yet been cloned), which mediates aggregation, at which ATP acts as a competitive antagonist. Recent evidence suggests that the platelet P$_{2T}$ receptor is a P2Y$_1$ receptor (Hechler et al. 1998). The P$_{2T}$ receptor couples to G$_{i2}$ proteins to mediate inhibition of adenylate cyclase activity, together with activation of PLC and an increase in intracellular Ca^{2+}, to promote aggregation (Hourani and Hall 1996). Platelets also express P2X$_1$ receptors, which, when activated, cause the opening of nonselective cation channels and the rapid influx of Ca^{2+} (MacKenzie et al. 1996) and may contribute to the platelet aggregatory process. Platelet A_{2A} receptors inhibit aggregation (Huttemann et al. 1984; Gurden et al. 1993; Monopoli et al. 1994; Cristalli et al. 1995), which may be significant in autoregulation of platelet aggregation by ADP and ATP released from platelets.

Nitric Oxide Mediation and Modulation of Purinergic Responses

As considered earlier, it is clear that NO has a crucial role as a mediator of vasodilatation following activation of P1 and P2 purinergic receptors on endothelial cells. EDHF may be significant in small vessels (Garland et al. 1995), and PGI_2, although generated in significant amounts following activation of endothelial P2 receptors, seems to make less of a contribution to the relaxation response; its primary role may be inhibition of platelet aggregation.

The release of NO and other EDRFs from endothelial cells in response to agonist activation of endothelial receptors has particular significance for those agonists which also have direct vasoconstrictor actions via receptors on the smooth muscle. Thus, removal of the endothelium, or blockade of endothelial cell NO formation with inhibitors of NOS, not only attenuates endothelium-dependent vasodilatation, but also facilitates vasoconstrictor actions mediated by receptors on the underlying smooth muscle. This applies not only to purines, but also to other substances, including acetylcholine and 5-HT, and highlights the importance of the endothelium, since vasoactive substances that produce vasodilatation in the presence of the endothelium may elicit vasoconstriction when endothelial function or integrity is impaired. It is interesting that among endogenous purines and pyrimidines, removal of the endothelium causes the most pronounced shift in the response to ATP, a reflection of the fact that ATP is both a potent vasodilator at endothelial P2 receptors and a potent vasoconstrictor at smooth muscle P2 receptors.

In addition to its role as a mediator of vasodilatation, NO can modulate vasoconstrictor responses to purines and other vasoactive agents. Although relatively little is known specifically about NO modulation of responses mediated by purinergic receptors and purine release, some general principles can be derived from related studies of NO modulation of responses mediated by other vasoactive compounds. The source of this NO may be endothelium or perivascular nerves, and an involvement of nonendothelial, non-neuronal NO has also been claimed. In most blood vessels, there is a continuous basal release of NO from the endothelium, which controls smooth muscle contractility and sets a basal vasodilator tone. Hence, removal of the endothelium or blockade of NO formation causes a nonspecific augmentation of vasoconstrictor responses, for example, to electrical field stimulation (EFS) of perivascular nerves, NA, ATP, and potassium chloride (see Ralevic and Burnstock 1996a). The postjunctional nature of this effect has been shown in the rat tail artery, where inhibition of NO potentiates vasoconstriction of EFS but does not affect the release of [^{3}H]NA (Bucher et al. 1992); basal release of endothelial NO is likely to be primarily responsible for inhibition of EFS-induced contractions, although direct actions of NA and ATP on endothelial receptors may be involved. In rabbit carotid artery, an increase in shear stress caused by an increase in perfusate viscosity attenuates sympathetic vasoconstriction only if the endothelium is present (Tesfamariam and Cohen 1988). This was attributed to a shear-stress-induced release of EDRF, but does not exclude a possible involvement of ATP coreleased with EDRF from endothelium. An important implication is that when vascular resistance falls, the increased blood flow dilates the blood vessels not only by a direct action of EDRF and ATP on the smooth muscle, but also by inhibition of sympathetic neurotransmission.

Prejunctional neuromodulation by the endothelium, which may involve NO, since this molecule can readily diffuse across cell membranes, is suggested by the demonstration that removal of the endothelium reduces [^{3}H]NA release from perivascular sympathetic nerves in rabbit carotid artery (Cohen and Weisbrod 1988) and dog pulmonary artery and vein (Greenberg et al. 1989). Although ATP release from sympathetic nerves was not assayed, a parallel inhibition might occur. On the other hand, prejunctional modulation by the endothelium may not involve NO, since maximally relaxing concentrations of NO have no effect on NA release from the rabbit carotid artery (Cohen et al. 1990). Modulation by endothelial prostaglandins seems unlikely, since blockade of prostaglandin synthesis with indomethacin fails to influence neurogenic contractions or the release of NA (Cohen and Weisbrod 1988). Although ATP released from the endothelium might not be expected to reach the adventitia intact, its breakdown by ectonucleotidases to generate adenosine is consistent with prejunctional inhibition of neurotransmitter release via A_1 receptors.

Nonendothelial NO may modulate agonist-evoked responses, since an inhibitor of NOS, N^G-nitro-L-arginine methylester (L-NAME), potentiates contractions in some endothelially denuded vessels: Endothelium-independent potentiation of vasoconstriction to EFS by NOS inhibition with N^G-nitro-L-arginine (L-NNA) in dog temporal arteries appears to be postjunctional, since the release of [^{3}H]NA is unaffected and is suggested to be due to elimination of neuronally released NO (Toda et al. 1991). Inhibition of neuronal NO formation has also been suggested to account for L-NNA potentiation of constriction to EFS in endothelium-denuded monkey mesenteric arteries (Tsuchiya et al. 1994). A non-neuronal, nonendothelial source of NO is implicated in the endothelium-independent enhancement by N^G-monomethyl-L-arginine (L-NMMA) of contractile responses to exogenous NA (and presumably to ATP as well) in guinea pig (Cederqvist et al. 1991; Cederqvist and Gustafsson 1994) and rabbit (MacLean et al. 1993) pulmonary arteries and in rabbit saphenous vein (Gordon et al. 1992). Endogenous NO has been suggested to prejunctionally inhibit sensorimotor vasodilatation in endothelium-denuded rat mesenteric arteries, although the source of NO was not identified (Li et al. 1993). By analogy with other perivascular nerves, where regulation of neurotransmitter release by autoreceptors is a common phenomenon, it is possible that NO coreleased with ATP from NANC nerves in rabbit portal vein (Brizzolara et al. 1993) modulates ATP and NO release from these nerves.

In addition to mediating smooth muscle vasodilatation, an important property of NO is its capacity to inhibit platelet adhesion to endothelial cell surfaces and platelet aggregation (Mellion et al. 1981; Radomski et al. 1987). One source of the NO is endothelial cells; it permeates platelets and activates soluble guanylate cyclase, leading to a marked accumulation of cGMP and inhibition of platelet aggregation (Mellion et al. 1981). EDRF/NO inhibition of platelet aggregation has been shown for a number of prothrombotic agents, including inhibition of platelet aggregation mediated by ADP in the pulmonary circulation of anesthetized rats (Bhardwaj et al. 1988). Platelets themselves posses an NO synthase that is activated during platelet aggregation and attenuates aggregation via an increase in cGMP (Radomski et al. 1990). Thus, control of platelet reactivity by NO will inhibit release of ADP, ATP, 5-HT, and thromboxane A_2 from platelets and will help to inhibit further aggregation and constriction of the underlying smooth muscle. In areas of damaged or denuded endothelium, the NO mechanism may be overwhelmed, leading to intravascular coagulation and vasoconstriction, but adjacent areas of intact endothelium will act to localize the platelet response to the damaged area.

Conclusions

Different purine receptor subtypes are expressed in characteristic patterns by vascular endothelium, smooth muscle, perivascular nerves, and platelets, where they have important and diverse effects on blood vessel tone. NO is implicated at several stages of the purinergic control of vascular tone: in addition to its important role as a mediator of vasodilatation in response to activation of specific endothelial purinergic receptors, NO modulates contractile responses mediated by smooth muscle purinergic receptors, and modulates the release of purines and pyrimidines from platelets and possibly from some perivascular nerves. The importance of NO in the purinergic control of blood flow is particularly apparent under conditions in which endothelial function or integrity is impaired, since vasodilatation mediated by purines at endothelial receptors may be converted to vasoconstriction at the underlying smooth muscle purine receptors, leading to local vasospasm.

Acknowledgments. The support of the Royal Society and the British Heart Foundation is gratefully acknowledged.

References

Abbracchio MP, & Burnstock G. 1994. Purinoceptors: are there families of P_{2X} and P_{2Y} purinoceptors? *Pharmacol Ther* 64:445–475.

Abebe W, Makujina SR, & Mustafa SJ. 1994. Adenosine receptor-mediated relaxation of porcine coronary artery in presence and absence of endothelium. *Am J Physiol* 266:H2018–H2025.

Akatsuka Y, Egashira K, Katsuda Y, Narishige T, Ueno H, Shimokawa H, & Takeshita A. 1994. ATP sensitive potassium channels are involved in adenosine A_2 receptor mediated coronary vasodilatation in the dog. *Cardiovasc Res 28:*906–911.

Angus JA, Broughton A, & Mulvany MJ. 1988. Role of α-adrenoceptors in constrictor responses of rat, guinea-pig and rabbit small arteries to neural activation. *J Physiol 403:*495–510.

Balwierczak JL, Sharif R, Krulan CM, Field FP, Weiss GB, & Miller MJS. 1991. Comparative effects of a selective adenosine A_2 receptor agonist, CGS 21680, and nitroprusside in vascular smooth muscle. *Eur J Pharmacol 196:*117–123.

Bean BP. 1992. Pharmacology and electrophysiology of ATP-activated ion channels. *Trends Pharmacol Sci 13:*87–90.

Bhardwaj R, Page CP, May GR, & Moore PK. 1988. Endothelium-derived relaxing factor inhibits platelet aggregation in human whole blood in vitro and in the rat in vivo. *Eur J Pharmacol 157:*83–91.

Bodin P, Milner P, Winter R, & Burnstock G. 1992. Chronic hypoxia changes the ratio of endothelin to ATP release from rat aortic endothelial cells exposed to high flow. *Proc R Soc Lond B 247:*131–135.

Born GVR, & Kratzer MAA. 1984. Source and concentration of extracellular adenosine triphosphate during haemostasis in rats, rabbits and man. *J Physiol 354:*419–429.

Bowden A, Patel V, Brown C, & Boarder MR. 1995. Evidence for requirement of tyrosine phosphorylation in endothelial P_{2Y}- and P_{2U}-purinoceptor stimulation of prostacyclin release. *Br J Pharmacol 116:*2563–2568.

Brizzolara A, & Burnstock G. 1991. Endothelium-dependent and endothelium-independent vasodilatation of the hepatic artery of the rabbit. *Br J Pharmacol 103:*1206–1212.

Brizzolara AL, Crowe R, & Burnstock G. 1993. Evidence for the involvement of both ATP and nitric oxide in non-adrenergic non-cholinergic inhibitory neurotransmission in the rabbit portal vein. *Br J Pharmacol 109:*606–608.

Bucher B, Ouedraogo S, Tschopl M, Paya D, & Stoclet JC. 1992. Role of the L-arginine pathway and of cyclic GMP in electrical field-induced noradrenaline release and vasoconstriction in the rat tail artery. *Br J Pharmacol 107:*976–982.

Burnstock G. 1987. Local control of blood pressure by purines. *Blood Vessels 24:*156–160.

Burnstock G. 1990. Co-transmission. The Fifth Heymans Memorial Lecture. *Arch Int Pharmacodyn Ther 304:*7–33.

Burnstock G, & Kennedy C. 1986. Purinergic receptors in the cardiovascular system. *Prog Pharmacol 6:*111–132.

Burnstock G, & King BF. 1996. Numbering of cloned P2 purinoceptors. *Drug Dev Res 38:*67–71.

Burnstock G, & Ralevic V. 1996. Cotransmission. In: Garland CJ, Angus JA, eds. *Pharmacology of Vascular Smooth Muscle*. Oxford: Oxford University Press, pp 210–232.

Burnstock G, & Warland JJI. 1987. A pharmacological study of the rabbit saphenous artery *in vitro:* a vessel with a large purinergic contractile response to sympathetic nerve stimulation. *Br J Pharmacol 90:*111–120.

Cederqvist B, & Gustafsson LE. 1994. Modulation of neuroeffector transmission in guinea-pig pulmonary artery and vas deferens by exogenous nitric oxide. *Acta Physiol Scand 150:*75–81.

Cederqvist B, Wiklund NP, Persson MG, & Gustafsson LE. 1991. Modulation of neuroeffector transmission in the guinea pig pulmonary artery by endogenous nitric oxide. *Neurosci Lett 127:*67–69.

Chen BC, Lee C-M, Lee YT, & Lin W-W. 1996. Characterization of signaling pathways of P_{2Y} and P_{2U} purinoceptors in bovine pulmonary artery endothelial cells. *J Cardiovasc Pharmacol 28:*192–199.

Chiang PH, Wu SN, Tsai EM, Wu CC, Shen MR, Huang CH, & Chiang CP. 1994. Adenosine modulation of neurotransmission in penile erection. *Br J Clin Pharmacol 38:*357–362.

Cohen RA, & Weisbrod RM. 1988. The endothelium inhibits norepinephrine release from adrenergic nerves of the rabbit carotid artery. *Am J Physiol 254:*H871–H878.

Cohen RA, Tesfamariam B, & Weisbrod RM. 1990. The endothelium inhibits adrenergic neurotransmission. In: Vanhoutte PM, Rubanyi GM, eds. *Proceedings of the First International Symposium on Endothelium-Derived Vasoactive Factors.* New York: Karger, pp 206–212.

Communi D, Raspe E, Pirotton S, & Boeynaems JM. 1995. Coexpression of P_{2Y} and P_{2U} receptors on aortic endothelial cells. Comparison of cell localization and signaling pathways. *Circ Res 76:*191–198.

Conti A, Monopoli A, Gamba M, Borea PA, & Ongini E. 1993. Effects of selective A_1 and A_2 adenosine receptor agonists on cardiovascular tissues. *Naunyn Schmiedebergs Arch Pharmacol 348:*108–112.

Cook SP, Vulchanova L, Hargreaves KM, Elde R, & McCleskey EW. 1997. Distinct ATP receptors on pain-sensing and stretch-sensing neurons. *Nature 387:*505–508.

Corr L, & Burnstock G. 1994. Analysis of P_2-purinoceptor subtypes on the smooth muscle and endothelium of rabbit coronary artery. *J Cardiovasc Res 23:*709–715.

Cristalli G, Camaioni E, Vittori S, Volpini R, Borea PA, Conti A, Dionisotti S, Ongini E, & Monopoli A. 1995. 2-Aralkynyl and 2-heteroalkynyl derivatives of adenosine-5′-*N*-ethyluronamide as selective A_{2A} adenosine receptor agonists. *J Med Chem 38:*1462–1472.

Drury AN, & Szent-Györgyi A. 1929. The physiological activity of adenine compounds with especial reference to their action upon the mammalian heart. *J Physiol 68:*213–237.

Dubyak GR, & El-Moatassim C. 1993. Signal transduction via P_2-purinergic receptors for extracellular ATP and other nucleotides. *Am J Physiol 265:*C577–C606.

Ellsworth ML, Forrester T, Ellis CG, & Dietrich HH. 1995. The erythrocyte as a regulator of vascular tone. *Am J Physiol 268:*H2155–H2161.

Eltze M, & Ullrich B. 1996. Characterization of vascular P_2 purinoceptors in the rat isolated perfused kidney. *Eur J Pharmacol 306:*139–152.

Evans RJ, & Kennedy C. 1994. Characterization of P_2-purinoceptors in the smooth muscle of the rat tail artery: a comparison between contractile and electrophysiological responses. *Br J Pharmacol 113:*853–860.

Evans RJ, & Suprenant A. 1992. Vasoconstriction of guinea-pig submucosal arterioles following sympathetic nerve stimulation is mediated by the release of ATP. *Br J Pharmacol 106:*242–249.

Forrester T. 1990. Release of ATP from heart: presentation of a release model using human erythrocyte. *Ann NY Acad Sci 603:*335–352.

Fredholm BB, Abbracchio MP, Burnstock G, Daly JW, Harden KT, Jacobson KA, Leff P, & Williams M. 1994. Nomenclature and classification of purinoceptors. *Pharmacol Rev 46:*143–156.

Fredholm BB, Burnstock G, Harden KT, & Spedding M. 1996. Receptor nomenclature. *Drug Dev Res 39:*461–466.

Fuder H, & Muth U. 1993. ATP and endogenous agonists inhibit evoked [^{3}H]-noradrenaline release in rat iris via A_1 and P_{2Y}-like purinoceptors. *Naunyn Schmiedebergs Arch Pharmacol 348:*352–357.

Furchgott RF, & Zawadzki JV. 1980. The obligatory role of endothelial cells in the relaxation of arterial smooth muscle by acetylcholine. *Nature 288:*373–376.

Gallinaro BJ, Reimer WJ, & Dixon SJ. 1995. Activation of protein kinase C inhibits ATP-induced $[Ca^{2+}]_i$ elevation in rat osteoblastic cells: selective effects on P_{2Y} and P_{2U} signaling pathways. *J Cell Physiol 162:*305–314.

Garland CJ, Plane F, Kemp BK, & Cocks TM. 1995. Endothelium-dependent hyperpolarization: a role in the control of vascular tone. *Trends Pharmacol Sci 16:*23–30.

Goetz V, Prada DA, & Pletscher MA. 1971. Adenine-, guanine- and uridine-5'-phosphonucleotides in blood platelets and storage organelles of various species. *J Pharmacol Exp Ther 178:*210–215.

Gonçalves J, & Queiroz G. 1993. Purinergic modulation of noradrenaline release in rat tail artery: tonic modulation mediated by inhibitory P_{2Y}- and facilitatory P_{2X}-purinoceptors. *Br J Pharmacol 117:*156–160.

Gordon JF, Baird M, Daly CJ, & McGrath JC. 1992. Endogenous nitric oxide modulates sympathetic neuroeffector transmission in the isolated rabbit lateral saphenous vein. *J Cardiovasc Pharmacol 20:*S68–S71.

Greenberg S, Diecke FPJ, Peevy K, & Tanaka TP. 1989. The endothelium modulates adrenergic neurotransmission to canine pulmonary arteries and veins. *Eur J Pharmacol 162:*67–80.

Gurden MF, Coates J, Ellis F, Evans B, Foster M, Hornby E, Kennedy I, Martin DP, Strong P, Vardey CJ, & Wheeldon A. 1993. Functional characterization of three adenosine receptor subtypes. *Br J Pharmacol 109:*693–698.

Hechler B, Léon C, Vial C, Vigne P, Frelin C, Cazenave J-P, & Gachet C. 1998. The P2Y$_1$ receptor is necessary for adenosine 5'-triphosphate-induced platelet aggregation. *Blood 92:*152–159.

Holton FA, & Holton P. 1953. The possibility that ATP is a transmitter at sensory nerve endings. *J Physiol 119:*50–51P.

Holton FA, & Holton P. 1954. The capillary dilator substances in dry powders of spinal roots: a possible role of ATP in chemical transmission. *J Physiol 126:*124–140.

Holton P. 1959. The liberation of ATP on antidromic stimulation of sensory nerves. *J Physiol 145:*494–504.

Holtz J, Förstermann U, Pohl U, Giesler M, & Bassenge E. 1984. Flow-dependent, endothelium-mediated dilation of epicardial coronary arteries in conscious dogs: effects of cyclooxygenase inhibition. *J Cardiovasc Pharmacol 6:*1161–1169.

Hourani SMO, & Hall DA, 1996. P2T purinoceptors: ADP receptors on platelets. In: *P2 Purinoceptors: Localization, Function and Transduction Mechanisms.* Chichester: Wiley, pp. 53–70.

Hull SS, Kaiser L, Jaffe MD, Sparks HV. 1986. Endothelium-dependent flow-induced dilation of canine femoral and saphenous arteries. *Blood Vessels 23:*183–198.

Huttemann E, Ukena D, Lenschow V, Schwabe U. 1984. R$_a$ adenosine receptors in human platelets: characterization by 5'-*N*- ethylcarboxamido[^{3}H]-adenosine binding in relation to adenylate cyclase activity. *Naunyn Schmiedebergs Arch Pharmacol 325:*226–233.

Ignarro LJ, Byrns RE, Buga GM, & Wood KS. 1987. Endothelium-derived relaxing factor from pulmonary artery and vein possess pharmacologic and chemical properties identical to those of nitric oxide radical. *Circ Res 61:*866–879.

Ishizaki M, Iizuka Y, Suzuki-Kusaba M, Kimura T, & Satoh S. 1997. Nonadrenergic contractile response of guinea pig portal vein to electrical field stimulation mimics response to UTP but not to ATP. *J Cardiovasc Pharmacol 29:*360–366.

Iwamoto T, Umemura S, Toya Y, Uchibori T, Kogi K, Takagi N, & Ishii M. 1994. Identification of adenosine A_2 receptor cAMP system in human aortic endothelial cells. *Biochem Biophys Res Commun 199:*905–910.

Juul B, Plesner L, & Aalkjaer C. 1992. Effects of ATP and UTP on $[Ca^{2+}]_i$, membrane potential and force in isolated rat small arteries. *J Vasc Res 29:*385–395.

Kaiser L, Hull SS, & Sparks HV. 1986. Methylene blue and ETYA block flow-dependent dilation in canine femoral artery. *Am J Physiol 250:*H974–H981.

Keefe KD, Pasco JS, & Eckman DM. 1992. Purinergic relaxation and hyperpolarization in guinea pig and rabbit coronary artery: role of the endothelium. *J Pharmacol Exp Ther 260:*592–600.

Kennedy C, & Burnstock G. 1985. Evidence for two types of P_2-purinergic receptor in the longitudinal muscle of the rabbit portal vein. *Eur J Pharmacol 111:*49–56.

Kurz K, von Kügelgen I, & Starke K. 1993. Prejunctional modulation of noradrenaline release in mouse and rat vas deferens: contribution of P_1- and P_2-purinoceptors. *Br J Pharmacol 110:*1465–1472.

Lagaud GJL, Stoclet JC, & Andriantsitohaina R. 1996. Calcium handling and purinoceptor subtypes involved in ATP-induced contraction in rat small mesenteric arteries. *J Physiol 492:*689–703.

Lewis CD, Hourani SM, Long CJ, & Collis MG. 1994. Characterization of adenosine receptors in the rat isolated aorta. *Gen Pharmacol 25:*1381–1387.

Lewis C, Neidhart S, Holy C, North RA, Buell G, & Surprenant A. 1995. Coexpression of $P2X_2$ and $P2X_3$ receptor subunits can account for ATP-gated currents in sensory neurones. *Nature 377:*432–435.

Li YJ, Yu XJ, & Deng HW. 1993. Nitric oxide modulates responses to sensory nerve activation of the perfused rat mesentery. *Eur J Pharmacol 239:*127–132.

Liu SF, McCormack DG, Evans TW, & Barnes PJ. 1989. Evidence for two P_2-purinoceptor subtypes in human small pulmonary arteries. *Br J Pharmacol 98:*1014–1020.

Liu SF, Crawley DE, Evans TW, & Barnes PJ. 1992. Endothelium-dependent nonadrenergic, noncholinergic neural relaxation in guinea pig pulmonary artery. *J Pharmacol Exp Ther 260:*541–548.

MacKenzie AB, Mahaut-Smith MP, & Sage SO. 1996. Activation of receptor-operated cation channels via P_{2X1} not P_{2T} receptors in human platelets. *J Biol Chem 271:* 2879–2881.

MacLean MR, McCulloch KM, MacMillan JB, & McGrath JC. 1993. Influences of the endothelium and hypoxia on neurogenic transmission in the isolated pulmonary artery of the rabbit. *Br J Pharmacol 108:*150–154.

Makujina SR, Sabouni MH, Bhatia S, Douglas FL, & Mustafa SF. 1992. Vasodilator effects of adenosine A_2 receptor agonists CGS 21680 and CGS 22492 in human vasculature. *Eur J Pharmacol 221:*243–247.

Martin PL, & Potts AA. 1994. The endothelium of the rat renal artery plays an obligatory role in A_2 adenosine receptor-mediated relaxation induced by 5'-N-ethylcarboxamidoadenosine and N^6-cyclopentyladenosine. *J Pharmacol Exp Ther 270:*893–899.

Martin PL, Ueeda M, & Olsson RA. 1993. 2-Phenylethoxy-9-methyladenosine: an adenosine receptor antagonist that discriminates between A_2 adenosine receptor in the aorta and the coronary vessels from the guinea pig. *J Pharmacol Exp Ther 265:*248–253.

Martin W, Cusack NJ, Carleton JS, & Gordon JL. 1985. Specificity of the P_2-purinoceptor that mediates endothelium-dependent relaxation of the pig aorta. *Eur J Pharmacol 108:*295–299.

Mathieson JJI, & Burnstock G. 1985. Purine-mediated relaxation and constriction of isolated rabbit mesenteric artery are not endothelium-dependent. *Eur J Pharmacol 118:*221–229.

Mellion BT, Ignarro LJ, Ohlstein EH, Pontecorvo EG, Hyman AL, & Kadowitz PJ. 1981. Evidence for the inhibitory role of guanosine 3′,5′-monophosphate in ADP-induced human platelet aggregation in the presence of nitric oxide and related vasodilation. *Blood 57:*946–955.

Merkel LA, Lappe RW, Rivera LM, Cox BF, & Perrone MH. 1992. Demonstration of vasorelaxant activity with an A_1-selective adenosine agonist in porcine coronary artery: involvement of potassium channels. *J Pharmacol Exp Ther 260:*437–443.

Milner P, Bodin P, Loesch A, & Burnstock G. 1990a. Rapid release of endothelin and ATP from isolated aortic endothelial cells exposed to increased flow. *Biochem Biophys Res Commun 170:*649–656.

Milner P, Kirkpatrick KA, Ralevic V, Toothill V, Pearson J, & Burnstock G. 1990b. Endothelial cells cultured from human umbilical vein release ATP, substance P and acetylcholine in response to increased flow. *Proc R Soc Lond B 241:*245–248.

Miyagi Y, Kobayashi S, Nishimura J, Fukui M, & Kanaide H. 1996. Dual regulation of cerebrovascular tone by UTP: P_{2U} receptor-mediated contraction and endothelium-dependent relaxation. *Br J Pharmacol 118:*847–856.

Monopoli A, Conti A, Zocchi C, Casati C, Volpini R, Cristalli G, & Ongini E. 1994. Pharmacology of the new selective A_{2A} adenosine receptor agonist 2-hexyl-5′-*N*-ethylcarboxamidoadenosine *Arzneimittelforschung 44:*1296–1304.

Olsson RA, & Pearson JD. 1990. Cardiovascular purinoceptors. *Physiol Rev 70:*761–845.

Palmer RMJ, Ferrige AG, & Moncada S. 1987. Nitric oxide release accounts for the biological activity of endothelium-derived relaxing factor. *Nature 327:*524–526.

Patel V, Brown C, Goodwin A, Wilkie N, & Boarder MR. 1996. Phosphorylation and activation of p42 and p44 mitogen-activated protein kinase are required for the P2 purinoceptor stimulation of endothelial prostacyclin production. *Biochem J 320:*221–226.

Pelleg A, & Hurt CM. 1996. Mechanism of action of ATP on canine pulmonary vagal C fibre nerve terminals. *J Physiol 490:*265–275.

Prentice DJ, & Hourani SMO. 1996. Activation of multiple sites by adenosine analogues in the rat isolated aorta. *Br J Pharmacol 118:*1509–1517.

Purkiss JR, Wilkinson GF, & Boarder MR. 1994. Differential regulation of inositol 1,4,5-trisphosphate by co-existing P_{2Y}-purinoceptors and nucleotide receptors on bovine aortic endothelial cells. *Br J Pharmacol 111:*723–728.

Qasabian RA, Schyvens C, Owe-Young R, Killen JP, Macdonald PS, Conigrave AD, & Williamson DJ. 1997. Characterization of the P_2 purinoceptors in rabbit pulmonary artery. *Br J Pharmacol 120:*553–558.

Radomski MW, Palmer RMJ, & Moncada S. 1987. Endogenous nitric oxide inhibits human platelet adhesion to vascular endothelium. *Lancet 2:*1057–1058.

Radomski MW, Palmer RMJ, & Moncada S. 1990. An L-arginine: nitric oxide pathway present in human platelets regulates aggregation. *Proc Natl Acad Sci USA 87:*5193–5197.

Ralevic V, & Burnstock G. 1991a. Effects of purines and pyrimidines on the rat mesenteric arterial bed. *Circ Res 69:*1583–1590.

Ralevic V, & Burnstock G. 1991b. Roles of P_2-purinoceptors in the cardiovascular system. *Circulation 84:*1–14.

Ralevic V, & Burnstock G. 1996a. Interactions between perivascular nerves and endothelial cells in control of local vascular tone. In: Bennett T, Gardiner SM, ed. *Nervous Control of Blood Vessels.* Chur, Switzerland: Harwood Academic, pp. 135–175.

Ralevic V, & Burnstock G. 1996b. Relative contribution of P_{2U}- and P_{2Y}-purinoceptors to endothelium-dependent vasodilatation in the golden hamster isolated mesenteric arterial bed. *Br J Pharmacol 117:*1797–1802.

Ralevic V, & Burnstock G. 1998. Receptors for purines and pyrimidines. *Pharmacol Rev 50:*413–492.

Ralevic V, Milner P, Kirkpatrick KA, & Burnstock G. 1992. Flow-induced release of adenosine 5′-triphosphate from endothelial cells of the rat mesenteric arterial bed. *Experientia 48:*31–34.

Rubanyi GM, Romero JC, & Vanhoutte PM. 1986. Flow-induced release of endothelium-derived relaxing factor. *Am J Physiol 250:*H1145–H1149.

Rubino A, & Burnstock G. 1996. Evidence for a P_2-purinoceptor mediating vasoconstriction by UTP, ATP and related nucleotides in the isolated pulmonary vasculature of the rat. *Br J Pharmacol 118:*1415–1420.

Saiag B, Milon D, Allain H, Rault B, & Driessche VD. 1990. Constriction of the smooth muscle of rat tail and femoral arteries and dog saphenous vein is induced by uridine triphosphate via 'pyrimidinoceptors', and by adenosine triphosphate via P_{2X} purinoceptors. *Blood Vessels 27:*352–364.

Saiag B, Milon D, Shacoori V, Allain H, Rault B, & Van Den Driessche J. 1992. Newly evidenced pyrimidinoceptors and the P_{2X} purinoceptors are present on the vascular smooth muscle and respectively mediate the UTP- and ATP-induced contractions of the dog maxillary internal vein. *Res Commun Chem Pathol Pharmacol 76:*89–94.

Saiag B, Bodin P, Shacoori V, Catheline M, Rault B, & Burnstock G. 1995. Uptake and flow-induced release of uridine nucleotides from isolated vascular endothelial cells. *Endothelium 2:*279–285.

Schiele JO, & Schwabe U. 1994. Characterization of the adenosine receptor in microvascular coronary endothelial cells. *Eur J Pharmacol 269:*51–58.

Simonsen U, Garcia-Sacritsàn A, & Prieto D. 1997. Involvement of ATP in the non-adrenergic non-cholinergic inhibitory neurotransmission of lamb isolated coronary small arteries. *Br J Pharmacol 120:*411–420.

Sjöblom-Widfeldt N. 1990. Neuromuscular transmission in blood vessels: phasic and tonic components: An in vitro study of mesenteric arteries of the rat. *Acta Physiol Scand-Suppl 138:*1–52.

Sneddon P, & Burnstock G. 1984. ATP as a co-transmitter in rat tail artery. *Eur J Pharmacol 106:*149–152.

Sobrevia L, Yudilevich DL, & Mann GE. 1997. Activation of A_2-purinoceptors by adenosine stimulates L-arginine transport (system y^+) and nitric oxide synthesis in human fetal endothelial cells. *J Physiol 499:*135–140.

Stoggall SM, & Shaw JS. 1990. The coexistence of adenosine A_1 and A_2 receptors in guinea-pig aorta. *Eur J Pharmacol 190:*329–335.

Strøbæk D, Olesen S-P, Christopersen P, & Dissing S. 1996. P_2-purinoceptor-mediated formation of inositol phosphates and intracellular Ca^{2+} transients in human coronary artery smooth muscle cells. *Br J Pharmacol 118:*1645–1652.

Su C. 1985. Extracellular functions of nucleotides in heart and blood vessels. *Annu Rev Physiol 47:*665–676.

Szentmiklósi AJ, Ujfalusi A, Cseppento A, Nosztray K, Kovacs P, & Szabo JZ. 1995. Adenosine receptors mediate both contractile and relaxant effects of adenosine in main pulmonary artery of guinea pigs. *Naunyn Schmiedebergs Arch Pharmacol 351:*417–425.

Tesfamariam B, & Cohen RA. 1988. Inhibition of adrenergic vasoconstriction by endothelial cell shear stress. *Circ Res 63:*720–725.

Toda N, Yoshida K, & Okamura T. 1991. Analysis of the potentiating actions of N^G-nitro-L-arginine on the contraction of the dog temporal artery elicited by transmural stimulation of noradrenergic nerves. *Naunyn Schmiedebergs Arch Pharmacol 343:* 221–224.

Tsuchiya K, Urabe M, Tamamoto R, Asada Y, & Lee TJ. 1994. Effects of N^ω-nitro-L-arginine and capsaicin on neurogenic vasomotor responses in isolated mesenteric arteries of the monkey. *J Pharm Pharmacol 46:*155–157.

Valera S, Hussy N, Evans RJ, Adami N, North RA, Surprenant A, & Buell G. 1994. A new class of ligand-gated ion channel defined by P_{2X} receptor for extracellular ATP. *Nature 371:*516–519.

Valera S, Talbot F, Evams RJ, Gos A, Antonarakis SE, Morris SA, & Buell GN. 1995. Characterization and chromosomal localisation of a human P2X receptor from the urinary bladder. *Receptor Channels 3:*283–289.

Vials A, & Burnstock G. 1993. A_2-purinoceptor-mediated relaxation in the guinea-pig coronary vasculature: a role for nitric oxide. *Br J Pharmacol 109:*424–429.

von Kügelgen I, Häussinger D, & Starke K. 1987. Evidence for a vasoconstriction-mediating receptor for UTP, distinct from the P_2 purinoceptor, in rabbit ear artery. *Naunyn Schmiedebergs Arch Pharmacol 336:*556–560.

von Kügelgen I, Schöffel E, & Starke K. 1989. Inhibition by nucleotides acting at presynaptic P_2-receptors of sympathetic neuroeffector transmission in the mouse vas deferens. *Naunyn Schmiedebergs Arch Pharmacol 340:*522–532.

von Kügelgen I, Bultmann R, & Starke K. 1990. Interaction of adenine nucleotides, UTP and suramin in mouse vas deferens: suramin-sensitive and suramin-insensitive components in the contractile effect of ATP. *Naunyn Schmiedebergs Arch Pharmacol 342:*198–205.

von Kügelgen I, Kurz K, & Starke K. 1993. Axon terminal P_2-purinoceptors in feedback control of sympathetic transmitter release. *Neuroscience 56:*263–267.

von Kügelgen I, Kurz K, & Starke K. 1994a. P_2-purinoceptor-mediated autoinhibition of sympathetic transmitter release in mouse and rat vas deferens. *Naunyn Schmiedebergs Arch Pharmacol 349:*125–134.

von Kügelgen I, Späth L, & Starke K. 1994b. Evidence for P_2-purinoceptor-mediated autoinhibition of noradrenaline release in rat brain cortex. *Br J Pharmacol 113:*815–822.

von Kügelgen I, Stoffel D, & Starke K. 1995. P_2-purinoceptor-mediated inhibition of noradrenaline release in rat atria. *Br J Pharmacol 115:*247–254.

Webb TE, Feolde E, Vigne P, Neary JT, Runberg A, Frelin C, & Barnard EA. 1996. The P2Y purinoceptor in rat brain microvascular endothelial cells couple to inhibition of adenylate cyclase. *Br J Pharmacol 119:*1385–1392.

Windscheif U, Ralevic V, Bäumert HG, Mutschler E, Lambrecht G, & Burnstock G. 1994. Vasoconstrictor and vasodilator responses to various agonists in the rat perfused mesenteric arterial bed: selective inhibition by PPADS of contractions mediated via P_{2X}-purinoceptors. *Br J Pharmacol 113:*1015–1021.

Yang S, Cheek DJ, Westfall DP, & Buxton IL. 1994. Purinergic axis in cardiac blood vessels. Agonist-mediated release of ATP from cardiac endothelial cells. *Circ Res 74:*401–407.

Yang S, Buxton ILO, Probert CB, Talbot JN, & Bradley ME. 1996. Evidence for a discrete UTP receptor in cardiac endothelial cells. *Br J Pharmacol 117:*1572–1578.

Ziganshin AU, Hoyle CHV, Lambrecht G, Mutschler E, Bäumert HG, & Burnstock G. 1994. Selective antagonism by PPADS at P_{2X}-purinoceptors in rabbit isolated blood vessels. *Br J Pharmacol 111:*923–929.

6
Nitric Oxide Derived from Perivascular Nerves and Endothelium

TOMIO OKAMURA AND NOBORU TODA

Introduction

Endothelium-derived relaxing factor (EDRF) was discovered by Furchgott and Zawadzki (1980), who observed that acetylcholine-induced relaxation of the isolated rabbit aorta was endothelium-dependent and that vascular smooth muscle directly responded to acetylcholine with slight contraction. The discovery brought us the marvelous idea that vascular endothelium influences not only the blood stream but also the smooth muscle cells, thus participating in the regulation of platelet aggregation and adhesion and of vascular tone. In 1988 EDRF was identified as nitric oxide (NO), a highly diffusible and short-lived free radical, synthesized by NO synthase from L-arginine (Palmer et al. 1988a). Specific inhibitors of NO synthase (NOS), introduced by Palmer et al. (1988b), enabled us to clarify the physiological roles of endogenous NO. This lipophilic gas molecule is now recognized to be a new intercellular messenger not only in the circulatory system but also in the central nervous and immune systems.

On the other hand, neurogenic relaxation of the isolated canine cerebral artery was discovered in 1975 (Toda 1975). A number of pharmacological studies revealed that the neurogenic response was nonadrenergic and noncholinergic. However, it took a long time to identify the neurotransmitter of this vasodilator nerve. By using specific antibodies and inhibitors of NOS, we demonstrated that the nerve containing NOS innervated the cerebral artery (Yoshida et al. 1993) and was responsible for the neurogenic relaxation, and the vasodilating neurotransmitter was concluded to be NO (Toda and Okamura 1990a). Histochemical and pharmacological studies revealed that NO-mediated vasodilator nerves are also present in peripheral arteries (Yoshida et al. 1993) and probably counteract the noradrenergic vasoconstrictor nerve function (Toda and Okamura 1992a).

This chapter reviews the physiological role of endogenous NO derived from endothelium and perivascular nerves in the control of vascular tone and blood pressure.

NO Derived from Vascular Endothelium

Endothelium-dependent relaxation has been demonstrated in many vascular preparations, including various arteries, veins, and microvessels, in response to a variety of substances, such as acetylcholine, substance P, bradykinin, adenosine diphosphate, thrombin, and calcium ionophore A23187 (Furchgott 1984; Angus and Cocks 1989). Other stimuli, such as hypoxia, heat, shear stress, and increase in blood flow also cause endothelium-dependent relaxation (Furchgott 1983; Rubanyi et al. 1986). Since it is hard to totally remove vascular endothelium in vivo, most studies on EDRF have been performed in vitro. Responses of endothelium-intact and endothelium-denuded vessels to the same stimuli have been compared and bioassay of EDRF has been performed (Griffith et al. 1984). The latter studies, in which the donor of EDRF (either a blood vessel with intact endothelium or vascular endothelial cells in culture) was separated from the detector (endothelium-denuded vascular preparations), allowed the effects of physical or chemical stimuli on the generation, stability, and actions of EDRF to be studied. By the use of these techniques, various properties of EDRF have been demonstrated, including its half-life basal release, its inactivation by superoxide anion (O_2^-), and inhibition of its function by methylene blue, a soluble guanylate cyclase inhibitor, and oxyhemoglobin. Combined use of the bioassay and mass spectrometry made it possible to identify EDRF as NO synthesized from the terminal guanidino nitrogen atom of L-arginine but not of D-arginine (Palmer et al. 1988a). Further, it was found that the release of NO from cultured endothelial cells was inhibited by N^G-monomethyl-L-arginine (L-NMMA) in an enantiomer-specific manner. The L-arginine analogue inhibits the synthesis of NO_2^- and NO_3^- and citrulline from L-arginine in macrophages (Hibbs et al. 1987). Since then, L-arginine analogues such as L-NMMA, N^G-nitro-L-arginine (L-NA), and N^G-nitro-L-arginine methylester (L-NAME) have been used as competitive and reversible inhibitors of NO synthase (Rees et al. 1990). These inhibitors produce endothelium-dependent contraction of rabbit aortic rings (Palmer et al. 1988b) and canine cerebral arteries (Figure 6.1) (Toda et al. 1993a), indicating that spontaneous release of NO maintains a decreased tone in these vessels. Further, they inhibit the release of NO induced by acetylcholine from the perfused rabbit aorta; the inhibition is reversed by L-arginine but not by D-arginine (Rees et al. 1989a).

Nitric oxide which synthase, synthesizes NO and L-citrulline from L-arginine and O_2, has been divided into two types, constitutive and inducible (Moncada et al. 1991). The former is present in vascular endothelial cells, platelets, and neurons, and its activity is calcium-dependent. A small amount of NO, on the order of picomoles, is released, and glucocorticoids do not affect the synthesis of the enzyme. On the other hand, the latter type is not present under normal conditions but is induced by cytokines such as interferon-γ and endotoxins such as lipopolysaccharide in macrophages, vascular smooth muscle, and glial cells. The activity of inducible NOS is calcium-independent. A large amount (nanomoles) of NO is released, and it is considered to be a pathogenetic factor in endotoxin shock, apop-

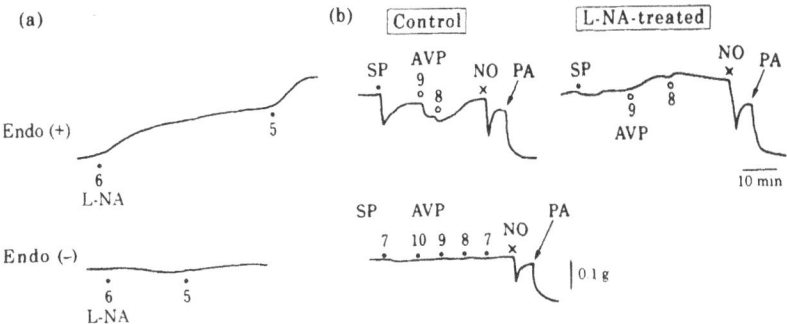

FIGURE 6.1. Functional evidence for basal (a) and stimulated (b) releases of EDRF/NO from cerebral artery strips. Endo (+) and (−), presence and absence of endothelium; L-NA, N^G-nitro-L-arginine; SP, substance P; AVP, arginine vasopressin; NO, nitric oxide (10^{-7} M acidified NaNO$_2$ solution); PA, 10^{-4} M papaverine. Numbers in the tracings represent concentrations of the drugs in $-\log[M]$.

tosis, etc. Glucocorticoids inhibit the induction of this enzyme. Endothelial NOS and neuronal NOS are both constitutive types, but they are different proteins. The endothelial enzyme has a molecular weight of 135 kDa and is considered to be membrane-bound, since it has a moiety that is N-myristoylated (Sessa et al. 1993). On the other hand, neuronal NOS, with a molecular weight of 168 kDa, seems to be a cytosolic enzyme. The different roles of these enzymes in the regulation of NO synthesis and/or release are not known. In pregnant or nonpregnant animals treated with female sex hormones, the activities of both endothelial and neuronal NOS are elevated compared with those in nonpregnant animals without hormone treatment. Therefore, an increase in uterine blood flow during pregnancy is postulated to be due to elevation of these constitutive NOS activities (Weiner et al. 1994). The NOS inhibitors of the L-arginine analogues described above specifically inhibit NOS of all types. Therefore, care has to be taken to conclude that the effect of these NOS inhibitors is associated with a depression of NO formation solely in the endothelium.

Recently, regional differences in EDRF- and endothelium-dependent relaxation have been reported. Acetylcholine-induced relaxation is very potent and is abolished by endothelial denudation and treatment with oxyhemoglobin or L-NA in the monkey temporal artery, whereas the relaxation is very weak and endothelium-independent in the monkey cerebral artery (Toda et al. 1991a) (Figure 6.2). On the other hand, relaxation caused by calcium ionophore A23187 is similar in magnitude and endothelium-dependent in both cerebral and temporal arteries (Toda et al. 1991a), indicating a heterogeneity in the density of endothelial muscarinic receptors. Such a regional difference has also been found in canine arteries (Wang et al. 1993). In pathological conditions, endothelium-dependent relaxation is not uniformly impaired in rats (Baggia et al. 1997). Second, relaxation induced by

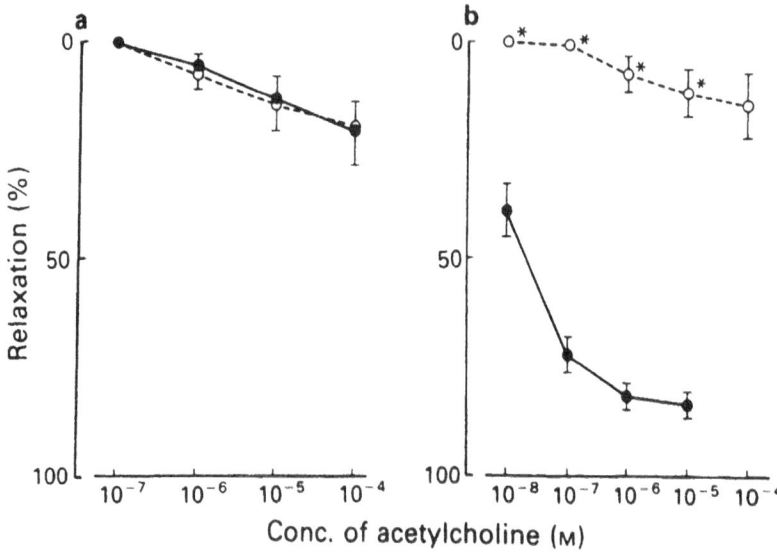

FIGURE 6.2. Concentration-relaxant response curves for acetylcholine in monkey cerebral (a) and temporal artery strips (b) with (●) and without (○) the endothelium. The strips were partially contracted with prostaglandin $F_{2\alpha}$. Relaxations induced by 10^{-4} M papaverine were taken as 100%. Significantly different from the value in the endothelium-intact arteries, *$P < 0.001$ (paired t-test). Bars indicate SE. From Toda et al. 1991a. Copyright 1991 by Stockton Press. Reprinted with permission.

acetylcholine is abolished by methylene blue (Okamura et al. 1989b) or NOS inhibitors in canine coronary artery but is not completely inhibited in the mesenteric artery. Combined treatment with cyclooxygenase inhibitors such as indomethacin abolished relaxation in the mesenteric artery (Okamura et al. 1989b), indicating that acetylcholine can stimulate the production of both EDRF and vasodilating prostanoids. Further, endothelium-derived hyperpolarizing factor(s) (EDHF) are reported to be involved in the relaxation caused by some EDRF-releasing agents, such as acetylcholine and substance P, since some endothelium-dependent relaxations induced by the agents are resistant to both cyclooxygenase and NOS inhibitors but are suppressed by K^+-induced depolarization or K^+-channel inhibitors (see review by Feletou and Vanhoutte 1996). Shimokawa et al. (1996) compared the contributions of prostacyclin (prostaglandin I_2, PGI_2), NO, and EDHF to the acetylcholine-induced, endothelium-dependent relaxation of isolated rat aorta, proximal mesenteric artery, and distal mesenteric artery. They demonstrated that the contribution of PGI_2 was minimal in three different-sized blood vessels. Nitric oxide contributed to the relaxation most prominently in the aorta, whereas the contribution of EDHF was largest in the distal mesenteric artery. Taken together, these findings suggest that the mediators of endothelium-dependent relaxation vary with the location and size of blood vessels.

The importance of endothelium-derived NO in the regulation of hemodynamics is evident. Cross-linked hemoglobin perfusion in anesthetized dogs causes an elevation of blood pressure and a decrease of blood flow to the renal, iliac, and mesenteric regions (Cases et al. 1997), and blood pressure is raised in mice with the endothelial NOS genes knocked out, (Huang et al. 1995). Reduction of endothelial cell function with aging or from damage by vascular disorders such as atherosclerosis, hypertension, and diabetes mellitus would diminish the peripheral circulation by decreasing the tonic vasodilating action of NO derived from the endothelium. However, NOS function is maintained in aged rats (Hill et al. 1997), and supplementation of L-arginine, a substrate of NOS, may restore impaired coronary blood flow in aged humans (Chauhan et al. 1996). It has also been demonstrated that treatment of hypertension (Lockette et al. 1986) and acute cardiac ischemia (Lee et al. 1996) may restore impaired endothelium-dependent relaxation in animal models.

NO Derived from Perivascular Nerves

Cerebral Arteries

Stimulation of perivascular nerves by electrical pulses applied transmurally or by nicotine produces frequency- or concentration-related relaxation of canine coronary and cerebral arteries. Relaxation of the coronary arteries is abolished by β-adrenoceptor antagonists, indicating that the neurogenic relaxation is due to noradrenaline released from adrenergic nerves, whereas the relaxation of the cerebral artery is resistant to adrenoceptor antagonists, P_1 purinoceptor antagonists, muscarinic and H_1 and H_2 histamine receptor antagonists, and inhibitors of Na^+-K^+-ATPase and cyclooxygenase (Toda 1978). Immunohistochemical studies have demonstrated the presence of dense networks of perivascular nerves containing polypeptides such as substance P (Edvinsson et al. 1981), vasoactive intestinal polypeptide (VIP) (Larsson et al. 1976), and calcitonin gene-related polypeptide (CGRP)(Wanaka et al. 1986). Because these peptides relax cerebral arteries, they were recognized as candidates for the vasodilator neurotransmitter. However, evidence for their release from perivascular nerves has not been obtained and neurogenic relaxation is observed even in vascular preparations made unresponsive to the peptides (Toda 1982; Toda and Okamura 1991) or treated with capsaicin (Okamura and Toda 1994a), which depletes neuronal substance P and CGRP (Kawasaki et al. 1988). Atrial natriuretic peptide has also been excluded as a neurotransmitter candidate (Okamura et al. 1989a). Since the neurotransmitter of the vasodilator nerves to the cerebral artery is unidentified, the nerves are called non-adrenergic, noncholinergic (NANC) nerves. However, hemolysate or methylene blue abolishes the neurogenic relaxation, suggesting that the properties of this neurotransmitter are similar to those of EDRF (Toda 1988). When the NOS inhibitors became available, the neurotransmitter of this vasodilator nerve was unmasked (Toda and Okamura 1990b). L-NMMA or L-NA does not affect neurogenic relaxation in canine coronary artery but abolishes cerebroarterial neurogenic relaxation in an enantiomer-specific manner (Toda et al. 1990). The response is restored by the addition of high concentrations of L-arginine but not D-arginine. Relaxation induced by exogenous NO is not affected by NOS inhibitors but is abolished by oxyhemoglobin and methylene blue. Similar results are obtained in human (Toda 1993), monkey (Toda and Okamura 1990b), sheep (Gaw et al. 1990), and porcine (Lee and Sarwinski 1991) cerebral arteries. Further, transmural electrical stimulation (TES) or nicotine increases NOx release from canine endothelium-denuded cerebral arteries and intracellular cyclic GMP production in the arteries (Toda and Okamura 1990a). Neurogenic relaxation by TES is not affected by L-type calcium antagonists such as nicardipine (Toda and Okamura 1992b) but is suppressed by N-type antagonists such as ω-conotoxin GVIA (Toda et al. 1995), calmodulin inhibitors such as W-7 and calmidazolium (Okamura and Toda 1994b), and calcium/calmodulin-dependent protein kinase II (CALCAM II) inhibitors such as KN62 (Toda et al. 1998). Immunohistochemical studies with nerve-derived NOS antibody (Bredt et al. 1990) have shown the presence of networks of positively stained nerve fibers and bundles in canine (Yoshida et al. 1993) and monkey (Yoshida et al. 1994) cerebral arteries and arterioles. The neurons are located mainly in the adventitia, and some fine fibers are located in the outer layer of the media. These results strongly support our hypothesis that NO or an NO-containing molecule synthesized by neural NOS from L-arginine in response to nerve stimulation is liberated from perivascular nerves innervating canine and primate cerebral arteries, and activates soluble guanylate cyclase in smooth muscle cells to elevate cyclic GMP production, resulting in vasorelaxation. Nitric oxide synthase is presumably activated by calcium that enters through N-type calcium channels upon nerve excitation, is bound to calmodulin, and activates CALCAM II in nerve cells. Neurotransmission through the NO-mediated vasodilator nerve (nitroxidergic nerve) is presented in the scheme (Figure 6.3). The possibilities have been excluded that NO derived from perivascular nerves liberates other vasodilator neurotransmitters and that unknown neurotransmitters stimulate NO production in smooth muscle cells (see review by Toda 1995).

Peripheral Arteries

In contrast to the response in cerebral arteries, nerve stimulation causes contraction in dog peripheral arteries, including the mesenteric, renal, temporal, and femoral arteries. The neurogenic contraction, which is sensitive to α-adrenoceptor antagonists, is potentiated by L-NMMA and L-NA (Figure 6.4), whereas noradrenaline-induced contraction is not enhanced in canine (Toda et al. 1991b) and monkey (Toda and Okamura 1992c) mesenteric arteries. TES-evoked noradrenaline release, estimated by overflow of ^{3}H from the isolated canine temporal

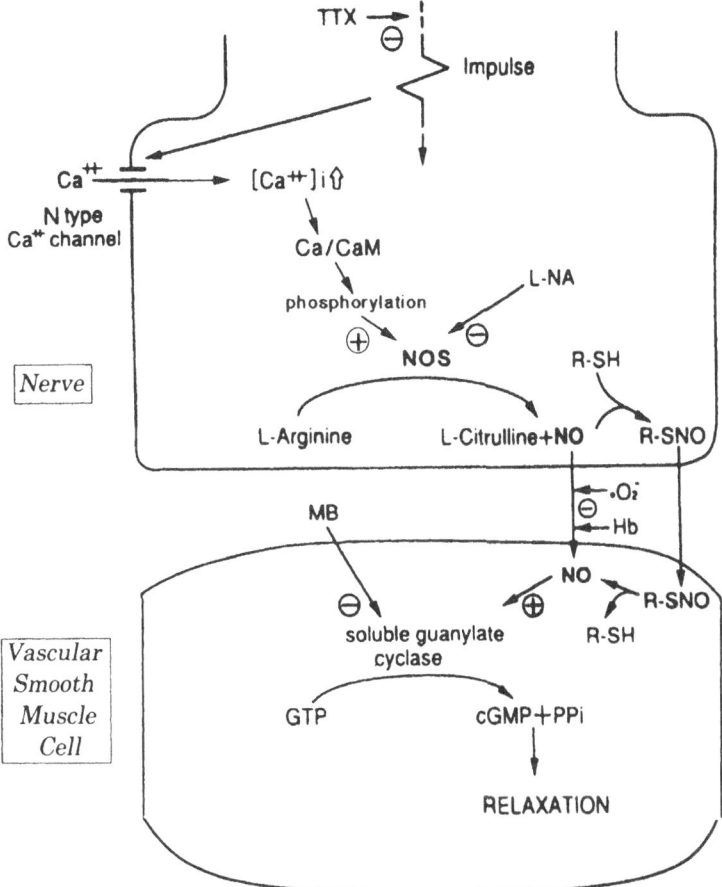

FIGURE 6.3. Hypothetical schema of the nitroxidergic nerve terminal and cerebral arterial smooth muscle cell. The nerve originates from the pterygopalatine ganglion in the dog. NOS, Nitric oxide (NO) synthase; CaM, calmodulin; R-SNO, S-nitrosothiol; Hb, oxyhemoglobin; MB, methylene blue; L-NA, N^G-nitro-L-arginine; GTP, guanosine triphosphate; TTX, tetrodotoxin.

(Toda et al. 1991c) and mesenteric arteries preincubated with [^{3}H]noradrenaline, is not increased by L-NA. These findings indicate that potentiation by NOS inhibitors of the response to adrenergic nerve stimulation is not due to increasing the responsiveness of smooth muscle to noradrenaline or release of noradrenaline from the nerves. When the arteries are precontracted and treated with an α-adrenoceptor antagonist alone or in combination with a P$_2$ purinoceptor antagonist, α β-methylene ATP, the neurogenic contraction is reversed to a relaxation, which is endothelium-independent and abolished by L-NA or L-NMMA in an enatiomer-specific manner (Figure 6.4). L-Arginine, but not D-arginine, restores the neurogenic relaxation. Similar results have also been reported in bovine mesenteric

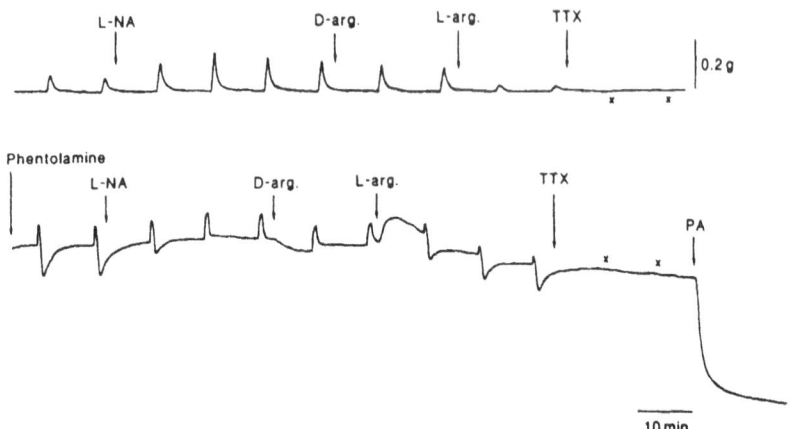

FIGURE 6.4. Representative tracings of responses to transmural electrical stimulation (5 Hz) of a monkey artery strip without the endothelium, before and after treatment with N^G-nitro-L-arginine (L-NA)(10^{-6} M), D-arginine (D-arg.) and L-arginine (L-arg.)(3×10^{-4} M), and tetrodotoxin (TTX)(3×10^{-7} M) in control (upper tracing) and phentolamine-containing (lower tracing) media. After the experimental sequence on the upper tracing was over, the strip was repeatedly washed and equilibrated. Then the strip was partially contracted with prostaglandin $F_{2\alpha}$ and treated with 10^{-7} M phentolamine. PA, 10^{-4} M papaverine to attain the maximal relaxation. x, Application of electrical stimulation under TTX-treated conditions. (From Toda and Okamura 1992c.) Copyright 1992 by Lippincott Williams & Wilkins. Reprinted with permission.

arteries (Leckstrom et al. 1993) and guinea pig pulmonary arteries (Liu et al. 1992). TES or nicotine stimulates the NO_x release from superfused canine temporal arteries without the endothelium and increases the intracellular content of cyclic GMP in the arteries. Therefore, the mechanisms of neurogenic relaxation in the peripheral arteries would be the same as those in the cerebral arteries. These findings strongly suggest that the potentiation of contraction by NOS inhibitors results from a suppression of NO-mediated vasodilator nerve function. Histochemical demonstration of the presence of immunoreactive NOS or NADPH diaphorase activity-containing nerve fibers and bundles in the adventitia of the arteries and arterioles supports this hypothesis (Yoshida et al. 1993). Taken together, our data strongly suggest that the vascular tone in canine and monkey peripheral arteries is reciprocally regulated by noradrenergic vasoconstrictor and nitroxidergic vasodilator nerves (Figure 6.5). Such an innervation is seen in renal (Okamura et al. 1996), gastroepiploic (Toda 1994), saphenous (Okamura and Toda 1994c), and uterine (Toda et al. 1994) arteries. It is well known that noradrenergic nerve function is the most important determinant in the regulation of peripheral arterial tone. Counteraction of the vasodilator nerve against noradrenergic nerve function would also be important in the regulation of local blood flow and systemic blood pressure. Functional impairment of the vasculature may be associated with an imbalance of these nerve functions.

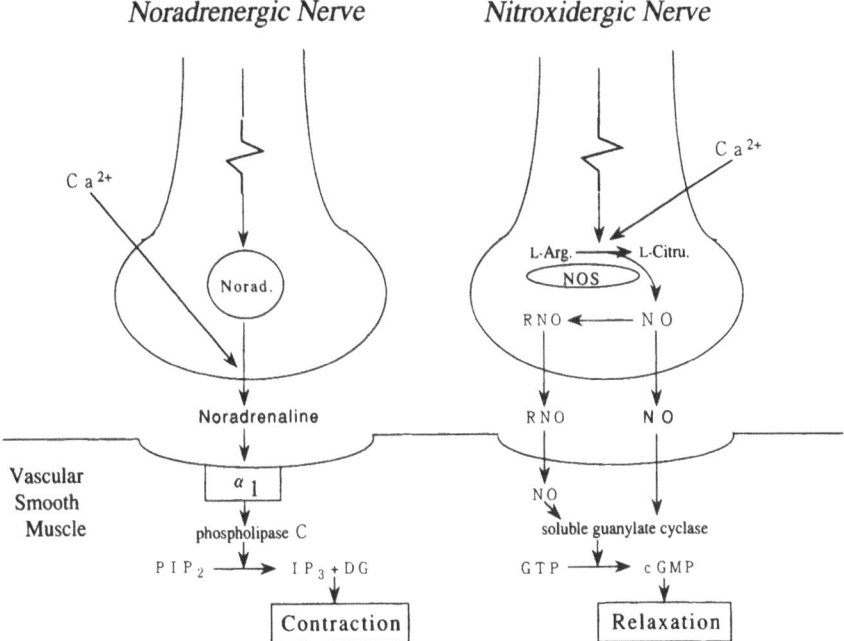

FIGURE 6.5. Schematic presentation of reciprocal regulation of the vascular smooth muscle by noradrenergic vasoconstrictor nerve and nitroxidergic vasodilator nerve. NOS, Nitric oxide (NO) synthase; L-Arg., L-arginine; RNO, NO-containing molecule; GTP, guanosine triphosphate; α_1 in square, α_1-adrenoceptor; PIP$_2$, phosphatidylinositol bisphosphate; IP$_3$, inositol trisphosphate; DG, diacylglycerol; Norad., noradrenaline.

Endogenous NO in the Regulation of Vascular Tone In Vivo

The physiological roles of endogenous NO in the regulation of cardiovascular function have been investigated by determining the in vivo effects of NOS inhibitors. Systemic administration of NOS inhibitors, including L-NMMA, L-NA, and L-NAME, increases systemic blood pressure and total vascular resistance in mammals, including rats (Gardiner et al. 1990), rabbits (Rees et al. 1989b), guinea pigs (Aisaka et al. 1989), dogs (Klabunde et al. 1991; Toda et al. 1993c), monkeys (Okamura et al. 1996), and humans (Haynes et al. 1993). The hypertension induced by NOS inhibitors is accompanied by a decrease in vascular conductance in various vascular beds in rats (Gardiner et al. 1990), dogs (Chu et al. 1991), and humans (Dijkhorst et al. 1997). D-Enantiomers of the arginine analogues are without effect. The changes caused by NOS inhibition are reversed by L-arginine, indicating that basal release of NO derived from L-arginine participates in hemodynamic control. A number of reports have suggested that NO involved in the regulation of blood pressure and organ blood flow is derived from the vascular endothelium, since the effects of EDRF-releasing agents such as acetylcholine and bradykinin are suppressed by treatment with NOS inhibitors under the same experimental conditions. However, the evidence is not sufficient to permit the conclusion that endothelium-derived NO is the sole cause of the hemodynamic changes elicited by NOS inhibitors. As previously described, these inhibitors also suppress the function of NO-mediated vasodilator nerves that supply arteries and arterioles. Our studies with anesthetized dogs (Toda et al. 1993c) and Japanese monkeys (Okamura et al. 1996) have demonstrated that elevation of blood pressure by L-NA is not influenced by treatment with phentolamine but is markedly suppressed by the ganglion-blocking agents hexamethonium and pentolinium, at doses sufficient to lower blood pressure to a similar magnitude to that caused by phentolamine and to reverse the bradycardia induced by a pressor dose of noradrenaline to a tachycardia. Togashi et al. (1992) have suggested that the hypertensive effect of the NOS inhibitor is due to a suppression of NOS activity in the central nervous system and a subsequent increase in the efferent discharge of peripheral sympathetic nerves. However, the pressor action of the NOS inhibitor was not reduced by treatment with phentolamine in our experiments. Further, intracisternal administration of L-NA does not significantly alter the systemic blood pressure in the dog (Toda et al. 1993c) and the monkey (Okamura et al. 1995), whereas the inhibitor constricts the basilar artery. Neurogenic relaxation of distal mesenteric arteries isolated from monkeys (outer diameter less than 200 μm) is abolished by L-NA, and the presence of NADPH diaphorase-positive fibers has been histologically demonstrated in monkey mesenteric arterioles (Okamura et al. 1996). These findings suggest that nitroxidergic nerves also participate in the physiological regulation of blood pressure and vascular tone in dogs and monkeys. Evidence for the importance of neuronal NO in the regulation of hemodynamics has recently been accumulated. Ayata et al. (1996) reported that augmentation of regional cerebral blood flow by mechanical stimulation of the whiskers in endothelial NOS knockout mice was inhibited by L-NA. The NO responsible for a sustained increase in cerebral blood flow during a seizure after systemic administration of kainate is considered to be of neural origin (Montecot et al. 1997). Nitric oxide-containing factors mediating hindlimb vasodilation caused by superior laryngeal nerve stimulation may also be derived from nerves (Possas and Lewis 1997).

Endothelial dysfunction has been related to several vascular disorders, such as vasospasm, atherosclerosis, hypertension, and diabetes mellitus, in experimental animals and humans. Although the contribution of nitroxidergic nerves to the regulation of vascular tone may be heterogeneous among animal species or vasculature region, the pathophysiological implication of neuronal NO deficiency in circulatory disorders is a matter of great interest for future investigation.

References

Aisaka K, Gross SS, Griffith OW, & Levi R. 1989. N^G-Methylarginine, an inhibitor of endothelium-derived nitric oxide synthesis, is a potent pressor agent in the guinea pig:

does nitric oxide regulate blood pressure in vivo? *Biochem Biophys Res Commun 160:*881–886.

Angus JA, & Cocks TM. 1989. Endothelium-derived relaxing factor. *Pharmacol Ther 41:*303–351.

Ayata C, Ma J, Meng W, Huang P, & Moskowitz MA. 1996. L-NA-sensitive rCBF augmentation during vibrissal stimulation in type III nitric oxide synthase mutant mice. *J Cereb Blood Flow Metab 16:*539–541.

Baggia S, Perkins K, & Greenberg B. 1997. Endothelium-dependent relaxation is not uniformly impaired in chronic heart failure. *J Cardiovasc Pharmacol 29:*389–396.

Bredt DS, Hwang PM, & Snyder SH. 1990. Localization of nitric oxide synthase indicating a neural role for nitric oxide. *Nature 347:*768–770.

Cases A, Stulak JM, Katusic Z, Villa E, & Romero JC. 1997. Hemodynamic and renal effects of cross-linked hemoglobin infusion. *Am J Physiol 272:*R793–R799.

Chauhan A, More RS, Mullins PA, Taylor G, Petch C, & Schofield PM. 1996. Aging-associated endothelial dysfunction in humans is reversed by L-arginine. *J Am Coll Cardiol 28:*1796–1804.

Chu A, Lin CC, Chambers DE, Kuehl WD, Palmer RMJ, Moncada S, & Cobb FR. 1991. Effects of inhibition of nitric oxide formation on basal tone and endothelium-dependent responses of the coronary arteries in awake dogs. *J Clin Invest 87:*1964–1968.

Dijkhorst O, Rabelink TJ, Boer P, & Koomans HA. 1997. Nifedipine attenuates systemic and renal vasoconstriction during nitric oxide inhibition in humans. *Hypertension 29:*1192–1198.

Edvinsson L, Muculloch J, & Uddman R. 1981. Immunohistochemical localization and effect upon cat pial arteries in vivo and in situ. *J Physiol 318:*251–258.

Feletou M, & Vanhoutte PM. 1996. Endothelium-derived hyperpolarizing factor. *Clin Exp Pharmacol Physiol 23:*1082–1090.

Furchgott RF. 1983. Role of endothelium in responses of vascular smooth muscle. *Circ Res 53:*557–573.

Furchgott RF. 1984. The role of endothelium in the responses of vascular smooth muscle to drugs. *Annu Rev Pharmacol Toxicol 24:*175–197.

Furchgott RF, & Zawadzki JV. 1980. The obligatory role of endothelial cells in the relaxation of arterial smooth muscle by acetylcholine. *Nature 288:*373–376.

Gardiner SM, Compton AM, Bennett T, Palmer RMJ, & Moncada S. 1990. Control of regional blood flow by endothelium-derived nitric oxide. *Hypertension 15:*486–492.

Gaw AJ, Wadsworth RM, & Humphrey PPA. 1990. Neurotransmission in the sheep middle cerebral artery: modulation of responses by 5-HT and haemolysate. *J Cereb Blood Flow Metab 10:*409–416.

Griffith TM, Edwards DH, Lewis MJ, Newby AC, & Henderson AH. 1984. The nature of endothelium-derived vascular relaxing factor. *Nature 329:*442–445.

Haynes WG, Noon JP, Walker BR, & Webb DJ. 1993. Inhibition of nitric oxide synthesis increases blood pressure in healthy humans. *J Hypertens 11:*1375–1380.

Hibbs JB Jr, Taintor RR, & Vavrin Z. 1987. Macrophage cytotoxicity: role for L-arginine deiminase activity and imino nitrogen oxidation to nitrite. *Science 235:*473–476.

Hill C, Lateef AM, Engels K, Samsell L, & Baylis C. 1997. Basal and stimulated nitric oxide in control of kidney function in the aging rat. *Am J Physiol, 272:*R1747–R1753.

Huang PL, Huang Z, Mashimo H, Bloch KD, Moskowitz MA, Bevan JA, & Fishman MC. 1995. Hypertension in mice lacking the gene for endothelial nitric oxide synthase. *Nature 377:*239–242.

Kawasaki H, Takasaki T, Saito A, & Goto K. 1988. Calcitonin gene-related peptide acts as a novel vasodilator neurotransmitter in mesenteric resistance vessels of the rat. *Nature* 335:164–167.

Klabunde RE, Ritger RC, & Helgren MC. 1991. Cardiovascular actions of inhibitors of endothelium-derived relaxing factor (nitric oxide) formation/release in anesthetized dogs. *Eur J Pharmacol 199:*51–59.

Larsson LI, Edvinsson L, Fahrenkrug J, Hakanson R., Owman C, Schaffalitzky-de-Muckadell O, & Sundler F. 1976. Immunohistochemical localization of a vasodilatory polypeptide (VIP) in cerebrovascular nerves. *Brain Res 113:*400–404.

Leckstrom A, Ahlner J, Grundstrom N, & Axelsson KL. 1993. Involvement of nitric oxide and peptides in the inhibitory non-adrenergic, non-cholinergic (NANC) response in bovine mesenteric artery. *Pharmacol Toxicol 72:*194–198.

Lee JJ, Oimos L, & Vanhoutte PM. 1996. Recovery of endothelium-dependent relaxations four weeks after ischemia and progressive reperfusion in canine coronary arteries. *Proc Assoc Am Physician 108:*362–367.

Lee TJ-F, Sarwinski SJ. 1991. Nitric oxidergic vasodilation in the porcine basilar artery. *Blood Vessels 28:*407–412.

Liu SF, Crawley DE, Rohde JAL, Evans TW, & Barnes PJ. 1992. Role of nitric oxide and guanosine 3′, 5′-cyclic monophosphate in mediating nonadrenergic, noncholinergic relaxation in guinea-pig pulmonary arteries. *Br J Pharmacol 107:*861–866.

Lockette W, Otsuka Y, & Carretero O. 1986. The loss of endothelium-dependent vascular relaxation in hypertension. *Hypertension 8(Suppl):*II61–II66.

Moncada S, Palmer RMJ, & Higgs EA. 1991. Nitric oxide: physiology, pathophysiology and pharmacology. *Pharmacol Rev 43:*109–142.

Montecot C, Borredon J, Seylaz J, & Pinard E. 1997. Nitric oxide of neuronal origin is involved in cerebral blood flow increase during seizures induced by kainate *J Cereb Blood Flow Metab 17:*94–99.

Okamura T, & Toda N. 1994a. Nitric oxide (NO)-mediated, vasodilator nerve function and its susceptibility to calcium antagonists. *J Auton Nerv Sys 49:*S55–S58.

Okamura T, & Toda N. 1994b. Inhibition by calmodulin antagonists of the neurogenic relaxation in cerebral arteries. *Eur J Pharmacol 256:*79–83.

Okamura T, & Toda N. 1994c. Mechanism underlying nicotine-induced relaxation in dog saphenous arteries. *Eur J Pharmacol 263:*85–91.

Okamura T, Inoue S, & Toda N. 1989a. Action of atrial natriuretic peptide (ANP) on dog cerebral arteries: evidence that neurogenic relaxation is not mediated by release of ANP. *Br J Pharmacol 97:*1258–1264.

Okamura T, Minami Y, Toda N. 1989b. Endothelium-dependent and independent mechanisms of action of acetylcholine in monkey and dog isolated arteries. *Pharmacology 38:*279–288.

Okamura T, Ayajiki K, & Toda N. 1995. Basilar arterial constriction caused by intracisternal N^G-nitro-L-arginine in anesthetized monkeys. *Cardiovasc Res 30:*663–667.

Okamura T, Ayajiki K, & Toda N. 1996. Neural mechanism of pressor action of nitric oxide synthase inhibition in anesthetized monkeys. *Hypertension 28:*341–346.

Palmer RMJ, Ashton DS, & Moncada S. 1988a. Vascular endothelial cells synthesize nitric oxide from L-arginine. *Nature 333:*664–666.

Palmer RMJ, Rees DD, Ashton DS, & Moncada S. 1988b. L-Arginine is the physiological precursor for the formation of nitric oxide in the endothelium-dependent relaxation. *Biochem Biophys Res Commun 153:*1251–1256.

Possas OS, & Lewis S J. 1997. NO-containing factors mediate hindlimb vasodilation produced by superior laryngeal nerve stimulation. *Am J Physiol 273*:H234–H243.

Rees DD, Palmer RMJ, Hodson HF, & Moncada S. 1989a. A specific inhibitor of nitric oxide formation from L-arginine attenuates endothelium-dependent relaxation. *Br J Pharmacol 96*:418–424.

Rees DD, Palmer RMJ, & Moncada S. 1989b. Role of endothelium-derived nitric oxide in the regulation of blood pressure. *Proc Natl Acad Sci USA 86*:3375–3378.

Rees DD, Schulz R, Hodson HF, Palmer RMJ, & Moncada S. 1990. Characterization of three inhibitors of endothelial nitric oxide synthase in vitro and in vivo. *Br J Pharmacol 101*:746–752.

Rubanyi GM, Romero JC, & Vanhoutte PM. 1986. Flow-induced release of endothelium-derived relaxing factor. *Am J Physiol 250*:H1145–H1149.

Sessa WC, Barber CM, & Lynch KR. 1993. Mutation of N-myristoylation site converts endothelial nitric oxide synthase from a membrane to a cytosolic protein. *Circ Res 72*:921–924.

Shimokawa H, Yasutake H, Fujii K, Owada MK, Nakaike R, Fukumoto Y, Takayanagi T, Nagao T, Egashira K, Fujishima M, & Takeshita A. 1996. The importance of the hyperpolarizing mechanism increases as the vessel size decreases in endothelium-dependent relaxations in rat mesenteric circulation. *J Cardiovasc Pharmacol 28*:703–711.

Toda N. 1975. Nicotine-induced relaxation in isolated canine cerebral arteries. *J Pharmacol Exp Ther 193*:376–384.

Toda N. 1978. Heterogeneity in the relaxation of vascular smooth muscle. In: Vanhoutte PM, ed. *Vasodilatation*. Basel: Karger, pp 129–136.

Toda N. 1982. Relaxant responses to transmural stimulation and nicotine of dog and monkey cerebral arteries. *Am J Physiol 243*:H145–H153.

Toda N. 1988. Hemolysate inhibits cerebral artery relaxation. *J Cereb Blood Flow Metab 8*:46–53.

Toda N. 1993. Mediation by nitric oxide of neurally-induced human cerebral artery relaxation. *Experientia 49*:51–53.

Toda N. 1994. Nitroxidergic innervation in smooth muscle. In: Matsuo Y, Kasuya Y, Tuchiya M, & Nagao F, eds. *Gastrointestinal Function; Regulation and Disturbances*, Vol 12. Tokyo: Excerpta Medica, pp3–7.

Toda N. 1995. Nitric oxide and regulation of cerebral arterial tone. In: Vincent R, ed. *Nitric Oxide in the Nervous System*. New York: Academic Press, pp 207–225.

Toda N, & Okamura T. 1990a. Possible role of nitric oxide in transmitting information from vasodilator nerve to cerebroarterial muscle. *Biochem Biophys Res Commun 170*:308–313.

Toda N, & Okamura T. 1990b. Mechanism underlying the response to vasodilator nerve stimulation in isolated dog and monkey cerebral arteries. *Am J Physiol 259*: H1511–H1517.

Toda N, & Okamura T. 1991. Suppression by N^G-monomethyl-L-arginine of cerebroarterial responses to nonadrenergic, noncholinergic vasodilator nerve stimulation. *J Cardiovasc Pharmacol 17 (Suppl 3)*:S234–S237.

Toda N, & Okamura T. 1992a. Regulation by nitroxidergic nerve of arterial tone. *News Physiol Sci 7*:148–152.

Toda N, & Okamura T. 1992b. Different susceptibility of vasodilator nerve, endothelium and smooth muscle functions to Ca^{++} antagonists in cerebral arteries. *J Pharmacol Exp Ther 261*:234–239.

Toda N, & Okamura T. 1992c. Mechanism of neurally induced monkey mesenteric artery relaxation and contraction. *Hypertension 19:*161–166.

Toda N, Minami Y, & Okamura T. 1990. Inhibitory effects of L-N^G-nitro-arginine on the synthesis of EDRF and the cerebroarterial response to vasodilator nerve stimulation. *Life Sci 47:*345–351.

Toda N, Kawakami M, Yamazaki M, & Okamura T. 1991a. Comparison of endothelium-dependent responses of monkey cerebral and temporal arteries. *Br J Pharmacol 102:*805–810.

Toda N, Kitamura Y, & Okamura T. 1991b. New idea on the mechanism of hypertension: suppression of nitroxidergic vasodilator nerve function. *J Vasc Med Biol 3:*235–241.

Toda N, Yoshida K, & Okamura T. 1991c. Analysis of the potentiating action of N^G-nitro-L-arginine on the contraction of the dog temporal artery elicited by transmural stimulation of noradrenergic nerves. *Naunyn Schmiedeberg's Arch Pharmacol 343:*221–224.

Toda N, Ayajiki K, & Okamura T. 1993a. Endothelial modulation of contractions caused by oxyhemoglobin and N^G-nitro-L-arginine in isolated dog and monkey cerebral arteries. *Stroke 24:*1584–1588.

Toda N, Ayajiki K, & Okamura T. 1993b. Neural mechanism underlying basilar arterial constriction by intracisternal L-NNA in anesthetized dogs. *Am J Physiol 265:*H103–H107.

Toda N, Kitamura Y, & Okamura T. 1993c. Neural mechanism of hypertension by nitric oxide synthase inhibitor in dogs. *Hypertension 21:*3–8.

Toda N, Kimura T, Yoshida K, Bredt DS, Snyder SH, Yoshida Y, & Okamura T. 1994. Human uterine arterial relaxation induced by nitroxidergic nerve stimulation. *Am J Physiol 266:*H1446–H1450.

Toda N, Uchiyama M, & Okamura T. 1995. Prejunctional modulation of nitroxidergic nerve function in canine cerebral arteries. *Brain Res 700:*213–218.

Toda N, Ayajiki K, & Okamura T. 1998. Effect of Ca^{2+}/calmodulin-dependent protein kinase II inhibitors on the neurogenic cerebroarterial relaxation. *Eur J Pharmacol 340:*59–65.

Togashi H, Sakuma I, Yoshioka M, Kobayashi T, Yasuda H, Kitabatake A, Saito H, Gross SS, & Levi R. 1992. A central nervous system action of nitric oxide in blood pressure regulation. *J Pharmacol Exp Ther 262:*343–347.

Wanaka K, Matsuyama T, Yoneda S, Kimura K, Kamada T, Girgis S, Macintyre I, Emson P, & Tohyama M. 1986. Origins and distribution of calcitonin gene-related peptide-containing nerves in the wall of the cerebral arteries of the guinea pig with special reference to the coexistence with substance P. *Brain Res 369:*185–192.

Wang Y, Okamura T, & Toda N. 1993. Mechanisms of acetylcholine-induced relaxation in dog external and internal ophthalmic arteries. *Exp Eye Res 57:*275–281.

Weiner CP, Lizasoain I, Baylis SA, Knowles RG, Charles IG, & Moncada S. 1994. Induction of calcium-dependent nitric oxide synthases by sex hormones. *Proc Natl Acad Sci USA 91:*5212–5216.

Yoshida K, Okamura T, Kimura H, Bredt DS, Snyder SH, & Toda N. 1993. Nitric oxide synthase-immunoreactive nerve fibers in dog cerebral and peripheral arteries. *Brain Res 629:*67–72.

Yoshida K, Okamura T, & Toda N. 1994. Histological and functional studies on the nitroxidergic nerve innervating monkey cerebral, mesenteric and temporal arteries. *Jpn J Pharmacol 65:*351–359.

7
Nitric Oxide and Hypertension

Roberto Zatz and Christine Baylis

Introduction

Since the pioneering work of Furchgott and Zawadzki (1980), it has become increasingly clear that the vascular endothelium plays a key role in the regulation of the cardiovascular system. The discovery in 1987 that nitric oxide (NO) accounts for the vasorelaxing action of the endothelium (Ignarro et al. 1987; Palmer et al. 1987), opening the possibility of specifically inhibiting NO synthesis, greatly enlarged our understanding of endothelial physiology and gave rise to a vast new area of biological knowledge: NO physiology.

Nitric oxide is produced in response to a variety of stimuli in the presence of the substrates L-arginine and oxygen as well as a number of essential cofactors. The synthesis of NO is crucially dependent on the action of three widely distributed isoforms of nitric oxide synthase (NOS): endothelial NOS (eNOS), brain (neural) NOS (bNOS), and inducible NOS (iNOS), eNOS and bNOS are constitutively expressed and widely distributed throughout the kidney (Bachmann and Mundel 1994), as well as the cardiovascular system and in strategic locations in the peripheral and central nervous system (Zhang et al. 1997). Nitric oxide synthase plays a major role in the physiologic control of vascular tone and kidney function (Moncada et al. 1991; Baylis and Qiu 1996; Kone and Baylis 1997). Inducible NOS can be activated in many cell types in response to a variety of immunologic stimuli and may play a key role in the process of immune defense. Limited constitutive iNOS expression in kidney and vascular smooth muscle has been described, although its role in the regulation of blood pressure and/or kidney function is still unclear (Kone and Baylis 1997).

Much of the evidence currently suggesting a role for NO deficiency in the pathogenesis of hypertension is derived from acute and chronic studies of NO inhibition, many of which will be discussed in the following sections.

Effects of Acute NOS Inhibition on Cardiovascular and Renal Hemodynamics

In 1989 Rees et al. (1989) showed that acute blockade of NO synthesis with the L-arginine analogue N^G-monomethyl- L-arginine (L-NMMA) markedly elevated blood pressure in rabbits, an effect that was reversed by administration of large amounts of L-arginine. This finding was corroborated by other studies in several species, including human (Haynes et al. 1993; Wolzt et al. 1997), indicating that by promoting constant vasorelaxation, NO must be an important modulator of circulatory function.

Acute systemic NOS inhibition with L-arginine analogues profoundly influences the renal circulation as well, producing intense vasoconstriction and markedly reducing renal plasma flow, although the glomerular filtration rate is relatively preserved. These effects, observed in several species, are apparent in both the awake and the anesthetized animal (Zatz and De-Nucci 1991; Baylis and Qiu 1996; Kone and Baylis 1997).

Effects of Chronic NO Inhibition on Cardiovascular and Renal Hemodynamics: The Chronic NO Synthase Inhibition Model

The finding that hypertension and renal vasoconstriction invariably follow the abrupt interruption of NO synthesis does not necessarily imply that NO is indispensable to the control of blood pressure and cardiovascular function on a long-term basis. Conceivably, other vasodilating systems might be recruited in the long run to compensate for NO blockade. This hypothesis was proven incorrect in 1992, when two independent groups showed that chronic oral administration of an inhibitor of NO synthesis, N^G-nitro-L-arginine methylester (L-NAME), promoted persistent hypertension and renal damage. Two other groups reported simultaneously or soon thereafter the development of hypertension associated with chronic L-NAME treatment (Arnal et al. 1992; Lahera et al. 1992). These findings, subsequently confirmed in several other studies, indicated that NO is a fundamental and irreplaceable element in the regulation of blood pressure, originating a new model of arterial hypertension, the chronic nitric oxide synthase inhibition (CNOSI) model.

Mechanisms of Hypertension in the CNOSI Model

Sustained hypertension requires the persistence of at least one of the following abnormalities: increased cardiac output, increased peripheral resistance, and impaired renal sodium excretory capacity (Guyton's hypothesis).

The available evidence unanimously suggests that cardiac output is decreased, not increased, in hypertension due to chronic NO inhibition (Takahashi et al.

1995; Kassab et al. 1998). Therefore, the mechanisms underlying the development of hypertension in this model must involve systemic vasoconstriction and/or salt retention.

Unopposed Vasoconstriction

Since the first studies of acute NO inhibition, the concept has emerged that the hemodynamic changes observed in this setting result from the abrupt withdrawal of a tonic vasodilator effect, leaving unopposed an equally tonic action of endogenous vasoconstrictors. Indeed, inhibition of vasoconstrictor compounds such as angiotensin II (Sigmon et al. 1992) or vasopressin (Manning et al. 1994) reverses at least in part the effects of *acute* NO inhibition. Accordingly, the renal vascular response to angiotensin II is amplified by acute NO inhibition (Baylis et al. 1994). There is now considerable evidence that at least some endogenous vasoconstrictor systems may contribute to the development of hypertension in the chronic NO inhibition model.

Renin–Angiotensin System

Experiments involving *acute* inhibition of angiotensin II in animals with chronic NO inhibition have yielded inconclusive results. Zanchi and co-workers (Zanchi et al. 1995) obtained a large blood pressure reduction with AT I receptor blockade, whereas Bank and co-workers (Bank et al. 1994) and Baylis and colleagues (Qiu et al. 1994) observed little effect of angiotensin II blockade on blood pressure or renal hemodynamics in CNOSI rats. However, combined, acute inhibition of *both* angiotensin II and α-adrenergic receptors almost completely reversed the L-NAME-induced hypertension (Qiu et al. 1994), raising the possibility that the state of generalized vasoconstriction characteristic of this model is maintained by an interaction between the sympathetic (see below) and the renin–angiotensin systems.

The effect of *chronic* inhibition of the renin–angiotensin system in the CNOSI model is more clear-cut. Ribeiro and co-workers (Ribeiro et al. 1992) showed that concomitant administration of the angiotensin II receptor antagonist losartan potassium to rats chronically treated with L-NAME prevented both the hypertension and the renal injury associated with this model. These findings have been corroborated in several subsequent studies involving chronic treatment with either angiotensin II receptor antagonists or angiotensin-converting enzyme inhibitors (Michel et al. 1996; Navarro-Cid et al. 1996), suggesting a key participation of the renin–angiotensin system in this model. In addition, chronic angiotensin II blockade reversed established hypertension (Pollock et al. 1993) and reduced persistent hypertension after discontinuation of NOS inhibition (Morton et al. 1993).

Sympathetic Nervous System

Acute α-adrenergic blockade in the CNOSI model has produced variable findings as to renal and systemic hemodynamics (Bank et al. 1994; Qiu et al. 1994; Zanchi et al. 1995). Acute ganglionic blockade, however, produced a large fall in blood pressure in CNOSI rats, suggesting that increased central sympathetic activity may underlie the hypertension observed in this model. Indeed, NOS inhibitors that are given systemically cross the blood–brain barrier and inhibit NOS in the central nervous system (Traystman et al. 1995), whereas direct administration of NOS inhibitors into some areas of the central nervous system produces hypertension (Harada et al. 1993; Tseng et al. 1996; Zhang et al. 1997). Studies in the awake rat have recently shown that the role of the sympathetic nervous system in CNOSI hypertension depends on the duration of inhibition (Sander et al. 1997), which may explain some of the conflicting findings obtained with acute α-adrenergic inhibition.

Additional evidence in support of a role for the sympathetic nervous system in this model was provided by the finding that chronic bilateral renal denervation prevented glomerular injury and coagulation (Nakashima et al. 1996) and delayed and attenuated chronic NOSI hypertension (Matsuoka et al. 1994). Other investigators, however, have reported evidence that renal nerve traffic plays little or no role in the hypertension of CNOSI (Granger et al. 1996; Reinhart et al. 1997).

Calcium Channels

Acute blockade of voltage-dependent calcium channels with verapamil in rats that had received L-NAME for 3 weeks reversed the hypertension (Bank et al. 1994). Chronic treatment with verapamil (Takase et al. 1996) prevented hypertension in rats subjected to CNOSI for 6 weeks, whereas chronic treatment with nifedipine (Ribeiro et al. 1995) attenuated hypertension and prevented renal injury in rats receiving L-NAME for 4 weeks. Erley and co-workers (Erley et al. 1995) reported attenuation of hypertension, but not of renal dysfunction, with the simultaneous use of felodipine in rats receiving L-NAME for 12 weeks. Together, these observations suggest that endogenous vasoconstrictors, especially angiotensin II, and an adequate operation of voltage-gated calcium channels are both necessary for the hemodynamic and cellular actions of chronic L-NAME to take place.

Endothelin

Neither acute (Bank et al. 1994) nor chronic endothelin inhibition with either a specific endothelin$_A$ or a combined endothelin$_A$/endothelin$_B$ antagonist (Fujihara et al. 1995; Sventek et al. 1997) exerted any effect on hypertension or renal injury in the CNOSI model. However, chronic endothelin$_A$/endothelin$_B$ blockade reduced the appearance of atherosclerotic-like changes in the preglomerular vessels, without affecting hypertension or glomerular damage (Bouriquet et al. 1996). Taken as a whole, these studies suggest that endothelin is unimportant in the pathogenesis of hypertension in this model.

Role of Salt Retention

According to Guyton's hypothesis (Guyton 1990), arterial hypertension reflects an impaired renal ability to excrete sodium. As a consequence, blood pressure will rise until renal sodium excretion again equals sodium intake. Nitric oxide may be central to this process of pressure natriuresis. Mattson and co-workers showed that dietary sodium excess overload largely increases medullary NO generation (Mattson and Higgins 1996), whereas chronic infusion of L-NAME in the rat renal medullary area promotes sustained arterial hypertension (Mattson et al. 1994). However, the results of studies of dietary salt influence on the CNOSI model have been conflicting. Excess dietary sodium has been shown to aggravate hypertension and renal injury in rats given chronic L-NAME (Fujihara et al. 1994; Tolins and Shultz 1994). Conversely, dietary salt restriction was shown to prevent the development of hypertension in this model (Lahera et al. 1992; Salazar et al. 1993), whereas other investigators observed no impact of varying salt intake on L-NAME-induced hypertension (Johnson and Freeman 1992; Jover et al. 1993; Fernandezrivas et al. 1995). These discrepancies may be due to the wide variation of inhibitor doses employed in these studies. Yamada and co-workers (Yamada et al. 1996) showed that chronic treatment with a low, nonpressor dose of L-NAME induced salt sensitivity akin to that observed in salt-sensitive Dahl rats, whereas rats treated with a much higher, pressor dose of L-NAME developed malignant hypertension with severe renal injury. Indeed, hypertension resulting from high-dose L-NAME treatment is associated with immediate and persistent volume depletion, rather than expansion (Qiu et al. 1998).

Organ Damage Secondary to NO Deficiency

At least three target organs—the kidneys, heart, and central nervous system—are known to be severely damaged in the CNOSI model. Although a large amount of injury can result from hypertension only, there is strong evidence that NO deficiency per se can lead to parenchymal damage as well.

Kidney

Five modalities of renal injury have been described in conjunction with CNOSI: classical glomerulosclerosis (Baylis et al. 1992; Ribeiro et al. 1992; Fujihara et al. 1994; Yamada et al. 1996); glomerular ischemia (Ribeiro et al. 1992; Fujihara et al. 1994; Yamada et al. 1996); glomerular segmental necrosis (Ribeiro et al. 1992; Fujihara et al. 1994; Yamada et al. 1996); interstitial expansion, and microvascular lesions (Ribeiro et al. 1992; Fujihara et al. 1994; Yamada et al. 1996). All types of renal injury are associated with progressive albuminuria, which can be amplified nearly to the nephrotic range by concomitant salt overload (Baylis et al. 1992; Fujihara et al. 1994; Yamada et al. 1996). These abnormalities probably explain the severe renal functional impairment observed at advanced stages of the CNOSI model, as well as the persistence of hypertension even after NO inhibition is discontinued.

Myocardium

Three definite types of myocardial damage have been recognized in association with the CNOSI model: cardiac hypertrophy, which appears to result from local activation of the renin–angiotensin system (Moreno et al. 1995; Takemoto et al. 1997), rather than directly from arterial hypertension (Numaguchi et al. 1995; Moreno et al. 1996); remodeling of the coronary circulation, also more consistently associated with activation of the renin–angiotensin system than with blood pressure elevation (Ito et al. 1995; Numaguchi et al. 1995; Takemoto et al. 1997); and focal myocardial necrosis and consequent fibrosis, which appear related neither to arterial hypertension nor to activation of the renin–angiotensin system, suggesting that NO inhibition may have a specific deleterious effect on the myocardium (Numaguchi et al. 1995; Moreno et al. 1996, 1997).

Central Nervous System

The role of NO as a mediator of central nervous system injury and in the pathogenesis of stroke is presently unclear. Cell culture studies suggested that NO might damage nervous tissue (Maiese et al. 1993), whereas previous L-NAME treatment was shown to attenuate cerebral damage in experimental stroke (Trifiletti 1992). Other studies, however, provided evidence that NO can actually be protective in experimental stroke (Morikawa et al. 1992; Prado et al. 1992). These apparently incongruent observations may reflect varying biological effects of NO originating from different NOS isoforms (Huang et al. 1996).

Is NO Deficiency a Cause of Hypertension?

Although the dramatic effects of chronic NO inhibition demonstrate beyond any doubt that pharmacological NO inhibition can severely elevate blood pressure, it is still unproven whether spontaneous NO deficiency is a cause of hypertension. Several clinical and experimental findings, however, seem to support this hypothesis. Chen and Sanders (1991) showed that, when challenged with dietary salt overload, Dahl salt-sensitive rats produced much less NO than salt-resistant rats. Reversal of this relative NO deficiency by excess L-arginine prevented salt-induced hypertension in these animals. In addition, L-arginine treatment ameliorates the pressure-natriuresis curve in the spontaneously hypertensive rat (Larson and Lockhart 1995). Moreover, deletion of the gene encoding the endothelial isoform of NO synthase leads to the development of systemic and pulmonary hypertension in mice (Huang et al. 1995; Shesely et al. 1996), whereas animals lacking the genes for the nNOS or iNOS isoforms are normotensive (Laubach et al. 1995; Nelson et al. 1995), indicating that the eNOS isoform is essential for the maintenance of circulatory homeostasis.

The results of several clinical studies are consistent with these experimental findings. Direct infusion of NOS inhibitors into the brachial artery largely increases local vascular resistance (Calver et al. 1994; Joannides et al. 1995), indicating that also in humans NO promotes moment-to-moment vasodilation. Accordingly, systemic administration of NOS inhibitors to healthy volunteers has caused a sharp blood pressure elevation (Haynes et al. 1993; Wolzt et al. 1997).

As in experimental animals, local NO production in humans can be enhanced by acetylcholine and other vasodilators (Imaizumi et al. 1992; Warren 1994), explaining at least in part the vasodilator action of these compounds. In persons with hypertension, these effects are largely attenuated, suggesting that vascular production of NO is impaired (Panza et al. 1993; Calver et al. 1994). Consistent with these findings, the vasoconstrictor response to local infusion of NO inhibitors is also attenuated (Lyons et al. 1994). However compelling, the present evidence is still far from constituting definitive proof that deficient NO production causes hypertension. Impaired endothelium-dependent vasodilation is not universally found among hypertensive patients (Cockcroft et al. 1994). Diminished NO production might conceivably result from hypertension itself or might simply reflect associated atherosclerosis (Creager et al. 1992). Direct vessel response to NO may be impaired in hypertensive patients (Preik et al. 1996), suggesting the existence of an intrinsic vascular defect, rather than impaired NO production. Moreover, attempts at finding a genetic linkage between eNOS and essential hypertension have been unsuccessful so far (Bonnardeaux et al. 1995; Hunt et al. 1996).

Summary

There is at present convincing evidence that NO deficiency can be an important element in the pathogenesis of hypertension. Most of these data come from studies of chronic inhibition of NOS (the CNOSI model), which is also a model of organ damage. Additional evidence has been obtained in clinical and experimental genetic and pharmacological studies. Further investigation is needed to establish definitively whether NO deficiency underlies any of the modalities of human or experimental hypertension and whether therapy with NO donors will have any beneficial impact.

References

Arnal JF, Warin L, & Michel JB. 1992. Determinants of aortic cyclic guanosine monophosphate in hypertension induced by chronic inhibition of nitric oxide synthase. *J Clin Invest 90:*647–652.

Bachmann S, & Mundel P. 1994. Nitric oxide in the kidney: synthesis, localization, and function. *Am J Kidney Dis 24:*112–129.

Bank N, Aynedjian HS, & Khan GA. 1994. Mechanism of vasoconstriction induced by chronic inhibition of nitric oxide in rats. *Hypertension 24:*322–328.

Baylis C, & Qiu CB. 1996. Importance of nitric oxide in the control of renal hemodynamics, *Kidney Int 49:* 1727–1731.

Baylis C, Mitruka B, & Deng A. 1992. Chronic blockade of nitric oxide synthesis in the rat produces systemic hypertension and glomerular damage. *J Clin Invest 90:*278–281.

Baylis C, Harvey J, & Engels K. 1994. Acute nitric oxide blockade amplifies the renal vasoconstrictor actions of angiotensin II. *J Am Soc Nephrol 5:*211–214.

Bonnardeaux A, Nadaud S, Charru A, Jeunemaitre X, Corvol P, & Soubrier F, 1995. Lack of evidence for linkage of the endothelial cell nitric oxide synthase gene to essential hypertension. *Circulation 91:*96–102.

Bouriquet N, Dupont M, Herizi A, Mimran A, & Casellas D. 1996. Preglomerular sudanophilia in L-NAME hypertensive rats: involvement of endothelin. *Hypertension 27:*382–391.

Calver A, Collier J, & Vallance P. 1994. Forearm blood flow responses to a nitric oxide synthase inhibitor in patients with treated essential hypertension. *Cardiovasc Res 28:*1720–1725.

Chen PY, & Sanders PW. 1991. L-Arginine abrogates salt-sensitive hypertension in Dahl/Rapp rats. *J Clin Invest 88:*1559–1567.

Cockcroft JR, Chowienczyk PJ, Benjamin N, & Ritter JM. 1994. Preserved endothelium-dependent vasodilation in patients with essential hypertension. *N Engl J Med 330:*1036–1040.

Creager MA, Gallagher SJ, Girerd XJ, Coleman SM, Dzau VJ, & Cooke JP. 1992. L-Arginine improves endothelium-dependent vasodilation in hypercholesterolemic humans. *J Clin Invest 90:*1248–1253.

Erley CM, Rebmann S, Strobel U, Schmidt T, Wehrmann M, Osswald H, & Risler T. 1995. Effects of antihypertensive therapy on blood pressure and renal function in rats with hypertension due to chronic blockade of nitric oxide synthesis. *Exp Nephrol 3:*293–299.

Fernandezrivas A, Garciaestan J, & Vargas F. 1995. Effects of chronic increased salt intake on nitric oxide synthesis inhibition-induced hypertension. *J Hypertens 13:*123–128.

Fujihara CK, Michellazzo SM, De Nucci G, & Zatz R. 1994. Sodium excess aggravates hypertension and renal parenchymal injury in rats with chronic NO inhibition. *Am J Physiol 266:*F697–F705.

Furchgott RF, & Zawadzki JV. 1980. The obligatory role of endothelial cells in the relaxation of arterial smooth muscle by acetylcholine. *Nature 288:*373–376.

Granger J, Novak J, Schnackenberg C, Williams S, & Reinhart GA. 1996. Role of renal nerves in mediating the hypertensive effects of nitric oxide synthesis inhibition. *Hypertension 27:*613–618.

Guyton AC. 1990. Long term arterial pressure control: an analysis from animal experiments and computer and graphic models. *Am J Physiol 259:*R865–R877.

Harada S, Tokunaga S, Momohara M, Masaki H, Tagawa T, Imaizumi T, & Takeshita A. 1993. Inhibition of nitric oxide formation in the nucleus tractus solitarius increases renal sympathetic nerve activity in rabbits. *Circ Res 72:*511–516.

Haynes WG, Noon JP, Walker BR, & Webb DJ. 1993. Inhibition of nitric oxide synthesis increases blood pressure in healthy humans. *J Hypertens 11:*1375–1380.

Huang PL, Huang ZH, Mashimo H, Bloch KD, Moskowitz MA, Bevan JA, & Fishman MC. 1995. Hypertension in mice lacking the gene for endothelial nitric oxide synthase. *Nature 377:*239–242.

Huang Z, Huang PL, Ma J, Meng W, Ayata C, Fishman MC, & Moskowitz MA. 1996. Enlarged infarcts in endothelial nitric oxide synthase knockout mice are attenuated by nitro-L-arginine. *J Cereb Blood Flow Metab 16:*981–987.

Hunt SC, Williams CS, Sharma AM, Inoue I, Williams RR, & Lalouel JM. 1996. Lack of linkage between the endothelial nitric oxide synthase gene and hypertension. *J Hypertens 10:*27–30.

Ignarro LJ, Buga GM, Wood KS, Byrns RE, & Chaudhuri G. 1987. Endothelium-derived relaxing factor produced and released from artery and vein is nitric oxide. *Proc Natl Acad Sci USA 84:*9265–9269.

Imaizumi T, Hirooka Y, Masaki H, Harada S, Momohara M, Tagawa T, & Takeshita A. 1992. Effects of L-arginine on forearm vessels and responses to acetylcholine. *Hypertension 20:*511–517.

Ito A, Egashira K, Kadokami T, Fukumoto Y, Takayanagi T, Nakaike R, Kuga T, Sueishi K, Shimokawa H, & Takeshita A. 1995. Chronic inhibition of endothelium-derived nitric oxide synthesis causes coronary microvascular structural changes and hyperreactivity to serotonin in pigs. *Circulation 92:*2636–2644.

Joannides R, Haefeli WE, Linder L, Richard V, Bakkali EH, Thuillez C, & Luscher TF. 1995. Nitric oxide is responsible for flow-dependent dilation of human peripheral conduit arteries in vivo. *Circulation 91:*1314–1319.

Johnson RA, & Freeman RH. 1992. Sustained hypertension in the rat induced by chronic blockade of nitric oxide production. *Am J Hypertens 5:*919–922.

Jover B, Herizi A, Ventre F, Dupont M, & Mimran A. 1993. Sodium and angiotensin in hypertension induced by long-term nitric oxide blockade. *Hypertension 21:*944-948.

Kassab S, Miller MT, Hester R, Novak J, & Granger JP. 1998. Systemic hemodynamics and regional blood flow during chronic nitric oxide synthesis inhibition in pregnant rats. *Hypertension 31:*315–320.

Kone BC, & Baylis C. 1997. Biosynthesis and homeostatic roles of nitric oxide in the kidney. *Am J Physiol 272:*F561–F578.

Lahera V, Salazar J, Salom MG, & Romero JC. 1992. Deficient production of nitric oxide induces volume-dependent hypertension. *J Hypertens Suppl 10:*S173–S177.

Larson TS, & Lockhart JC. 1995. Restoration of vasa recta hemodynamics and pressure natriuresis in SHR by L-arginine. *Am J Physiol Renal Fluid Electrolyte Physiol 37:*F907–F912.

Laubach VE, Shesely EG, Smithies O, & Sherman PA. 1995. Mice lacking inducible nitric oxide synthase are not resistant to lipopolysaccharide-induced death. *Proc Natl Acad Sci USA 92:*10688–10692.

Lyons D, Webster J, & Benjamin N. 1994. The effect of antihypertensive therapy on responsiveness to local intra-arterial N-G-monomethyl-L-arginine in patients with essential hypertension. *J Hypertens 12:*1047–1052.

Maiese K, Boniece IR, Skurat K, & Wagner JA. 1993. Protein kinases modulate the sensitivity of hippocampal neurons to nitric oxide toxicity and anoxia. *J Neurosci Res 36:*77–87.

Manning RD, Hu LF, & Williamson TD. 1994. Mechanisms involved in the cardiovascular-renal actions of nitric oxide inhibition. *Hypertension 23:*951–956.

Matsuoka H, Nishida H, Nomura G, Vanvliet BN, & Toshima H. 1994. Hypertension induced by nitric oxide synthesis inhibition is renal nerve dependent. *Hypertension 23:*971–975.

Mattson D, & Higgins DJ. 1996. Influence of dietary sodium intake on renal medullary nitric oxide synthase. *Hypertension 27:*688–692.

Mattson DL, Lu S, Nakanishi K, Papanek PE, & Cowley AW Jr. 1994. Effect of chronic renal medullary nitric oxide inhibition on blood pressure. *Am J Physiol 266:*H1918–H1926.

Michel JB, Xu YC, Blot S, Philippe M, & Chatellier G. 1996. Improved survival in rats administered N-G-nitro L-arginine methyl ester due to converting enzyme inhibition. *J Cardiovasc Pharmacol 28:*142–148.

Moncada S, Palmer RMJ, & Higgs EA. 1991. Nitric oxide: physiology, pathophysiology and pharmacology. *Pharmacol Rev 43:*109–141.

Moreno H, Nathan LP, Costa SKP, Metze K, Antunes E, Zatz R, & De Nucci G. 1995. Enalapril does not prevent the myocardial ischemia caused by the chronic inhibition of nitric oxide synthesis. *Eur J Pharmacol 287:*93–96.

Moreno H, Metze K, Bento AC, Antunes E, Zatz R, & De Nucci G. 1996. Chronic nitric oxide inhibition as a model of hypertensive heart muscle disease. *Basic Res Cardiol 91:*248–255.

Moreno H, Nathan LP, Metze K, Costa SKP, Antunes E, Hyslop S, Zatz R, De Nucci G. 1997. Non-specific inhibitors of nitric oxide synthase cause myocardial necrosis in the rat. *Clin Exp Pharmacol Physiol 24:*349–352.

Morikawa E, Huang Z, & Moskowitz MA. 1992. L-Arginine decreases infarct size caused by middle cerebral arterial occlusion in SHR. *Am J Physiol 263:*H1632–H1635.

Morton JJ, Beattie EC, Speirs A, & Gulliver F. 1993. Persistent hypertension following inhibition of nitric oxide formation in the young Wistar rat: role of renin and vascular hypertrophy. *J Hypertens 11:*1083–1088.

Nakashima A, Matsuoka H, Yasukawa H, Kohno K, Nishida H, Nomura G, Imaizumi T, & Morimatsu M. 1996. Renal denervation prevents intraglomerular platelet aggregation and glomerular injury induced by chronic inhibition of nitric oxide synthesis. *Nephron 73:*34–40.

Navarro-Cid J, Maeso R, Rodrigo E, Munoz-Garcia R, Ruilope LM, Lahera V, & Cachofeiro V. 1996. Renal and vascular consequences of the chronic nitric oxide synthase inhibition. Effects of antihypertensive drugs. *Am J Hypertens 9:*1077–1083.

Nelson RJ, Demas GE, Huang PL, Fishman MC, Dawson VL, Dawson TM, & Snyder SH. 1995. Behavioral abnormalities in male mice lacking neuronal nitric oxide synthase. *Nature 23:*383–386.

Numaguchi K, Egashira K, Takemoto M, Kadokami T, Shimokawa H, Sueishi K, & Takeshita A. 1995. Chronic inhibition of nitric oxide synthesis causes coronary microvascular remodeling in rats. *Hypertension 26:*957–962.

Palmer RMJ, Ferrige AG, & Moncada S. 1987. Nitric oxide release accounts for the biologic activity of endothelium-derived relaxing factor. *Nature 327:*524–526.

Panza JA, Casino PR, Kilcoyne CM, & Quyyumi AA. 1993. Role of endothelium-derived nitric oxide in the abnormal endothelium-dependent vascular relaxation of patients with essential hypertension. *Circulation 87:*1468–1474.

Pollock DM, Polakowski JS, Divish BJ, & Opgenorth TJ. 1993. Angiotensin blockade reverses hypertension during long-term nitric oxide synthase inhibition. *Hypertension 21:*660–666.

Prado R, Watson BD, Kuluz J, & Dietrich WD. 1992. Endothelium-derived nitric oxide synthase inhibition. Effects on cerebral blood flow, pial artery diameter, and vascular morphology in rats. *Stroke 23:*1118–1123.

Preik M, Kelm M, Feelisch M, & Strauer BE. 1996. Impaired effectiveness of nitric oxide-donors in resistance arteries of patients with arterial hypertension. *J Hypertens 14:*903–908.

Qiu C, Engels K, & Baylis C. 1994. Angiotensin II and alpha 1-adrenergic tone in chronic nitric oxide blockade-induced hypertension. *Am J Physiol 266:*R1470–R1476.

Qiu C, Beierwaltes W, Racusen L, Muchant D, & Baylis C. 1998. Evolution of chronic nitric oxide inhibition hypertension: relationship to renal function. *Hypertension 31:*21–26.

Rees DD, Palmer RMJ, & Moncada S. 1989. Role of endothelium-derived nitric oxide in the regulation of blood pressure. *Proc Natl Acad Sci USA 86:*3375–3378.

Reinhart GA, Lohmeier TE, & Mizelle HL. 1997. Temporal influence of the renal nerves on renal excretory function during chronic inhibition of nitric oxide synthesis. *Hypertension 29:*199–204.

Ribeiro MO, Antunes E, De Nucci G, Lovisolo SM, & Zatz R. 1992. Chronic inhibition of nitric oxide synthesis: a new model of arterial hypertension. *Hypertension 20:*298–303.

Ribeiro MO, Antunes E, Muscara MN, De Nucci G, & Zatz R. 1995. Nifedipine prevents renal injury in rats with chronic nitric oxide inhibition. *Hypertension 26:*150–155.

Salazar FJ, Alberola A, Pinilla JM, Romero JC, & Quesada T. 1993. Salt-induced increase in arterial pressure during nitric oxide synthesis inhibition. *Hypertension 22:*49–55.

Sander M, Hansen J, & Victor RG. 1997. The sympathetic nervous system is involved in the maintenance but not initiation of the hypertension induced by *N*-omega-nitro-L-arginine methyl ester. *Hypertension 30:*64–70.

Shesely EG, Maeda N, Kim HS, Desai KM, Krege JH, Laubach VE, Sherman PA, Sessa WC, & Smithies O. 1996. Elevated blood pressures in mice lacking endothelial nitric oxide synthase. *Proc Natl Acad Sci USA 93:*13176–13181.

Sigmon DH, Carretero OA, & Beierwaltes WH. 1992. Plasma renin activity and the renal response to nitric oxide synthesis inhibition. *J Am Soc Nephrol 3:*1288–1294.

Sventek P, Turgeon A, & Schiffrin EL. 1997. Vascular endothelin-1 gene expression and effect on blood pressure of chronic ET(A) endothelin receptor antagonism after nitric oxide synthase inhibition with L-NAME in normal rats. *Circulation 95:*240–244.

Takahashi H, Hara K, Komiyama Y, Masuda M, Murakami T, Nishimura M, Nambu A, & Yoshimura M. 1995. Mechanism of hypertension induced by chronic inhibition of nitric oxide in rats. *Hypertens Res 18:*319–324.

Takase H, Moreau P, Kung CF, Nava E, & Luscher TF. 1996. Antihypertensive therapy prevents endothelial dysfunction in chronic nitric oxide deficiency: effect of verapamil and trandolapril. *Hypertension 27:*25–31.

Takemoto M, Egashira K, Usui M, Numaguchi K, Tomita H, Tsutsui H, Shimokawa H, Sueishi K, & Takeshita A. 1997. Important role of tissue angiotensin-converting enzyme activity in the pathogenesis of coronary vascular and myocardial structural changes induced by long-term blockade of nitric oxide synthesis in rats. *J Clin Invest 99:*278–287.

Tolins JP, & Shultz PJ. 1994. Endogenous nitric oxide synthesis determines sensitivity to the pressor effect of salt. *Kidney Int 46:*230–236.

Traystman RJ, Moore LE, Helfaer MA, Davis S, Banasiak K, Williams M, & Hurn PD. 1995. Nitro-L-arginine analogues. Dose and time related nitric oxide inhibition in brain. *Stroke 26:*864–869.

Trifiletti RR. 1992. Neuroprotective effects of N^G-nitro-L-arginine in focal stroke in the 7-day old rat. *Eur J Pharmacol 218:*197–198.

Tseng CJ, Liu HY, Lin HC, Ger LP, Tung CS, & Yen MH. 1996. Cardiovascular effects of nitric oxide in the brain stem nuclei of rats. *Hypertension 27:*36–42.

Warren JB. 1994. Nitric oxide and human skin blood flow responses to acetylcholine and ultraviolet light. *FASEB J 8:*247–251.

Wolzt M, Schmetterer L, Ferber W, Artner E, Mensik C, Eichler HG, & Krejcy K. 1997. Effect of nitric oxide synthase inhibition on renal hemodynamics in humans: reversal by L-arginine. *Am J Physiol Renal Physiol 41:* F178–F182.

Yamada SS, Sassaki AL, Fujihara CK, Malheiros DMAC, De Nucci G, & Zatz R. 1996. Effect of salt intake and inhibitor dose on arterial hypertension and renal injury induced by chronic nitric oxide blockade. *Hypertension 27:*1165–1172.

Zanchi A, Schaad NC, Osterheld MC, Grouzmann E, Nussberger J, Brunner HR, & Waeber B. 1995. Effects of chronic NO synthase inhibition in rats on renin–angiotensin system and sympathetic nervous system. *Am J Physiol 268:*H2267–H2273.

Zatz R, & De Nucci G. 1991. Effects of acute nitric oxide inhibition on rat glomerular microcirculation. *Am J Physiol 261:*F360–F363.

Zhang K, Mayhan WG, Patel KP. 1997. Nitric oxide within the paraventricular nucleus mediates changes in renal sympathetic nerve activity. *Am J Physiol 273:*R864–R872.

Part II
Vascular Bed

A. Cerebral

8
Nitric Oxide in Control of the Cerebral Circulation

Charles W. Leffler

Introduction

Contributions of nitric oxide (NO) to the control of cerebral hemodynamics have been studied extensively over the past decade. Because of the voluminous literature on this topic, recent reviews will be cited when possible.

Because the mediating role of NO was discovered only recently, most data on the subject have been obtained from studies in adult rodents. In the present chapter, I will try to include data from other groups, as the possibility must be considered that some roles of NO in cerebral hemodynamic control may be modified by species and age.

Cerebral arteries resist intense vasoconstriction, and NO may contribute to this resistance (Toda and Okamura 1996). Generalized intense sympathetic outflow may markedly decrease blood flow to the skin, gastrointestinal tract, skeletal muscle, and kidneys. In contrast, although cerebral vascular resistance can be increased by intense sympathetic stimulation, the elevation is extremely modest under physiological conditions, and the flow usually will not be affected if the arterial pressure is increased, or it will be affected much less than in peripheral vascular beds if the sympathetic discharge is the result of hypotension. The blood–brain barrier characteristics of the cerebral vascular endothelium isolate the vascular smooth muscle from many systemic perturbations. Whereas lipid-soluble mediators, particularly respiratory gases, can readily transverse these cells, circulating hormonal signaling requires endothelial transduction or transport, or reflex relay from areas without a blood–brain barrier.

Sources of NO in the Cerebral Circulation

Under physiological conditions, there are four potential sources of NO that can affect the cerebrovascular circulation: the endothelium, neurons, glia, and blood.

With pathology, additional sources producing much greater amounts of nitric oxide from inducible nitric oxide synthase (iNOS) can be added: vascular smooth muscle, leukocytes, macrophages, and microglia (Brian et al. 1996; Pelligrino et al. 1996a; Tanaka 1996).

Cerebral vascular endothelial cells, as in other tissues, constituitively express endothelial nitric oxide synthase (eNOS) (Brain et al. 1996) that may contribute a tonic dilator influence, affect reactivity (see below), and inhibit platelet adhesion and aggregation (Pelligrino et al. 1996a).

Perivascular neurons are a rich supply of NO from neuronal nitric oxide synthase (nNOS). Nerve fibers-positive for nNOS are found in the adventitia and outer media of arterioles and arteries of numerous species, including humans. Transmural electrical stimulation dilates cerebral arteries from monkeys, humans, sheep, rats, and pigs. These dilations are blocked by NOS inhibitors (Toda and Okamura 1996). Parasympathetic nonadrenergic, noncholinergic (NANC) neurons (nitroxidergic) originating in the pterygopalatine (sphenopalatine) ganglion produce and release NO (Brian et al. 1996; Goadsby et al. 1996; Tanaka 1996; Toda and Okamura 1996; Ignacio et al. 1997). In addition, to a lesser extent, the nasociliary nerve from the trigeminal ganglion appears to contribute perivascular nitroxidergic nerve fibers, along with fibers from the otic, internal carotid, and nodose ganglia (Tanaka 1996). Release of NO concomitantly with constrictor neurotransmitters can attenuate or reverse to vasodilation the constrictor effects. Such action appears to be an important contributor to the marked resistance of cerebral arteries and arterioles in vivo to vasoconstriction (Toda and Okamura 1996). Furthermore, the excitatory amino acid glutamate via *N*-methyl-D-aspertate (NMDA) receptors activates nNOS in perivascular neurons to provide NO that may contribute to vasodilation in response to increased neuronal activity.

Astrocytes can express low levels of nNOS constitutively (Brian et al. 1996) and iNOS in conditions of injury. Contributions of NO from the glia to the control of the cerebral vascular circulation under physiological conditions require considerably more investigation.

Nitric oxide carried in blood can also act to affect cerebral vascular tone. Oxygenated hemoglobin can act as a transporter of NO in the form of *S*-nitrosohemoglobin (Jia et al. 1996; Stamler et al. 1997). Countercurrent exchange of oxygen between arteries on arterioles and veins or venules results in marked reduction of PO_2 in small arterioles and capillaries. Such reduction certainly promotes oxygen delivery by reducing PO_2 in exchange vessels to the steep portion of the oxygen dissociation curve (brain tissue microvascular PO_2 in the range of 19–32 mmHg). In addition, deoxygenation of hemoglobin results in the release of NO from *S*-nitrosohemoglobin, which was loaded particularly in the pulmonary circulation. This NO can contribute to the maintenance of cerebral microvascular blood flow and encourage vasodilation in response to falling PO_2 in the tissues (see Hypoxia page 118) and may thus contribute to matching microvascular blood flow to oxygen consumption.

Basal Cerebral Blood Flow

Pharmacological studies suggest that NO contributes to the maintenance of appropriate cerebral hemodynamics under resting conditions (Tanaka 1996; Thompson et al. 1996; Yang 1996). There is evidence implicating endothelium (Tanaka 1996), neurons (Toda et al. 1993b), and S-nitrosohemoglobin (Stamler et al. 1997) as sources of NO under basal conditions. The contributions of nNOS to basal cerebral vascular tone remain a bit controversial (Tanaka 1996; Toda and Okamura 1996), and Nω-nitro-L-arginine (L-NNA) (topical) decreased regional cerebral blood flow (CBF) (laser doppler) in wild-type but not in eNOS knockout mice (Ma et al. 1996). Systemically administered L-arginine analogues that inhibit NOS increase cerebral vascular resistance and can decrease CBF (Tanaka 1996). These inhibitors tend to increase arterial pressure, at times markedly, resulting in autoregulatory vasoconstriction that can complicate data interpretation. However, when CBF is reduced below the before-treatment level or blood pressure is controlled, the increase in tone beyond that necessary to return flow to the basal level must be considered to result from removal of a tonic dilator influence. Systemic application of NOS inhibitors would remove all the NO sources. In contrast, application of NOS inhibitors on the brain side of the blood–brain barrier should not eliminate NO from S-nitrosohemoglobin. Such treatment, application to pial vessels either via a cranial window or by cisternal injections, does reduce pial arteriolar diameter and CBF, respectively. In fact, when applied via cranial windows, N^G-monomethyl-L-arginine (L-NMMA) has been shown to cause constriction of vessels from the basilar artery (Faraci 1990) to the pial arterioles (Rosenblum et al. 1990). Inhibition of NOS does not affect cerebral oxygen or glucose consumption (Tanaka 1996), indicating that the increase in cerebrovascular resistance cannot be explained by a decrease in metabolism.

Innumerable inputs may be responsible for providing the stimulation that results in basal NO production. One factor that has been clearly established is shear stress on endothelial cells (Hecker et al. 1993; Rossitti et al. 1995). Physiological shear stresses increase eNOS expression and NO production from endothelial cells in culture and from intact vessels (Ngai and Winn 1995; Uematsu et al. 1995). Such stimulation may be an important component of baseline NO release by endothelium, and increasing NO with increasing shear stress appears to function to counterbalance vasoconstriction.

Mechanisms of Effects of NO on Reactivity

Conventionally, dilator mechanisms are perceived in the agonist, receptor, second-messenger, and action model. In the case of NO, the receptor is actually the enzyme (soluble guanylyl cyclase) responsible for generating the second messenger, cGMP (Brian et al. 1996; Moro et al. 1996; Pelligrino et al. 1996a; Toda and Okamura 1996). In such a model, the amount of second messenger generated and thus the degree of dilation depends upon the quantity of agonist delivered (i.e., dose–response relationship). In the case of NO, the second messenger produced, cGMP, could produce a decrease in vascular tone via multiple mechanisms. Cyclic GMP, via G kinase, can decrease cytosolic Ca^{2+} (Ishikawa et al. 1993; Lincoln and Cornwell 1993; Lorenz et al. 1994). It can also decrease the Ca^{2+} sensitivity of the contractile machinery by activating myosin light-chain phosphatase to decrease myosin light-chain phosphorylation (Lee et al. 1997). G kinase phosphorylation can inhibit phosphodiesterase with high affinity for cAMP (Degerman et al. 1997), resulting in increased cAMP and activation of A kinase, and producing crosstalk between the two second-messenger pathways. G kinase can also produce a decrease in vascular tone via hyperpolarization (Robertson et al. 1993; Nelson et al. 1995). Phosphorylation of Ca^{2+} channels on the endoplasmic reticulum increases the open probability, resulting in localized discrete releases of Ca^{2+} (Ca^{2+} sparks). Intracellular organization to locate Ca^{2+} release sites in close proximity to Ca^{2+} activated K^+ channels (K^+_{Ca}) increases the open probability of K^+_{Ca} in response to calcium sparks, causing hyperpolarization and dilation.

Another mechanism by which NO can influence the cerebral vasculature is by activating other systems to allow response to stimulation. This mechanism is referred to as "permissive." Continually accumulating evidence suggests that permissive contributions of specific paracrine mediators may far outweigh conventional mechanisms in their roles in control of vascular tone. The permissive concept of paracrine cerebrovascular communication developed from studies of a different paracrine system, prostanoids. In the newborn pig and, apparently, in the newborn human, prostanoids, rather than NO, are the principal paracrine mediators of vascular tone from the endothelium-derived relaxing factor (EDRF) group (Leffler 1997). As a result, acetylcholine does not cause endothelial-dependent dilation of pial arterioles but instead causes vasoconstriction (Wagerle and Busija 1989). Surprisingly, this vasoconstriction is accompanied by massive release of dilator prostanoids (Busija et al. 1988) but is blocked, not enhanced, by inhibition of prostanoid synthesis with indomethacin (Wagerle and Busija 1989). This puzzle was solved with the observation that activation of the thromboxane receptor (TP) was necessary to permit vasoconstriction secondarily to muscarinic-1 receptor activation (Armstead et al. 1989). The large quantities of dilator prostanoids produced actually do attenuate the constriction to acetylcholine. We and another group also studying newborn pigs wondered if such a permissive mechanism could be involved in the prostanoid dependence of cerebral vasodilation to hypercapnia (Leffler et al. 1994a, b; Wagerle and Degiulio 1994). Indeed, increasing concentrations of prostacyclin are not needed in response to hypercapnia to produce dilation, but, instead occupation of the prostacyclin receptor (IP) is obligatory. Further, other prostanoid-dependent dilator responses are permissive rather than conventional (Leffler et al. 1995; Leffler and Fedinec 1997). Contemporaneously, it was discovered not only that prostanoid-dependent dilation in the newborn circulation can involve permissive signaling, but also that the NO-dependent vasodilation to hypercapnia in the adult rat involves permissive rather than conventional signals (Iadecola et al. 1994). Thus, the inhibitory effect of NOS inhibition on cerebrovascular dilation can be reversed by low, constant levels of NO (Iadecola et al. 1994; Iadecola and Zhang 1996; Okamoto et al. 1997; Smith et al. 1997), so increasing NO production above a basal level is not needed. Interestingly, in the adult rat the permissive contribution of NO appears to involve the necessity for a basal cGMP "tone" (Yang and Iadecola 1997). This is in contrast to the "IP permission" in the newborn pig, which appears to involve protein kinase C rather than the more commonly associated IP receptor-coupled pathway to cAMP (Rama et al. 1997). Such permissive paracrine mechanisms may provide intercellular communication of the functional integrity of the local vasculature (endothelium, vascular smooth muscle, and perivascular neurons) (Leffler 1997). The underlying primary mechanism by which hypercapnia, for example, ultimately causes a decrease in vascular tone is probably the same across species, age, and maybe even vascular beds. Differences reside in the permissive signals to allow activation of that mechanism.

Hypercapnia

The contributions of NO to hypercapnia-induced cerebrovascular dilation continue to be debated (Fabricius et al. 1996; Tanaka 1996). The majority of studies in several species, including the monkey (Thompson et al. 1996), rat (Iadecola and Zhang 1996; Ma et al. 1996; Estevez and Phillis 1997; Okamoto et al. 1997; Smith et al. 1997), pig (Zuckerman et al. 1996), and cat (Sandor et al. 1994), suggest that NO is an important contributor to cerebral vasodilation in response to CO_2. However, with chronic inhibition of NOS and in NOS knockouts, alternative mediators seem to replace the function of NO (Wang et al. 1994a, b; Irikura et al. 1995). Further, it has been suggested that NO contributes to dilation to milder hypercapnia (PCO_2, 50–80 mmHg) but not to that in response to extreme hypercapnia (100 mmHg) (Pelligrino et al. 1996a). However, this dichotomy is also debated (Toda and Okamura 1996). Important crosstalk occurs between the NO/cGMP system and the prostaglandin I_2 (PGI) cAMP system in the control of vascular tone in response to hypercapnia (Pelligrino et al. 1996a).

The mechanisms by which NO can influence cerebrovascular responses to PCO_2 are beginning to be examined. Although progressive vasodilation in response to CO_2 can be demonstrated from extreme hypocapnia to extreme hypercapnia, the mechanisms involved in maintaining high tone at levels of CO_2 less than normal may be different from (i.e., not simply the opposite of) mechanisms involved in vasodilation to CO_2 at levels above normal (Mirro et al. 1993). In the newborn pig, constriction in response to hypocapnia appears to involve elevation of cytosolic calcium mediated by phospholipase C and inositol phosphate/diacylglycerol signaling (Albuquerque et al. 1995). In contrast, dilation in response to hypercapnia is mediated by cAMP (Parfenova et al. 1994). Whether a similar pattern exists in the adult animal remains to be determined.

Virtually all available evidence indicates that the role of NO in cerebrovasodilation to hypercapnia is a permissive (see above) rather than a classic one (Iadecola et al. 1994; Iadecola and Zhang 1996; Okamoto et al. 1997; Smith et al. 1997). The permissive function of NO seems to result from the fact that a minimal level of vascular smooth muscle cGMP is necessary before the cell will respond to elevated PCO_2 by decreasing tone (Iadecola et al. 1994; Iadecola and Zhang 1996). The actual mediator of the vasodilation may be cAMP rather than cGMP, although this mechanism has not been clearly established in the adult animal.

The NO involved in hypercapnic cerebral vasodilation does not appear to be from endothelium. Several studies seem to show clearly that the dilation is independent of endothelium (Suzuki et al. 1987; Toda et al. 1993a; Lassen and Edvinsson 1994; Wang et al. 1994a; Toda and Okamura 1996). This is in contrast to the newborn pig (see below), in which endothelial injury blocks dilation to hypercapnia (Leffler et al. 1994b, 1995). Several studies implicate nNOS (Wang et al. 1994a; Ma et al. 1996; Okamoto et al. 1997), but others report that the NO involved in hypercapnia dilation is not from nerves (Iadecola et al. 1993; Sandor et al. 1994). It appears, however, that at least in adult rats, the source of the NO providing the necessary permissive signal for vasodilation is neuronal.

A third potential source of NO during hypercapnia is S-nitrosohemoglobin. Decreasing the affinity of hemoglobin for O_2 as PCO_2 rises would decrease the O_2 saturation of hemoglobin, adding to the release of NO from S-nitrosohemoglobin in arterioles as O_2 falls due to countercurrent exchange (Stamler et al. 1997) (see above).

Finally, the signal involved in permitting cerebral vasodilation in response to hypercapnia may change during development. In neonatal pigs and humans, prostacyclin appears to be necessary for cerebral vasodilation to hypercapnia (Leffler 1997). In contrast, in the juvenile pig (Zuckerman et al. 1996) and the adult rat (Wang et al. 1994b), both prostacyclin and NO appear to contribute. In the newborn, the role of prostacyclin is a permissive one (Leffler et al. 1994a, b). Conversely, in the adult rat, NO provides a permissive signal whereas prostacyclin may act in a classical fashion (Pelligrino et al. 1996a). Considerably more research is needed on the mechanisms involved in the prostanoid component of hypercapnic vasodilation in the juvenile and adult animal.

Hypoxia

The evidence for a role of NO in cerebrovasodilation in response to hypoxia is mixed. Considerable data are available to suggest that NO does not contribute to decreased cerebral vascular tone or increased CBF in response to hypoxia (Pelligrino et al. 1993; McPherson et al. 1994a, b; Iadecola and Zhang 1996; Tanaka 1996). However, equally convincing data, often from the same species, seem to clearly include NO as a mediator of hypoxic cerebrovascular dilation (Isozumi et al. 1994; Audibert et al. 1995; Reid et al. 1995; van Bel et al. 1995). The question of the endothelial dependence of dilation to hypoxia has not been studied much, but some data do suggest a potential role of endothelium in cerebrovasodilation to hypoxia (Pearce et al. 1989; Leffler et al. 1997). In newborn pigs, endothelial injury reduces but does not block vasodilation to hypoxemia (Leffler et al. 1997). However, the mediator of this endothelial-dependent component of dilation to hypoxemia is not NO (Leffler et al. 1997). Superficially in contrast, Armstead (1995a) finds a clear NO component to vasodilation to hypoxemia in the same model. The only apparent difference between these studies involves the duration of hypoxemia. In the first study, which could not detect an NO component (Leffler et al. 1997), the hypoxic stimulus was 5 minutes. In the second study, which did detect a role for NO (Armstead 1995a), hypoxia was given for 10 minutes. In piglets opioids have been shown to contribute to vasodilation during hypoxemia (Armstead 1995b, Shankar and Armstead 1995). Opioids and adenosine appear to interact with NO in their dilator mechanisms (Armstead 1995a, 1997; Wilderman and Armstead 1996). Therefore, hypoxia could produce dilation independently of NO initially, with maintenance of vasodilation having an NO component related to opioids and adenosine. As noted above, some studies in rats find a role for NO (Pelligrino et al. 1995), whereas others do not (Buchanan and Phillis 1993). It has been reported that NO contributes to dilation to severe hypoxemia (~35 mmHg) but not to modest hypoxemia (~45 mmHg) (Pelligrino et al. 1995). These data also indicate that the source of the NO during severe hypoxia is neuronal, not endothelial (Pelligrino et al. 1995). In rats, the source of this nitrodergic innervation activated by hypoxia is the rostral ventrolateral medulla (Underwood et al. 1994).

As noted above under Sources of NO in the Cerebral Circulation, S-nitrosohemoglobin is certainly a potential contributor with hypoxia (Jia et al. 1996; Stamler et al. 1997). Particularly with localized tissue hypoxia and systemic normoxia, release of NO from S-nitrosohemoglobin within hypoxic tissue could provide localized vasodilation to restore adequate flow to the hypoxic tissue.

The contributions of NO to cerebrovasodilation caused by hypoxia appear to be classical rather than permissive. No evidence has been presented that a minimal constant level of NO is necessary for dilation to hypoxia (i.e., permissive). However, this subject has not been thoroughly investigated.

Autoregulation

The contributions of NO to cerebrovascular adjustments to changing perfusion pressure have not been clearly established. Somewhat surprisingly, far less research has been conducted on the autoregulatory role of NO in the cerebral circulation than on gas tensions and neuronal mechanisms. Those data that are available appear to be in conflict. Evidence has been presented from studies in anesthetized rats to indicate that NO contributes to dilation in response to hypotension in such a way that NOS inhibition raises the lower autoregulatory pressure limit (Preckel et al. 1996; Tanaka 1996). Conversely, several other studies in rats and monkeys were unable to detect any effect of NOS inhibition on cerebral vasodilation to hypotension (Wang et al. 1992; Kelly et al. 1994; Thompson et al. 1996). Similarly, in newborn pigs, one study detected attenuation of autoregulatory vasodilation by NOS inhibition (O'Neil et al. 1995) and another could not

(Hayden and Leffler 1996). In piglets, endothelial injury inhibits autoregulatory vasodilation (Eidson et al. 1995). However, the endothelial-dependent dilator(s) involved have not been demonstrated with certainty. Hypotension does cause an increase in cerebral prostanoid production (Leffler and Busija 1987). Further, administration of indomethacin during hypotension markedly reduces CBF and produces coma in unanesthetized piglets (Leffler et al. 1986). However, although indomethacin markedly constricts pial arterioles of hypotensive piglets, pial arterioles dilate normally to hypotension in piglets pretreated with indomethacin (Hayden and Leffler 1997). Such data suggest that endothelially derived prostanoids cause or maintain dilation in the intact cerebral circulation, but that one or more alternative endothelium-dependent dilators can substitute if prostanoid production is blocked. Whether NO is a dilator or one of the dilators under such conditions remains to be determined.

It has been reported that the contribution of NO to cerebral autoregulatory vasodilation in rats is regionally specific (Tanaka 1996). It has also been reported that inhibition of NOS can raise the upper limit of autoregulation in rats (Wang et al. 1992), as might be expected as a consequence of removal of a tonic vasodilator input.

Nerves

The cerebral arterial vasculature is heavily innervated with nNOS-positive fibers (Goadsby et al. 1996; Tanaka 1996). The contributions of NO of nervous origin to the control of cerebral hemodynamics have received considerable attention.

Much evidence suggests that neuronal NO can contribute to the coupling of CBF with metabolism locally. Numerous studies have used whisker stimulation in rats as a model to produce localized activation of specific loci in the somatosensory cortex. Coincident with the increased brain activity is an increase of blood flow to the activated region. Most investigations have found that L-arginine derivatives that inhibit NOS markedly attenuate the increase in blood flow in response to whisker stimulation (Northington et al. 1992, Dirnagl et al. 1993, 1994; Irikura et al. 1994; Lindauer et al. 1996). Some questions have been raised regarding the origin of the NO (endothelial versus neuronal), but vibrissal stimulation produces equivalent cortical barrel field hyperemia in eNOS knockout and wild-type mice (Ayata et al. 1996). However, others have not been able to detect attenuation of the active hyperemia response to whisker stimulation after NOS inhibition (Wang et al. 1993; Adachi et al. 1994). The reasons for these differing results remain to be determined.

Nitric oxide-induced elevation of local CBF has also been demonstrated by stimulation of the ganglia of origin of the nitrodergic nerves. Stimulation of the cat facial nerve produces cortical hyperemia. Parasympathetic nerves from the cat or rat sphenopalatine (pterygopalatine) ganglion utilize NO as a neurotransmitter, along with vasoactive intestinal polypeptide, as suggested by inhibition with topical N^G-nitro-L-arginine methylester (L-NAME) (Morita-Tsuzuki et al. 1993; Morita et al. 1994; Goadsby et al. 1996; Toda and Okamura 1996). In rats, NO-mediated cortical vasodilation has been demonstrated in response to electrical stimulation of the nucleus basalis of Meynert, basal forebrain, or rostral ventro-lateral medulla (Adachi et al. 1992; Raszklewicz et al. 1992; Golanov and Reis 1994).

As noted above, neuronal production of NO in rats appears critical to the appropriate response to hypercapnia, possibly playing a permissive role to communicate the functional integrity of vascular innervation. With the exception of hypercapnia, whether nNOS-derived NO effects on cerebral hemodynamics are classical or permissive has not been determined.

Activation of NMDA glutamate receptors produces cerebral vasodilation that appears to involve neuronal NO production (Faraci and Breese 1993; Faraci and Brian 1995; Meng et al. 1995; Pelligrino et al. 1996). NMDA-induced dilation can be blocked by tetrodotoxin or 7-nitroindazole (Faraci and Breese 1993). These data strongly suggest that neuronal NO is an important mechanism for coupling blood flow to metabolism. Cerebellar parallel fiber stimulation produces active hyperemia involving glutamate-induced NO release (Akgoren et al. 1994; Li and Iadecola 1994). Increases in cerebral metabolism via excitatory amino acid release during cortical spreading depression produce concomitant vasodilation, effectively matching perfusion to metabolism. Most investigations, but not all (Zhang et al. 1994), implicate neuronal NO in this dilation, but not the neuronal depolarization (Wahl et al. 1994).

Nitric oxide may also be an important mediator of vasodilation during focal epileptic seizures (Pereira de Vasconcelos et al. 1995), but it does not appear to be important in the global hyperemia associated with grand mal seizures (Wang et al. 1994c).

Summary and Conclusion

Nitric oxide is unquestionably integral to normal cerebrovascular function. However, the mechanisms involved in the contributions of NO to vascular tone and reactivity are diverse and, at present, incompletely understood. In physiological conditions, the sources of NO affecting the cerebrovasculature are the endothelium, neurons, glia, and blood. NO appears to contribute to the maintenance of appropriate cerebral hemodynamics under resting conditions. NO can affect cerebrovascular reactivity by either conventional or permissive contributions. NO is a potential contributor to cerebral vasodilation in response to hypercapnia, hypoxia, and increased metabolism/neuronal activity. Considerably more research into NO and the cerebral vasculature is certainly warranted.

References

Adachi T, Inanami O, & Sato A. 1992. Nitric oxide (NO) is involved in increased cerebral cortical blood flow following stimulation of the nucleus basalis of Meynert in anesthetized rats. *Neurosci Lett 139:*201–204.

Adachi K, Takahashi S, Melzer P, Campos KL, Nelson T, Kennedy C, & Sokoloff L. 1994. Increases in local cerebral blood flow associated with somatosensory activation are not mediated by NO. *Am J Physiol 267:*H2155–H2162.

Akgoren N, Fabricius M, & Lauritzen M. 1994. Importance of nitric oxide for local increases of blood flow in rat cerebellar cortex during electrical stimulation, *Proc Natl Acad Sci USA 91:*5903–5907.

Albuquerque MLC, Lowery-Smith L, Hsu P, Parfenova H, & Leffler CW. 1995. Low CO_2 stimulates inositol phosphate turnover and increased $Ins(1,4,5)P_3$ levels in piglet cerebral microvascular smooth muscle cells. *Proc Soc Exp Biol Med 209:*14–19.

Armstead WM. 1995a. Opioids and nitric oxide contribute to hypoxia-induced pial arterial vasodilation in newborn pigs. *Am J Physiol 268:*H226–H232.

Armstead WM. 1995b. The contribution of delta-1 and delta-2 opioid receptors to hypoxia-induced pial artery dilation in the newborn pig. *J Cereb Blood Flow Metab 15:*539–546.

Armstead WM. 1997. Role of nitric oxide, cyclic nucleotides, and the activation of ATP-sensitive K^+ channels in the contribution of adenosine to hypoxia-induced pial artery dilation. *J Cereb Blood Flow Metab 17:*100–108.

Armstead WM, Mirro R, Busija DW, & Leffler CW. 1989. Permissive role of prostanoids in acetylcholine-induced cerebral constriction. *J Pharmacol Exp Ther 251:*1012–1019.

Audibert G, Saunier CG, Siat J, Hartemann D, & Lambert J. 1995. Effect of the inhibitor of nitric oxide synthase, N^G-nitro-L-arginine methyl ester, on cerebral and myocardial blood flows during hypoxia in the awake dog. *Anesth Analg 81:*945–951.

Ayata C, Ma J, Meng W, Huang P, & Moskowitz MA. 1996. L-NA-sensitive rCBF augmentation during vibrissal stimulation in type III nitric oxide synthase mutant mice. *J Cereb Blood Flow Metab 16:*539–541.

Brian JE Jr, Faraci FM, & Heistad DD. 1996. Recent insights into the regulation of cerebral circulation. *Clin Exp Pharm Physiol 23:*449–457.

Buchanan JE, & Phillis JW. 1993. The role of nitric oxide in the regulation of cerebral blood flow. *Brain Res 610:*248–255.

Busija DW, Pourcyrous M, Leffler CW, & Wagerle LC. 1988. Acetylcholine dramatically increases prostanoid synthesis in piglet parietal cortex. *Brain Res 439:*122–126.

Degerman E, Belfrage P, & Manganiello VC. 1997. Structure, localization, and regulation of cGMP-inhibited phosphodiesterase (PDE_3). *J Biol Chem 272:*6823–6826.

Dirnagl U, Lindauer U, & Villringer A. 1993. Role of nitric oxide in the coupling of cerebral blood flow to neuronal activation in rats. *Neurosci Lett 149:*43–46.

Dirnagl U, Niwa K, Lindauer U, & Villringer A. 1994. Coupling of cerebral blood flow to neuronal activation: role of adenosine and nitric oxide. *Am J Physiol 267:*H296–H301.

Eidson TH, Edrington JL, Albuquerque MLC, Zuckerman SL, & Leffler CW. 1995. Light/dye microvascular injury eliminates pial arteriolar dilation in hypotensive piglets. *Pediatr Res 37:*10–15.

Estevez AY, & Phillis JW. 1997. Hypercapnia-induced increases in cerebral blood flow: roles of adenosine, nitric oxide and cortical arousal. *Brain Res 758:*1–8.

Fabricius M, Rubin I, Bundgaard M, & Lauritzen M. 1996. NOS activity in brain and endothelium: relation to hypercapnic rise of cerebral blood flow in rats. *Am J Physiol 271:*H2035–H2044.

Faraci FM. 1990. Role of nitric oxide in regulation of basilar artery tone in vivo. *Am J Physiol 259:*H1216–H1221.

Faraci FM, & Breese KR. 1993. Nitric oxide mediates vasodilation in response to activation of N-methyl-D-aspartate receptors in brain. *Circ Res 72:*476–480.

Faraci FM, & Brian JE Jr. 1995. 7-Nitroindazole inhibits brain nitric oxide synthase and cerebral vasodilation in response to *N*-methyl-D-aspartate. *Stroke 26:*2172–2175.

Goadsby PJ, Uddman R, & Edvinsson L. 1996. Cerebral vasodilatation in the cat involves nitric oxide from parasympathetic nerves. *Brain Res 707:*110–118.

Golanov EV, & Reis DJ. 1994. Nitric oxide and prostanoids participate in cerebral vasodilation elicited by electrical stimulation of the rostral ventrolateral medulla. *J Cereb Blood Flow Metab 14:*492–502.

Hayden JE, & Leffler CW. 1996. Role of nitric oxide in cerebral vascular tone during hypotension in newborn pigs. *Pediatr Res 39:*215A.

Hayden JE, & Leffler CW. 1997. The effects of treatment with indomethacin on the cerebral vasculature of newborn piglets prior to and during hemorrhagic hypotension. *Pediatr Res 41:*78–82.

Hecker M, Mulsch A, Bassenge E, & Busse R. 1993. Vasoconstriction and increased flow: two principal mechanisms of shear stress-dependent endothelial autacoid release. *Am J Physiol 265:*H828–H833.

Iadecola C, & Zhang F. 1996. Permissive and obligatory roles of NO in cerebrovascular responses to hypercapnia and acetylcholine. *Am J Physiol 271:*R990–R1001.

Iadecola C, Zhang F, & Xu X. 1993. Role of nitric oxide synthase-containing vascular nerves in cerebrovasodilatation elicited from cerebellum. *Am J Physiol 264:*R738–R746.

Iadecola C, Zhang F, & Xu X. 1994. SIN-1 reverses attenuation of hypercapnic cerebrovasodilation by nitric oxide synthase inhibitors. *Am J Physiol 267:*R228–R235.

Ignacio CS, Curling PE, Childers WF, & Bryan RH Jr, 1997. Nitric oxide-synthesizing perivascular nerves in the rat middle cerebral artery. *Am J Physiol 273:*R661–R668.

Irikura K, Maynard KI, & Moskowitz MA, 1994, Importance of nitric oxide synthase inhibition to the attenuated vascular responses induced by topical L-nitroarginine during vibrissal stimulation. *J Cereb Blood Flow Metab 14:*45–48.

Irikura K, Huang PL, Ma JY, Lee WS, Dalkara T, Fishman MC, Dawson TM, Snyder SH, & Moskowitz MA. 1995. Cerebrovascular alterations in mice lacking neuronal nitric oxide synthase gene expression. *Proc Natl Acad Sci USA 92:*6823–6827.

Ishikawa T, Hume JR, & Keef KD. 1993. Regulation of Ca^{2+} channels by cAMP and cGMP in vascular smooth muscle cells. *Circ Res 73:*1128–1137.

Isozumi K, Fukuuchi Y, Takeda H, & Itoh Y. 1994. Mechanisms of CBF augmentation during hypoxia in cats: probable participation of prostacyclin, nitric oxide, and adenosine. *Keio J Med 43:*31–36.

Jia L, Bonaventura C, Bonaventura J, & Stamler JS. 1996. *S*-Nitrosohaemoglobin: a dynamic activity of blood involved in vascular control. *Nature 380:*221–226.

Kelly PA, Thomas CL, Ritche IM, & Arbuthnott GW. 1994. Cerebrovascular autoregulation in response to hypertension induced by N^G-nitro-L-arginine methyl ester. *Neuroscience 59:*13–20.

Lassen NA, & Edvinsson L. 1994. Hypercapnic vasodilatation in isolated rat basilar arteries is exerted via low pH and does not involve nitric oxide synthase stimulation or cyclic GMP production. *Acta Physiol Scand 152:*391–397.

Lee MR, Li LL, & Kitazawa T. 1997. Cyclic GMP causes Ca^{2+} desensitization in vascular smooth muscle by activating the myosin light chain phosphatase. *J Biol Chem 272:*5063–5068.

Leffler CW. 1997. Prostanoids: intrinsic modulators of cerebral circulation. *NIPS 12:*72–77.

Leffler CW, & Busija DW. 1987 Prostanoids and pial arteriolar diameter in hypotensive newborn pigs. *Am J Physiol 252:*H687–H691.

Leffler CW, & Fedinec AL. 1997. Newborn piglet cerebral microvascular responses to epoxyeicosatrienoic acids. *Am J Physiol 273:*H333–H338.

Leffler CW, Busija DW, Beasley DG, & Fletcher AM, 1986. Maintenance of cerebral circulation during hemorrhagic hypotension in newborn pigs: role of prostanoids, *Circ Res 59:*562–567.

Leffler CW, Mirro R, Pharris LJ, & Shibata M. 1994a. Permissive role of prostacyclin in cerebral vasodilation to hypercapnia in newborn pigs. *Am J Physiol 267:*H285–H291.

Leffler CW, Mirro R, Shanklin DR, Armstead WM, & Shibata M, 1994b. Light/dye microvascular injury selectively eliminates hypercapnia-induced pial arteriolar dilation in newborn pigs. *Am J Physiol 266:*H623–H630.

Leffler CW, Fedinec A, & Shibata M. 1995. Prostacyclin receptor activation and pial arteriolar dilation after endothelial injury in piglets. *Stroke 26:*2103–2111.

Leffler CW, Smith JS, Edrington JL, Zuckerman SL, & Parfenova H, 1997. Mechanisms of hypoxia-induced cerebrovascular dilation in the newborn pig. *Am J Physiol 272:*H1323–H1332.

Li J, & Iadecola C. 1994. Nitric oxide and adenosine mediate vasodilation during functional activation in cerebellar cortex. *Neuropharmacology 33:*1453–1461.

Lincoln TM, & Cornwell TL. 1993. Intracellular cyclic GMP receptor proteins. *FASEB J 7:*328–338.

Lindauer U, Megow D, Schultze J, Weber JR, & Dirnagl U. 1996. Nitric oxide synthase inhibition does not affect somatosensory evoked potentials in the rat. *Neurosci Lett 216:*207–210.

Lorenz JN, Bielefeld DR, & Sperelakis N. 1994. Regulation of calcium channel current in A7r5 vascular smooth muscle cells by cyclic nucleotides. *Am J Physiol 266:*C1656–C1663.

Ma J, Meng W, Ayata C, Huang PL, Fishman MC, & Moskowitz MA. 1996, L-NNA-sensitive regional cerebral blood flow augmentation during hypercapnia in type III NOS mutant mice. *Am J Physiol 271:*H1717–H1719.

McPherson RW, Koehler RC, & Traystman RJ. 1994. Hypoxia, α_2-adrenergic, and nitric oxide-dependent interactions on canine cerebral blood flow. *Am J Physiol 266:*H476–H482.

Meng W, Tobin JR, & Busija DW. 1995. Glutamate-induced cerebral vasodilation is mediated by nitric oxide through N-methyl-D-aspartate receptors. *Stroke 26:*857–863.

Mirro R, Pharris LJ, Armstead WM, Shibata M, & Leffler CW. 1993. Effects of indomethacin on newborn pig pial arteriolar responses to PCO_2. *J Appl Physiol 75:*1300–1305.

Morita Y, Hardebo JE, & Bouskela E. 1994. The role of nitric oxide in the cerebrovascular flow response to stimulation of postganglionic parasympathetic nerves in the rat. *J Auton Nerv Syst (Netherlands) 49 (Suppl):*S77–S81.

Morita-Tsuzuki Y, Hardebo JE, & Bouskela E. 1993. Inhibition of nitric oxide synthase attenuates the cerebral blood flow response to stimulation of postganglionic parasympathetic nerves in the rat. *J Cereb Blood Flow Metab 13:*993–997.

Moro MA, Russell RJ, Cellek S, Lizasoain I, Su Y, Darley-Usmar VM, Radomski, MW, & Moncada S. 1996. cGMP mediated the vascular and platalet actions of nitric oxide confirmation using an inhibitor of the soluble gangly cyclase. *Proc Natl Acad Sci USA 93:*1480–1485.

Nelson MT, Cheng H, Rubart M, Santana LF, Bonev AD, Knot HJ, & Lederer WJ. 1995. Relaxation of arterial smooth muscle by calcium sparks. *Science 270:*633–637.

Ngai AC, & Winn HR. 1995. Modulation of cerebral arteriolar diameter, by intraluminal flow and pressure. *Circ Res* 77:832–840.

Northington FJ, Metherne GP, & Berne RM. 1992. Competitive inhibition of nitric oxide synthase prevents the cortical hyperemia associated with peripheral nerve stimulation. *Proc Natl Acad Sci USA* 89:6649–6652.

Okamoto H, Hudetz AG, Roman RJ, Bosnjak ZJ, & Kampine JP. 1997. Neuronal NOS-derived NO plays permissive role in cerebral blood flow response to hypercapnia. *Am J Physiol* 272:H559–H566.

O'Neil JT, Sogn A, Hunt T, & Palacino J. 1995. Inhibition of nitric oxide synthase alters the cerebral blood flow response to hemorrhagic hypotension in piglets. *Pediatr Res* 37:227A.

Parfenova H, Shibata M, Zuckerman S, & Leffler CW. 1994. Carbon dioxide and cerebral circulation in newborn pigs: cyclic nucleotides and prostanoids in vascular regulation. *Am J Physiol* 266:H1494–H1501.

Pearce WJ, Ashwall S, & Cuevas J. 1989. Direct effects of graded hypoxia on intact and de-nuded rabbit cranial arteries. *Am J Physiol* 257:H824–H833.

Pelligrino DA, Koenig HM, & Albrecht RF. 1993. Nitric oxide synthesis and regional cerebral blood flow responses to hypercapnia and hypoxia in the rat. *J Cereb Blood Flow Metab* 13:80–87.

Pelligrino DA, Wang Q, Koenig HM, & Albrecht RF. 1995. Role of nitric oxide, adenosine, N-methyl-D-aspartate receptors, and neuronal activation in hypoxia-induced pial arteriolar dilation in rats. *Brain Res* 704:61–70.

Pelligrino DA, Baughman VL, & Koenig HM. 1996a. Nitric oxide and the brain. *Int Anesthesiol Clin* 34:113–132.

Pelligrino DA, Gay RL III, Baughman VL, & Wang Q. 1996b. NO synthase inhibition modulates NMDA-induced changes in cerebral blood flow and EEG activity. *Am J Physiol* 271:H990–H995.

Pereira de Vasconcelos A, Baldwin RA, & Wasterlain CG. 1995. Nitric oxide mediates the increase in local cerebral blood flow during focal seizures. *Proc Natl Acad Sci USA* 92:3175–3179.

Preckel MP, Leftheriotis G, Ferber C, Degoute CS, Banssillon V, & Saumet JL. 1996. Effect of nitric oxide blockade on the lower limit of the cortical cerebral autoregulation in pentobarbital-anaesthetized rats. *Int J Microcirc Clin Exp* 16:277–283.

Rama GP, Parfenova H, & Leffler CW. 1997. Protein kinase Cs and tyrosine kinases in permissive action of prostacyclin on cerebrovascular regulation in newborn pigs. *Pediatr Res* 41:83–89.

Raszkiewicz JL, Linville DG, Kerwin JFJ, Wagenaar F, & Arneric SP. 1992. Nitric oxide synthase is critical in mediating basal forebrain regulation of cortical cerebral circulation. *J Neurosci Res* 33:129–135.

Reid JM, Davies AG, Ashcroft FM, & Paterson DJ. 1995. Effect of L-NMMA, cromakalim, and glibenclamide on cerebral blood flow in hypercapnia and hypoxia. *Am J Physiol* 269:H916–H922, 1995.

Robertson BE, Schubert R, Hescheler J, & Nelson MT. 1993. cGMP-dependent protein kinase activates Ca-activated K channels in cerebral artery smooth muscle cells. *Am J Physiol* 265:C299–C303.

Rosenblum WI, Nishimura H, & Nelson GH. 1990. Endothelium-dependent L-arg- and L-NMMA-sensitive mechanisms regulate tone of brain microvessels. *Am J Physiol* 259:H1369–H1401.

Rossitti S, Frangos J, Girard PR, & Bevan J. 1995. Regulation of vascular tone. *Can J Physiol Pharmacol* 73:544–550.

Sandor P, Komjati K, Reivich M, & Nyrary I. 1994. Major role of nitric oxide in the mediation of regional CO_2 responsiveness of the cerebral and spinal cord vessels in the cat. *J Cereb Blood Flow Metab 14*:49–58.

Shankar V, & Armstead WM. 1995. Opioids contribute to hypoxia-induced pial artery dilation through activation of ATP-sensitive K^+ channels. *Am J Physiol 269*:H997–H1002.

Smith JJ, Lee JG, Hudetz AG, Hillard CJ, Bosnjak ZJ, & Kampine JP. 1997. The role of nitric oxide in the cerebrovascular response to hypercapnia. *Anesth Analg 84*:363–369.

Stamler JS, Jia L, Eu JP, McMahon TJ, Demchenko IT, Bonaventura J, Gernert K, & Piantadosi CA. 1997. Blood flow regulation by *S*-nitrosohemoglobin in the physiological oxygen gradient. *Science 276*:2034–2037.

Suzuki N, Kawamura J, Yamawaki T, Itoh N, & Obara K. 1987. Role of endothelium in responses of pial vessels to changes in blood pressure and to carbon dioxide in cats. *J Cereb Blood Flow Metab 7*:S275.

Tanaka K. 1996. Is nitric oxide really important for regulation of the cerebral circulation? Yes or no? 1996. *Keio J Med 45*:14–27.

Thompson BG, Pluta RM, Girton ME, & Oldfield EH. 1996. Nitric oxide mediation of chemoregulation but not autoregulation of cerebral blood flow in primates. *J Neurosurg 84*:71–78.

Toda N, & Okamura T. 1996. Nitroxidergic nerve: regulation of vascular tone and blood flow in the brain. *J Hypertens 14*:423–434.

Toda N, Ayajiki K, Enokibori M, & Okamura T. 1993a. Monkey cerebral arterial relaxation caused by hypercapnic acidosis and hypertonic bicarbonate. *Am J Physiol 265*:H929–H933.

Toda N, Ayajiki K, & Okamura T. 1993b. Neural mechanism underlying basilar arterial constriction by intracisternal L-NNA in anesthetized dogs. *Am J Physiol 265*:H103–H107.

Uematsu M, Ohara Y, Navas JP, Nishida K, Murphy TJ, Alexander RW, Nerem RM, & Harrison DG. 1995. Regulation of endothelial cell nitric oxide synthase mRNA expression by shear stress. *Am J Physiol 269*:C1371–C1378.

Underwood MD, Iadecola C, & Reis DJ. 1994. Lesions of the rostral ventrolateral medulla reduce the cerebrovascular response to hypoxia. *Brain Res 635*:217–223.

van Bel F, Sola A, Roman C, & Rudolph AM. 1995. Role of nitric oxide in the regulation of the cerebral circulation in the lamb fetus during normoxemia and hypoxemia. *Biol Neonate 68*:200–210.

Wagerle LC, Busija DW. 1989. Cholinergic mechanisms in cerebral circulation of the newborn piglet: effect of inhibitors of arachidonic acid metabolism. *Circ Res 64*:1030–1036.

Wagerle LC, & DeGiulio PA. 1994. Indomethacin-sensitive CO_2-reactivity of cerebral arterioles is restored by dilator prostaglandin. *Am J Physiol 266*:H1332–H1338.

Wahl M, Schilling L, Parsons AA, & Kaumann A. 1994. Involvement of calcitonin generelated peptide (CGRP) and nitric oxide (NO) in the pial artery dilatation elicited by cortical spreading depression. *Brain Res 637*:204–210.

Wang Q, Paulson OB, & Lassen NA. 1992. Is autoregulation of cerebral blood flow in rats influenced by nitro-L-arginine, a blocker of the synthesis of nitric oxide? *Acta Physiol Scand 145*:297–298.

Wang Q, Kjaer T, Jorgensen MB, Paulson OB, Lassen NA, Diemer NH, & Lou HC. 1993. Nitric oxide does not act as a mediator coupling cerebral blood flow to neural activity following somatosensory stimuli in rats. *Neurol Res 15*:33–36.

Wang Q, Pelligrino DA, Koenig HM, & Albrecht RF. 1994a. The role of endothelium and nitric oxide in rat pial arteriolar dilatory responses to CO_2 in vivo. *J Cereb Blood Flow Metab 14:*944–951.

Wang Q, Pelligrino DA, Paulson OB, & Lassen NA. 1994b. Comparison of the effects of N^G-nitro-L-arginine and indomethacin on the hypercapnic cerebral blood flow increase in rats. *Brain Res 641:*257–264.

Wang Q, Theard MA, Pelligrino DA, Baughman VL, Hoffman WE, Albrecht RF, Cwik M, Paulson OB, & Lassen NA. 1994c. Nitric oxide (NO) is an endogenous anticonvulsant but not a mediator of the increase in cerebral blood flow accompanying bicuculline-induced seizures in rats. *Brain Res 658:*192–198.

Wilderman MJ, & Armstead WM. 1996. Relationship between nitric oxide and opioids in hypoxia-induced pial artery vasodilation. *Am J Physiol 270:*H869–H874.

Yang G, Iadecola C. 1997. Obligatory role of NO in glutamate-dependent hyperemia evoked from cerebellar parallel fibers. *Am J Physiol 272:*R1155–R1161.

Yang ST. 1996. Role of nitric oxide in the maintenance of resting cerebral blood flow during chronic hypertension. *Life Sci 58:*1231–1238.

Zhang ZG, Chopp M, Maynard KI, & Moskowitz MA. 1994. Cerebral blood flow changes during cortical spreading depression are not altered by inhibition of nitric oxide synthesis. *J Cereb Blood Flow Metab 14:*939–943.

Zuckerman SL, Armstead WM, Hsu P, Shibata M, & Leffler CW. 1996. Age-dependence of cerebrovascular response mechanisms in the domestic pig. *Am J Physiol 271:*H535–H540.

B. Pulmonary

9
Intravenous Anesthetics in the Pulmonary Circulation

ALAN D. KAYE, BOBBY D. NOSSAMAN, AND PHILIP J. KADOWITZ

Intravenous anesthetics in either bolus or constant infusion may be used to produce induction or maintenance of anesthesia. Examples of these anesthetics are ultra-short-acting thiobarbiturates, propofol, etomidate, and ketamine. Of the four, propofol and ketamine possess significant pulmonary effects. Propofol, a diisopropylphenol derivative, has been used for the induction of general anesthesia and is considered to be a short-acting anesthetic agent with a rapid onset of action and rapid elimination (Smith et al. 1994). Propofol initially decreases the respiratory rate and has an immediate temporary hypotensive action secondary to a reduction in total peripheral resistance. Propofol can produce a decrease in the ventilatory response to hypoxia, suggesting the need for supplementary oxygen in all uses of propofol (Smith et al. 1994). Propofol decreases both systemic arterial and pulmonary arterial pressures without causing a significant reduction in cardiac output in normal humans and children with congenital heart disease (Smith et al. 1994; Williams et al. 1994). The decreases in blood pressure associated with propofol are secondary to a reduction in both systemic and pulmonary vascular resistance; however, little if any shunt has been demonstrated (Mendoza et al. 1992).

Over the past few years, numerous studies have found varying results as to the mechanism of propofol activity. In hyperoxic and hypoxic dogs, the hemodynamic response of propofol was not affected by cylooxygenase inhibition with aspirin (Naeije et al. 1989). In cultured porcine aortic endothelial cells, propofol stimulated the production of nitric oxide (NO) (Petros et al. 1993). Propofol has also been shown to have vasodilator properties in isolated pulmonary arteries by endothelium-dependent mechanisms (Park et al. 1992; Rich et al. 1994; Gacar et al. 1995). Other studies, however, have demonstrated propofol-induced vasodilation via an endothelium-independent pathway (Chang et al. 1993; Miyawaki et al. 1995; Moreno et al. 1997).

Additional mechanisms have been implicated in the vasodilatory effect of propofol. Propofol-induced vasodilation was inhibited by the cyclooxygenase inhibitor indomethacin in the rat (Park et al. 1992). In rat thoracic aorta and pulmonary artery precontracted with phenylephrine, propofol induced concentration-dependent relaxation, which was significantly altered after incubation with the synthase (NOS) inhibitor N^G-nitro-L-arginine-methylester (L-NAME) (Park et al. 1992). Studies in canine coronary arteries have found no significant effect on propofol-induced vasodilation after administration of L-NAME (Moore et al. 1994). The development of arginine analogues that inhibit NO synthesis have facilitated the investigation of the role of NO in regulating vascular tone and mediating responses in the pulmonary circulation (Nishiwaki et al. 1992). Nitric oxide synthesis inhibitors decrease pulmonary vasodilator responses to acetylcholine, enhance pressor responses to angiotensin II and hypoxia, and have a small effect on baseline pressures in the isolated perfused rat lung (Feng et al. 1994).

Recent studies in the rat and chick have suggested that other pathways may be involved in propofol-mediated responses. In precontracted isolated guinea pig basilar arteries, propofol-induced vasodilation was associated with a decrease in calcium influx and a reduction in intracellular calcium release (Gelb et al. 1996). A related study found that in cultured chick sensory neurons, propofol inhibited L-type and T-type calcium currents (Olcese et al. 1994). In the isolated rat lung, glibenclamide, the K^+-ATP channel blocker, significantly inhibited the vasodilator effect of propofol, suggesting that activation of K^+-ATP channels mediates propofol responses (Erdemli et al. 1995). Additional studies in human T lymphocytes and clonal pheochromocytoma cells have also suggested a role for potassium channels in propofol-induced effects (Magnelli et al. 1992, Mozrzymas et al. 1996). In canine myocardial cells, propofol has been demonstrated to inhibit both calcium and potassium currents (Bulijubasic et al. 1996).

Ketamine is a potent analgesic at subanesthetic plasma concentrations, and its analgesic and anesthetic effects may be mediated by different mechanisms (Platt et al. 1991). Ketamine is a potent cerebrovasodilator and produces a dissociative amnesia, evident on the electroencephalogram as a dissociation of the thalamocortical and limbic systems (Takeshita et al. 1972; Reich et al. 1989). Ketamine alters systemic arterial pressure, with significant increases in heart rate, cardiac output, cardiac work, and myocardial oxygen requirements in normal humans (Kreuscher and Gauch 1967; Tweed et al. 1972). The increases in blood pressure and cardiac output associated with ketamine are secondary to a direct increase in central sympathetic nervous system outflow (Tweed et al. 1972). Ketamine does not produce a significant decrease in ventilation, and the response to carbon dioxide is maintained initially. In the lung, ketamine is as effective as halothane or enflurane in preventing experimentally induced bronchospasm in dogs (Hirshman et al. 1979).

Over the past few years, studies have demonstrated conflicting results with regard to the mechanism of action of ketamine-induced responses in the lung. As previously mentioned, The development of arginine analogues that inhibit NO synthesis has facilitated the investigation of the role of NO in regulating vascular tone and responses in the pulmonary circulation. Nitric oxide synthesis inhibitors decrease pulmonary vasodilator responses to acetylcholine, enhance pressor responses to angiotensin II and hypoxia, and have a small effect on baseline pressures in the isolated perfused rat lung, suggesting that NO plays a role in mediating or modulating pulmonary vascular responses in the rat (Lippton et al. 1992; Nishiwaki et al. 1992; Cheng et al. 1994; Feng et al. 1994). In one study in rat pulmonary arteries with chronic hypoxic pulmonary hypertension, ketamine had no effect on NO mediated relaxation in either normal or pulmonary hypertensive rats (Maruyama et al. 1995). Ketamine has vasodilator properties in the rat and guinea pig via endothelium- and epithelium-independent pathways (Rich et al. 1994; Gratton et al. 1995a, b; Sato et al. 1997). In the isolated rabbit portal vein, ketamine-induced vasorelaxation was not endothelium-dependent but rather was due to blockade of the voltage-gated influx of extracellular calcium (Yamazaki et al. 1995).

Ketamine relaxed isolated guinea pig trachea after constriction with carbachol, histamine, prostaglandin F_2, and potassium (Pedersen et al. 1993). A possible mechanism for the relaxation of the porcine trachealis muscle is a decrease in the excitability of postsynaptic nicotinic receptors of the intramural ganglia of the peripheral vagus nerve and the muscarinic receptors of the nerve and the smooth muscle cell (Wilson et al. 1993). In canine distal and proximal coronary arteries in vitro, after preconstriction with endothelin, relaxation responses to increasing concentrations of ketamine were observed (Coughlan et al. 1992).

Other pathways may be involved in the vasodilatory effect of ketamine in the lung. Although ketamine-induced vasorelaxant effects have been reported to be unaffected after treatment with the cyclooxgenase inhibitor indomethacin in the trachea of guinea pigs, recent studies in rat and guinea pig ventricular myocytes have shown ketamine-inhibited activity of the K^+-ATP channel in single rat ventricular myocytes (Ko S et al. 1997; Sato T et al. 1997). In isolated guinea pig ventricular myocytes treated with ketamine, there was a significant inhibition of the inward potassium rectifier current, suggesting that inhibition of this type of potassium channel may be involved in ketamine-induced responses (Baum 1993). Ketamine has also been shown to decrease the frequency of open-state, long-lived, calcium-activated potassium currents in GH3 cells by a competitive interaction between calcium and ketamine (Denson et al. 1994).

Although much is known about the beneficial effects of propofol and ketamine, the mechanism of each agent in the pulmonary circulation remains controversial. In our laboratory, we investigated the direct effects of ketamine, 2,6-diisopropyl phenol, disodium edetate, and its intralipid emulsion in the rat pulmonary vascular bed. In addition, we studied the influence of the NO synthesis blocker N^G-nitro-L-arginine benzylester (L-NABB), U-37883A, an ATP-sensitive potassium channel opener, and the cyclooxygenase blocker, meclofenamate on responses to propofol and to ketamine in the pulmonary vascular bed under conditions of controlled blood flow and ventilation and constant left atrial pressure in the isolated blood-perfused rat lung and when tone was increased experimentally.

All experimental protocols were approved by the Advisory Committee for Animal Resources at Tulane University Medical Center. In these experiments, 96 male Sprague-Dawley rats weighing 300–350 g (Hill Top Laboratories,

Scottdale, PA) were anesthetized with pentobarbital sodium (50 mg/kg) intraperitoneally. After stable anesthesia had been obtained, the trachea was approached, cannulated with a short section of polyethylene tubing and connected to a rodent ventilator (Harvard Apparatus, South Natick, MA), and ventilated with air enriched with oxygen, a mixture of 30% O_2/5% CO_2, with a tidal volume of 4-5 ml/kg and 2 cm H_2O positive end respiratory pressure. The rats were heparinized with 1,000 units of heparin (Sigma Chemical Co., St. Louis, MO) and rapidly exsanguinated by withdrawing blood from the carotid artery.

The lungs were exposed by median sternotomy, and a ligature was placed around the aorta to prevent systemic loss of blood. The main pulmonary artery was catheterized, and the lungs were rapidly isolated from the animal and suspended in a warmed (38°C), humidified (100%) water-jacketed chamber. An external heat exchanger (Haake D1 Heat Exchanger, Baxter Instrument Co., Harahan, LA) maintained the temperature of the perfusate and the isolated-lung chamber constant throughout the experiment. The perfusate solution (15 ml of heparinized blood and 5 ml of modified Krebs-Heinsleit solution) was placed into a reservoir and mixed constantly by a magnetic stirrer (Thermolyne, Cimarec II, Dubuque, IA). The lungs were perfused with a peristaltic roller pump (Cole-Palmer Instrument Co., Berrington, IL), and blood from the left atrium was allowed to collect in the reservoir, so that left atrial pressure was equal to atmospheric pressure. Once the isolated lung perfusion circuit had been established, the flow rate was set at 8-14 ml/min to maintain physiologic baseline pulmonary arterial perfusion pressure, which averaged approximately 15 mm Hg and was not changed during the experiment. The isolated blood-perfused lung preparation was allowed to stabilize for 30 minute before the experiments were begun. All vascular pressures were measured with Viggo-Spectramed transducers (Viggo-Spectramed, Oxnard, CA) zeroed at the level of the pulmonary arterial cannula. Pulmonary arterial perfusion pressure, airway pressure, and reservoir blood level were continuously monitored, electronically averaged, and recorded with a Grass polygraph, Model 7 (Grass Instrument Co., Quincy, MA). The modified Krebs-Heinseleit solution had the following composition (g/l): NaCl, 66.37; KCl, 3.58; $CaCl_2 \cdot H_2O$, 3.68; KH_2PO_4, 1.63; $MgSO_4 \cdot H_2O$, 1.45; $NaHCO_3$, 1.6; dextrose, 1.0 (pH 7.4). The solution was diluted (1:10) using double distilled water and made fresh daily.

L-NABE, acetylcholine hydrochloride, sodium meclofenamate (Warner-Lambert-Park-Davis, Ann Arbor, MI), sodium arachidonate, disodium edetate, and isoproterenol hydrochloride (Sigma Chemical Co., St. Louis, MO) were dissolved in normal saline. Pure 2,6-diisopropylphenol and propofol were a generous gift of Zeneca Pharmaceuticals (Wilmington, DE). Levcromakalim (SmithKline Beecham, Sussex, UK) was dissolved in 20% ethanol–saline solution at a concentration of 1 mg/ml and was diluted in 0.9% saline immediately before use. Nitroglycerin (American Regent Laboratories, Shirley, NY) was prepared by the manufacturer in a 30% alcohol, 30% propylene glycol solution. The thromboxane receptor agonist U46619 (11α, 9α-epoxymethano-9α, 11β-dideoxyprostaglandin F2α, Upjohn, Kalamazoo, MI) was dissolved in 100% ethanol at a concentration of 10 mg/ml, and further dilutions were made in normal saline. U-37883A (Upjohn, Kalamazoo, MI) was sonicated in a normal saline solution. BAY K8644 was dissolved in a 1:4 solution of cremophor EL and tris (hydroxymethyl) aminomethane (Tris) and Tris-HCl, pH 7.4. The resulting suspension was warmed, and polyethylene glycol and Tris, pH 7.4, were added to make a stock solution that was stored in a brown bottle in a freezer at $-20°C$. Verapamil was prepared by the manufacturer in solution. Working solutions were prepared frequently (every 2 or 3 days) by diluting the stock solution in normal saline solution, stored in brown, stoppered bottles, and kept on crushed ice during experiments.

Arterial blood gas tensions and pH were measured using a Corning Model 178 (Corning Instruments, Medfield, MA) at the beginning and at intervals during an experiment. Blood pH was maintained between 7.35 and 7.45 by adding small amounts of $NaHCO_3$ solution to the perfusion reservoir. To serve as a control before agonist injections, equivalent volumes of saline or distilled water were injected directly into the perfusion circuit. All injections were made in small volumes in a random sequence, and sufficient time was permitted between agonist injections for pressures to return to baseline values. Since pulmonary blood flow and outflow pressure were maintained constant, changes in perfusion pressure in this preparation reflect changes in pulmonary vascular resistance. All vascular pressures are expressed in absolute units (mm Hg) as means $\pm$ SEM. The data were analyzed using analysis of variance with post-hoc Scheffe's F test for repeated measures (StatView, Abascus Concepts, Berkeley, CA) on a Power Macintosh 6100/66 computer. A P value of less than 0.05 was used as the criterion for statistical significance.

Because the pulmonary vascular bed of the rat has little if any vasoconstrictor tone under resting conditions when FiO_2 is greater than 0.21, pulmonary arterial pressure must be actively increased so vasodilator responses can be expressed. In all of the present experiments, the tone was raised in the control period to values of 31–39 mm Hg with an infusion of U46619. Under conditions of elevated tone in the control period, pulmonary vascular responses of each agonist were obtained. The agonists were injected in small volumes directly into the perfusion circuit distal to the pump during the control period. Afterward, the blocker was administered and the agonists were again injected in a random sequence. Since L-NABE modestly increases tone, U46619 infusion was initially terminated when the NO synthase inhibitor was administered and after the peak increase in pulmonary arterial pressure in response to L-NABE (100 mg/kg i.a.) was attained, U46619 infusion was resumed if necessary to raise pulmonary vascular tone to a level similar to that attained during the control period. In some experiments, however, L-NABE administration alone was sufficient to raise pulmonary vascular tone to a level equal to the control level, and in these experiments, U46619 infusion was resumed only later, when lobar arterial pressure had fallen below 30 mm Hg.

In separate experiments with meclofenamate and U-37883A, responses to the agonists were again studied during the control period with U46619. Prior to meclofenamate or U-37883A infusion, U46619 was terminated and pulmonary arterial pressure was permitted to return to near control values. After the peak increase in pulmonary arterial pressure in response to meclofenamate (2.5 mg/kg i.a.) or U-37883A (2 mg/kg i.a.), the U46619 infusion was resumed if necessary to raise pulmonary vascular tone to a level similar to that attained during the control period. In some experiments, meclofenamate administration alone was sufficient to raise pulmonary vascular tone to a level equal to the control level, and in these experiments, U46619 infusion was resumed only later, when pulmonary arterial pressure had fallen below 30 mm Hg.

In the last series of experiments, the effects of ketamine and propofol on the pulmonary pressor response to ventilatory hypoxia were investigated. In these experiments, the lungs were ventilated with a 3% O_2, 5% CO_2, 92% N_2 gas mixture until a stable pressor response was obtained. Ketamine or propofol was then added to the perfusion reservoir in order to evaluate the effects of ketamine on the pressor response to hypoxia in the isolated perfused rat lung.

Because changes in pulmonary blood flow induce passive changes in pulmonary vascular resistance, the direct effects of propofol on the pulmonary vascular bed were investigated in the rat under constant-flow conditions. Under low resting tone conditions in four preparations, intraarterial injections of propofol in doses of 0.03–1.0 mg produced only small decreases in pulmonary arterial pressure (data not shown). However, when the baseline tone in the pulmonary vascular bed was raised to a high steady value with an infusion of U46619, injections of propofol into the perfusion circuit in doses of 0.03–1.0 mg caused significant dose-related decreases in pulmonary arterial perfusion pressure (Figure 9.1). Decreases in pulmonary arterial pressure in response to propofol were rapid in onset, and pressure returned to control values over a 3- to 5-minute period depending on the dose injected. Pure 2,6-diisopropylphenol and a lipid emulsion identical to the one used in propofol were administered, and neither solution caused significant changes in pulmonary arterial perfusion pressure (data not shown). Under low resting tone conditions, injections of ketamine in doses of 0.01–1.0 mg produced only small decreases in pulmonary arterial pressure (Figure 9.2). However, when the baseline tone in the pulmonary vascular bed was raised to a high steady value with an infusion of U46619 or in response to ventilatory hypoxia, injections of ketamine into the perfusion circuit in doses of 0.01–1.0 mg caused significant dose-related decreases-in-pulmonary arterial perfusion pressure (Figure 9.2). Decreases in pulmonary arterial pressure in response to ketamine were rapid in onset, and pressure returned to control values over a 3- to 5-minute period, depending on the dose injected.

When the baseline tone in the pulmonary vascular bed was raised to a high steady value with an infusion of U46619 or ventilatory hypoxia, injections of propofol (30–300 μg i.a.), ketamine (30–300 μg i.a.), acetylcholine (300–1000 ng i.a.), nitroglycerin (0.3–3 ng i.a.), and isoproterenol (0.03–0.3 ng i.a.) into the perfusion circuit caused significant dose-related decreases in pulmonary arterial pressure (Figure 9.3). Decreases in pulmonary arterial pressure in response to intraarterial injections of acetylcholine were significantly attenuated after administration of L-NABE (Figure 9.3). Decreases in pulmonary arterial pressure in response to intraarterial in-

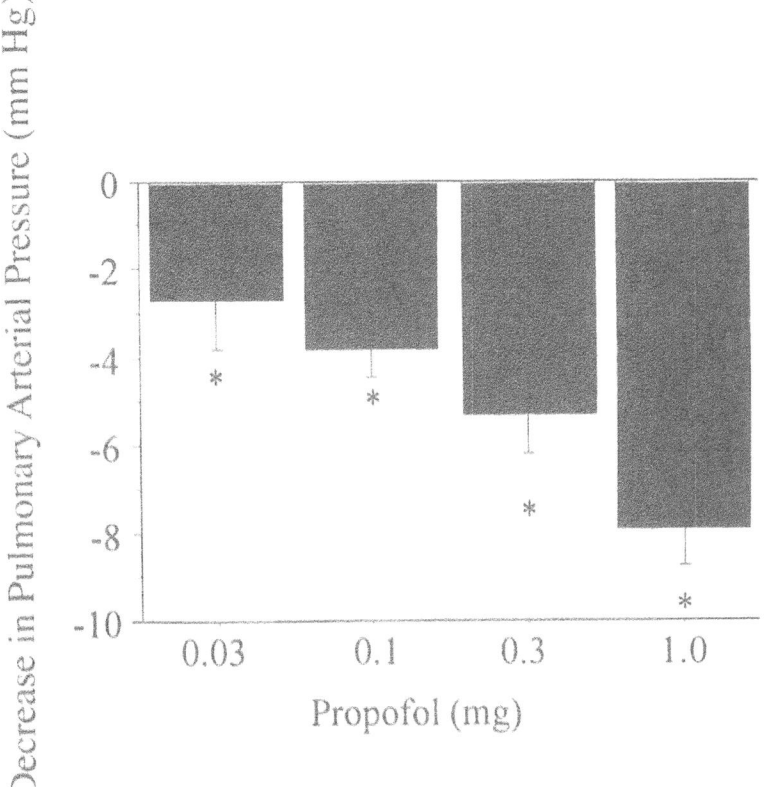

FIGURE 9.1. Influence of propofol under conditions of elevated tone with U46619 or ventilatory hypoxia, $N = 12-14$. (Redrawn from data presented at the International Anesthesia Research Society, Orlando, Florida, March, 1998; submitted to *Acta Anaesthesiologica Scandinavica* for publication.)

jections of propofol, ketamine, nitroglycerin, and isoproterenol were not significantly reduced after administration of L-NABE (Figure 9.3).

When the baseline tone in the pulmonary vascular bed was raised to a high steady value with an infusion of U46619, injections of propofol, ketamine, and levcromakalim into the perfused pulmonary artery in doses of $10-100$ μg i.a. caused significant dose-related decreases in pulmonary arterial pressure (Figure 9.4). Decreases in pulmonary arterial pressure in response to intrapulmonary injections of propofol were not significantly attenuated after administration of U-37883A, whereas the vasodilator responses to the K$^+$-ATP channel opener levcromakalim were significantly reduced (Figure 9.4). The effects of propofol and levcromakalim with and without U46619 were studied, and there was no significant difference with regard to the blocker effects of U-37883A on the vasodilator response to levcromakalim.

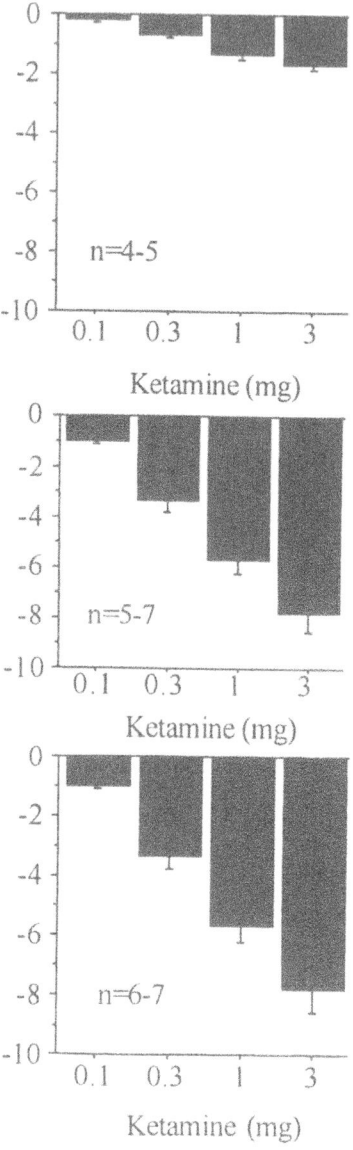

FIGURE 9.2. Influence of ketamine under low tone (upper) and conditions of elevated tone with U46619 (middle) or ventilatory hypoxia (lower). (Redrawn from data submitted to the American Society of Anesthesiology National Meeting, Orlando, Florida, September 1998; accepted for publication by *Anesthesia and Analgesia*.)

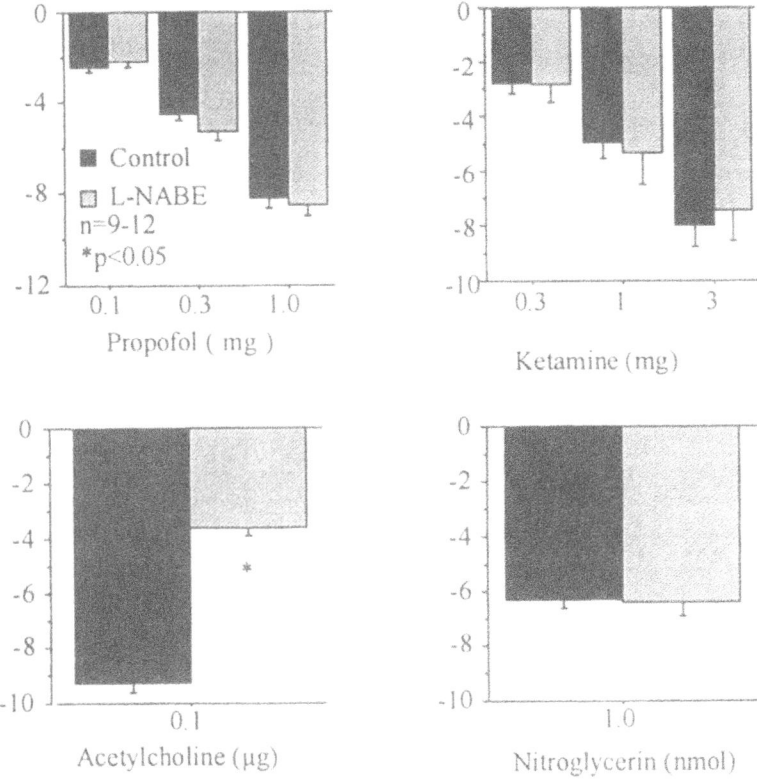

FIGURE 9.3. Influence of N^G-nitro-L-arginine benzylester (L-NABE) on vasodilator responses to propofol, ketamine, isoproterenol, acetylcholine, and nitroglycerin under conditions of elevated tone with U46619 infusion. n indicates number of animals. *$P < 0.05$. (Redrawn from data submitted to the American Society of Anesthesiology National Meeting and presented at the International Anesthesia Research Society, 1998; portions of data accepted for publication by *Anesthesia and Analgesia* and submitted to *Acta Anaesthesiologica Scandinavica* for publication.)

When pulmonary arterial pressure was increased by the infusion of U46619, decreases in pulmonary arterial pressure in response to intraarterial injections of arachidonic acid were significantly smaller than responses obtained when arachidonic acid was injected during the control U46619 infusion (Figure 9.5). Decreases in pulmonary arterial pressure in response to intraarterial injections of propofol or ketamine were not significantly attenuated after administration of meclofenamate (Figure 9.5).

When the baseline tone in the pulmonary vascular bed was raised to a high steady value with an infusion of U46619, injections of disodium edetate into the perfusion circuit in doses of 10–100 μl caused significant dose-related decreases

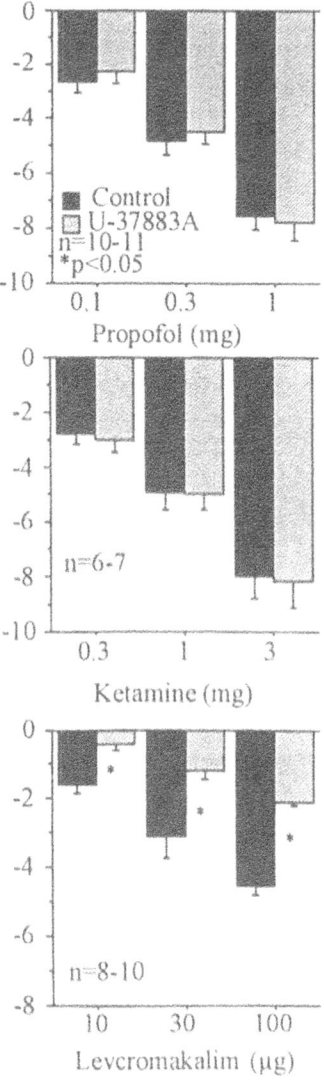

FIGURE 9.4. Influence of U-37883A on vasodilator responses to propofol, ketamine, and levcromakalim under elevated tone conditions. *n* indicates number of animals. $*P < 0.05$. (Redrawn from data submitted to the American Society of Anesthesiology National Meeting and presented at the International Anesthesia Research Society, 1998; portions of data accepted for publication by *Anesthesia and Analgesia* and submitted to *Acta Anaesthesiologica Scandinavica* for publication.)

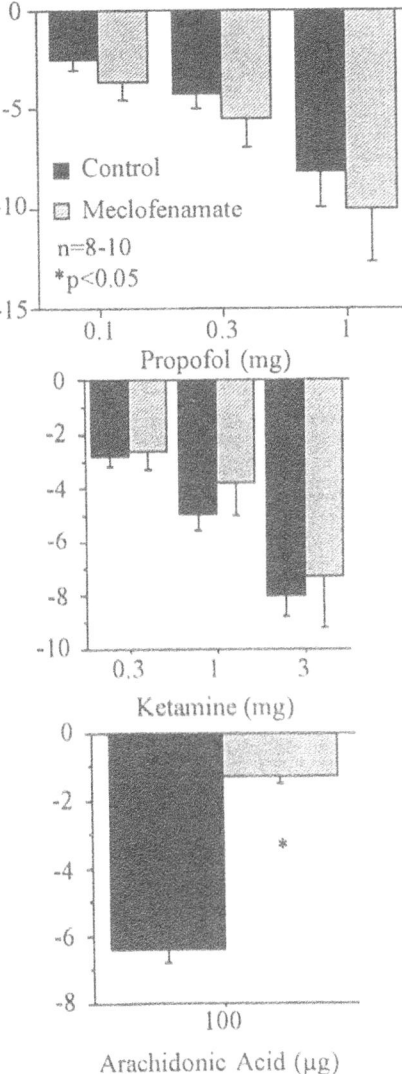

FIGURE 9.5. Influence of meclofenamate on vasodilator responses to propofol, ketamine, and arachidonic acid. n indicates number of animals. *$P < 0.05$. (Redrawn from data submitted to the American Society of Anesthesiology National Meeting and presented at the International Anesthesia Research Society, 1998; portions of data accepted for publication by *Anesthesia and Analgesia* and submitted to *Acta Anaesthesiologica Scandinavica* for publication.)

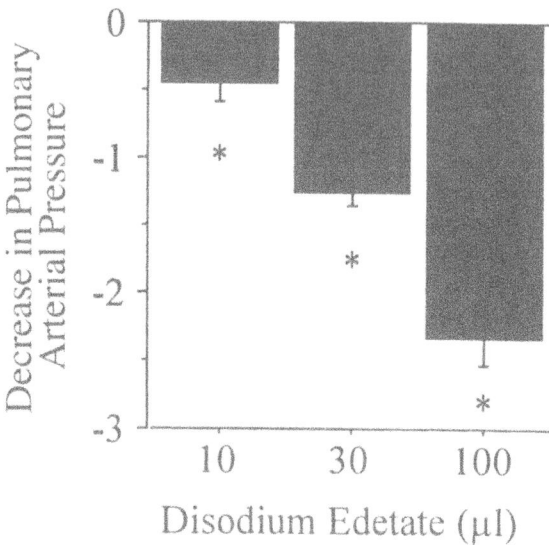

FIGURE 9.6. Influence of disodium edetate under conditions of elevated tone with U46619. *$P < 0.05$. (Redrawn from data presented at the International Anesthesia Research Society, Orlando, Florida, March, 1998; submitted to *Acta Anaesthesiologica Scandinavica* for publication.)

in pulmonary arterial perfusion pressure (Figure 9.6). Decreases in pulmonary arterial pressure in response to disodium edetate were rapid in onset, and pressure returned to control values over a 3- to 5-minute period, depending on the dose injected.

Verapamil significantly decreased BAY K8644-induced pressor responses in the pulmonary vascular bed of the rat (Figure 9.7). When pulmonary arterial pressure was increased by the infusion of U46619, decreases in pulmonary arterial pressure in response to intraarterial injections of ketamine were significantly smaller than responses obtained when ketamine was injected during the control period when the tone was increased with an U46619 infusion (Figure 9.7).

The present investigation shows that propofol, ketamine, and a constituent of the propofol solution, disodium edetate, decreased pulmonary arterial pressure when the tone in the pulmonary vascular bed was increased to a high steady level with U46619. Inasmuch as pulmonary blood flow and left atrial pressure were maintained constant, the decreases in pulmonary pressure reflect decreases in pulmonary vascular resistance. The decreases in pulmonary arterial pressure were dose-dependent and were not reduced by L-NABE, meclofenamate, or U-37883A, an ATP-sensitive potassium channel antagonist. These results suggest that decreases in pulmonary vascular resistance in response to propofol appear to occur by a mechanism independent of the release of endothelium-derived NO the release of vasodilator products in the cyclooxygenase pathway, or activation of K^+-ATP channels.

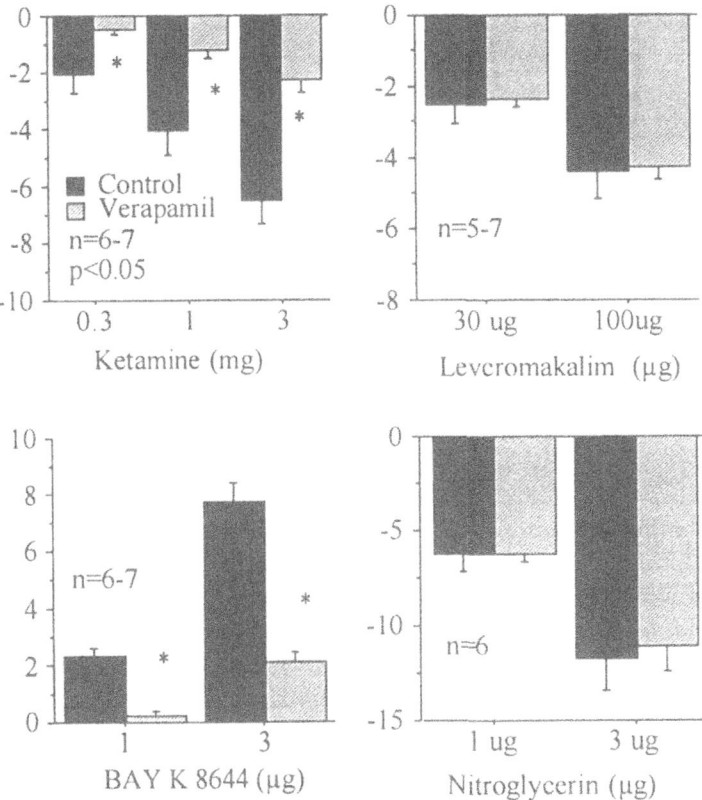

FIGURE 9.7. Influence of verapamil on vasodilator responses to ketamine, nitroglycerin, levcromakalim, and BAY K 8644. n indicates number of animals. *$P < 0.05$. (Redrawn from data submitted to the American Society of Anesthesiology National Meeting, Orlando, Florida, September 1998; accepted for publication by *Anesthesia and Analgesia*.)

Although the exact mechanism by which propofol and ketamine dilate the pulmonary vascular bed is not well understood, these induction agents have significant pulmonary vasodilator activity. Propofol and ketamine produced dose-related decreases in perfusion pressure in the rat pulmonary vascular bed that were not significantly decreased in the presence of meclofenamate, suggesting a minimal role for vasodilator prostaglandins. The observation that vasodilator responses to propofol are not inhibited by L-NABE or meclofenamate suggests that the release of NO and the release of cyclooxygenase products are not the major mechanisms of action by which propofol or ketamine decreases pulmonary vascular resistance.

The present investigation showed that U-37883A, a novel vascular selective nonsulfonylurea K^+-ATP channel antagonist, attenuated decreases in pulmonary arterial perfusion pressure in response to the K^+-ATP channel opener, levcromakalim, without significantly altering responses to propofol or to ketamine.

These data suggest that activation of K^+-ATP channels does not play a role in the mediation of the responses to propofol or ketamine.

The use of L-arginine analogues such as L-NABE is complicated since some studies have demonstrated that there is potentiation of vasoconstrictor responses to U46619, and these authors have questioned the use of these probes in studies on the role of epithelium-derived relaxing factor (EDRF) in the pulmonary vascular bed in vivo (Lippton et al. 1992). However, we have used L-NABE and N^G-nitro-L-arginine (L-NA) with consistent results (Feng et al. 1994; Kaye et al. 1996a, b).

Pure 2,6-diisopropyl phenol and an intralipid emulsion, key components of the commercial propofol preparation, had no significant effect in the pulmonary vascular bed of the rat. These data are similar to findings when normal healthy human volunteers were injected with 500 ml of 10% intralipid solution (Greene et al. 1976). In another study, intralipid was demonstrated to raise pulmonary vascular resistance in anesthetized rabbits (Uezono and Clarke 1995). The difference in the results of the present experiments may reflect differences in experimental methods or species.

The effects of ketamine may involve an effect on calcium influx or release from the sarcoplasmic reticulum. In rabbit and guinea pig preparations, ketamine inhibits L-type calcium channels, reduces calcium influx, and/or activates a Ca^{2+}-ATPase (Yamazaki et al. 1992; Abdalla et al., 1994; Lee and Hou 1995). However, in vascular smooth muscle cells, ketamine had no significant effect on calcium uptake into intracellular stores or on calcium extrusion (Kanmura et al. 1996). Ketamine was also found to inhibit agonist-induced synthesis of inositol 1,4,5-trisphosphate in rabbit mesenteric artery and inhibit the release of calcium (Kanmura et al. 1993). A similar mechanism for the inhibition of calcium release was found in the rabbit femoral artery, where ketamine decreased the release of calcium and reduced phospholipase C activity (Ratz et al. 1993).

In studies in GH3 cells, calcium-activated potassium channels were blocked and calcium levels decreased in a dose-dependent manner (Denson et al. 1994). A follow-up study in these cells indicated a significant disruption of the phospholipase A_2 arachidonic acid signal transduction pathway (Denson et al. 1996). In the present study, vasodilator responses to ketamine were not altered after administration of the ATP-sensitive potassium channel blocker U-37883A in a dose that significantly decreased responses to the ATP-sensitive potassium channel opener levcromakalim, suggesting that ketamine-induced vasodilation does not involve activation of these potassium channels.

The results of the present investigation suggest that calcium is involved in ketamine responses in the lung, since ketamine-induced vasodilator responses were attenuated by the L-type calcium channel blocker verapamil. Verapamil inhibited responses to BAY K 8644, a nifedipine analogue that promotes calcium entry, showing that L-type calcium channels were blocked. Additional mechanisms for the ketamine-induced vasodilator response may play a role mediating responses in the pulmonary vascular bed of the rat. We will investigate these in future studies.

In conclusion, the results of the present study show that propofol and ketamine have significant vasodilator activity in the pulmonary vascular bed of the rat when the tone is increased experimentally. The results suggest that that the vasodilator response to propofol and ketamine is independent of the release of endothelial-derived NO and cyclooxygenase products, or activation of K^+-ATP channels. Results also suggest that pure 2,6-diisopropyl phenol requires the lipid emulsion present with disodium edetate in the commercial propofol preparation for the full expression of pulmonary vasodilator activity in the rat. Ketamine-induced responses were attenuated after administration of the L-type calcium channel blocker verapamil when the tone in the pulmonary vascular bed was increased to a high steady level. Additional studies are needed to define exactly the mechanism of vasodilator responses to propofol in the pulmonary vascular bed of the rat.

References

Abdalla SS, Laravuso RB, & Will JA. 1994. Mechanisms of inhibitory effects of ketamine on guinea pig isolated pulmonary artery. *Anesth Analg 78:*17–22.

Baum V. 1993. Distinctive effects of three intravenous anesthetics on the inward rectifier and the delayed rectifier potassium currents in myocardium: implications for the mechanism of action. *Anesth Analg 76:*18–23.

Bulijubasic N, Marijic J, Berczi V, Supan D, Kampine J, & Bosnjak Z. 1996. Differential effects of etomidate, propofol, and midazolam on calcium and potassium channel currents in canine myocardial cells. *Anesthesiology 85:*1092–1099.

Chang K, & Davis R. 1993. Propofol produces endothelium-independent vasodilation and may act as a calcium channel blocker. *Anesth Analg 76:*24–32.

Cheng D, Dewitt B, McMahon T, & Kadowitz P. 1994. Comparative effects of L-NNA and alkyl esters of L-NNA on pulmonary vasodilator responses to acetylcholine, bradykinin, and substance P. *Am J Physiol 266:*H2416–H2423.

Coughlan MG, Flynn NM, Kenny D, et al. 1992. Differential relaxant effect of high concentrations of intravenous anesthetics on endothelin-constricted proximal and distal canine coronary arteries. *Anesth Analg 74:*378–383.

Denson DD, & Eaton DC. 1994. Ketamine inhibition of large conductance calcium-activated potassium channels is modulated by intracellular calcium. *Am J Physiol 267:*C1452–1458.

Denson DD, Duchatelle P, & Eaton DC. 1994. The effect of racemic ketamine on the large conductance Ca(+2)-activated potassium (BK) channels in GH3 cells. *Brain Res 638:*61–68.

Denson DD, Worrell RT, & Eaton DC. 1996. A possible role for phospholipase A2 in the action of general anesthetics. *Am J Physiol 270:*C636–644.

Erdemli O, Gumusel B, & Sahin-Erdemli I. 1995. The pulmonary vascular response to propofol in the isolated perfused rat lung. *Eur J Anaesthesiol 6:*617–623.

Feng C, Cheng D, Kaye A, Kadowitz P, & Nossaman B. 1994. Influence of N-LAME, LY83583, glibenclamide and L158809 on pulmonary circulation. *Eur J Pharm 263:*133–140.

Gacar N, Gok S, Kalyoncu I, Ozen I, Suokan N, & Akturk G. 1995. The effect of endothelium on the response to propofol on bovine coronary artery rings. *Acta Anaesthesiol Scand 39:*1080–1083.

Gelb A, Zhang C, & Hamilton J. 1996. Propofol induces dilation and inhibits constriction in guinea pig basilar arteries. *Anesth Analg 83:*472–476.

Gratton JP, Maurice MC, Rae GA, & D'Orleans-Juste P. 1995a. Pharmacological properties of endothelins and big endothelins in ketamine/xylazine or urethane anesthetized rats. *Am J Hypertens 8:*1121–1127.

Gratton JP, Rae GA, Claing A, et al. 1995b. Different pressor and bronchoconstrictor properties of human big-endothelin-1, 2 (1–38) and 3 in ketamine/xylazine-anaesthetized guinea-pigs. *Br J Pharmacol 114:*720–726.

Greene HL, Hazlett D, & Demaree R. 1976. Relationship between intralipid-induced hyperlipemia and pulmonary function. *Am J Clin Nutr 29:*127–135.

Hirshman CA, Downes H, Farbood A, & Bergman NA. 1979. Ketamine block of bronchospasm in experimental canine asthma. *Br J Anesth 51:*713–718.

Kanmura Y, Kajikuri J, Itoh T, & Yoshitake J. 1993. Effects of ketamine on contraction and synthesis of inositol 1,4,5-trisphosphate in smooth muscle of the rabbit mesenteric artery. *Anesthesiology 79:*571–579.

Kanmura Y, Missiaen L, & Casteels R. 1996. The effects of ketamine on calcium movements in A7r5 vascular smooth muscle cells. *Anesth Analg 83:*1105–1109.

Kaye A, Ibrahim I, Kadowitz P, & Nossaman B. 1996a. Analysis of responses to pentoxifylline in the pulmonary vascular bed of the cat. *Crit Care Med 24:*263–267.

Kaye A, Feng C, Smith D, Kang B, Kadowitz P, & Nossaman B. 1996b. Analysis of responses of pentoxifylline in the pulmonary vascular bed of the rat. *Am J Ther 3:*640–646.

Ko S, Lee S, & Han Y, et al. 1997. Blockade of myocardial ATP-sensitive potassium channels by ketamine. *Anesthesiology 87:*68–74.

Kreuscher H, & Gauch H. 1967. The effect of phencylidine derivatives ketamine (CI 581) on the cardiovascular system of the man. *Anaesthesist 16:*229–233.

Lee TS, & Hou X. 1995. Vasoactive effects of ketamine on isolated rabbit pulmonary arteries. *Chest 107:*1152–1155.

Lippton HL, Hao Q, & Hyman Ä. 1992. L-NAME enhances pulmonary vasoconstriction without inhibiting EDRF-dependent vasodilation. *J Appl Physiol 73:*2432–2439.

Magnelli V, Nobile M, & Maestrone E. 1992. K^+ channels in pc12 cells are affected by propofol. *Pflugers Arch 420:*393–398.

Maruyama K, Maruyama J, Yokochi A, Muneyuki M, & Miyasaka K. 1995. Vasodilatory effects of ketamine on pulmonary arteries in rats with chronic hypoxic pulmonary hypertension. *Anesth Analg 80:*786–792.

Mendoza CU, Suarez M, Castaneda R, Hernandez A, & Sanchez R. 1992. Comparative study between the effects of total intravenous anesthesia with propofol and balanced anesthesia with halothane on the alveolar-arterial oxygen tension difference and on the pulmonary shunt. *Arch Med Res 23:*139–142.

Miyawaki I, Nakamura K, Terasako K, Toda H, Kakuyama M, & Mori K. 1995. Modification of endothelium-dependent relaxation by propofol, ketamine, and midazolam. *Anesth Analg 81:*474–479.

Moore P, Nguyen D, Boldy R, & Reitan J. 1994. Comparative effects of nitric oxide on the coronary vasomotor responses to etomidate, propofol, and thiopental in anesthetized dogs. *Anesth Analg 79:*439–446.

Moreno L, Martinez-Cuesta M, Muedra V, Beltran B, & Esplugues J. 1997. Role of endothelium in the relaxation induced by propofol and thiopental in isolated arteries from man. *J Pharm Pharmacol 49:*430–432.

Mozrzymas J, Teisseyre A, & Vittur F. 1996. Propofol blocks voltage-gated potassium channels in human T lymphocytes. *Biochem Pharm 52:*843–849.

Naeije R, Lejeune P, Leeman M, Melot C, & Deloof T. 1989. Effects of propofol on pulmonary and systemic arterial pressure-flow relationships in hyperoxic and hypoxic dogs. *Br J Anesth 62:*532–539.

Nishiwaki K, Nyhan D, Rock P, Desai P, Peterson W, Pribble C, & Murray P. 1992. N^{ω}-l-nitro-*l*-arginine and pulmonary vascular-flow relationship in conscious dogs. *Am J Physiol 262:*H1331–H1337.

Olcese R, Usai C, Maestrone E, & Nobile M. 1994. The general anesthetic propofol inhibits transmembrane calcium current in chick sensory neurons. *Anesth Analg 78:* 955–960.

Park W, Lynch C, & Johns R. 1992. Effects of propofol and thiopental in isolated rat aorta and pulmonary artery. *Anesthesiology 77:* 956–963.

Pedersen CM, Thirstrup S, & Nielsen-Kudsk JE. 1993. Smooth muscle relaxant effects of propofol and ketamine in isolated guinea-pig trachea. *Eur J Pharmacol 238:*75–80.

Petros A, Bogle R, & Pearson J. 1993. Propofol stimulates nitric oxide release from cultured porcine aortic endothelial cells. *Br J Pharmacol 109:*6–7.

Platt OS, Thorington BD, & Brambilla DJ. 1991. Pain in sickle cell disease: rates and risk factors. *N Engl J Med 325:*11–15.

Ratz PH, Callahan PE, & Lattanzio FA Jr. 1993. Ketamine relaxes rabbit femoral arteries by reducing $[Ca^{2+}]_i$ and phospholipase C activity. *Eur J Pharmacol 236:*433–441.

Reich DL, & Silvay G. 1989. Ketamine: an update on the first twenty-five years of clinical experience. Can J Anesth *36:*186–197.

Rich G, Roos C, Anderson S, Daugherty M, & Uncles D. 1994. Direct effects of intravenous anesthetics on pulmonary vascular resistance in the isolated rat lung. *Anesth Analg 78:*961–966.

Sato T, Hirota K, Matsuki A, et al. 1997. The relaxant effect of ketamine on guinea pig airway smooth muscle is epithelium-independent. *Anesth Analg 84:*641–647.

Smith I, White P, Natahnson M, & Gouldson R. 1994. Propofol: an update on its clinical use. *Anesthesiology 81:*1005–1043.

Takeshita H, Okuda Y, & Sari A. 1972. The effects of ketamine on cerebral circulation and metabolism in man. *Anesthesiology 36:*69–75.

Tweed WA, Minuck MS, & Mymin D. 1972. Circulatory response to ketamine anesthesia. *Anesthesiology 37:*613–619.

Uezono S, & Clarke WR. 1995. The effect of propofol on normal and increased pulmonary vascular resistance in isolated perfused rabbit lung. *Anesth Analg 80:*577–582.

Williams G, Hanson K, Geiduschek J, Jones T, Baptiste B, & Morray J. 1994. Systemic and pulmonary vascular effects of propofol in children with congenital heart disease. *Anesthesiology 81:*A1372.

Wilson LE, Hatch DJ, & Rehder K. 1993. Mechanisms of the relaxant action of ketamine on isolated porcine trachealis muscle. *Br J Anaesth 71:*544–550.

Yamazaki M, Ito Y, Kuze S, et al. 1992. Effects of ketamine on voltage-dependent Ca^{2+} currents in single smooth muscle cells from rabbit portal vein. *Pharmacology 45:*162–169.

Yamazaki M, Momose Y, Shakunaga K, et al. 1995. The vasodilatory effects of ketamine on isolated rabbit portal veins. *Pharmacol Toxicol 76:*3–8.

10
Inhaled Nitric Oxide Therapy for Acute Respiratory Failure

WILLIAM E. HURFORD, WOLFGANG STEUDEL, AND WARREN M. ZAPOL

Introduction: A Selective Pulmonary Vasodilator

Pulmonary hypertension and severe hypoxemia complicate the care of patients with diseases such as the acute respiratory distress syndrome (ARDS). Numerous vasodilator therapies aimed at reducing pulmonary hypertension have been tested in these patients. All of the currently available intravenous vasodilators produce systemic vasodilation and hypotension at dosages sufficient to reduce the pulmonary artery pressure. In addition, intravenous infusions of systemic vasodilators such as nitroprusside or prostacyclin (prostaglandin I_2; PGI_2) markedly increase the venous admixture (Radermacher et al. 1990). In 1991, inhaled nitric oxide (NO) was reported to selectively vasodilate the pulmonary circulation (Fratacci et al. 1991; Frostell et al. 1991). The use of this novel therapy as an adjunct in the treatment of ARDS has attracted immense interest.

Inhaled NO Is a Selective Pulmonary Vasodilator

When administered by inhalation, NO diffuses into the pulmonary vasculature of ventilated lung regions and produces relaxation of pulmonary vascular smooth muscle, thereby decreasing pulmonary hypertension (Fratacci et al. 1991; Frostell et al. 1991). Since the NO is inhaled, the gas is distributed predominantly to well-ventilated alveoli and not to collapsed or fluid-filled areas of the lung. In the presence of increased vasomotor tone, selective vasodilation of well-ventilated lung regions causes a "steal" or diversion of pulmonary artery blood flow toward well-ventilated alveoli, and often improves the matching of ventilation to perfusion and increases arterial oxygenation (Pison et al. 1993).

It is unclear whether inhaled NO has clinical utility in addition to its properties as a selective pulmonary vasodilator. Laboratory studies suggest that inhaled NO has important effects in reducing some forms of lung and tissue injury. These effects include the ability to scavenge oxygen free radicals (Kanner et al. 1991; Clancy et al. 1992; Wink et al. 1993; Kavanagh et al. 1994; Hassoun et al. 1995), reduce oxygen toxicity (Garat et al. 1994, 1995; Nelin et al. 1995), and inhibit platelet and leukocyte aggregation (May et al. 1991; Högman et al. 1994; Samama et al. 1995). If these effects are clinically significant, early and continued therapy with inhaled NO could reduce the severity of some forms of lung injury.

Evaluation of Inhaled NO in Laboratory Models

Endotoxin Administration

The effects of inhaled NO on endotoxin-induced pulmonary hypertension are complex. In a study by Weitzberg and colleagues, inhaled NO at a concentration of 10 ppm selectively decreased the acute pulmonary hypertension occurring at least 30 minutes after the intravenous administration of *Escherichia coli* endotoxin in anesthetized pigs (Weitzberg et al. 1993). Arterial oxygenation and pH were also improved during NO inhalation. The early increase of pulmonary artery pressure (within 30 minutes after endotoxin administration) was unaffected by 10 ppm of inhaled NO.

Ogura and co-workers reported that NO inhalation (40 ppm) decreased the late-phase pulmonary hypertension following the infusion of *E. coli* endotoxin in anesthetized swine (Ogura et al. 1994b). Nitric oxide inhalation improved arterial oxygenation by redistributing blood flow from true shunt to ventilated lung regions, as measured by the multiple inert gas elimination technique. The amount of pulmonary edema, as assessed by the blood-free wet-to-dry lung weight ratio, was decreased in animals breathing 40 ppm NO. Dahm and co-workers similarly reported selective pulmonary vasodilation following NO inhalation in an anesthetized swine endotoxemia model (Dahm et al. 1994).

Oleic Acid-Induced Lung Injury

A common animal model of ARDS is produced by intravenous injection of oleic acid. An injection of this 18-carbon unsaturated fatty acid produces a syndrome of acute endothelial and alveolar epithelial cell necrosis, resulting in proteinaceous alveolar edema that mimics the acute phase of ARDS. In a dog model of acute lung injury induced by oleic acid injections, Putensen and co-workers examined the role of NO by giving inhaled NO and/or intravenous L-NMMA (N^G-monomethyl-L-arginine) (Putensen et al. 1994a). After the induction of lung injury, inhaled NO at 40 ppm improved gas exchange by redistributing blood flow from shunting regions to lung units with a nearly ideal ratio of ventilation to perfusion ($\dot{V}/\dot{Q}$). The improvement of $\dot{V}/\dot{Q}$ matching and gas exchange was most pronounced when NO was inhaled in the presence of systemic L-NMMA. Systemic L-NMMA administration alone increased pulmonary and systemic vascular resistance but did not affect $\dot{V}/\dot{Q}$ mismatch and gas exchange. Inhaled NO reversed the pulmonary but not the systemic vasoconstriction caused by L-NMMA. This effect may be clinically important, because the infusion of nitric oxide synthase (NOS) inhibitors is being considered as treatment for sepsis-induced hypotension (Putensen et al. 1994a; Klemm et al. 1995).

The effect of inhaled NO on pulmonary vascular resistance in a canine ARDS model was reported by Romand and co-workers (Romand et al. 1994). Nitric oxide inhalation (up to 145 ppm) reduced hypoxic pulmonary vasoconstriction, but not the mild pulmonary hypertension (mean pulmonary artery pressure, 23 mmHg) induced by oleic acid infusion. In a pig model of acute lung injury induced by oleic acid injection, Shah et al. (1994) reported that inhaled NO (10–80 ppm) reduced mean pulmonary artery pressure and shunt fraction in a concentration-dependent manner and improved oxygenation. Because almost the same dose of oleic acid (0.08–0.1 ml/kg) was used to induce acute lung injury in these studies, these varying results are most likely due to differences between the experimental species.

The degree of lung inflation may be an important determinant of the effects of inhaled NO. Putensen and colleagues reported that the recruitment of lung units by the application of 10 cm H_2O continuous positive airway pressure (CPAP) augmented the improvement of oxygenation caused by inhaling 40 ppm NO in anesthetized dogs with oleic acid-induced lung injury (Putensen et al. 1994b). Application of CPAP converted extensive shunting regions [shunt, 48 ± 7% (SD) of cardiac output] to regions of more normal $\dot{V}/\dot{Q}$. Inhaled NO also selectively reduced the pulmonary artery pressure in this study, but the extent of pulmonary vasodilation was not dependent upon the application of CPAP.

Inhaled NO mediates pulmonary vasodilation during lung injury by increasing cGMP levels within vascular smooth muscle. This increase is reflected by increased plasma cGMP concentrations. Rovira and co-workers studied a model of acute lung injury induced by bilateral lung lavage in anesthetized lambs (Rovira et al. 1994). When endogenous NO production was inhibited by infusing N^G-nitro-L-arginine methylester (L-NAME), a consistent increase of aortic, as compared with pulmonary arterial plasma cGMP concentration could be measured within 5 minutes of breathing 60 ppm NO. Increased aortic plasma cGMP levels were associated with selective pulmonary vasodilation, reduced venous admixture, and increased PaO_2. Levels of plasma cGMP returned to baseline within 10 min of discontinuation of NO breathing.

Inhalation Injury

The effect of inhaled NO on lung injury caused by smoke inhalation has been studied in sheep (Ogura et al. 1994a). Compared with air-breathing controls, pulmonary artery hypertension was reduced and oxygenation was improved in sheep breathing 20 ppm NO in air for 48 hours. There were no significant differences between the two groups in lung wet-to-dry weight ratio, compliance, or histologic changes. Nitric oxide inhalation neither improved nor worsened the tracheobronchial or alveolar pathologic changes occurring after inhalation injury.

Oxidant-Induced Acute Lung Injury

Oxygen-derived free radicals are important in the pathogenesis of ARDS and may contribute to leukocyte adherence and emigration and subsequent endothelial cell injury (Kubes et al. 1993). Depending on the species or model used, endogenous NO appears to enhance (Hughes et al. 1990; Mulligan et al. 1991; Ialenti et al. 1992; Mayhan 1992) or attenuate (Hutcheson et al. 1990; Boughton et al. 1992; Kubes 1992) the acute inflammatory response.

Nitric oxide combines with reactive oxygen species to produce intermediates, such as the peroxynitrite anion, which can exacerbate cellular injury. Nozik and co-workers, for example, reported that the administration of 1 mM L-arginine to isolated buffer-perfused rabbit lungs produced significant pulmonary hypertension and edema when the lungs were ventilated with 95% oxygen or in the presence of a H_2O_2 generating system (Nozik et al. 1995). This injury was attenuated by the administration of L-NAME or by pretreatment with catalase. The authors postulated that the administration of L-arginine increased the synthesis of NO within the lungs and that the NO reacted with H_2O_2 to cause lung injury via reactive intermediates.

Nitric oxide may also act as a superoxide scavenger. In the isolated buffer-perfused rabbit lung, Kavanagh and co-workers investigated the effects of inhaled NO and endogenous NOS inhibition on oxidant-induced acute lung injury (Kavanagh et al. 1994). Superoxide radicals were produced by the combination of purine and xanthine oxidase. Pretreatment with inhaled NO (90–120 ppm) prevented the increase in pulmonary artery pressure and capillary permeability as measured by the pulmonary capillary filtration coefficient. Inhibition of endogenous NOS by the infusion of L-NAME increased pulmonary vascular tone without affecting capillary permeability. In addition, Guidot and co-workers reported that inhaled NO prevented neutrophil-mediated permeability edema in isolated rat lungs and speculated that NO may have antiinflammatory properties (Guidot et al. 1995). Although the mechanism by which inhaled NO protects against lung injury is unclear, it may involve alterations in cellular cGMP levels and cytoskeletal changes.

Inhibition of cGMP-Specific Phosphodiesterases

The selectivity of inhaled NO is primarily due to the inactivation of NO by its rapid combination with hemoglobin within the pulmonary circulation (Rimar and Gillis 1993). Although this selectivity is a unique characteristic of inhaled NO, its short duration of action could be a disadvantage, because many patients with chronic pulmonary hypertension or severe ARDS require continuous vasodilator therapy. Although there is little evidence for acute pulmonary toxicity at low concentrations (<40 ppm) of inhaled NO, the toxicity of prolonged NO exposure in humans is unclear (Zapol et al. 1994). Conceivably, pharmacological agents that potentiate or prolong the pulmonary vasodilator effects of inhaled NO could reduce the effective dose of NO and reduce the toxicity of prolonged exposures.

Zaprinast (M&B 22948; 2-o-propoxyphenyl-8-azapurin-6-one, Rhone-Poulenc Rorer, Dagenham, Essex, UK) is a type 5 phosphodiesterase inhibitor that selectively inhibits the hydrolysis of cGMP with minimal effects on the breakdown of adenosine 3′,5′-cyclic monophosphate (cAMP) (Harris et al. 1989). Vasodilator responses to infusions of endothelium-dependent vasodilators and nitrosova-sodilators are modified by zaprinast in newborn lambs (Braner et al. 1993) and in the isolated cat lobar artery (McMahon et al. 1993). Ichinose and co-workers investigated the effects of zaprinast on the pulmonary vasodilating effects of inhaled NO in awake spontaneously breathing lambs with pharmacologically induced pulmonary hypertension (Ichinose et al. 1995). The reduction of pulmonary vascular resistance was significantly greater during zaprinast infusion. The duration of the vasodilator response to inhaled nitric oxide ($t_{1/2}$) was also markedly increased by zaprinast infusion (Fig. 10.1). Similar results have been reported in the ovine transitional circulation by Ziegler and co-workers (Ziegler et al. 1994).

Combination of Inhaled NO Therapy with Partial Liquid Ventilation

Perfluorocarbons are inert liquids that have the ability to lower surface tension in surfactant-depleted lungs, without being absorbed by the respiratory epithelium, and transport large concentrations of respiratory gases. Zobel et al. (1997) demonstrated that after induction of acute lung injury in piglets by repeated bilateral lung lavage, inhaled NO (1, 10, and 20 ppm) enhanced pulmonary gas exchange during partial liquid ventilation with perfluorocarbon. An additive effect of inhaled NO and partial liquid ventilation on pulmonary gas exchange in acute lung injury was confirmed by others (Houmes et al. 1997).

Nitric Oxide in Poorly Ventilated Lung Regions

Hopkins et al. (1997) studied the effects of NO on gas exchange by selectively creating areas of shunt or areas of low $\dot{V}/\dot{Q}$. Nitric oxide (80 ppm) reduced the blood flow to shunt areas. In the areas with ventilation/perfusion inequality, NO caused an inconsistent response. When the pulmonary vascular resistance of partially obstructed areas was decreased by NO inhalation, $\dot{V}/\dot{Q}$ inequality increased, since the blood flow to a relatively poorly ventilated area was increased by vasodilation. When NO did not reach the lung distal to the partial obstruction, and thus did not affect local vascular resistance $\dot{V}/\dot{Q}$ matching was improved. These results indicate that the effects of NO on $\dot{V}/\dot{Q}$ mismatching are dependent on the distribution of NO within lung regions. Nitric oxide improves oxygenation when administered to well-ventilated lung regions that are vasoconstricted. Nitric oxide can worsen oxygenation if it reaches and increases perfusion to poorly ventilated lung regions.

Acute Respiratory Distress Syndrome

Acute pulmonary hypertension consistently occurs in severe ARDS. In survivors, pulmonary vascular resistance progressively decreases over time. Nonsurvivors tend to have a persistently increased pulmonary vascular resistance. The increased

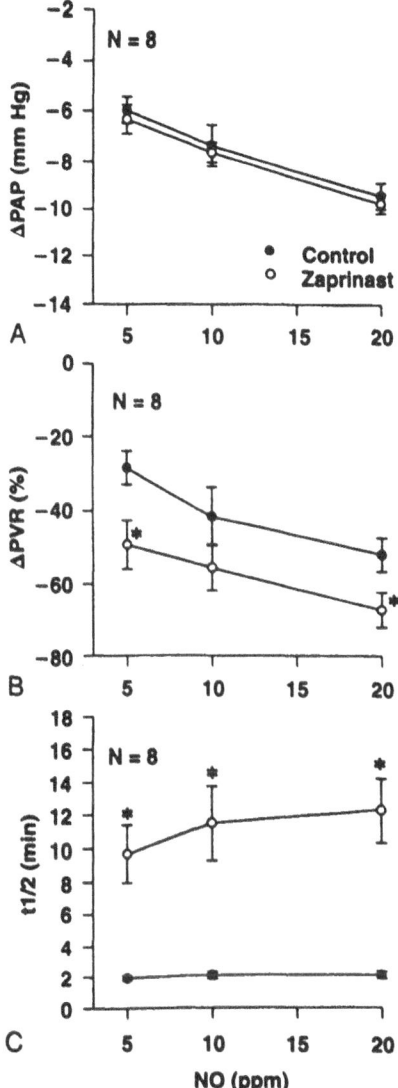

FIGURE 10.1. Influence of continuous intravenous infusion of zaprinast (0.1 mg/kg/min), a selective phosphodiesterase inhibitor, on magnitude of peak decreases of (A) mean pulmonary arterial pressures (PAP), (B) percent change of pulmonary vascular resistance (PVR), and (C) half times of vasodilating effect ($t_{1/2}$) in response to NO inhalation during pulmonary hypertension induced by U-46619 infusion. Values are means $\pm$ SE; N, number of lambs, * $P < 0.05$ compared with controls. From Ichinose et al. (1995). Copyright 1995 by the American Physiological Society. Reproduced with permission.

pulmonary arterial pressure is independent of changes of cardiac output and persists after the correction of systemic hypoxemia (Zapol and Snider 1977).

The pulmonary vascular changes in ARDS are produced by a complex combination of primary lung injury (e.g., from aspiration, trauma, or infection), the consequences of the pulmonary inflammatory response to injury (hypoxia, acidosis, and release of cytokines and components of the complement system and the arachidonic acid pathway, as well as inhibitors of fibrinolysis), and the iatrogenic complications of intensive care therapy (oxygen toxicity and barotrauma). In severe ARDS, thromboembolic occlusion of the pulmonary vasculature is also common (Zapol and Jones 1987).

Rossaint and co-workers compared the effects of NO inhalation (18 and 36 ppm) and intravenously infused prostacyclin in nine patients with ARDS (Rossaint et al. 1993). Nitric oxide selectively reduced mean pulmonary artery pressure (Fig. 10.2). Oxygenation improved because of decreased venous admixture ($\dot{Q}_{VA}/\dot{Q}_t$). Although the intravenous infusion of prostacyclin also reduced pulmonary artery pressure, mean arterial pressure, and arterial partial pressure of oxygen (PaO_2) decreased as $\dot{Q}_{VA}/\dot{Q}_t$ increased.

Very low concentrations of inhaled NO may effectively reduce pulmonary artery pressure and improve oxygenation in many patients with ARDS. Inhaled NO concentrations of less than 2 ppm have been effective in some patients (Fig. 10.3) (Gerlach et al. 1993; Rossaint et al. 1993, Puybasset et al. 1994, 1995). Right ventricular ejection fraction may increase in some patients responding to inhaled NO, suggesting that the observed decreases of pulmonary artery pressure may be hemodynamically important (Fierobe et al. 1995; Rossaint et al. 1995).

It is attractive to speculate that decreasing pulmonary capillary pressure by breathing NO may reduce the severity of lung injury in patients with ARDS. Benzing and co-workers (1995) reported that breathing 40 ppm NO decreased pulmonary capillary pressure and the pulmonary transvascular flux of albumin (an index of fluid efflux into the pulmonary interstitium). It is unclear whether such changes are clinically significant or persist during long-term NO therapy.

If inhaled NO can improve the matching of ventilation to perfusion, it might be expected that (V_D/V_T) and arterial partial pressure of carbon dioxide ($PaCO_2$) would decrease during NO inhalation. Although most studies have reported no change in $PaCO_2$ during NO inhalation, several have shown small but statistically significant decreases in V_D/V_T and $PaCO_2$ (Puybasset et al. 1994; Fierobe et al. 1995). These changes were too small to permit minute ventilation to be reduced.

Some patients do not respond to NO inhalation with pulmonary vasodilation and improved oxygenation. In studies of critically ill patients with ARDS who were ventilated with inhaled NO (<40 ppm), approximately 35% of patients had minimal or no response to inhaled NO (Bigatello et al. 1994; Lundin et al. 1997; Manktelow et al. 1997; Dellinger et al. 1998). Several hypotheses have been raised to explain the mechanisms of nonresponsiveness. Manktelow et al. (1997) and Krafft et al. (1996) reported that ARDS patients without sepsis were more likely to respond to inhaled NO than those patients with septic shock. Holzmann et al. (1996) tested the effects of sepsis on NO responsiveness in an isolated lung

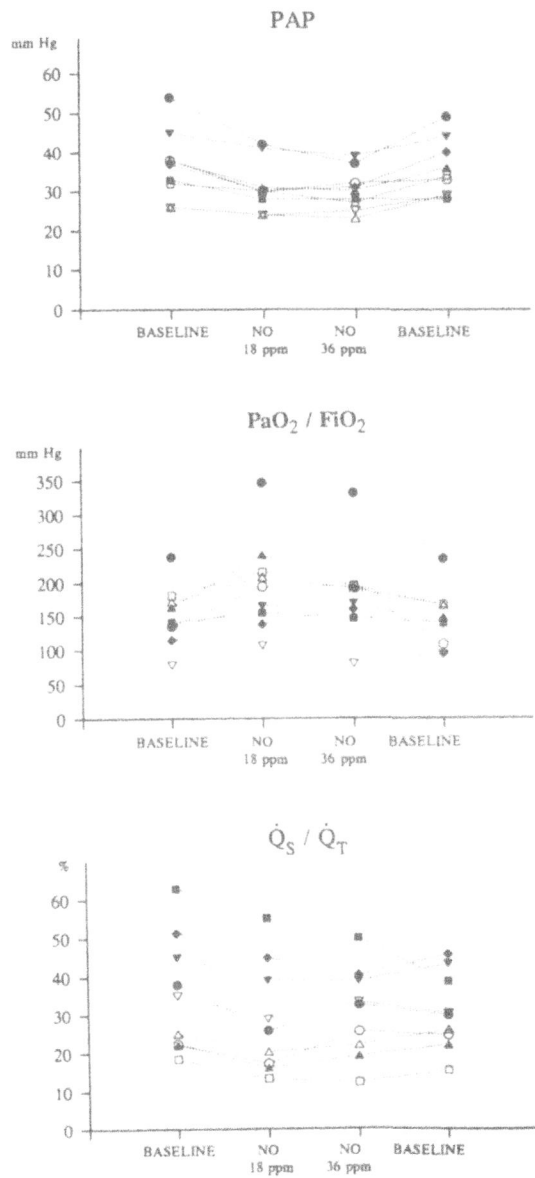

FIGURE 10.2. Mean pulmonary artery pressure (PAP), arterial oxygenation efficiency (PaO_2/FiO_2), and venous admixture ($\dot{Q}_s/\dot{Q}_T$) during inhalation of 18 and 36 ppm NO in nine patients with ARDS. Solid symbols represent patients treated with extracorporeal membrane oxygenation. From Roissant et al. (1993). Copyright 1993 by Lippincott-Raven Publishers. Reproduced with permission.

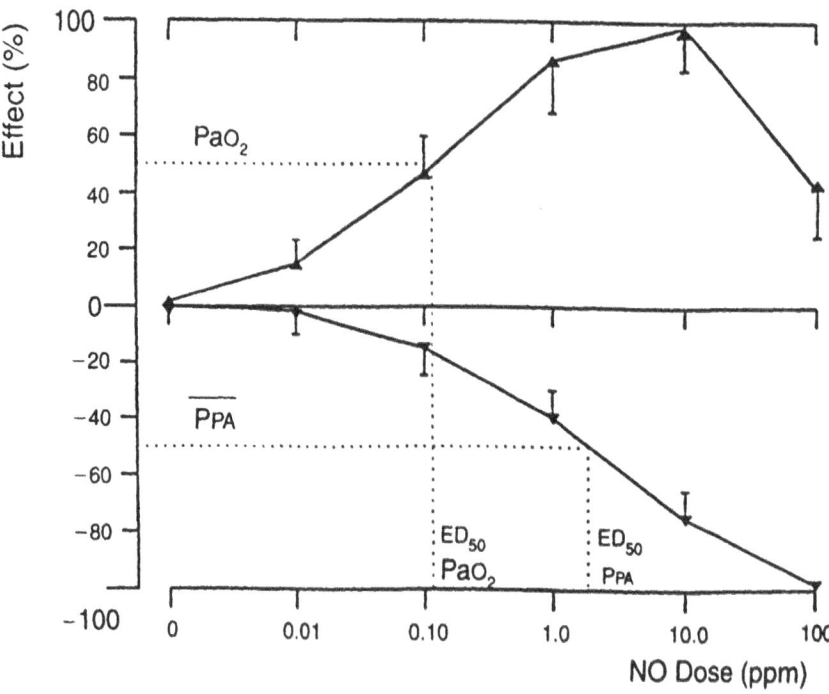

FIGURE 10.3. Dose-response of inhaled NO for (upper) PaO_2 and (lower) mean pulmonary artery pressure (P_{PA}). Values are means ± SD for 12 patients with ARDS, expressed as percentage of maximal change. The estimated ED_{50} values of NO for the PaO_2 increase and for the pulmonary artery decrease are indicated above the x axis. From Gerlach et al. (1993). Copyright 1993 by Blackwell Science Ltd. Reproduced with permission.

model. They reported that hyporesponsiveness was associated with a decrease in cGMP release, suggesting that the signal transduction of the NO pathway is down-regulated in sepsis. This was attributable to increased PDE activity and, therefore, increased cGMP metabolism and breakdown. Other factors are undoubtedly important. In general, the baseline level of pulmonary vascular resistance appears to predict the degree of pulmonary vasoconstriction reversible by NO inhalation. Those with the greatest degree of pulmonary hypertension appear to respond best to NO inhalation (Bigatello et al. 1994; Young et al. 1994; Puybasset et al. 1995). The concomitant intravenous infusion of a novel vasoconstricting drug, almitrine, which increases the degree of hypoxic vasoconstriction in the lung, has been reported to enhance the beneficial effect of inhaled NO on PaO_2 (Payen et al. 1993; Wysocki et al. 1994). Almitrine, however, also increased mean pulmonary artery pressure and cardiac output. A favorable response to inhaled NO may also be related to the degree of alveolar recruitment (Putensen et al. 1994b; Puybasset et al. 1995).

   Tachyphylaxis to NO inhalation has not been observed, even when NO inhalation is continued for up to 53 days (Rossaint et al. 1993). Continuous inhalation of NO, however, is necessary. Pulmonary artery pressure and PaO_2 quickly return to baseline values after discontinuation of the gas. In some cases, rapid withdrawal of inhaled NO therapy has been associated with the sudden worsening of hypoxemia and pulmonary hypertension (Gerlach et al. 1993; Bigatello et al. 1994; Miller et al. 1995). This deterioration appears to be transient, but the hemodynamic consequences of this "rebound" may be catastrophic. The severity of the rebound may be minimized by transiently increasing inspired fraction of oxygen (FiO$_2$), administering intravenous vasoactive agents as necessary, and avoiding sudden discontinuation of NO inhalation in patients with hemodynamic instability or severe hypoxemia. The mechanism for this acute pulmonary vasoconstriction is unknown, and its occurrence in patients is unpredictable. It may be due to reduced production of endogenous NO caused by inhaling high levels of exogenous NO, a phenomenon known as product inhibition (Assreuy et al. 1993; Rengasamy and Johns 1993; Kiff et al. 1994). It could also be due to enhanced release of vasoconstrictor substances by the pulmonary circulation or enhanced phosphodiesterase activity.

Several randomized, double-blind studies of the efficacy of NO inhalation in patients with ARDS have been performed. In a multicenter phase 2 trial, Dellinger and colleagues reported the effects of inhaled NO (0, 1.25, 5, 20, 40 ppm) on 177 patients with nonseptic ARDS (Dellinger et al. 1998). Sixty percent of patients receiving inhaled NO had at least a 20% improvement of PaO$_2$ after 4 hours of treatment, compared with 24% of patients receiving placebo. The 28-day mortality rate of all patients was 30% and was unaffected by NO therapy. Although most outcome variables were equivalent in both groups, ARDS patients who received 5 ppm NO appeared to have a more rapid resolution of their respiratory failure (Fig. 10.4). A randomized, double-blind, multicenter phase 3 study investigating the long-term effects of 5 ppm inhaled NO on ARDS is currently in progress.

A European phase 3 trial studied 260 medical and surgical patients with acute lung injury (Lundin et al. 1997). A positive response was defined as an increase of PaO$_2$ greater than 25%. Sixty-six percent of the patients were responders, and the treatment was continued at the lowest effective dose of inhaled NO or placebo by random allocation up to 30 days. An initial abstract of the study reported a mortality rate of 45% in patients receiving inhaled NO treatment, 38% in placebo patients, and 45% in nonresponders (Lundin et al. 1997). The complete results have not yet been published.

Neonatal Respiratory Failure

At birth there is a sustained decrease in pulmonary vascular resistance and an increase in pulmonary blood flow, in part due to increased oxygen tensions. If this does not occur, persistent pulmonary hypertension of the newborn (PPHN) may result. Persistent pulmonary hypertension of the newborn is characterized by increased pulmonary vascular resistance and severe hypoxemia, which is unresponsive to oxygen therapy. Extracorporeal membrane oxygenation (ECMO) is often

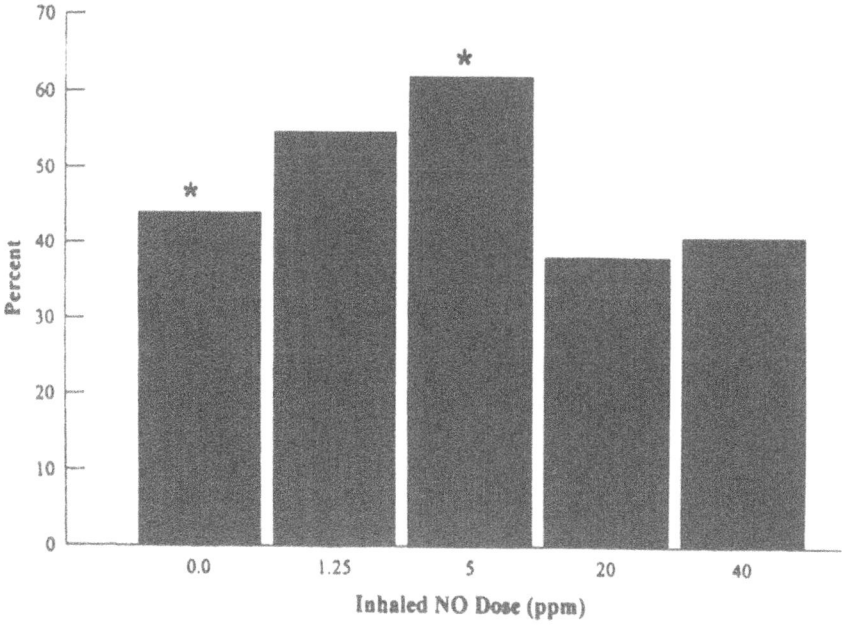

FIGURE 10.4. Percentage of patients with ARDS in each dose group who were alive and off mechanical ventilation at day 28 following randomization to 0, 1.25, 5, 20, or 40 ppm NO breathing. *$P < 0.05$ for the difference between the 0 ppm and 5 ppm groups. From Dellinger et al. (1998). Copyright 1998 by Lippincott Williams & Wilkens. Reproduced with permission.

used to support these infants; however, anticoagulation and cannulation of large vessels are associated with complications. It has been hypothesized that endogenous production of NO by the pulmonary vasculature may be decreased in PPHN. If so, then inhaled NO might provide an effective therapy for these severely ill babies (Kinsella et al. 1992; Roberts et al. 1992).

Laboratory Studies

Laboratory studies have documented that inhaled NO is an effective vasodilator of the neonatal pulmonary circulation (Roberts et al. 1993; DeMarco et al. 1994; Etches et al. 1994; Kinsella et al. 1994; Nelin et al. 1994; Skimming et al. 1994; Karamanoukian et al. 1995; Rosenberg et al. 1995). Zayek and co-workers studied the effects of inhaled NO (6 to 100 ppm) in a model of PPHN created by prenatal ligation of the ductus arteriosus in lambs (Zayek et al. 1993a). Inhaled NO caused dose-dependent decreases in pulmonary artery pressure and pulmonary vascular resistance. Decreased right-to-left shunting of blood flow through the foramen ovale during NO inhalation increased PaO_2 and arterial oxygen saturation and decreased $PaCO_2$. Systemic blood pressure was unaffected by breathing NO (Zayek et al. 1993a). Using this model, Zayek and co-workers subsequently documented an increased survival rate with NO therapy (Zayek et al. 1993b). Inhalation of NO also can attenuate hypoxic pulmonary vascular remodeling of the pulmonary circulation (Kouyoumdjian et al. 1994; Roberts et al. 1995). Conceivably, inhaled NO therapy might be used to limit the chronic pulmonary vascular changes that accompany neonatal acute respiratory failure.

Clinical Studies

Several large, randomized, multicenter trials on the effect of inhaled NO in near-term and term newborn patients have been published recently (Anonymous 1997a,b; Kinsella et al. 1997; Roberts et al. 1997). In the majority of PPHN patients, NO had beneficial effects and reduced the requirement for ECMO. In follow-up of children who received NO treatment for PPHN as newborns, neurodevelopment, the incidence of airway disease, and the need for supplemental oxygen one to two years after NO inhalation were comparable to those in conventionally or ECMO-treated patients (Rosenberg et al. 1997).

Preterm Neonates with Respiratory Distress Syndrome

Respiratory distress syndrome, or hyaline membrane disease, of the premature infant is characterized by surfactant deficiency or dysfunction and, in many cases, with pulmonary hypertension. Several clinical series of inhaled NO in premature infants with respiratory distress syndrome have been reported (Abman et al. 1993; Peliowski et al. 1995; Skimming et al. 1997). In general, inhaled NO has improved systemic oxygenation and pulmonary hypertension and reduced the need for ECMO in large populations of patients with diverse causes of neonatal respiratory failure (Hoffman et al. 1997). Overall mortality has been unaffected, however, and 40% to 50% of patients did not respond to NO inhalation.

Summary

Nitric oxide inhalation has been used successfully to treat pulmonary hypertension and hypoxemia in thousands of patients worldwide since its clinical introduction in 1991. The intense interest in inhaled NO is indicative of the great need for simple and effective therapies for these critically ill patients. The clinical utility of inhaled NO therapy is being currently evaluated in prospective randomized clinical trials. As we understand why inhaled NO works, and why it is sometimes ineffective, we will greatly expand our knowledge of the physiology and pathophysiology of the pulmonary circulation.

References

Abman SH, Kinsella JP, Schaffer MS, & Wilkening RB. 1993. Inhaled nitric oxide in the management of a premature newborn with severe respiratory distress and pulmonary hypertension. *Pediatrics* 92:606–609.

Anonymous. 1997a. Inhaled nitric oxide and hypoxic respiratory failure in infants with congenital diaphragmatic hernia. The Neonatal Inhaled Nitric Oxide Study Group (NINOS). *Pediatrics 99:*838–845.

Anonymous. 1997b. Inhaled nitric oxide in full-term and nearly full-term infants with hypoxic respiratory failure. The Neonatal Inhaled Nitric Oxide Study Group. *N Engl J Med 336:*597–604.

Assreuy J, Cunha FQ, Liew FY, & Moncada S. 1993. Feedback inhibition of nitric oxide synthase activity by nitric oxide. *Br J Pharmacol 108:*833–837.

Benzing A, Bräutigam P, Geiger K, Loop T, Beyer U, & Moser E. 1995. Inhaled nitric oxide reduces pulmonary transvascular albumin flux in patients with acute lung injury. *Anesthesiology 83:*1153–1161.

Bigatello LM, Hurford WE, Kacmarek RM, Roberts JD Jr, & Zapol WM. 1994. Prolonged inhalation of low concentrations of nitric oxide in patients with severe adult respiratory distress syndrome. Effects on pulmonary hemodynamics and oxygenation. *Anesthesiology 80:*761–770.

Boughton SN, Deakin AM, & Whittle BJ. 1992. Actions of nitric oxide on the acute gastrointestinal damage induced by PAF in the rat. *Ann NY Acad Sci 664:*126–139.

Braner DAV, Fineman JR, Chang R, & Soifer SJ. 1993. M&B 22948, a cGMP phosphodiesterase inhibitor, is a pulmonary vasodilator in lambs. *Am J Physiol 264:* H252–H258.

Clancy RM, Leszynska-Piziak J, & Abramson SB. 1992. Nitric oxide, an endothelial cell relaxation factor, inhibits neutrophil superoxide anion production via a direct action on the NADPH oxidase. *J Clin Invest 90:*1116–1121.

Dahm P, Blomquist S, Martensson L, Thorne J, & Zoucas E. 1994. Circulatory and ventilatory effects of intermittent nitric oxide inhalation during porcine endotoxemia. *J Trauma 37:*769–777.

Dellinger RP, Zimmerman JL, Taylor RW, Straube RC, Hauser DL, Criner GJ, Davis K, Hyers TM, Papadakos P, and the Inhaled Nitric Oxide in ARDS Study Group. 1998. Effects of inhaled nitric oxide in patients with acute respiratory distress syndrome: results of a randomized phase II trial. *Crit Care Med 26:*15–23.

DeMarco V, Skimming J, Ellis TM, & Cassin S. 1994. Nitric oxide inhalation: effects on the ovine neonatal pulmonary and systemic circulations. *Chest 105:*91S–92S.

Etches PC, Finer NN, Barrington KJ, Graham AJ, & Chan WK. 1994. Nitric oxide reverses acute hypoxic pulmonary hypertension in the newborn piglet. *Pediatr Res 35:*15–19.

Fierobe L, Brunet F, Dhainaut JF, Monchi M, Belghith M, Mira JP, Dall'ava-Santucci J, & Dinh-Xuan AT. 1995. Effect of inhaled nitric oxide on right ventricular function in adult respiratory distress syndrome. *Am J Respir Crit Care Med 151:*1414–1419.

Fratacci MD, Frostell CG, Chen TY, Wain JC Jr, Robinson DR, & Zapol WM. 1991. Inhaled nitric oxide. A selective pulmonary vasodilator of heparin-protamine vasoconstriction in sheep. *Anesthesiology 75:*990–999.

Frostell C, Fratacci MD, Wain JC, Jones R, & Zapol WM. 1991. Inhaled nitric oxide. A selective pulmonary vasodilator reversing hypoxic pulmonary vasoconstriction. *Circulation 83:*2038–2047.

Garat C, Adnot S, Reyaiguia S, Meignan M, & Jayr C. 1994. Effect of inhaled nitric oxide on 100% oxygen induced lung injury in rats. *Anesthesiology 81:*A1454.

Garat C, Adnot S, Rezaiguia S, Kouyoumdjian C, Meignan M, & Jayr C. 1995. Effect of inhaled NO or treatment with L-NAME on 100% oxygen induced lung injury in rats. *Am J Respir Crit Care Med 151:*A757.

Gerlach H, Rossaint R, Pappert D, & Falke KJ. 1993. Time-course and dose-response of nitric oxide inhalation for systemic oxygenation and pulmonary hypertension in patients with adult respiratory distress syndrome. *Eur J Clin Invest 23:*499–502.

Guidot DM, Repine MJ, Hybertson BM, & Repine JE. 1995. Inhaled nitric oxide prevents neutrophil-mediated, oxygen radical-dependent leak in isolated rat lungs. *Am J Physiol 269*(1 Pt 1):L2–L5.

Hatris AL, Lemp BM, Bentley RG, Perrone MH, Hamel LT, & Silver PJ. 1989. Phosphodiesterase isozyme inhibition and the potentiation by zaprinast of endothelium-derived relaxing factor and guanylate cyclase stimulating agents in vascular smooth muscle. *J Pharmacol Exp Ther 249:*394–400.

Hassoun PM, Yu FS, Zulueta JJ, White AC, & Lanzillo JJ. 1995. Effect of nitric oxide and cell redox status on the regulation of endothelial cell xanthine dehydrogenase. *Am J Physiol 268:*L809–L817.

Hoffman GM, Ross GA, Day SE, Rice TB, & Nelin LD. 1997. Inhaled nitric oxide reduces the utilization of extracorporeal membrane oxygenation in persistent pulmonary hypertension of the newborn. *Crit Care Med 25:*352–359.

Högman M, Frostell C, Arnberg H, Sandhagen B, & Hedenstierna G. 1994. Prolonged bleeding time during nitric oxide inhalation in the rabbit. *Acta Physiol Scand 151:*125–129.

Holzmann A, Bloch KD, Sanchez LS, Filippov G, & Zapol WM. 1996. Hyporesponsiveness to inhaled nitric oxide in isolated, perfused lungs from endotoxin-challenged rats. *Am J Physiol 271:*L981–L986.

Hopkins SR, Johnson EC, Richardson RS, Wagner H, De Rosa M, & Wagner PD. 1997. Effects of inhaled nitric oxide on gas exchange in lungs with shunt or poorly ventilated areas. *Am J Respir Crit Care Med 156:*484–491.

Houmes RJ, Hartog A, Verbrugge SJ, Bohm S, & Lachmann B. 1997. Combining partial liquid ventilation with nitric oxide to improve gas exchange in acute lung injury. *Intensive Care Med 23:*163–169.

Hughes SR, Williams TJ, & Brain SD. 1990. Evidence that endogenous nitric oxide modulates oedema formation induced by substance P. *Eur J Pharmacol 191:*481–484.

Hutcheson IR, Whittle BJ, & Boughton SN. 1990. Role of nitric oxide in maintaining vascular integrity in endotoxin-induced acute intestinal damage in the rat. *Eur J Pharmacol 191:*485–488.

Ialenti A, Ianaro A, Moncada S, & Di RM. 1992. Modulation of acute inflammation by endogenous nitric oxide. *Eur J Pharmacol 211:*177–182.

Ichinose F, Adrie C, Hurford WE, & Zapol WM. 1995. Prolonged pulmonary vasodilator action of inhaled nitric oxide by zaprinast in awake lambs. *J Appl Physiol 78:*1288–1295.

Kanner J, Harel S, & Granit R. 1991. Nitric oxide as an antioxidant. *Arch Biochem Biophys 289:*130–136.

Karamanoukian HL, Glick PL, Wilcox DT, Rossman JE, Holm BA, & Morin FC III, 1995. Pathophysiology of congenital diaphragmatic hernia. VIII: inhaled nitric oxide requires exogenous surfactant therapy in the lamb model of congenital diaphragmatic hernia. *J Pediat Surg 30:*1–4.

Kavanagh BP, Mouchawar A, Goldsmith J, & Pearl RG. 1994. Effects of inhaled NO and inhibition of endogenous NO synthesis in oxidant-induced acute lung injury. *J Appl Physiol 76:*1324–1329.

Kiff RJ, Moss DW, & Moncada S. 1994. Effect of nitric oxide gas on the generation of nitric oxide by isolated blood vessels: implications for inhalation therapy. *Br J Pharmacol 113:*496–498.

Kinsella JP, Ivy DD, & Abman SH. 1994. Inhaled nitric oxide improves gas exchange and lowers pulmonary vascular resistance in severe experimental hyaline membrane disease. *Pediatr Res 36:*402–408.

Kinsella JP, Shaffer E, Neish SR, & Abman SH. 1992. Low-dose inhalational nitric oxide in persistent pulmonary hypertension of the newborn. *Lancet 340:*8819–8820.

Kinsella JP, Truog WE, Walsh WF, Goldberg RN, Bancalari E, Mayock DE, Redding GJ, deLemos RA, Sardesai S, McCurnin DC, et al. 1997. Randomized, multicenter trial of inhaled nitric oxide and high-frequency oscillatory ventilation in severe, persistent pulmonary hypertension of the newborn. *J Pediat 131:*55–62.

Klemm P, Thiemermann C, Winkmaier G, Martorana PA, & Henning R. 1995. Effects of nitric oxide synthase inhibition combined with nitric oxide inhalation in a porcine model of endotoxin shock. *Br J Pharmacol 224:*363–368.

Kouyoumdjian C, Adnot S, Levame M, Eddahibi S, Bousbaa H, & Raffestin B. 1994. Continuous inhalation of nitric oxide protects against development of pulmonary hypertension in chronically hypoxic rats. *J Clin Invest 94:*578–584.

Krafft P, Fridrich P, Fitzgerald RD, Koc D, & Steltzer H. 1996. Effectiveness of nitric oxide inhalation in septic ARDS. *Chest 109:*486–493.

Kubes P. 1992. Nitric oxide modulates epithelial permeability in the feline small intestine. *Am J Physiol 262 (Gastrointestin 25):*G1138–G1142.

Kubes P, Kanwar S, Niu X-F, & Gaboury JP, 1993. Nitric oxide synthesis inhibition induces leukocyte adhesion via superoxide and mast cells. *FASEB J 7:*1293–1299.

Lundin S, Mang H, Smithies M, et al. 1997. Inhalation of nitric oxide in acute lung injury: preliminary results of a European multicenter study. *Intensive Care Med 23 (suppl 1):*S2.

Manktelow C, Bigatello LM, Hess D, & Hurford WE. 1997. Physiological determinants of the response to inhaled nitric oxide in patients with the acute respiratory distress syndrome. *Anesthesiology 87:*297–307.

May GR, Crook P, Moore PK, & Page CP. 1991. The role of nitric oxide as an endogenous regulator of platelet and neutrophil activation within the pulmonary circulation of the rabbit. *Br J Pharmacol 102:*759–763.

Mayhan WG. 1992. Role of nitric oxide in modulating permeability of hamster cheek pouch in response to adenosine 5′-diphosphate and bradykinin. *Inflammation 16:*295–305.

McMahon TJ, Ignarro LJ, & Kadowitz PJ. 1993. Influence of zaprinast on vascular tone and vasodilator responses in the cat pulmonary vascular bed. *J Appl Physiol 74:*1704–1711.

Miller OI, Tang SF, Keech A, & Celermajer DS. 1995. Rebound pulmonary hypertension on withdrawal from inhaled nitric oxide. *Lancet 346:*51–52.

Mulligan MS, Hevel JM, Marletta MA, & Ward PA. 1991. Tissue injury caused by deposition of immune complexes is L-arginine dependent. *Proc Natl Acad Sci USA 88:*6338–6342.

Nelin LD, Moshin J, Thomas CJ, Sasidharan P, & Dawson CA. 1994. The effect of inhaled nitric oxide on the pulmonary circulation of the neonatal pig. *Pediatr Res 35:*20–24.

Nelin L, Dolinski S, Morrisey J, & Dawson C. 1995. The effect of inhaled nitric oxide on survival of rats in oxygen. *Am J Respir Crit Care Med 151:*A757.

Nozik ES, Huang Y-CT, & Piantadosi CA. 1995. L-Arginine enhances injury in the isolated rabbit lung during hyperoxia. *Respir Physiol 100*:63–74.

Ogura H, Cioffi WG Jr, Jordan BS, Okerberg CV, Johnson AA, Mason AD Jr, & Pruitt BA Jr. 1994a. The effect of inhaled nitric oxide on smoke inhalation injury in an ovine model. *J Trauma 37*:294–302.

Ogura H, Cioffi WG, Offner PJ, Jordan BS, Johnson AA, & Pruitt BA Jr. 1994b. Effect of inhaled nitric oxide on pulmonary function after sepsis in a swine model. *Surgery 116*:313–321.

Payen DM, Gatecel C, & Plaisance P. 1993. Almitrine effect on nitric oxide inhalation in adult respiratory distress syndrome. *Lancet 341*:1164.

Peliowski A, Finer NN, Etches PC, Tierney AJ, & Ryan CA. 1995. Inhaled nitric oxide for premature infants after prolonged rupture of the membranes. *J Pediat 126*:450–453.

Pison U, Lopez FA, Heidelmeyer CF, Roissant R, & Falke KJ. 1993. Inhaled nitric oxide reverses hypoxic pulmonary vasoconstriction without impairing gas exchange. *J Appl Physiol 74*:1287–1292.

Putensen C, Rasanen J, & Downs JB. 1994a. Effect of endogenous and inhaled nitric oxide on the ventilation-perfusion relationships in oleic-acid lung injury. *Am J Respir Crit Care Med 150*:330–336.

Putensen C, Rasanen J, Lopez FA, & Downs JB. 1994b. Continuous positive airway pressure modulates effect of inhaled nitric oxide on the ventilation-perfusion distributions in canine lung injury. *Chest 106*:1563–1569.

Puybasset L, Rouby JJ, Mourgeon E, Stewart TE, Cluzel P, Arthaud M, Poete P, Bodin L, Korinek AM, & Viars P. 1994. Inhaled nitric oxide in acute respiratory failure: dose-response curves. *Intensive Care Med 20*:319–327.

Puybasset L, Rouby JJ, Mourgeon E, Cluzel P, Souhil Z, Law-Koune JD, Stewart T, Devilliers C, Lu Q, Roche S, et al. 1995. Factors influencing cardiopulmonary effects of inhaled nitric oxide in acute respiratory failure. *Am J Respir Crit Care Med 152*:318–328.

Radermacher P, Santak B, Wust HJ, Tarnow J, & Falke KJ. 1990. Prostacyclin for the treatment of pulmonary hypertension in the Adult Respiratory Distress Syndrome: effects on pulmonary capillary pressure and ventilation-perfusion distributions. *Anesthesiology 72*:238–244.

Rengasamy A, & Johns RA. 1993. Regulation of nitric oxide synthase by nitric oxide. *Mol Pharmacol 44*:124–128.

Rimar S, & Gillis CN. 1993. Selective pulmonary vasodilation by inhaled nitric oxide is due to hemoglobin inactivation. *Circulation 88*:2884–2887.

Roberts JD Jr, Polaner DM, Lang P, & Zapol WM. 1992. Inhaled nitric oxide in persistent pulmonary hypertension of the newborn. *Lancet 340*:818–819.

Roberts JD Jr, Chen TY, Kawai N, Wain J, Dupuy P, Shimouchi A, Bloch K, Polaner D, & Zapol WM. 1993. Inhaled nitric oxide reverses pulmonary vasoconstriction in the hypoxic and acidotic newborn lamb. *Circ Res 72*:246–254.

Roberts JD Jr, Roberts CT, Jones RC, & Zapol WM. 1995. Continuous nitric oxide inhalation reduces pulmonary arterial structural changes, right ventricular hypertrophy, and growth retardation in the hypoxic newborn rat. *Circ Res 76*:215–222.

Roberts JD Jr, Fineman JR, Morin FC III, Shaul PW, Rimar S, Schreiber MD, Polin RA, Zwass MS, Zayek MM, Gross I, Heymann MA, & Zapol WM, for the Inhaled Nitric Oxide Study Group. 1997. Inhaled nitric oxide and persistent pulmonary hypertension of the newborn. *N Engl J Med 336*:605–610.

Romand JA, Pinsky MR, Firestone L, Zar HA, & Lancaster JR Jr. 1994. Effect of inhaled nitric oxide on pulmonary hemodynamics after acute lung injury in dogs. *J Appl Physiol 76:*1356–1362.

Rosenberg AA, Kinsella JP, & Abman SH. 1995. Cerebral hemodynamics and distribution of left ventricular output during inhalation of nitric oxide. *Crit Care Med 23:* 1391–1397.

Rosenberg AA, Kennaugh JM, Moreland SG, Fashaw LM, Hale KA, Torielli FM, Abman SH, & Kinsella JP. 1997. Longitudinal follow-up of a cohort of newborn infants treated with inhaled nitric oxide for persistent pulmonary hypertension. *J Pediat 131:*70–75.

Rossaint R, Falke KJ, Lopez F, Slama K, Pison U, & Zapol WM. 1993. Inhaled nitric oxide for the adult respiratory distress syndrome. *N Engl J Med 328:*399–405.

Rossaint R, Slama K, Steudel W, Gerlach H, Pappert D, Veit S, & Falke K. 1995. Effects of inhaled nitric oxide on right ventricular function in severe acute respiratory distress syndrome. *Intensive Care Med 21:*197–203.

Rovira I, Chen TY, Winkler M, Kawai N, Bloch KD, & Zapol WM, 1994. Effects of inhaled nitric oxide on pulmonary hemodynamics and gas exchange in an ovine model of ARDS. *J Appl Physiol 76:*345–355.

Samama CM, Diaby M, Fellahi JL, Mdhafar A, Eyraud D, Arock M, Guillosson JJ, Coriat P, & Rouby JJ. 1995. Inhibition of platelet aggregation by inhaled nitric oxide in patients with acute respiratory distress syndrome. *Anesthesiology 83:*56–65.

Shah NS, Nakayama DK, Jacob TD, Nishio I, Imai T, Billiar TR, Exler R, Yousem SA, Motoyama EK, & Peitzman AB. 1994. Efficacy of inhaled nitric oxide in a porcine model of adult respiratory distress syndrome. *Arch Surg 129:*158–164.

Skimming JW, deMarco VG, & Cassin S. 1994. The effects of nitric oxide inhalation on the pulmonary circulation of preterm lambs. *Pediatr Res 37:*35–40.

Skimming JW, Bender KA, Hutchison AA, & Drummond WH. 1997. Nitric oxide inhalation in infants with respiratory distress syndrome. *J Pediat 130:*225–230.

Weitzberg E, Rudehill A, & Lundberg JM. 1993. Nitric oxide inhalation attenuates pulmonary hypertension and improves gas exchange in endotoxin shock. *Eur J Pharmacol 233:*85–94.

Wink DA, Hanbauer I, Krishna MC, DeGraff W, Gamson J, & Mitchell JB, 1993. Nitric oxide protects against cellular damage and cytotoxicity from reactive oxygen species. *Proc Natl Acad Sci USA 90:*9813–9817.

Wysocki M, Delclaux C, Roupie E, Langeron O, Liu N, Herman B, Lemaire F, & Brochard L. 1994. Additive effect on gas exchange of inhaled nitric oxide and intravenous almitrine bismesylate in the adult respiratory distress syndrome. *Intensive Care Med 20:*254–259.

Young JD, Brampton WJ, Knighton JD, & Finfer SR. 1994. Inhaled nitric oxide in acute respiratory failure in adults. *Br J Anaesth 73:*499–502.

Zapol WM, Jones R. 1987. Vascular components of ARDS: clinical pulmonary hemodynamics and morphology. *Am Rev Respir Dis 136:*471–474.

Zapol WM, & Snider MT. 1977. Pulmonary hypertension in severe acute respiratory failure. *N Engl J Med 296:*476–480.

Zapol WM, Rimar S, Gillis N, Marletta M, & Bosken CH. 1994. Nitric oxide and the lung. NHLBI workshop summary. *Am J Respir Crit Care Med 149:*1375–1380.

Zayek M, Cleveland D, & Morin FC III. 1993a. Treatment of persistent pulmonary hypertension in the newborn lamb by inhaled nitric oxide. *J Pediatrics 122:*743–750.

Zayek M, Wild L, Roberts JD, & Morin FC III. 1993b: Effect of nitric oxide on the survival rate and incidence of lung injury in newborn lambs with persistent pulmonary hypertension. *J Pediat 123:*947–952.

Ziegler JW, Ivy DD, Kinsella JP, Clark WR, & Abman SH. 1994. Dypyridamole, a cGMP phosphodiesterase inhibitor, augments inhaled nitric oxide induced pulmonary vasodilation in the ovine transitional circulation. *Pediatr Res 90A:*527.

Zobel G, Urlesberger B, Dacar D, Rodl S, Reiterer F, & Friehs I. 1997. Partial liquid ventilation combined with inhaled nitric oxide in acute respiratory failure with pulmonary hypertension in piglets. *Pediatr Res 41:*172–177.

11
Clinical Applications of Inhaled Nitric Oxide

Elie Haddad, Lesley J. Millatt, and Roger A. Johns

Introduction

The concept of using inhaled nitric oxide (NO) as a selective pulmonary vasodilator arose from the recent realization that NO is a potent endogenous vasodilator that plays a major role in modulating vascular smooth muscle tone in both the systemic and pulmonary circulations (Moncada and Higgs 1993). Importantly, inhaled NO does not dilate the systemic circulation because of its avid binding and inactivation by oxyhemoglobin upon exposure to the pulmonary circulation (Rimar and Gillis 1993). Therefore, inhaled NO holds great promise as the long-sought selective pulmonary vasodilator. Inhaled NO is distributed only to those areas of the lung that are ventilated, and dilates only those vessels directly adjacent to the ventilated alveoli (Figure 11.1). Areas of the lung that are collapsed do not receive NO and do not vasodilate. Ventilation–perfusion matching is thus preserved or improved by NO. Inhaled NO is a unique therapy, in that it is delivered directly to the desired site of action. Its action as a selective pulmonary vasodilator in pediatric and adult patients may make it an ideal therapeutic agent for numerous diseases associated with increased pulmonary vascular resistance (PVR).

Therapeutic Applications

Initial animal studies demonstrated the selective pulmonary vasodilator effect of inhaled NO (Frostell et al. 1991), which was confirmed in later clinical studies. Pepke-Zaba et al. (1991), in the first reported use in humans, showed that inhaled NO decreased PVR in eight patients with severe pulmonary hypertension. The vasodilating action of inhaled NO occurred within the first minutes of inhalation and was immediately reversible; there were no systemic effects (Pepke-Zaba et al. 1991). The clinical use of inhaled NO has now been reported in a broad spectrum of situations that may benefit from a selective increase in pulmonary flow and therefore improved arterial oxygenation and/or right heart function (Table 11.1).

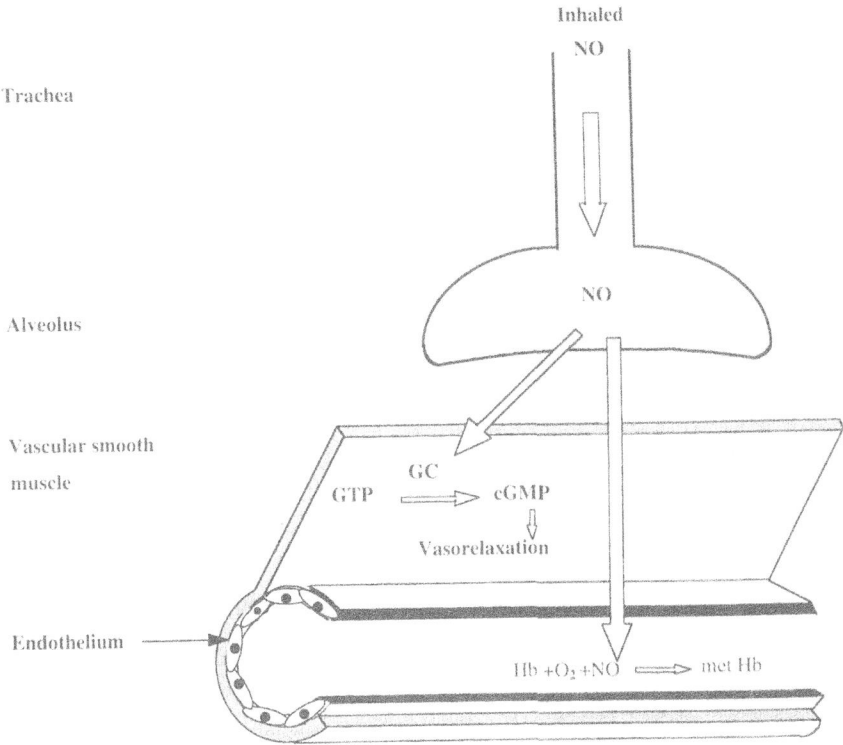

FIGURE 11.1. Selective pulmonary vasodilator action of inhaled NO. Inhaled NO enters ventilated alveoli and diffuses into the smooth muscle cells of the pulmonary blood vessels, where it activates guanylate cyclase (GC). GC converts guanosine triphosphate (GTP) to cyclic guanosine monophosphate (cGMP), resulting in pulmonary vasodilation. Inhaled NO, entering the pulmonary circulation, is rapidly inactivated by combining with hemoglobin (Hb) and O_2 to form Fe (III) methemoglobin (MetHb). In this way, the vasodilator activity of inhaled NO is restricted to the pulmonary circulation.

Adult Patients

Acute Lung Injury or Its Aggravated Form, Acute Respiratory Distress Syndrome

Acute respiratory distress syndrome (ARDS) is characterized by a sudden generalized progressive inflammation of the lung, with noncardiogenic pulmonary edema, pulmonary hypertension, a reduction in total lung compliance, and systemic hypoxemia. Rossaint et al. (1993) were the first to report the effects of short-term administration of inhaled NO (18 and 36 ppm) to patients with severe ARDS. In this study, inhaled NO improved arterial oxygenation and decreased pulmonary arterial pressure without altering systemic hemodynamics. Inert gas

TABLE 11.1. Therapeutic and diagnostic applications of inhaled NO.

Therapeutic Applications			
Adult-70 total		Pediatric-76 total 91	
Acute respiratory distress syndrome	54.3%	PPHN	23.7%
Chronic obstructive pulmonary disease, pulmonary fibrosis	7.1%	SRDS, respiratory failure	22.4%
Asthma	4.3%	Congenital heart disease	6.6%
Heart surgery	18.6%	Congenital diaphragmatic hernia	3.9%
Lung surgery	8.6%	Postoperative pulmonary	
hypertension	36.8%		
Primary pulmonary hypertension	2.9%	Acute chest syndrome	1.3%
Pulmonary embolism	2.9%	Primary pulmonary hyper-tension	2.6%
High-altitude pulmonary edema	1.4%	Secondary pulmonary hyper-tension	2.6%
Diagnostic Applications-35 total			
Reversibility studies	9		
Pulmonary diffusion capacity	3		
Exhaled NO	23		

Therapeutic applications are expressed as the percentage of total reports on the therapeutic use of inhaled NO in humans, for adult and pediatric patients, searched in the Medline database. Diagnostic applications of inhaled NO are expressed as the number of reports found for each application in the Medline database. PPHN, Persistent pulmonary hypertension of the newborn; SRDS, severe respiratory distress syndrome.

analysis in these patients revealed that NO also improved the ventilation/perfusion ratio by reducing right-to-left shunting. In the same study, a continued long-term benefit of inhaled NO at 5 to 20 ppm for 3 to 53 days was observed in some patients. There was no observed tolerance to inhaled NO in this study. Gerlach et al. (1993a) demonstrated that in ARDS patients, the dose for half-maximal effect (ED_{50}) of inhaled NO that improved oxygenation (0.1 ppm) was lower than the ED_{50} concentration of inhaled NO that reduced PVR (>2 ppm).

The beneficial effects of inhaled NO on oxygenation and PVR in ARDS patients have been confirmed in multiple studies. However, which patients will respond and the magnitude of the response remains unpredictable. Bigatello et al. (1994) reported that the reduction in PVR in response to inhaled NO was proportional to baseline abnormalities. In addition, Puybasset et al. (1995) demonstrated that the improvement in oxygenation was greater if positive end-expiratory pressure produced alveolar recruitment. However, other reports have not been able to demonstrate any variables that predict those patients who will respond to inhaled NO therapy (Rossaint et al. 1995; Lowson et al. 1996). The percentage of responders to NO may be influenced by the etiology of ARDS (Krafft et al. 1996). The response to inhaled NO may also vary in the same patient at different times (Gerlach et al. 1993b). This was also noted in a study by Rossaint et al. (1995), in which most of the nonresponders became responders on subsequent days.

Taken together, these studies have important clinical implications for the use of inhaled NO in this patient population: (1) The optimal dose of inhaled NO should be determined for each patient repetitively during the course of the treatment. (2) Low doses of inhaled NO, similar to environmental concentrations, may exert a beneficial effect on systemic oxygenation. (3) Improvement of systemic oxygenation does not necessarily correlate with a reduction in pulmonary artery pressure. (4) The long-term benefits of inhaled NO on morbidity and mortality in this patient population remain unknown and require further clinical investigation; existing studies have failed to demonstrate such benefit.

Inhaled NO can also be used, and may be more efficacious, in combination with other therapies for ARDS. Inhaled NO prevents exacerbation of pulmonary hypertension during the use of acute permissive hypercapnia in ARDS, while continuing to improve oxygenation (Roberts et al. 1992). In ARDS patients, inhaled NO combined with almitrine bismesylate enhanced the improvement in oxygenation resulting from either of these therapies alone (Payen et al. 1993). In 60% of patients, prone ventilation led to a further improvement in systemic oxygenation when combined with inhaled NO and almitrine (Jolliet et al. 1997). Both inhaled NO and inhaled prostacyclin improve oxygenation in ARDS (Pappert et al. 1995; Walmrath et al. 1996), although it is not clear whether there is any long-term benefit from combination therapy.

Rebound hypoxemia and pulmonary hypertension following withdrawal of inhaled NO may be a significant risk (Lavoie et al. 1996). Inhaled NO has to be slowly weaned, suggesting that it may suppress the endogenous NO–guanylyl cyclase pathway, as has been shown by in vitro studies (Rengasamy and Johns 1993; Ravichandran et al. 1995).

Primary Pulmonary Hypertension

Pepke-Zaba et al. (1991) showed that patients with severe primary pulmonary hypertension (PPHT) responded to inhaled NO (40 ppm). In this study, PVR decreased by 5% to 68% from baseline, with no change in systemic resistance. Snell et al. (1995) reported the successful use of continuous administration of inhaled NO (40 ppm) for 68 days as a bridge to heart-lung transplantation in a woman with end-stage PPHT. Importantly, there was no evidence of significant NO toxicity in the explanted lungs. Under light microscopy, the lungs did not show any parenchymal abnormalities.

Chronic Obstructive Pulmonary Disease and Pulmonary Fibrosis

Studies have shown that inhaled NO selectively reduces PVR and improves hypoxemia in patients with chronic obstructive pulmonary disease (COPD) (Adnot et al. 1993) and end-stage pulmonary fibrosis (Channick et al. 1994). Other authors observed a decrease in PVR but reported no improvement (Moinard et al. 1994) or a worsening (Barbera et al. 1996) in gas exchange in COPD patients. The response to inhaled NO may depend on the concentration of NO or the severity of the disease. Recently, Yoshida et al. (1997) assessed the effect of combined NO

(2 ppm) and O_2 inhalation in COPD patients. They showed that this combination decreased mean pulmonary artery pressure and increased oxygenation to a greater extent than O_2 inhalation alone. However, given the inconsistencies in responses to NO, more convincing studies are needed before inhaled NO becomes a standard therapy for chronic lung disease.

Asthma

In animal models of induced bronchospasm, inhaled NO appeared to be a powerful bronchodilator, comparable to β_2-agonists (Dupuy et al. 1992). However, inhaled NO at 80–100 ppm had no effect on airway tone in healthy volunteers and COPD patients (Hogman et al. 1993b), and only a weak bronchodilatory effect in asthmatic patients (Kacmarek et al. 1996).

Heart Surgery

Pulmonary hypertension may be observed after valvular or coronary artery bypass surgery or heart transplantation. It has been suggested that the inflammatory response associated with cardiopulmonary bypass induces endothelial dysfunction, which may contribute to the postoperative elevation in PVR (Wessel et al. 1993). This may in turn lead to right ventricular dysfunction or failure. Once again, inhaled NO demonstrated its efficacy in reducing pulmonary hypertension in such situations (Girard et al. 1993; Rich et al. 1993), thereby selectively lowering right ventricular afterload. Moreover, the use of inhaled NO may avoid or reduce the need for right ventricular assist devices and/or pharmacological circulatory support in the pre- or postoperative period in patients with pulmonary hypertension associated with right ventricular failure (George and Boscoe 1997). By alleviating pulmonary hypertension, inhaled NO may also be useful in increasing the number of patients eligible for left ventricular assist devices (Hare et al. 1997).

Lung Surgery

After lung transplantation, transient graft dysfunction may occur and may be complicated by a reperfusion syndrome manifested by pulmonary hypertension and edema, and by right ventricular and respiratory failure. Adatia et al. (1994) first suggested that inhaled NO has great potential in the treatment of this manifestation of lung graft dysfunction. Date et al. (1996) reported a retrospective study among 243 patients who underwent lung transplantation. Of these patients, 13.2% (32 patients) developed an immediate severe allograft dysfunction, with an arterial oxygen tension/inspired oxygen fraction ratio of less than 150. Inhaled NO (30–60 ppm) was used to treat 15 of the 32 patients for 15 to 217 hours. Nitric oxide administration resulted in a rapid improvement of systemic oxygenation and pulmonary arterial pressure, without affecting systemic pressure and cardiac output. However, in this report, inhaled NO did not significantly improve the outcome in terms of requirement for extracorporeal membrane oxygenation (ECMO) duration of mechanical ventilation, airway complications, and mortality.

Pediatric Patients

Persistent Pulmonary Hypertension of the Newborn, Severe Respiratory Distress Syndrome in Premature Infants, and Hypoxic Respiratory Failure in Term Infants

Persistent pulmonary hypertension of the newborn (PPHN) is characterized by increased PVR after birth, with anatomic right-to-left shunting through a patent foramen ovale and/or patent ductus arteriosus. This results in a marked hypoxemia associated with high mortality and morbidity. Conventional therapy has included hyperventilation, alkalosis, hyperoxia, inotropic support, intravenous dilators, and ECMO. ECMO is an expensive and invasive technique associated with significant morbidity. Studies have indicated that inhaled NO can be used as a highly effective therapy for PPHN (Roberts et al. 1992; Kinsella et al. 1993). A prospective randomized multicenter study involving 58 full-term infants with PPHN associated with severe hypoxemia assessed the effect of inhaled NO on systemic oxygenation (Roberts et al. 1997). In this study, inhaled NO significantly improved systemic oxygenation in 53% of treated neonates. In addition, the need for ECMO was significantly lower in the NO-treated group (40%) than in the control group (71%). The longest period of NO administration was 8.5 days, with no side effects observed; NO doses started at 80 ppm and were successfully decreased within 2 days to 20 ppm.

Severe respiratory distress syndrome (SRDS) occurs in premature infants and is characterized by a primary physiological abnormality consisting of a deficiency or dysfunction of surfactant. Surfactant dysfunction decreases lung compliance and lung volume, causing abnormal gas exchange. Some neonates with severe hyaline membrane disease also have marked pulmonary hypertension, which is associated with severe lung disease and poor outcome (Walther et al. 1992). In this setting, pulmonary hypertension may be due in part to the mechanical effects of low lung volume. In addition, it seems that pulmonary vasoconstriction contributes to the increased PVR observed in some premature infants. Abman et al. (1993) first reported the beneficial effect of inhaled NO on PVR and oxygenation in a premature infant with sepsis associated with SRDS. Skimming et al. (1997) recently reported the effects of administering 5 or 20 ppm inhaled NO to 23 infants with SRDS. During the 15-minute treatment period, all infants showed a significant increase in arterial blood oxygenation, which was independent of the dose of inhaled NO additional clinical trials are required to define the potential role of inhaled NO in the management of this disease.

A recent prospective multicenter study evaluated whether inhaled NO reduces mortality or the need for ECMO in 236 infants born at or near term with hypoxic respiratory failure (Stork et al. 1997). The primary outcome (defined as death by 120 days of age or the initiation of ECMO therapy) was significantly lower (46%)

in the infants treated with inhaled NO than in the control group (64%). However, there were no differences in secondary outcomes (overall mortality, length of hospitalization, days of respiratory support, air leakage, or bronchopulmonary dysplasia) between the NO-treated and the control groups. The conclusion of this study was that inhaled NO therapy was safe, well tolerated, and easy to administer and reduced the need for ECMO in these infants.

Postoperative Pulmonary Hypertension

Pulmonary hypertension may occur in pediatric patients after surgery for congenital heart defects, congenital diaphragmatic hernia, and heart and/or lung transplantation. In this context, it is a major source of postoperative morbidity and mortality. Conventional therapy for this pulmonary hypertension includes hyperventilation, supplemental oxygen, inotropic support, and vasodilators. However, pharmacological vasdilators are not selective for the pulmonary vasculature and often cause systemic hypotension. Many reports have shown inhaled NO to be efficient in selectively reducing postoperative pulmonary hypertension (Haydar et al. 1992; Frostell et al. 1993; Sellden et al. 1993; Beghetti et al. 1995; Goldman et al. 1996). However, controlled clinical trials have not yet been performed to rigorously evaluate this therapy.

Congenital Heart Disease

Many forms of congenital heart disease are associated with progressive pulmonary hypertension and a reversal of the left-to-right shunt (the latter resulting from a defect at the atrial, ventricular, or aortopulmonary level), leading to cyanosis and chronic hypoxemia. Inhaled NO has been shown to significantly reduce PVR and improve oxygenation in infants with congenital heart disease (Roberts et al. 1993; Winberg et al. 1994). In addition, Ichida et al. (1997) recently examined the hemodynamic effects of inhaled NO (20 ppm) and a newly developed prostacyclin analogue, beraprost, in children with congenital heart disease complicated by pulmonary hypertension. They showed that beraprost and NO achieved similar reductions in PVR, and that combined beraprost-NO therapy had an additive effect on pulmonary vasodilation. They concluded that this combination may provide a new therapy for pulmonary hypertension in patients with congenital heart disease.

Congenital Diaphragmatic Hernia

Congenital diaphragmatic hernia may be associated with hypoxemia induced by reversible pulmonary hypertension. Finer et al. (1994) demonstrated an improvement in arterial oxygenation in five of seven infants with congenital diaphragmatic hernia who were treated with inhaled NO. However, in a controlled multicenter study involving 53 newborns with congenital diaphragmatic hernia, inhaled NO failed to reduce the need for ECMO or the mortality associated with this disease (Finer et al. 1997). Thus, the efficacy of inhaled NO in improving the outcome of patients with congenital diaphragmatic hernia is not yet proven.

Acute Chest Syndrome

Patients with sickle cell disease may develop acute chest syndrome, characterized by fever, chest pain, and radiographic evidence of pulmonary infiltrate, effusion, or edema. There is evidence that pulmonary vascular occlusion with ischemia and infarction plays an important role in this syndrome. Atz and Wessel (1997) recently reported the successful use of inhaled NO to decrease PVR and improve arterial oxygenation in two children with acute chest syndrome. More extensive studies may lead to the use of inhaled NO as an additional therapy for such patients, diminishing the need for more traditional invasive therapies.

Diagnostic Applications

Evaluation of Pulmonary Hypertension Reversibility

The decision to operate on a patient with pulmonary hypertension may depend on the reversibility of pulmonary vasoconstriction. In adult candidates for heart transplantation and in children with congenital heart disease, it is often desirable to determine the reversibility of the pulmonary hypertension during preoperative cardiac catheterization, to aid prediction of the postoperative PVR. Hyperoxic breathing has been used to dilate the pulmonary vasculature, as have intravenous vasodilators (prostacyclin, tolazoline, prostaglandin E_1, or sodium nitroprusside), which are nonselective and dilate both the systemic and the pulmonary circulations. Inhaled NO has been shown to provide a useful and safe means of selectively evaluating the pulmonary vasodilatory capacity of patients with heart failure (Semigran et al. 1994) and infants with congenital heart disease (Roberts et al. 1993). The latter study demonstrated that inhaled NO has a greater selective pulmonary vasodilator effect than hyperoxic breathing. Adatia et al. (1995) tested the reversibility of pulmonary hypertension in 11 patients, using inhaled NO (80 ppm) to establish the indications for heart or heart–lung transplantation. Inhaled NO was also used after surgery, in neonates with congenital heart disease and proximal pulmonary artery hypertension or marked cyanosis, as a short challenge to differentiate between pulmonary vasoconstriction and anatomical obstruction of pulmonary flow (Adatia et al. 1996). However, pulmonary edema has been reported in candidates for heart transplantation in whom inhaled NO was used for preoperative screening of the reversibility of pulmonary hypertension (Bocchi et al. 1994). This may therefore limit the utility of inhaled NO for such evaluations.

Determination of the Diffusion Capacity of the Alveolocapillary Membrane

Carbon monoxide (CO) has traditionally been used to measure the pulmonary diffusion capacity (D_L) because of it has a very high affinity for hemoglobin and because its uptake is almost independent of capillary blood flow. However, inhaled NO seems to be a better tool than CO for measuring D_L because of its greater affinity for hemoglobin and its higher water solubility at body temperature (Borland and Higenbottam 1989).

Exhaled NO

Nitric oxide production from the nasal passages and lungs of normal human subjects can be detected in exhaled gas, and may be used in the assessment of respiratory tract diseases. Nitric oxide concentrations in end-expiratory "alveolar" gas in healthy subjects are very low, in the range of 8 to 20 parts per billion (Borland et al. 1993; Persson et al. 1993). However, during oral breathing, NO concentrations in mixed exhaled air are higher than in the trachea (Gerlach et al. 1994; Lundberg et al. 1994b; Lundberg et al. 1995). This suggests that the upper airways, in particular the sinuses, are the main source of exhaled NO. Moreover, Lundberg et al. (1995) showed that endogenous NO produced by the upper airways, nasal passages, and sinuses may reach concentrations commonly used in the therapeutic administration of NO (>10 ppm). They suggested that this endogenous inhaled NO may be important in controlling normal pulmonary tone. In addition, the reduction of nitrite to NO in the acidic fluid (Benjamin et al. 1994; Lundberg et al. 1994a) and the mucosa (Rachmilewitz et al. 1994) of the stomach produces intestinal NO regurgitation. Accurate measurement of pulmonary NO formation may therefore be problematic because of potential contamination from gastric and upper airway NO production. However, increased exhaled NO has been found in patients with bronchiectasis (Kharitonov et al. 1995a), asthma (Kharitonov et al. 1994; Nelson et al. 1997), and upper respiratory tract infections (Kharitonov et al. 1995b). In contrast, patients with PPHT or Kartagener's syndrome, a ciliary disease in which mucus engorges the paranasal sinuses, have a decreased exhaled NO (Cremona et al. 1994; Lundberg et al. 1994b).

Despite their use for several years, measurement techniques for exhaled NO have still not been standardized. Different techniques have been reported, such as mixed exhaled NO, peak exhaled NO, and NO production over time in patients blowing into tubes or bags, with or without nose clips and with varying amounts of oropharyngeal pressure (Persson et al. 1993; Massaro et al. 1996; Silkoff et al. 1997). A standardized breath technique may be useful in the future as a noninvasive tool for determining exhaled NO in patients with respiratory diseases. This could be used as a marker of the severity of the disease or to monitor treatment efficacy (Baraldi et al. 1997; Nelson et al. 1997).

Toxicity

Currently there is little evidence to suggest that low concentrations of inhaled NO are toxic. Animal studies in which mice were exposed to 2.4 ppm NO for 2 years revealed no significant side effects (Nakajima et al. 1980). Furthermore, clinical studies that exposed humans to inhaled NO at low concentrations (5–20 ppm) for up to 53 days did not demonstrate any toxic effects (Rossaint et al. 1993). However, NO and higher oxides of nitrogen were responsible for a death that occurred during anesthesia as a result of contamination of nitrous oxide cylinders (Clutton-Brock 1967). Death was due to acute pulmonary injury, methemoglobinemia, and asphyxia and occurred at an estimated NO concentration of 10,000 ppm.

In the presence of oxygen, NO is rapidly oxidized to nitrogen dioxide (NO_2), which is known to be highly toxic (Rimar and Gillis 1993). The rate of oxidation of NO to NO_2 is dependent on the initial concentration of NO and the inspired O_2 concentration (Greenbaum et al. 1967). In a study involving 123 patients inhaling 10–80 ppm NO for several hours, NO_2 levels only exceeded 3 ppm in 4 patients (Wessel et al. 1994). However, even low concentrations of NO_2 (2 ppm) have been shown to be extremely toxic and to cause changes in lung histopathology, including loss of cilia, hypertrophy, and focal hyperplasia in the epithelium of terminal bronchioles in the rat (Evans et al. 1972). It is therefore important to keep NO_2 formation during inhaled NO therapy to a minimum, by the use of adequate filters or scavengers and high-quality gas mixtures.

The most commonly reported side effect of inhaled NO therapy is methemoglobinemia. Inhaled NO combines with hemoglobin in red blood cells and is converted to methemoglobin (Figure 11.1), reducing the capacity of hemoglobin for oxygen. A study in normal adults showed that 3 hours of inhalation of up to 128 ppm NO (much higher than clinical doses) did not result in clinically significant methemoglobinemia (Young et al. 1994). However, there is evidence to suggest that more prolonged NO inhalation may result in significant levels of methemoglobinemia in some patients (Adatia et al. 1994; Hovenga et al. 1996).

Concern exists regarding the potential for toxic effects of chronic low doses of inhaled NO, such as surfactant inactivation (Haddad et al. 1994), lung injury, and inflammation (Kooy et al. 1995). These effects are believed to be caused by the reaction of NO with superoxide, leading to the formation of peroxynitrite. The formation of this highly toxic strong oxidant molecule may be a significant side effect of inhaled NO. Indeed, tissue superoxide formation is known to be increased in a number of situations in which inhaled NO may be used, such as sepsis, inflammatory diseases, and ischemia-reperfusion injury during lung graft dysfunction (Freeman 1994). However, the progression of the underlying disease itself makes it difficult to identify potential NO-related toxicities. Detailed studies of lung pathologic states and pulmonary function, particularly with chronic exposure to inhaled NO, are lacking. In addition, potential mutagenic effects (Nguyen et al. 1992) and specific effects on cell function independent of structural changes need to be evaluated. The ability of NO to inhibit enzyme activity via stimulation of ADP-ribosylation (Dimmeler et al. 1992) and to inactivate or alter the function of a number of iron- and heme-based proteins, including cyclooxygenase, lipoxygenase (Kanner et al. 1992), and oxidative cytochromes, makes such concerns significant, particularly with regard to long-term inhalation therapy.

Inhaled NO-related toxicity should have a special consideration in premature and term newborns, as there are several specific concerns related to these patients: (1) Inhaled NO has been shown to prolong bleeding time in human volunteers (Hogman et al. 1993a). It is therefore possible that inhaled NO may increase the

TABLE 11.2. Toxicity of inhaled NO.

		Toxicity
Methemoglobinemia	11	Kinsella (1993), Adatia (1994), Ahluwalia (1994), Lonnqvist (1994), Breuer (1995), Heal (1995), Date (1996), Hovenga (1996), Dotsch (1997), Nakajima (1997)
Oxygen desaturation	1	Tulleken (1997)
Systemic hypotension	1	Henrichsen (1996)
Pulmonary edema	1	Bocchi (1994)
Impairment of neutrophil respiratory burst	1	Gessler (1996)

Adverse effects are expressed as the number of reports of each toxic effect found among the 146 citations of the use of inhaled NO in humans in the Medline database.

preexisting risk of intracranial hemorrhage in sick premature newborns. (2) A rapid reversal of predominantly right-to-left shunting across the ductus arteriosus by NO could markedly increase pulmonary blood flow, thus leading to pulmonary edema, a worsening of lung injury, and, potentially, severe systemic hypotension (Henrichsen et al. 1996). (3) The premature lung may have fewer antioxidant defenses, making it more sensitive to potential adverse effects of NO and reactive oxygen species. (4) Neonates are at increased risk of developing methemoglobinemia because of their lower levels of methemoglobin reductase activity (Nilsson et al. 1990). Although studies indicate that chronic low doses of inhaled NO do not cause significant methemoglobinemia in the newborn (Roberts et al. 1992, 1997), a recent clinical case described the occurrence of methemoglobinemia in a newborn with PPHN after 2 days of administration of 20–80 ppm inhaled NO (Nakajima et al. 1997).

In summary, there are a limited number of reports of inhaled NO-related toxicity in humans (Table 11.2). However, in most clinical situations in which inhaled NO therapy may be appropriate, the benefits are likely to outweigh the risks.

Delivery, Monitoring, and Dosage

Inhaled NO therapy must provide a steady, accurate NO concentration at appropriate flow rates, while keeping NO_2 formation to a minimum. Stock cylinders should contain medical-grade NO in an inert carrier gas such as nitrogen, since pure NO is corrosive and spontaneously degrades into highly toxic nitrogen oxides. Nitric oxide should ideally be mixed with the carrier gas (usually oxygen and/or air) immediately before inhalation to decrease the contact time between NO and oxygen. This minimizes the generation of toxic NO_2 (Foubert et al. 1992).

The concentrations NO and NO_2 must be continuously monitored on the inspiration limb of the ventilator circuit. Samples of gas may be assayed by chemiluminescence, infrared analysis, Raman spectroscopy, and electrochemical methods. Chemiluminescence is the most accurate method, but it is expensive. Electrochemical analyzers are less expensive and probably more suitable for routine clinical use. Commercial NO delivery systems with NO flow regulators and

NO/NO$_2$ monitoring are now becoming available. Inhaled NO can be administered to patients during mechanical ventilation or, less often, in spontaneously breathing patients by the use of a close-fitting, nonrebreathing mask.

Proposed courses of treatment of patients with inhaled NO vary widely, from 0.1 to 40 ppm for periods of a few hours to a few weeks (Gerlach et al. 1993a; Rossaint et al. 1993; Bigatello et al. 1994; Puybasset et al. 1994). In some cases (Gerlach et al. 1993a), long-term inhalation (up to 2 weeks) of NO concentrations as low as 0.06 to 0.25 ppm improved systemic oxygenation. In contrast, higher doses (10 ppm) have been shown to increase intrapulmonary right-to-left shunting, thus decreasing systemic oxygenation (Gerlach et al. 1993b). Therefore, the minimum effective inhaled NO concentrations should be determined for each patient.

Preventing Toxicity of Inhaled NO

There is still insufficient information to come to a definitive statement on the safety of inhaled NO, and further investigations are needed to study potential side effects. In the meantime, a variety of measures can be taken to ensure safe clinical use and to avoid the contamination of the immediate environment with the gas.

Ethical Considerations

Until the risk/benefit ratio has been established, it would seem logical to use inhaled NO only in a clinical trial that meets the requirements of the Declaration of Helsinki and the rules for Good Clinical Practice, and in which the approval of the hospital ethics committee and the informed consent of the patient or the family are obtained.

Environmental Safety

Only adequate NO delivery systems with feedback controls and alarm systems should be used. Nitric oxide and NO$_2$ should be continuously monitored to ensure patient safety. Finally, methemoglobin levels in the plasma should be measured daily. Nitric oxide cylinders should be restrained to prevent tilting, and repeated NO measurements should be performed in the patient's room. In patients undergoing mechanical ventilation the use of cuffed endotracheal tubes or close-fitting uncuffed tubes is recommended to avoid environmental contamination. Whenever the patient is disconnected from the ventilator circuit, the outlet of the ventilator should be plugged. When needed, endotracheal suction should be performed through a suction bullet or a bronchoscopy adaptor. In poorly ventilated units, exhaled gases and gases spilled from the release valve of the ventilator should be scavenged to avoid buildup in the atmosphere, Squire et al. (1996) recently compared three methods of scavenging NO and NO$_2$. Both soda lime and activated charcoal were found to be ineffective, whereas a commercially available industrial filter (ABEK HgCONO-P3) efficiently scavenged these gases.

Conclusion

Inhaled NO shows great potential as a therapy for a number of disease states in which elevated pulmonary vascular resistance is a problem. Nevertheless, before allowing inhaled NO to be widely used in daily clinical practice, long-term toxicological evaluation of inhaled NO therapy is required. In addition, the benefits of inhaled NO to overall patient outcome have not been proven in terms of mortality or morbidity (Girard et al. 1992; Adatia et al. 1993, 1995; Rich et al. 1993). Thus, the use of inhaled NO should still be be thought of as experimental, and not as a standard clinical therapeutic or diagnostic tool. However, significant evidence suggests that inhaled NO has a promising future in the clinical arena.

References

Abman SH, Kinsella JP, Schaffer MS, & Wilkening RB. 1993. Inhaled nitric oxide in the management of a premature newborn with severe respiratory distress and pulmonary hypertension. *Pediatrics 92:*606–609.

Adatia I, Thompson J, Landzberg M, & Wessel DL. 1993. Inhaled nitric oxide in chronic obstructive lung disease. *Lancet 341:*307–308.

Adatia I, Lillehei C, Arnold JH, Thompson JE, Palazzo R, Fackler JC, & Wessel DL. 1994. Inhaled nitric oxide in the treatment of postoperative graft dysfunction after lung transplantation, *Ann Thorac Surg 57:*1311–1318.

Adatia I, Perry S, Landzberg M, Moore P, Thompson JE, & Wessel DL. 1995. Inhaled nitric oxide and hemodynamic evaluation of patients with pulmonary hypertension before transplantation. *J Am Coll Cardiol 25:*1656–1664.

Adatia I, Atz AM, Jonas RA, & Wessel DL. 1996. Diagnostic use of inhaled nitric oxide after neonatal cardiac operations. *J Thorac Cardiovasc Surg 112:*1403–1405.

Adnot S, Kouyoumdjian C, Defouilloy C, Andrivet P, Sediame S, Herigault R, & Fratacci MD. 1993. Hemodynamic and gas exchange responses to infusion of acetylcholine and inhalation of nitric oxide in patients with chronic obstructive lung disease and pulmonary hypertension. *Am Rev Respir Dis 148:*310–316.

Ahluwalia JS, Kelsall AW, Raine J, Rennie JM, Mahmood M, Oduro A, Latimer R, Pickett J, & Higenbottam TW. 1994. Safety of inhaled nitric oxide in premature neonates. *Acta Paediatr 83:*347–348.

Atz AM, & Wessel DL. 1997. Inhaled nitric oxide in sickle cell disease with acute chest syndrome. *Anesthesiology 87:*988–990.

Baraldi E, Azzolin NM, Zanconato S, Dario C, & Zacchello F. 1997. Corticosteroids decrease exhaled nitric oxide in children with acute asthma. *J Pediatr 131:*381–385.

Barbera JA, Roger N, Roca J, Rovira I, Higenbottam TW, & Rodriguez-Roisin R. 1996. Worsening of pulmonary gas exchange with nitric oxide inhalation in chronic obstructive pulmonary disease. *Lancet 347:*436–440.

Beghetti M, Habre W, Friedli B, & Berner M. 1995. Continuous low dose inhaled nitric oxide for treatment of severe pulmonary hypertension after cardiac surgery in paediatric patients. *Br Heart J 73*:65–68.

Benjamin N, O'Driscoll F, Dougall H, Duncan C, Smith L, Golden M, & McKenzie H. 1994. Stomach NO synthesis. *Nature 368*:502.

Bigatello LM, Hurford WE, Kacmarek RM, Roberts JD Jr, & Zapol WM. 1994. Prolonged inhalation of low concentrations of nitric oxide in patients with severe adult respiratory distress syndrome. Effects on pulmonary hemodynamics and oxygenation. *Anesthesiology 80*:761–770.

Bocchi EA, Bacal F, Auler Junior JO, Carmone MJ, Bellotti G, & Pileggi F. 1994. Inhaled nitric oxide leading to pulmonary edema in stable severe heart failure. *Am J Cardiol 74*:70–72.

Borland C, Cox Y, & Higenbottam T. 1993. Measurement of exhaled nitric oxide in man. *Thorax 48*:1160–1162.

Borland CD, & Higenbottam TW, 1989. A simultaneous single breath measurement of pulmonary diffusing capacity with nitric oxide and carbon monoxide. *Eur Respir J 2*:56–63.

Breuer J, von Brenndorff CI, Baden W, Sieverding L, Steil E, Haberle L, Fenchel G, & Apitz J. 1995. Improvement of perioperative hemodynamics and gas exchange by inhalation of nitric oxide in children with congenital cardiopulmonary defects. *Z Kardiol 84*:1009–1017.

Channick RN, Hoch RC, Newhart JW, Johnson FW, & Smith CM. 1994. Improvement in pulmonary hypertension and hypoxemia during nitric oxide inhalation in a patient with end-stage pulmonary fibrosis. *Am J Respir Crit Care Med 149*:811–814.

Clutton-Brock J. 1967. Two cases of poisoning by contamination of nitrous oxide with higher oxides of nitrogen during anaesthesia. *Br J Anaesth 39*:388–392.

Cremona G, Higenbottam T, Borland C, & Mist B. 1994. Mixed expired nitric oxide in primary pulmonary hypertension in relation to lung diffusion capacity. *QJM 87*:547–551.

Date H, Triantafillou AN, Trulock EP, Pohl MS, Cooper JD, & Petterson GA. 1996. Inhaled nitric oxide reduces human lung allograft dysfunction. *J Thorac Cardiovasc Surg 111*:913–919.

Dimmeler S, Lottspeich F, & Brune B. 1992. Nitric oxide causes ADP-ribosylation and inhibition of glyceraldehyde-3-phosphate dehydrogenase. *J Biol Chem 267*:16771–16774.

Dotsch J, Demirakca S, Hamm R, Knothe C, Bauer J, Kuhl PG, & Rascher W. 1997. Extracorporeal circulation increases nitric oxide-induced methemoglobinemia in vivo and in vitro. *Crit Care Med 25*:1153–1158.

Dupuy PM, Shore SA, Drazen JM, Frostell C, Hill WA, & Zapol WM. 1992. Bronchodilator action of inhaled nitric oxide in guinea pigs. *J Clin Invest 90*:421–428.

Evans MJ, Stephens RJ, Cabral LJ, & Freeman G. 1972. Cell renewal in the lungs of rats exposed to low levels of NO_2. *Arch Environ Health 24*:108–188.

Finer NN, Etches PC, Kamstra B, Tiemey AJ, Peliowski A, & Ryan CA. 1994. Inhaled nitric oxide in infants referred for extracorporeal membrane oxygenation: dose response. *J Pediatr 124*:302–308.

Finer N, Solimano A, Germain F, Walker R, Ramirez AM, Singhal N, Bourcier L, Fajardo C, Cook V, Kirpalani H, Monkman S, Johnston A, Mullahoo K, Peliowski A, Etches P, Kamstra B, Wearden M, Gomez M, & Moon, Y. 1997. Inhaled nitric oxide and hypoxic respiratory failure in infants with congenital diaphragmatic hernia. *Pediatrics 99*:838–845.

Foubert L, Fleming B, Latimer R, Jonas M, Oduro A, Borland C, & Higenbottam T. 1992. Safety guidelines for use of nitric oxide. *Lancet 339:*1615–1616.

Freeman B. 1994. Free radical chemistry of nitric oxide. Looking at the dark side. *Chest 105:*79S–84S.

Frostell C, Fratacci MD, Wain JC, Jones R, & Zapol WM. 1991. Inhaled nitric oxide. A selective pulmonary vasodilator reversing hypoxic pulmonary vasoconstriction [published erratum appears in *Circulation* 1991 Nov;*84*(5):2212]. *Circulation 83:*2038–2047.

Frostell CG, Lonnqvist PA, Sonesson SE, Gustafsson LE, Lohr G, & Noack G. 1993. Near fatal pulmonary hypertension after surgical repair of congenital diaphragmatic hernia. Successful use of inhaled nitric oxide. *Anaesthesia 48:*679–683.

George SJ, & Boscoe MJ. 1997. Inhaled nitric oxide for right ventricular dysfunction following cardiac transplantation. *Br J Clin Pract 51:*53–55.

Gerlach H, Pappert D, Lewandowski K, Rossaint R, & Falke KJ. 1993a. Long-term inhalation with evaluated low doses of nitric oxide for selective improvement of oxygenation in patients with adult respiratory distress syndrome. *Intensive Care Med 19:*443–449.

Gerlach H, Rossaint R, Pappert D, & Falke KJ. 1993b. Time-course and dose-response of nitric oxide inhalation for systemic oxygenation and pulmonary hypertension in patients with adult respiratory distress syndrome. *Eur J Clin Invest 23:*499–502.

Gerlach H, Rossaint R, Pappert D, Knorr M, & Falke KJ. 1994. Autoinhalation of nitric oxide after endogenous synthesis in nasopharynx. *Lancet 343:*518–519.

Gessler P, Nebe T, Birle A, Mueller W, & Kachel W. 1996. A new side effect of inhaled nitric oxide in neonates and infants with pulmonary hypertension: functional impairment of the neutrophil respiratory burst. *Intensive Care Med 22:*252–258.

Girard C, Lehot JJ, Pannetier JC, Filley S, Ffrench P, & Estanove S. 1992. Inhaled nitric oxide after mitral valve replacement in patients with chronic pulmonary artery hypertension. *Anesthesiology 77:*880–883.

Girard C, Durand PG, Vedrinne C, Pannetier JC, Estanove S, Falke K, Adnot S, & Lemaire F. 1993. Inhaled nitric oxide for right ventricular failure after heart transplantation. *J Cardiothorac Vasc Anesth 7:*481–485.

Goldman AP, Delius RE, Deanfield JE, de Leval MR, Sigston PE, & Macrae DJ. 1996. Nitric oxide might reduce the need for extracorporeal support in children with critical postoperative pulmonary hypertension. *Ann Thorac Surg 62:*750–755.

Greenbaum R, Bay J, Hargreaves MD, Kain ML, Kelman GR, Nunn JF, Prys-Roberts C, & Siebold K. 1967. Effects of higher oxides of nitrogen on the anaesthetized dog. *Br J Anaesth 39:*393–404.

Haddad IY, Crow JP, Hu P, Ye Y, Beckman J, & Matalon S. 1994. Concurrent generation of nitric oxide and superoxide damages surfactant protein A. *Am J Physiol 267:*L242–L249.

Hare JM, Shernan SK, Body SC, Graydon E, Colucci WS, & Couper GS. 1997. Influence of inhaled nitric oxide on systemic flow and ventricular filling pressure in patients receiving mechanical circulatory assistance. *Circulation 95:*2250–2253.

Haydar A, Malhere T, Mauriat P, Journois D, Pouard P, Denis N, Lefebvre D, Safran D, & Vouhe P. 1992. Inhaled nitric oxide for postoperative pulmonary hypertension in patients with congenital heart defects. *Lancet 340:*1545.

Heal CA, & Spencer SA. 1995. Methaemoglobinaemia with high-dose nitric oxide administration. *Acta Paediatr 84:*1318–1319.

Henrichsen T, Goldman AP, & Macrae DJ. 1996. Inhaled nitric oxide can cause severe systemic hypotension. *J Pediatr 129:*183.

Hogman M, Frostell C, Arnberg H, & Hedenstierna G. 1993a. Bleeding time prolongation and NO inhalation. *Lancet 341:*1664–1665.

Hogman M, Frostell CG, Hedenstrom H, & Hedenstierna G. 1993b. Inhalation of nitric oxide modulates adult human bronchial tone. *Am Rev Respir Dis 148:*1474–1478.

Hovenga S, Koenders ME, van der Werf TS, Moshage H, & Zijlstra JG. 1996. Methaemoglobinaemia after inhalation of nitric oxide for treatment of hydrochlorothiazide-induced pulmonary oedema. *Lancet 348:*1035–1036.

Ichida F, Uese K, Hashimoto I, Hamamichi Y, Tsubata S, Fukahara K, Murakami A, & Miyawaki T. 1997. Acute effect of oral prostacyclin and inhaled nitric oxide on pulmonary hypertension in children. *J Cardiol 29:*217–224.

Jolliet P, Bulpa P, Ritz M, Ricou B, Lopez J, & Chevrolet JC. 1997. Additive beneficial effects of the prone position, nitric oxide, and almitrine bismesylate on gas exchange and oxygen transport in acute respiratory distress syndrome. *Crit Care Med 25:*786–794.

Kacmarek RM, Ripple R, Cockrill BA, Bloch KJ, Zapol WM, & Johnson DC. 1996. Inhaled nitric oxide. A bronchodilator in mild asthmatics with methacholine-induced bronchospasm. *Am J Respir Crit Care Med 153:*128–135.

Kanner J, Harel S, & Granit R. 1992. Nitric oxide, an inhibitor of lipid oxidation by lipoxygenase, cyclooxygenase and hemoglobin. *Lipids 27:*46–49.

Kharitonov SA, Yates D, Robbins RA, Logan-Sinclair R, Shinebourne EA, & Barnes PJ. 1994. Increased nitric oxide in exhaled air of asthmatic patients. *Lancet 343:*133–135.

Kharitonov SA, Wells AU, O'Connor BJ, Cole PJ, Hansell DM, Logan-Sinclair RB, & Barnes, PJ. 1995a. Elevated levels of exhaled nitric oxide in bronchiectasis. *Am J Respir Crit Care Med 151:*1889–1893.

Kharitonov SA, Yates D, & Barnes PJ. 1995b. Increased nitric oxide in exhaled air of normal human subjects with upper respiratory tract infections. *Eur Respir J 8:*295–297.

Kinsella JP, Neish SR, Ivy DD, Shaffer E, & Abman SH. 1993. Clinical responses to prolonged treatment of persistent pulmonary hypertension of the newborn with low doses of inhaled nitric oxide. *J Pediatr 123:*103–108.

Kooy NW, Royall JA, Ye YZ, Kelly DR, & Beckman JS. 1995. Evidence for in vivo peroxynitrite production in human acute lung injury. *Am J Respir Crit Care Med 151:*1250–1254.

Krafft P, Fridrich P, Fitzgerald RD, Koc D, & Steltzer H. 1996. Effectiveness of nitric oxide inhalation in septic ARDS. *Chest 109:*486–493.

Lavoie A, Hall JB, Olson DM, & Wylam ME. 1996. Life-threatening effects of discontinuing inhaled nitric oxide in severe respiratory failure. *Am J Respir Crit Care Med 153:*1985–1987.

Lonnqvist PA, Winberg P, Lundell B, Sellden H, & Olsson GL. 1994. Inhaled nitric oxide in neonates and children with pulmonary hypertension. *Acta Paediatr 83:*1132–1136.

Lowson SM, Rich GF, McArdle PA, Jaidev J, & Morris GN. 1996. The response to varying concentrations of inhaled nitric oxide in patients with acute respiratory distress syndrome. *Anesth Analg 82:*574–581.

Lundberg JO, Weitzberg E, Lundberg JM, & Alving K. 1994a. Intragastric nitric oxide production in humans: measurements in expelled air. *Gut 35:*1543–1546.

Lundberg JO, Weitzberg E, Nordvall SL, Kuylenstierna R, Lundberg JM, & Alving K. 1994b. Primarily nasal origin of exhaled nitric oxide and absence in Kartagener's syndrome. *Eur Respir J 7:*1501–1504.

Lundberg JO, Farkas-Szallasi T, Weitzberg E, Rinder J, Lidholm J, Anggaard A, Hokfelt T, Lundberg JM, & Alving K. 1995. High nitric oxide production in human paranasal sinuses. *Nat Med 1:*370–373.

Massaro AF, Mehta S, Lilly CM, Kobzik L, Reilly JJ, & Drazen JM. 1996. Elevated nitric oxide concentrations in isolated lower airway gas of asthmatic subjects. *Am J Respir Crit Care Med 153:*1510–1514.

Moinard J, Manier G, Pillet O, Castaing Y. 1994. Effect of inhaled nitric oxide on hemodynamics and VA/Q inequalities in patients with chronic obstructive pulmonary disease. *Am J Respir Crit Care Med 149:*1482–1487.

Moncada S, & Higgs A. 1993. The L-arginine-nitric oxide pathway. *N Engl J Med 329:*2002–2012.

Nakajima T, Oda H, Kusumoto S, & Nogami H. 1980. Biological effects of nitrogen dioxide and nitric oxide. In: Lee SD, ed. *Nitrogen Oxides and Their Effects on Health.* Ann Arbor Mich: Ann Arbor Science Publishers, pp 121–141.

Nakajima W, Ishida A, Arai H, & Takada G. 1997. Methaemoglobinaemia after inhalation of nitric oxide in infant with pulmonary hypertension. *Lancet 350:*1002–1003.

Nelson BV, Sears S, Woods J, Ling CY, Hunt J, Clapper LM, & Gaston B. 1997. Expired nitric oxide as a marker for childhood asthma. *J Pediatr 130:*423–427.

Nguyen T, Brunson D, Crespi CL, Penman BW, Wishnok JS, & Tannenbaum SR. 1992. DNA damage and mutation in human cells exposed to nitric oxide in vitro. *Proc Natl Acad Sci USA 89:*3030–3034.

Nilsson A, Engberg G, Henneberg S, Danielson K, & De Verdier CH. 1990. Inverse relationship between age-dependent erythrocyte activity of methaemoglobin reductase and prilocaine-induced methaemoglobinaemia during infancy. *Br J Anaesth 64:*72–76.

Pappert D, Busch T, Gerlach H, Lewandowski K, Radermacher P, & Rossaint R. 1995. Aerosolized prostacyclin versus inhaled nitric oxide in children with severe acute respiratory distress syndrome. *Anesthesiology 82:*1507–1511.

Payen DM, Gatecel C, & Plaisance P. 1993. Almitrine effect on nitric oxide inhalation in adult respiratory distress syndrome. *Lancet 341:*1664.

Pepke-Zaba J, Higenbottam TW, Dinh-Xuan AT, Stone D, & Wallwork J. 1991. Inhaled nitric oxide as a cause of selective pulmonary vasodilatation in pulmonary hypertension. *Lancet 338:*1173–1174.

Persson MG, Wiklund NP, & Gustafsson LE. 1993. Endogenous nitric oxide in single exhalations and the change during exercise. *Am Rev Respir Dis 148:*1210–1214.

Puybasset L, Stewart T, Rouby JJ, Cluzel P, Mourgeon E, Belin MF, Arthaud M, Landault C, & Viars P. 1994. Inhaled nitric oxide reverses the increase in pulmonary vascular resistance induced by permissive hypercapnia in patients with acute respiratory distress syndrome. *Anesthesiology 80:*1254–1267.

Puybasset L, Rouby JJ, Mourgeon E, Cluzel P, Souhil Z, Law-Koune JD, Stewart T, Devilliers C, Lu Q, Roche S, et al. 1995. Factors influencing cardiopulmonary effects of inhaled nitric oxide in acute respiratory failure. *Am J Respir Crit Care Med 152:*318–328.

Rachmilewitz D, Karmeli F, Eliakim R, Stalnikowicz R, Ackerman Z, Amir G, & Stamler JS. 1994. Enhanced gastric nitric oxide synthase activity in duodenal ulcer patients. *Gut 35:*1394–1397.

Ravichandran LV, Johns RA, & Rengasamy A. 1995. Direct and reversible inhibition of endothelial nitric oxide synthase by nitric oxide. *Am J Physiol 268:*H2216–H2223.

Rengasamy A, & Johns RA. 1993. Regulation of nitric oxide synthase by nitric oxide. *Mol Pharmacol 44:*124–128.

Rich GF, Murphy GD Jr, Roos CM, & Johns RA. 1993. Inhaled nitric oxide. Selective pulmonary vasodilation in cardiac surgical patients. *Anesthesiology 78:*1028–1035.

Rimar S, & Gillis CN. 1993. Selective pulmonary vasodilation by inhaled nitric oxide is due to hemoglobin inactivation. *Circulation 88:*2884–2887.

Roberts JD, Polaner DM, Lang P, & Zapol WM. 1992. Inhaled nitric oxide in persistent pulmonary hypertension of the newborn. *Lancet 340:*818–819.

Roberts JD Jr, Lang P, Bigatello LM, Vlahakes GJ, & Zapol WM. 1993. Inhaled nitric oxide in congenital heart disease. *Circulation 87:*447–453.

Roberts JD Jr, Fineman JR, Morin FC, Shaul PW, Rimar S, Schreiber MD, Polin RA, Zwass MS, Zayek MM, Gross I, Heymann MA, & Zapol WM. 1997. Inhaled nitric oxide and persistent pulmonary hypertension of the newborn. The Inhaled Nitric Oxide Study Group. *N Engl J Med 336:*605–610.

Rossaint R, Falke KJ, Lopez F, Slama K, Pison U, & Zapol WM. 1993. Inhaled nitric oxide for the adult respiratory distress syndrome. *N Engl J Med 328:*399–405.

Rossaint R, Gerlach H, Schmidt-Ruhnke H, Pappert D, Lewandowski K, Steudel W, & Falke K. 1995. Efficacy of inhaled nitric oxide in patients with severe ARDS. *Chest 107:*1107–1115.

Sellden H, Winberg P, Gustafsson LE, Lundell B, Book K, & Frostell CG. 1993. Inhalation of nitric oxide reduced pulmonary hypertension after cardiac surgery in a 3.2-kg infant. *Anesthesiology 78:*577–580.

Semigran MJ, Cockrill BA, Kacmarek R, Thompson BT, Zapol WM, Dec GW, & Fifer MA. 1994. Hemodynamic effects of inhaled nitric oxide in heart failure. *J Am Coll Cardiol 24:*982–988.

Silkoff PE, McClean PA, Slutsky AS, Furlott HG, Hoffstein E, Wakita S, Chapman KR, Szalai JP, & Zamel N. 1997. Marked flow-dependence of exhaled nitric oxide using a new technique to exclude nasal nitric oxide. *Am J Respir Crit Care Med 155:*260–267.

Skimming JW, Bender KA, Hutchison AA, & Drummond WH. 1997. Nitric oxide inhalation in infants with respiratory distress syndrome. *J Pediatr 130:*225–230.

Snell GI, Salamonsen RF, Bergin P, Esmore DS, Khan S, & Williams TJ. 1995. Inhaled nitric oxide used as a bridge to heart-lung transplantation in a patient with end-stage pulmonary hypertension. *Am J Respir Crit Care Med 151:*1263–1266.

Squire S, Kightley R, & Petros AJ. 1996. An effective method of scavenging nitric oxide. *Br J Anaesth 77:*432–434.

Stork E, Gorjanc E, Verter J, Younes N, Stenzel BA, Powers T, Sokol G, Wright LL, Yaffe SJ, Catz C, Vanmeurs K, Rhine W, Ball B, Brilli R, Moles L, Crowley M, Backstrom C, Crouse D, Hudson T, Konduri G, Bara R, Kleinman M, Hensman A, Rothstein RW, Ehrenkranz RA, Solimano A, et al. 1997. Inhaled nitric oxide in full-term and nearly full-term infants with hypoxic respiratory failure. *N Engl J Med 336:*597–604.

Tulleken JE, Zijlstra JG, Evers K, & van der Werf TS. 1997. Oxygen desaturation after treatment with inhaled nitric oxide for obstructive shock due to massive pulmonary embolism. *Chest 112:*296–298.

Walmrath D, Schneider T, Schermuly R, Olschewski H, Grimminger F, & Seeger W. 1996. Direct comparison of inhaled nitric oxide and aerosolized prostacyclin in acute respiratory distress syndrome. *Am J Respir Crit Care Med 153:*991–996.

Walther FJ, Benders MJ, & Leighton JO. 1992. Persistent pulmonary hypertension in premature neonates with severe respiratory distress syndrome. *Pediatrics 90:*899–904.

Wessel DL, Adatia I, Giglia TM, Thompson JE, & Kulik TJ. 1993. Use of inhaled nitric oxide and acetylcholine in the evaluation of pulmonary hypertension and endothelial function after cardiopulmonary bypass. *Circulation 88:*2128–2138.

Wessel DL, Adatia I, Thompson JE, & Hickey PR. 1994. Delivery and monitoring of inhaled nitric oxide in patients with pulmonary hypertension. *Crit Care Med 22:*930–938.

Winberg P, Lundell BP, & Gustafsson LE. 1994. Effect of inhaled nitric oxide on raised pulmonary vascular resistance in children with congenital heart disease. *Br Heart J* 71:282–286.

Yoshida M, Taguchi O, Gabazza EC, Kobayashi T, Yamakami T, Kobayashi H, Maruyama K, & Shima T. 1997. Combined inhalation of nitric oxide and oxygen in chronic obstructive pulmonary disease. *Am J Respir Crit Care Med* 155:526–529.

Young JD, Dyar O, Xiong L, & Howell S. 1994. Methaemoglobin production in normal adults inhaling low concentrations of nitric oxide. *Intensive Care Med* 20:581–584.

12
Nitric Oxide and the Perinatal Pulmonary Circulation

SIDNEY CASSIN

In fetal sheep close to term, pulmonary blood flow is approximately 35–40 ml/min/kg body weight (approximately 8%–10% of the total combined ventricular output), as measured by microsphere techniques in unanesthetized fetal sheep with indwelling catheters (Rudolph and Heymann 1970). The pulmonary arterial pressure is suprasystemic. The elevated vascular resistance to pulmonary blood flow in fetal lungs may be due to (a) arterioles that may have a very thick smooth muscle layer; (b) relative hypoxia in utero (normal PO_2 in pulmonary arteries is 18–22 mmHg); (c) vessels that are extremely reactive to changes in pH, PO_2, and PCO_2; and (d) circulating vasoactive substances. Since the postnatal survival of the fetus depends on a very rapid and appropriate establishment of pulmonary blood flow and alveolar ventilation, a remarkable adaptation of the pulmonary circulation must occur with the initiation of respiration. The circulatory pattern must change toward the adult pattern, in which the ventricles work in series rather than in parallel, and the lungs become for the first time organs of gaseous exchange. During the transition from an in utero liquid-breathing fetus to an air-breathing newborn, pulmonary blood flow increases approximately fivefold when measured in the lamb or kid. The factors responsible for this change in pulmonary blood flow have been studied in detail (Tod and Cassin 1997).

The basic mechanisms for this alteration in pulmonary circulation have not been elucidated. In term lambs and kids, pulmonary arterial pressure decreases to almost adult levels within several hours after birth; however, in humans, mean pulmonary arterial pressure may be only half of the aortic pressure within the first 60 hours of life (Moss et al. 1963). At this time it appears that multiple factors must be involved in the decline of pulmonary vascular resistance: expansion of alveoli with gas and/or increases in arterial oxygen tension, decreases in carbon dioxide tension, and increases in arterial pH. These changes could activate synthesis and/or release of a host of vasoactive substances, including those derived from endothelial cells, that could be involved in the decrease in pulmonary vascular resistance upon initiation of breathing and in the subsequent maintenance of a low pulmonary vascular resistance.

Many researchers have investigated vasoactive mediators that may be responsible for maintaining a high pulmonary vascular tone in the fetus. Others have sought to determine which mediators are responsible for the hypoxic pulmonary vasoconstriction in the newborn and adult. At this time, there does not appear to be a unique substance capable of accounting for all of the manifestations involved in regulation of the perinatal pulmonary circulation. Instead, it has become apparent that many substances may contribute in various ways to modulating pulmonary vascular responses in the fetus and newborn, and that the net interaction among these substances eventually results in what we term regulation of the pulmonary circulation.

One of the most intensively studied mediators in the last decade has been the endothelium-derived relaxing factor (EDRF). This work stems from an initial observation by Furchgott and Zawadzki (1980) that vascular relaxation induced by acetylcholine was dependent on the presence of an intact endothelium. They also provided evidence that acetylcholine-induced smooth muscle relaxation resulted from an endothelium mediated release of a labile relaxing substance, later known as EDRF. Endothelium-dependent relaxation has been observed in a wide variety of vascular preparations, including arteries, veins, and microvessels of several vascular regions, in response to a variety of substances and physiological stimuli. These include acetylcholine, adenine nucleotides, calcium ionophore A23187, bradykinin, and substance P. Stimuli included are hypoxia, increase in blood flow, and electrical stimulation. Furchgott (Furchgott and Vanhoutte 1989) and Ignarro (Ignarro et al. 1988) independently suggested that EDRF might be nitric oxide (NO) on the basis of pharmacological similarities between EDRF and NO generated from either acidified NO_2^- or NO gas. Evidence currently supports the identification of EDRF as NO (Moncada et al. 1991; Nathan 1992). Another proposed nonprostanoid EDRF that differs from NO has been identified as endothelium-dependent hyperpolarizing factor (EDHF) (Komori and Vanhoutte 1990). It is possible that EDHF is another EDRF distinct from NO (Nagao and Vanhoutte 1993) that causes opening of potassium channels and leads to smooth muscle hyperpolarization. The exact nature of EDHF remains unknown, but its action is not inhibited by hemoglobin or methylene blue. A major difficulty in trying to understand EDHF is the demonstration that NO itself can open potassium channels and cause hyperpolarization (Tare et al. 1990). Both exogenous NO and native EDRF-NO can directly activate single Ca^{2+}-activated potassium channels in cell-free membrane patches without requiring cGMP (Bolotina et al. 1994). Finally, a specific inhibitor of Ca^{2+}-activated potassium channels (K^+_{Ca}), charybdotoxin, seems to abolish the methylene blue-resistant component and the NO-induced relaxation of rabbit aorta (Bolotina et al. 1994). The physiological role of EDHF is largely unexplored, but it seems to contribute to endothelium-dependent relaxant responses to some vasodilators, such as ACh (Nagao and Vanhoutte 1993).

Prostacyclin (prostaglandin I_2, PGI_2) is a major product of the arachidonic acid metabolic pathway in pulmonary endothelium. It appears to be a very powerful vasodilator in the lung. PGI_2 is also an endothelium-derived relaxing factor and may play an important role in maintaining the pulmonary vascular bed in a dilated state. PGI_2 is a potent dilator of the fetal and perinatal pulmonary circulations (Leffler and Hessler 1979; Leffler et al. 1980; Cassin et al. 1981). Leffler et al. (1978, 1984) demonstrated that it was released into the pulmonary circulation at the time of birth. Inhibition of cyclooxygenase with indomethacin and the resultant block in PGI_2 synthesis diminishes the reduction in pulmonary arterial pressure at the time of birth (Leffler et al. 1978). Frangos et al. (1985) suggested that endothelial cells may regulate fetal pulmonary vascular tone through the release of PGI_2. The role of endothelium in the regulation of the fetal pulmonary vascular tone has been investigated in detail. Thus, in isolated pig pulmonary arteries, the relaxant response to acetylcholine is negligible at birth (5 minutes to 2 hours) but develops rapidly thereafter, becoming maximal at 3 to 10 days. The response decreases gradually at 3 to 8 weeks and becomes even lower in mature adults (Zellers and Vanhoutte 1991). In contrast, Dawes and Mott (1962) demonstrated that acetylcholine caused marked pulmonary dilation in fetal sheep prior to term. In addition, Cassin et al. (1964a) demonstrated that in the very immature fetal sheep (80 days gestational age), acetylcholine produced a pronounced vasodilation. It is possible that these different responses to the same agent are due to species differences and/or the use of isolated rings versus an intact lung preparation.

Production and Action of NO

Activation of receptors (such as acetylcholine and bradykinin) and/or shear stress initiates a calcium influx into endothelial cells that results in stimulation of nitric oxide synthase (NOS). Nitric oxide synthase in turn stimulates production of NO with the conversion of L-arginine to citruline. The NO rapidly diffuses small distances to smooth muscle cells and activates soluble guanylate cyclase (sGC), which in turn leads to increased synthesis of cyclic guanosine monophosphate (cGMP) from guanosine triphosphate (GTP). Elevation of cGMP results in a regulation of protein phosphorylation, ion channel conductivity, and phosphodiesterase activity with consequent smooth muscle relaxation.

Pulmonary Effects of NO

The effects of NO on the adult and perinatal pulmonary circulation have been studied in detail (Shaul 1995). Clearly there is a release of NO in response to a number of agonists, including ACH, bradykimn, and endothelin (Cassin et al. 1991; Ignarro et al. 1986, 1987; Liu et al. 1991; Namiki et al. 1992; Reader et al. 1994). All of these substances have been shown to dilate high-tone pulmonary arteries in vivo and in vitro (Ignarro et al. 1986, 1987) partly via NO release. NO has been shown to counteract hypoxic pulmonary vasoconstriction as well as vasoconstriction caused by specific agonists such as prostaglandin F_2 (PGF_2) (Liu et al. 1991). Also, injection of a variety of NOS inhibitors has been shown to elevate pulmonary arterial tone, suggesting involvement of NOS in the normal regulation of pulmonary vascular tone (Stamler et al. 1994).

Shaul et al. (1993) have presented data to indicate that chronic hypoxia in rats leads to depressed NO-mediated effects on isolated pulmonary arterial segments in vitro. It appears that these effects are due to diminished endothelial production in the pulmonary circulation.

In the near term, in situ pump-perfused fetal lung intrapulmonary infusion of $N\omega$-nitro-L-arginine (L-NA) caused an increase in pulmonary arterial pressure at constant flow (Cassin 1993). Similarly, in the intact unanesthetized near-term fetal sheep, L-NA produces a fall in pulmonary blood flow and an increase in pulmonary arterial pressure (Abman et al. 1990). These studies suggest that there is a basal EDRF-NO production that is involved in regulating the pulmonary vascular resistance of fetal sheep late in gestation. Acetylcholine (Abman et al. 1990; Tiktinsky et al. 1992) and bradykinin (Banerjee et al. 1994; Glasgow et al. 1997), both endothelium-dependent dilators, increase pulmonary blood flow and decrease pulmonary arterial pressure in the intact near-term fetal sheep. These changes are blocked with NOS inhibitors. Furthermore, endothelin-1 is a potent fetal pulmonary dilator (Cassin 1993; Wong et al. 1994), and circulatory levels of endothelin-1 are elevated in the newborn period (Wong et al. 1994). Similar studies in the isolated neonatal guinea pig lung (Davidson and Eldemerdash 1991) have provided data to indicate that EDRF-NO also modulates pulmonary vascular tone in the newborn period, and Wang and Coceani (1994) have demonstrated that isolated fetal pulmonary resistance vessels of sheep possess an EDRF-NO relaxing mechanism stimulated by bradykinin.

As indicated previously, with the initiation of ventilation (100% oxygen) in term or near-term fetal sheep, there is a dramatic increase (5- to 10-fold) in pulmonary blood flow and rearrangement of the circulation (Dawes et al. 1954; Cassin et al. 1964b). More recently, Abman et al. (1990) have repeated these studies in the intact, chronically instrumented near-term fetal sheep and obtained similar results. However, when fetal sheep were treated with a NOS inhibitor, L-NA, prior to ventilation, the dilation was attenuated. In 1964 Cassin et al. (1964b) demonstrated that rhythmic distension of fetal lungs with 7% CO_2 in N_2, 3% O_2, or room air, all resulted in pulmonary vasodilation. At that time the mechanism for dilation was thought to be due to a direct mechanical effect upon the lungs, or a neural reflex. It was also suggested that changes in PaO_2 and $PaCO_2$ could affect the pulmonary vessels directly. Since then, investigators (Cornfield et al. 1992) have repeated the studies of rhythmic distension and have demonstrated similar increases in pulmonary flow, which are diminished in the presence of L-NA inhibition of NOS activity. Thus, it appears that EDRF-NO modulates not only the pulmonary dilator responses to increased oxygen tension, but also those in response to ventilation without change in arterial oxygen tension.

Since phosphodiesterase is clearly involved in the mechanism(s) of action of EDRF-NO, studies have been carried out to test the effects of selective phosphodiesterase inhibitors on pulmonary smooth muscle relaxation. Ziegler et al. (1995) recently demonstrated that two selective phosphodiesterase-5 (PDE-5) inhibitors, dipyridamole and zaprinast, both induced pulmonary vasodilation in intact, unanesthetized mature fetal sheep in utero. The authors speculated that these PDE inhibitors might be useful as pulmonary vasodilators, either alone or along with cGMP-dependent dilators. Using 8-phenyltheophylline, they demonstrated that the dipyridamole dilation was not due to its effects on adenosine. Skimming et al. (1997) tested the hypothesis that A2 adenosine receptors mediated hemodynamic responses to intravenous dipyridamole. Their studies led to the conclusion, also, that dipyridamole dilates both pulmonary and systemic vasculature via a predominantly adenosine-independent mechanism. More recently, Ziegler et al. (1998) tested their hypothesis that PDE-5 opposes pulmonary vasodilation and maintains the high pulmonary vascular resistance in normal fetal lungs. Following administration of Ach or gaseous NO, dipyridamole potentiated the vasodilator responses of these endothelium-dependent and-independent dilators. In a similar study, Skimming et al. (1996) demonstrated that neither the magnitude nor the duration of the response of the fetal ovine pulmonary circulation was affected by simultaneous infusions of zaprinast and U46619 along with acetylcholine endothelin-1 (a fetal pulmonary vasodilator (Cassin et al. 1991), and ethanol saturated with NO. This led to the conclusion that endogenously produced NO may involve mechanisms other than raising cGMP concentrations to modulate pulmonary vascular tone. Bolotina et al. (1994) reported that exogenous NO as well as native EDRF-NO could directly activate single Ca^{2+}-dependent K^+ channels in cell-free membrane patches without requiring cGMP. It appears, therefore, that cyclic nucleotide-independent activation of K^+ channels could account, at least in part, for NO-induced smooth muscle relaxation. The series of experiments described above related to cyclic nucleotides are clearly complex and quite difficult to perform. There are obvious differences and contradictions in the results of the multiple investigations. However, it is not clear why differences exist. Perhaps it is due to the ages at which the fetal sheep studies were carried out, the protocols used (anesthetized versus unanesthetized fetal preparations), or the mechanisms involved other than cytoplasmic cGMP that cause vascular dilation in response to endogenously produced NO.

Nitric Oxide, cGMP, Pulmonary Vasodilators, and Fetal Lung Liquid Production

Recently, experiments (Cummings 1995, 1997) have demonstrated that several pulmonary vasodilators [acetylcholine, prostaglardin D_2 (PGD_2), FPL 55712, and NO solubilized in saline] not only cause a decrease in pulmonary vascular resistance but also affect lung liquid production. These very interesting experiments suggest that mechanisms causing the profound decrease in pulmonary vascular resistance at birth are also responsible for drying the lungs. In an effort to explain the mechanism responsible for these critical events at the time of birth, Kabbani and Cassin (1998) studied the effects of cGMP on fetal sheep pulmonary blood flow and lung liquid production. Intrapulmonary injection of 8-bromo-cGMP for 1 hour resulted in a marked decrease in lung liquid production (70% in the hour after injection and 44% in the second hour after injection), while pulmonary blood flow was elevated from 56 ml/min to 167 and 187 ml/min. Animals infused with saline showed no significant changes in lung liquid production or blood flow. Although pulmonary blood flow increased in the first hour of injection, lung liquid production did not decrease at this time, suggesting a time dissociation between changes in blood flow and lung liquid production. Thus, it appears that blood flow elevation and lung liquid reduction may not be causally related. However, a common transduction pathway involving cGMP may be responsible for both lung liquid reduction and elevation of pulmonary blood flow. The demonstration by these authors, in preliminary studies, that the dihydropyridine-type Ca^{2+}-channel blocker (nitrendipene) caused a significant rise in blood flow without change in lung liquid production supports the view that pulmonary blood flow and lung liquid production may be independent processes that depend on a common transduction pathway involving cGMP. It is suggested that elevated cGMP (which is responsible for elevated flow) may inhibit a cAMP phosphodiesterase (PDE III), which in turn would tend to elevate plasma cAMP and result in reabsorption of lung liquid by the same mechanism described for epinephrine-induced fluid reabsorption (Olver et al. 1986; Strang 1991).

Factors Affecting NO-Mediated Responses

Evidence has been presented that inhibition of NOS leads to an increase in pulmonary vascular resistance in the adult (Stamler et al. 1994) as well as in the fetal (Cassin 1993) pulmonary circulation. Thus, it appears that NOS regulates normal pulmonary vascular tone in the immature as well as the mature pulmonary circulations under normoxic conditions. Recently, evidence has been presented that NO-mediated processes are diminished in arteries obtained from patients with end-stage chronic obstructive pulmonary disease (Dinh-Xuan et al. 1991). Studies were also carried out on isolated lung preparations from adult rats exposed to prolonged hypoxia (Adnot et al. 1991). Similar studies on the effects of NO dependent pulmonary responses have been performed in newborn pigs (Fike et al. 1998). Control newborn pigs (1–3 days of age) raised in room air were compared with those kept in 10% oxygen for 10–12 days. The exhaled NO output and plasma nitrites and nitrates were measured in anesthetized piglets. Nitrites and nitrates were also measured in perfusates of isolated lungs. Finally, the amounts of endothelial NOS (eNOS) were estimated in whole-lung homogenates. Nitrites, nitrates, NO output, and eNOS were constantly lower in the chronically hypoxic piglets than in the controls. In another study, this same group (Fike and Kaplowitz 1996) explored the pulmonary vascular responses to agents that stimulate or inhibit NO synthesis in control and chronically hypoxic lungs of newborn piglets. The pulmonary vascular responses to acetylcholine were blunted in chronically hypoxic piglets. In addition, inhibition of NOS in chronically hypoxic piglets (10–12 days) did not cause constriction of either arteries or veins. The data suggest the possibility that depressed NO production in chronically hypoxic piglets might contribute to the pathogenesis of pulmonary hypertension in the newborn. In contrast, pulmonary endothelial cells showed a marked increase in NO production at a PO_2 of 680 mm Hg when compared with cells at PO_2 values of 150 and

40 mmHg. Thus, there appears to be an acute modulation of NO production and pulmonary blood flow when arterial oxygen tension is elevated (Shaul and Wells 1994).

Recently, evidence has been presented to suggest that estrogen has a direct effect on vascular beds other than the perinatal pulmonary circulation. These effects are the result of stimulation of prostacyclin as well as NO (Mikkola et al. 1998). Weiner and Thompson (1997) in their recent review point out that pregnancy, with its associated elevation in plasma estradiol concentration, increases Ca^{2+}-dependent NOS activity early in gestation. There is also an increase in cGMP with changes in blood flow (Weiner and Thompson 1997). Similarly, it has been shown that estrogen increases coronary blood flow in nonpregnant sheep (Lang et al. 1997). This increase in blood flow is prevented by an inhibitor of NOS. Since estrogen has been shown to have NO-mediated effects in peripheral vascular beds, and estrogen levels are elevated with parturition, it has been suggested that estrogen may, in fact, be involved in NO-mediated pulmonary vasodilation at birth (Lantin-Hermoso et al. 1997). Additional studies have demonstrated that estrogen upregulates endothelial NOS (eNOS) gene expression by activation of pulmonary arterial endothelial cell estrogen receptors, with a consequent increase in intracellular Ca^{2+} (MacRitchie et al. 1997).

Inhaled NO

There are multiple reasons for considering the use of inhaled NO as a therapeutic approach for pulmonary hypertension in both the adult and the newborn. Of primary significance is the fact that as a gas, NO may be administered directly to the pulmonary system with relative ease. Also of particular importance is the fact that NO is able to diffuse directly into the vascular smooth muscle of the pulmonary circulation and produce local effects. In contrast to a host of other potent pulmonary vasodilators, NO does not produce systemic vasodilation (Frostell et al. 1993). This is so because the NO that enters the vascular system binds with hemoglobin and other hemoproteins to form nitrosohemoglobin, which eliminates its dilator activity (Moncada et al. 1991; Knowles and Moncada 1994). As detailed by DeMarco et al. (1996), many investigators have shown that inhaled NO acts as a highly selective pulmonary vasodilator in perinatal, juvenile, and adult animals as well as in humans. Pepke-Zaba et al. (1991) noted a small reduction in pulmonary arterial pressure in several patients with primary pulmonary hypertension treated with NO. Frostell et al. (1991) demonstrated a reversal of pulmonary arterial hypertension induced in sheep with the thromboxane mimic U46619 and normalization of pulmonary vascular resistance after sheep inhaled NO (80 ppm). In these experiments, no systemic vascular effect of NO was noted. Other experiments were carried out by Fratacci et al. (1991), who demonstrated that NO could be used as a selective pulmonary vasodilator in sheep made pulmonary hypertensive following a heparin-protamine reaction. Zayek et al. (1993) addressed the question of the usefulness of NO inhalation in newborn near-term lambs with experimentally induced pulmonary hypertension as a result of ductal ligation. They compared lambs that inhaled NO for 23 hours with a control group and noted a significant increase in survival of those inhaling NO. Studies by Roberts et al. (1993) in newborn sheep demonstrated no further vasodilation of the hypertensive pulmonary circulation when NO concentrations were elevated from 80 to 160 ppm. They also showed that cGMP in lung tissue and preductal blood was elevated, indicating that vasodilation involved the NO-arginine signal transduction pathway.

DeMarco et al. (1996) tested the hypothesis that inhalation of NO would result in selective, dose-dependent dilation of the pulmonary circulation in unanesthetized, spontaneously breathing newborn lambs made hypoxic or infused with the thromboxane mimic U46619. These studies demonstrated that inhalation of 10 to 80 ppm NO decreased the pulmonary vascular resistance of neonates treated with U46619. Similarly, treatment with 80 ppm NO decreased pulmonary arterial pressure and pulmonary vascular resistance when lambs breathed 8% oxygen.

Studies by this same group (Skimming et al. 1995) were also carried out on preterm lambs used as a model of the respiratory distress syndrome. The goals of these studies were to determine whether inhaled NO could cause a significant elevation of systemic PaO_2, and to explore the mechanisms responsible for improvement of oxygenation. Fetal sheep (126 or 127 days of gestation) were delivered under chloralose anesthesia and maintained on this anesthesia for the duration of the experiments. After placement of endotracheal tubes and vascular catheters, the animals were ventilated with 95% O_2. After 100 minutes of ventilation on this gas, NO (5 and 10 ppm) was added to the ventilatory system for 20 minutes. Compared with control periods, ventilation with added NO resulted in improvement in PaO_2 as well as in physiologic intrapulmonary shunting. There was also a selective decrease in pulmonary arterial pressure. There was further improvement in blood oxygenation and pulmonary hemodynamics when NO was elevated from 10 to 20 ppm. Many clinical studies have now been done on infants with persistent pulmonary hypertension to determine the usefulness of inhaled NO in this condition. Roberts et al. (1992) demonstrated improvement in oxygen exchange, which was confirmed by Kinsella et al. (1992).

Although there are clearly beneficial effects on NO utilization in animal models and newborn infants, investigators have been cautions in the use of NO because of potential adverse side effects. Thus, NO or reactive intermediates formed by the reaction of NO with reduced oxygen species may cause injury to the mammalian lung that may not be easily detectable. Recently, Matalon et al. (1996) studied the potential toxicity of NO in newborn lambs exposed to 0, 20, 80, and 200 ppm NO in either 21% or 60% oxygen for 6 hours. The results indicated that NO inhaled at 80 and 200 ppm damaged the surfactant system. Although these concentrations of NO are considerably greater than those normally used in human infants or adults, the product of NO concentration and time inhaled is comparable to those seen in patients breathing NO at 20 ppm for 3 days. Inhalation of NO generates a complex series of events that may be deleterious to the organism. Clearly, these depend on multiple factors, including the concentrations of reactive oxygen and nitrogen, appropriate antioxidant defense, and the presence of lung inflammation.

References

Abman SH, Chatfield BA, Hall SL, & McMurtry IF. 1990. Role of endothelium-derived relaxing factor during transition of pulmonary circulation at birth. *Am J Physiol* 259:H1921–H1927.

Adnot S, Raffestin B, Eddahibi S, Braquet P, & Chabrier PE. 1991. Loss of endothelium-dependent relaxant activity in the pulmonary circulation of rats exposed to chronic hypoxia. *J Clin Invest* 87:155–162.

Banerjee A, Roman C, & Heymann MA. 1994. Bradykinin receptor blockade does not affect oxygen-mediated pulmonary vasodilation in fetal lambs. *Pediatr Res 36:* 474–480.

Bolotina VM, Najibi S, Palacino JJ, Pagano PJ, & Cohen RA. 1994. Nitric oxide directly activates calcium-dependent potassium channels in vascular smooth muscle. *Nature 368:*850–853.

Cassin S, Winikor I, Tod M, Philips J, Frisinger J, Jordan J, & Gibbs C. 1981. Effects of prostacyclin on the fetal pulmonary circulation. *Pediatr Pharmacol 1:*197–207.

Cassin S, Kristova V, Davis T, Kadowitz P, & Gause G. 1991. Tone-dependent responses to endothelin in the isolated perfused fetal sheep pulmonary circulation in situ. *J Appl Physiol 70:*1228–1234.

Cassin S. 1993. The role of eicosanoids and endothelium-dependent factors in regulation of the fetal pulmonary circulation. *J Lipid Mediat 6:*477–485.

Cassin S, Dawes GS, & Ross BB. 1964a. Pulmonary blood flow and vascular resistance in immature foetal lambs. *J Physiol (Lond) 171:*80–89.

Cassin S, Dawes GS, Mott JC, Ross BB, & Strang LB. 1964b. The vascular resistance of the foetal and newly ventilated lung of the lamb. *J Physiol (Lond) 171:*61–79.

Cornfield DN, Chatfield BA, McQueston JA, McMurtry IF, & Abman SH. 1992. Effects of birth-related stimuli on L-arginine-dependent pulmonary vasodilation in ovine fetus. *Am J Physiol 262:*H1474–H1481.

Cummings JJ. 1995. Pulmonary vasodilator drugs decrease lung liquid production in fetal sheep. *J Appl Physiol 79:*1212–1218.

Cummings JJ. 1997. Nitric oxide decreases lung liquid production in fetal lambs. *J Appl Physiol 83:*1538–1544.

Davidson D, & Eldemerdash A. 1991. Endothelium-derived relaxing factor: evidence that it regulates pulmonary vascular resistance in the isolated neonatal guinea pig lung. *Pediatr Res 29:*538–542.

Dawes GS, Mott JC, & Widdicombe JG. 1954. The foetal circulation in the lamb. *J Physiol (Lond) I 26:*563–587.

Dawes GS, & Mott JC. 1962. The vascular tone of the foetal lung. *J Physiol (Lond) 164:*465–477.

DeMarco V, Skimming JW, Ellis TM, & Cassin S. 1996. Nitric oxide inhalation: effects on the ovine neonatal pulmonary and systemic circulations. *Reprod Fertil Dev 8:*431–438.

Dinh-Xuan AT, Higenbottam TW, Clelland CA, Pepke-Zaba J, Cremona G, Butt AY, Large SR, Wells FC, & Wallwork J. 1991. Impairment of endothelium-dependent pulmonary-artery relaxation in chronic obstructive lung disease. *N Engl J Med 324:*1539–1547.

Fike CD, & Kaplowitz MR. 1996. Chronic hypoxia alters nitric oxide-dependent pulmonary vascular responses in lungs of newborn pigs. *J Appl Physiol 81:*2078–2087.

Fike CD, Kaplowitz MR, Thomas CJ, & Nelin LD. 1998. Chronic hypoxia decreases nitric oxide production and endothelial nitric oxide synthase in newborn pig lungs. *Am J Physiol 274:*L517–L526.

Frangos JA, Eskin SG, McIntire LV, & Ives CL. 1985. Flow effects on prostacyclin production by cultured human endothelial cells. *Science 227:*1477–1479.

Fratacci MD, Frostell CG, Chen TY, Wain JCJ, Robinson DR, & Zapol WM. 1991. Inhaled nitric oxide. A selective pulmonary vasodilator of heparin-protamine vasoconstriction in sheep. *Anesthesiology 75:*990–999.

Frostell CG, Fratacci MD, Wain JC, Jones R, & Zapol WM. 1991. Inhaled nitric oxide. A selective pulmonary vasodilator reversing hypoxic pulmonary vasoconstriction [published erratum appears in Circulation 1991 Nov; 84(5):2212]. *Circulation 83:* 2038–2047.

Frostell CG, Blomqvist H, Hedenstierna G, Lundberg J, & Zapol WM. 1993. Inhaled nitric oxide selectively reverses human hypoxic pulmonary vasoconstriction without causing systemic vasodilation [see comments]. *Anesthesiology 78:*427–435.

Furchgott RF, & Vanhoutte PM. 1989. Endothelium-derived relaxing and contracting factors. *FASEB J 3:*2007–2018.

Furchgott RF, & Zawadzki JV. 1980. The obligatory role of endothelial cells in the relaxation of arterial smooth muscle by acetylcholine. *Nature 288:*373–376.

Glasgow RE, Buga GM, Ignarro LJ, Chaudhuri G, & Heymann MA. 1997. Endothelium-derived relaxing factor as a mediator of bradykinin-induced perinatal pulmonary vasodilatation in fetal sheep. *Reprod Fertil Dev 9:*213–216.

Ignarro LJ, Harbison RG, Wood KS, & Kadowitz PJ. 1986. Activation of purified soluble guanylate cyclase by endothelium-derived relaxing factor from intrapulmonary artery and vein: stimulation by acetylcholine, bradykinin and arachidonic acid. *J Pharmacol Exp Ther 237:*893–900.

Ignarro LJ, Byrns RE, Buga GM, & Wood KS. 1987. Endothelium-derived relaxing factor from pulmonary artery and vein possesses pharmacologic and chemical properties identical to those of nitric oxide radical. *Circ Res 61:*866–879.

Ignarro LJ, Buga GM, Byrns RE, Wood KS, & Chaudhuri G. 1988. Endothelium-derived relaxing factor and nitric oxide possess identical pharmacologic properties as relaxants of bovine arterial and venous smooth muscle. *J Pharmacol Exp Ther 246:*218–226.

Kabbani MS, & Cassin S. 1998. The effects of cGMP on fetal sheep pulmonary blood flow and lung liquid production. *Pediatr Res 43:*325–330.

Kinsella JP, Neish SR, Shaffer E, & Abman SH. 1992. Low-dose inhalation nitric oxide in persistent pulmonary hypertension of the newborn [see comments]. *Lancet 340:* 819–820.

Knowles RG, & Moncada S. 1994. Nitric oxide synthases in mammals. *Biochem J 298:* 249–258.

Komori K, & Vanhoutte PM. 1990. Endothelium-derived hyperpolarizing factor. *Blood Vessels 27:*238–245.

Lang U, Baker RS, & Clark KE. 1997. Estrogen-induced increases in coronary blood flow are antagonized by inhibitors of nitric oxide synthesis. *Eur J Obstet Gynecol Reprod Biol 74:*229–235.

Lantin-Hermoso RL, Rosenfeld CR, Yuhanna IS, German Z, Chen Z, & Shaul PW. 1997. Estrogen acutely stimulates nitric oxide synthase activity in fetal pulmonary artery endothelium. *Am J Physiol 273:*L119–L126.

Leffler CW, Hessler JR. 1979. Pulmonary and systemic vascular effects of exogenous prostaglandin I₂ in fetal lambs. *Eur J Pharmacol 54:*37–42.

Leffler CW, Tyler TL, & Cassin S. 1978. Effect of indomethacin on pulmonary vascular response to ventilation of fetal goats. *Am J Physiol 234:*H346–H351.

Leffler CW, Hessler JR, & Terragno NA. 1980. Ventilation-induced release of prostaglandinlike material from fetal lungs. *Am J Physiol 238:*H282–H286.

Leffler CW, Hessler JR, & Green RS. 1984. The onset of breathing at birth stimulates pulmonary vascular prostacyclin synthesis. *Pediatr Res 18:*938–942.

Liu SF, Crawley DE, Evans TW, & Barnes PJ. 1991. Endogenous nitric oxide modulates adrenergic neural vasoconstriction in guinea-pig pulmonary artery. *Br J Pharmacol 104:*565–569.

MacRitchie AN, Jun SS, Chen Z, German Z, Yuhanna IS, Sherman TS, & Shaul PW. 1997. Estrogen upregulates endothelial nitric oxide synthase gene expression in fetal pulmonary artery endothelium. *Circ Res 81:*355–362.

Matalon S, DeMarco V, Haddad IY, Myles C, Skimming JW, Schurch S, Cheng S, & Cassin S. 1996. Inhaled nitric oxide injures the pulmonary surfactant system of lambs in vivo. *Am J Physiol 270:*L273–L280.

Mikkola T, Viinikka L, & Ylikorkala O. 1998. Estrogen and postmenopausal estrogen/progestin therapy: effect on endothelium-dependent prostacyclin, nitric oxide and endothelin-1 production [In Process Citation]. *Eur J Obstet Gynecol Reprod Biol 79:*75–82.

Moncada S, Palmer RM, & Higgs EA. 1991. Nitric oxide: physiology, pathophysiology, and pharmacology. *Pharmacol Rev 43:*109–142.

Moss AJ, Emmanouilides GC, & Duffie GR. 1963. Closure of the ductus arteriosus in the newborn infant. *Pediatrics 32:*25–30.

Nagao T, & Vanhoutte PM. 1993. Endothelium-derived hyperpolarizing factor and endothelium-dependent relaxations. *Am J Respir Cell Mol Biol 8:*1–6.

Namiki A, Hirata Y, Ishikawa M, Moroi M, Aikawa J, & Machii K. 1992. Endothelin-1- and endothelin-3-induced vasorelaxation via common generation of endothelium-derived nitric oxide. *Life Sci 50:*677–682.

Nathan C. 1992. Nitric oxide as a secretory product of mammalian cells. *FASEB J 6:*3051–3064.

Olver RE, Ramsden CA, Strang LB, & Walters DV. 1986. The role of amilorideblockable sodium transport in adrenaline-induced lung liquid reabsorption in the fetal lamb. *J Physiol (Lond) 376:*321–340.

Pepke-Zaba J, Higenbottam TW, Dinh-Xuan AT, Stone D, & Wallwork J. 1991. Inhaled nitric oxide as a cause of selective pulmonary vasodilatation in pulmonary hypertension. *Lancet 338:*1173–1174.

Reeder LB, Yang LH, & Ferguson MK. 1994. Modulation of lymphatic spontaneous contractions by EDRF. *J Surg Res 56:*620–625.

Roberts JD, Polaner DM, Lang P, & Zapol WM. 1992. Inhaled nitric oxide in persistent pulmonary hypertension of the newborn [see comments]. *Lancet 340:*818–819.

Roberts JDJ, Chen TY, Kawai N, Wain J, Dupuy P, Shimouchi A, Bloch K, Polaner D, & Zapol WM. 1993. Inhaled nitric oxide reverses pulmonary vasoconstriction in the hypoxic and acidotic newborn lamb. *Circ Res 72:*246–254.

Rudolph AM, & Heymann MA. 1970. Circulatory change during growth in the fetal lamb. *Circ Res 26:*289–299.

Shaul PW. 1995. Nitric oxide in the developing lung. *Adv Pediatr 42:*367–414.

Shaul PW, & Wells LB. 1994. Oxygen modulates nitric oxide production selectively in fetal pulmonary endothelial cells. *Am J Respir Cell Mol Biol 11:*432–438.

Shaul PW, Wells LB, & Horning KM. 1993. Acute and prolonged hypoxia attenuate endothelial nitric oxide production in rat pulmonary arteries by different mechanisms. *J Cardiovasc Pharmacol 22:*819–827.

Skimming JW, DeMarco VG, & Cassin S. 1995. The effects of nitric oxide inhalation on the pulmonary circulation of preterm lambs. *Pediatr Res 37:*35–40.

Skimming JW, DeMarco VG, Kadowitz PJ, & Cassin S. 1996. Effects of zaprinast and dissolved nitric oxide on the pulmonary circulation of fetal sheep. *Pediatr Res 39:* 223–228.

Skimming JW, DeMarco VG, & Cassin S. 1997. Effects of dipyridamole and adenosine infusions on ovine pulmonary and systemic circulations. *Am J Physiol 272:*H921–H926.

Stamler JS, Loh E, Roddy MA, Currie KE, & Creager MA. 1994. Nitric oxide regulates basal systemic and pulmonary vascular resistance in healthy humans. *Circulation 89:*2035–2040.

Strang LB. 1991. Fetal lung liquid: secretion and reabsorption. *Physiol Rev 71:*991–1016.

Tare M, Parkington HC, Coleman HA, Neild TO, & Dusting GJ. 1990. Hyperpolarization and relaxation of arterial smooth muscle caused by nitric oxide derived from the endothelium. *Nature 346:*69–71.

Tiktinsky MH, Cummings JJ, & Morin FC. 1992. Acetylcholine increases pulmonary blood flow in intact fetuses via endothelium-dependent vasodilation. *Am J Physiol 262:*H406–H410.

Tod ML, & Cassin S. 1997. Fetal and neonatal pulmonary circulation. In: Crystal RG, West JB, Barnes PJ, Cherniak NS, & Weibel ER, eds. The Lung: Scientific Foundations (Second Edition), New York: Lippincott-Raven, 2129–2139.

Wang Y, & Coceani F. 1994. EDRF in pulmonary resistance vessels from fetal lamb: stimulation by oxygen and bradykinin. *Am J Physiol 266:*H936–H943.

Weiner CP, & Thompson LP. 1997. Nitric oxide and pregnancy. *Semin Perinatol 21:*367–380.

Wong J, Fineman JR, & Heymann MA. 1994. The role of endothelin and endothelin receptor subtypes in regulation of fetal pulmonary vascular tone. *Pediatr Res 35:* 664–670.

Zayek M, Wild L, Roberts JD, & Morin FC. 1993. Effect of nitric oxide on the survival rate and incidence of lung injury in newborn lambs with persistent pulmonary hypertension. *J Pediatr 123:*947–952.

Zellers TM, & Vanhoutte PM. 1991. Endothelium-dependent relaxations of piglet pulmonary arteries augment with maturation. *Pediatr Res 30:*176–180.

Ziegler JW, Ivy DD, Fox JJ, Kinsella JP, Clarke WR, & Abman SH. 1995. Dipyridamole, a cGMP phosphodiesterase inhibitor, causes pulmonary vasodilation in the ovine fetus. *Am J Physiol 269:*H473–H479.

Ziegler JW, Ivy DD, Fox JJ, Kinsella JP, Clarke WR, & Abman SH. 1998. Dipyridamole potentiates pulmonary vasodilation induced by acetylcholine and nitric oxide in the ovine fetus. *Am J Respir Crit Care Med 157:*1104–1110.

13
Nitric Oxide and the Pulmonary Circulation in the Adult

Bobby D. Nossaman, Alan D. Kaye, and Philip J. Kadowitz

Introduction

Despite its very potent vasodilating action in vivo, acetylcholine does not always produce vasorelaxation in isolated blood vessel preparations (Furchgott and Bhadrakom 1953; Furchgott 1955). In 1980 Furchgott and Zawadski reported that the relaxing effect of acetylcholine on isolated arteries was dependent upon the presence of the vascular endothelium (Furchgott and Zawadzki 1980). They also suggested that the relaxing effect of acetylcholine was mediated by an unstable humoral factor, later named endothelium-dependent relaxing factor (EDRF) (Furchgott 1984, Martin et al. 1986). On the basis of the similar pharmacological responses of EDRF and nitric oxide (NO), two research groups independently published results in 1987 suggesting that NO accounts for the action of EDRF (Furchgott 1988; Ignarro et al. 1988). Further studies in vascular strips and platelets demonstrated that the activities of EDRF and NO were largely indistinguishable (Furchgott 1984; Griffith et al. 1984; Ignarro et al. 1987a,b). The similarity was finally confirmed, although questioned by others (Myers et al. 1990).

Until recently, NO was discussed only for its contribution to environmental pollution, e. g., as an internal combustion engine pollutant, and in formation of smog and acid rain. However, from its apparent widespread distribution in mammalian biology (Nathan 1992), NO played an important role in the evolution of living systems (Anbar 1996). Nitric oxide was honored by receiving the title of "Molecule of the Year" in 1992 (Culotta and Koshland 1992).

Biology

The generation of endogenous NO results in the oxidation of one of the two terminal guanidino nitrogen atoms of L-arginine, with the subsequent cleavage of the oxidized L-arginine into NO and L-citrulline (Figure 13.1) (Palmer et al. 1988a; Palmer et al, 1988).

This process requires the constitutive form of the enzyme, nitric oxide synthase (NOS) (Pollock et al. 1991; Janssens et al. 1992). When NO is inhaled, the highly

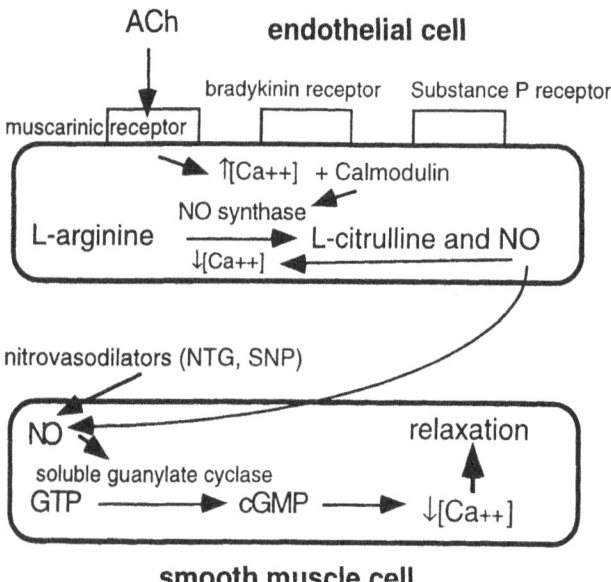

ACh   **endothelial cell**

FIGURE 13.1. Simplified diagram of the endogenous generation and endogenous or exogenous action of nitric oxide in the endothelial and smooth muscle cell. Ach, Acetylcholine; NTG, nitroglycerin; SNP, sodium nitroprusside.

lipid-soluble nature of NO allows for rapid penetration into the airway and vascular smooth muscle cells. When NO is delivered via the nitrosovasodilator vehicles sodium nitroprusside and nitroglycerin, they release NO molecules that also traverse the endothelial cells and enter the vascular smooth muscle cells. Once in vascular smooth muscle, NO leads to the activation of soluble guanylate cyclase, which catalyzes the conversion of magnesium guanosine triphosphate to cyclic guanosine monophosphate (cGMP) (Fig. 13.1). Cyclic guanosine monophosphate facilitates the phosphorylation of several proteins by the cGMP-dependent protein kinases, which indirectly leads to the dephosphorylation of myosin light chains and the relaxation of vascular smooth muscle cells (Holzmann 1982; Ignarro et al. 1983; Rapoport and Murad 1983; Griffin et al. 1985; MacLeod et al. 1987; Ignarro 1989; Monoada et al. 1991).

When NO diffuses into the intravascular space, it rapidly binds with hemoglobin, forming nitrosylhemoglobin, and is inactivated (Gibson and Roughton 1957; Oda et al. 1975). Nitrosylhemoglobin is rapidly oxidized to methemoglobin, forming nitrite and nitrate (Chiuodi and Mohler 1985). NO may also form nitrate, while oxidizing hemoglobin to methemoglobin (Doyle and Hoekstra 1981). Methemoglobin is subsequently reduced by methemoglobin reductase. Therefore, the short biological half-life of NO limits its vasodilatory activity to the localized area (Frostell et al. 1991, 1993). Nitric oxide produced by endothelial cells is a major determinant of vascular tone, since it controls blood pressure and blood flow distribution in each organ system (Rees et al. 1989; Stamler et al. 1994).

In 1993 a novel use of inhaled NO gas to selectively reduce pulmonary vaso-constriction induced by either hypoxia or thromboxane analogues was described, leading to an explosion of clinical interest in inhaled NO (Frostell et al. 1993). NO has been used to treat acute respiratory distress syndrome (ARDS) (Grover et al. 1992; Zapol and Hurford 1993; Fierobe et al. 1995; Rossetti et al. 1996; Walmrath et al. 1996; Zwissler et al. 1996), persistent pulmonary hypertension in congenital heart disease (Roberts et al. 1993; Atz et al. 1996; Berner et al. 1996), idiopathic pulmonary hypertension (Pepka-Zaba et al. 1991; Radermacher and Ramnos 1994), acute pneumonia (Blomqvist et al. 1993; Yaghi et al. 1993), and chronic ob-structive airway disease (Adatia et al. 1993; Moinard et al. 1994; Barbera et al. 1996). Ventilatory hypoxia is a consequence of many lung diseases, and the re-sponse of the pulmonary circulation occupies an interface between the normal and abnormal pulmonary circulation in the transition from health to disease. The ef-fects of NO on the cardiovascular and pulmonary systems have been the most stud-ied, and the effects observed in the pulmonary system are reviewed in this chapter.

Pulmonary Circulation

The healthy pulmonary circulation operates at a resistance one-fifth of that of the systemic circulation, even when pulmonary blood flow markedly increases with exercise (Stamler et al. 1994). High levels of NO are continuously produced in the human upper airway and are inhaled at each inspiration (Gerlach et al. 1994; Lundberg et al. 1994, 1995). Pulmonary-produced NO maintains low pulmonary arterial pressure at rest or during exercise (Stamler et al. 1994) and can oppose hy-poxic pulmonary vasoconstriction (HPV) (Frostell et al. 1993).

Hypoxic Pulmonary Vasoconstriction

In response to acute airway hypoxia, the pulmonary circulation increases in arte-rial pressure and vascular resistance (Fishman 1990). This acute hypoxic pressor response is a unique physiologic feature that has attracted much attention since it was first described in 1946 by Von Euler and Liljestrand. The pulmonary pressor response begins within seconds of ventilation with a hypoxic gas mixture, reaches a maximum within minutes (Eauge 1968), and is sustained for long periods of time (Malik and Kidd 1973). Although alveolar, or airway, hypoxia alone is suffi-cient to stimulate hypoxic vasoconstriction, an additional contribution to the ini-tiation of the response can come from the PO_2 of mixed venous blood (Marshall et al. 1981; Marshall and Marshall 1983). Hypoxic pulmonary vasoconstriction is a key determinant of regional pulmonary blood flow. By diverting blood flow from areas of alveolar hypoxia, HPV minimizes the consequences of regional hy-poventilation, thereby promoting regional ventilation–perfusion matching and improving gas exchange (Von Euler and Liljestrand 1946). A common hemody-namic feature of acute lung injury associated with multisystem trauma, sepsis, major general surgery, and cardiovascular surgery is ventilatory hypoxia. Therefore, alveolar hypoxia is a common feature of many forms of severe respiratory disease (Rubin 1995).

Mechanism of HPV

Although the mechanism of HPV is not known, two possible mechanisms have been proposed. Hypoxic pulmonary vasoconstruction could be initiated by endothelial release of a vasoconstrictor following detection of hypoxia. This release may explain differences observed in the response to hypoxia in pulmonary resistance vessels (Murray et al. 1990; Yuan et al. 1990; Madden et al. 1992) systemic resistance vessels (Yuan et al. 1990; Madden et al. 1992). However, the search for these mediators has been thorough but unsatisfactory. Some of the mediators investigated include histamine, serotonin, angiotensin II, catecholamines (Hauge 1968; Porcelli et al. 1977; Grover et al. 1983; McMurty et al. 1988), neurotransmitters (Silove et al. 1968a,b; Kazemi et al. 1972; Szidon and Flint 1977; Hales and Westphal 1979; Nandiwada et al. 1983), and vasoconstricting products of arachidonic acid metabolism, (Weir et al. 1976; Hales et al. 1978; Voekel et al. 1981; Morganroth et al. 1985; Schuster and Dennis 1987; Lonigro et al. 1988).

Alternatively, HPV may be due to a decrease in endothelium-derived relaxing factor (Weir 1978), Nitric oxide is an obvious candidate, and there is some evidence that hypoxia reduces NO synthesis (Jones et al. 1989; Rodman et al. 1990). However, if NO is released tonically, an inhibitor of NO synthesis given during normoxia should reproduce HPV. In fact, augmentation of resting pulmonary vascular tone by NO synthesis antagonists has generally been modest both in vivo (McCormick and Paterson 1993) and in vitro (Archer et al. 1989; Mazmanian et al. 1989; Liu et al. 1991; Bansal et al. 1993; Oka et al. 1993), and the pattern of the vasoconstrictor response to NO inhibitors is different from that for HPV. In most experimental systems, NO synthesis inhibitors augment HPV (Brashers et al. 1988; Archer et al. 1989; Persson et al. 1990; Liu et al. 1991; Ogata et al. 1992; Ohe et al. 1992). Thus, at the present time, the most likely physiologic role of NO in the normal pulmonary circulation is as a modulator of HPV. Experiments with isolated-perfused ventilated lungs from several species have demonstrated that a mechanism within the lung itself elicits pulmonary vasoconstriction during acute alveolar hypoxia (Voelkel 1986; Archer et al. 1989). Archer and co-workers (Archer et al. 1989) have hypothesized that during hypoxia, there is synthesis of NO, which tends to counteract HPV. Hyman et al. (1991) demonstrated that methylene blue, an inhibitor of soluble guanylate cyclase, prevented the increase in lobar arterial pressure in response to ventilatory hypoxia. The mechanism underlying HPV remains unknown.

In our laboratory (Feng et al. 1994) we investigated the effects of N^G-nitro-L-argininemethyl ester (L-NAME), an inhibitor of NO synthase; 6-anilino-5,8-quinolinedione (LY83583), an inhibitor of soluble guanylate cyclase; glybenclamide, a ATP-sensitive K^+-channel blocking agent; and 5,7-dimethyl-2-ethyl-3-[[2'-(1H-tetrazol-5yl)-[1,1']-biphenyl-4-yl]methyl]-3H-imidazo[4,5-b] pyridine

(L158809), an angiotensin II type I receptor antagonist, on the response to venti-latory hypoxia in the isolated blood-perfused rat lung.

In these experiments, 45 male Sprague-Dawley rats weighing 300 – 350 g (Hill Top Laboratories, Scottdale, PA) were intraperitoneally anesthetized with pento-barbital sodium (50 mg/kg). After stable anesthesia had been obtained, the trachea was surgically approached, cannulated with a short section of polyethylene tub-ing, connected to a rodent ventilator (Harvard Apparatus, South Natick, MA), and ventilated with room air enriched with 95% O_2/5% CO_2, with a tidal volume of 5 – 7 ml/kg and 2 cm H_2O positive end-expiratory pressure. The rats were hep-arinized with 1,000 units of heparin i.v., (Sigma Chemical Co., St. Louis, MO) and rapidly exsanguinated by withdrawing blood from the carotid artery.

The lungs were exposed by median sternotomy, and a ligature was placed around the aorta to prevent systemic loss of blood. The main pulmonary artery was catheterized, and the lungs were removed en bloc and suspended in a warmed (38°C), humidified (100%), water-jacketed chamber. An external heat exchanger (Haake D1 Heat Exchanger, Baxter Instrument Co., Harahan, LA) maintained the temperature of the perfusate and the isolated-lung chamber constant throughout the experiment. The perfusate solution (15 ml of heparinized blood and 5 ml of modified Krebs-Heinseleit solution) was placed in a reservoir and constantly mixed by a magnetic stirrer (Thermolyne, Cimarec II, Dubuque, IA). The lungs were perfused with a peristaltic roller pump (Cole-Palmer Instrument Co., Berrington, IL). Once the isolated lung perfusion circuit was established, the flow rate was set at 8 – 14 ml/min to maintain physiologic baseline pulmonary arterial perfusion pressure of 15 ± 0.5 mm Hg. The flow rate was confirmed in some ex-periments by timed collection of blood, using a stop watch and graduated cylinder at the end of the experiment. All vascular pressures were measured with Viggo-Spectramed transducers (Viggo-Spectramed, Oxnard, CA) zeroed at the level of the pulmonary arterial cannula. Pulmonary arterial perfusion pressure, airway pressure, and reservoir blood level were continuously monitored, electronically averaged, and recorded with a Grass polygraph, Model 7 (Grass Instrument Co., Quincy, MA). The modified Krebs-Heinseleit solution had the following compo-sition (g/l): NaCl, 66.37; KCl, 3.58: $CaCl_2 \cdot 2 H_2O$, 3.68; KH_2PO_4, 1.63; $MgSO_4 \cdot 7 H_2O$, 1.45; $NaHCO_3$, 2.0; Ficoll (type 70, Sigma Chemical Co., St. Louis, MO), 2.0; (pH 7.35 – 7.45). The solution was made fresh daily in double-distilled water.

Glybenclamide (Sigma Chemical Co., St. Louis, MO) was dissolved in a solu-tion containing 5 ml of 100% ethanol, 0.3 ml of 10% NaOH, and 4.7 ml of 0.9% NaCl. The resulting solution was diluted with 10 ml of 50 mM Tris, pH 8.4, to make a 5 mg/ml solution. LY83583 (6-anilino-5,8-quinolinedione; Calbiochemi-cal Co., La Jolla, CA) was dissolved in methanol and then diluted with distilled water. 5,7-Dimethyl-2-ethyl-3-2'-(1H-tetrazol-5yl)]-[1,1']-biphenyl-4-yl]methyl]-3H-imidazo[4,5-b] pyridine (L158809) (Merck-Sharp & Dohme, West Point, PA) was dissolved in saturated $NaHCO_3$ in 0.9% NaCl. Levcromakalim (Smith Kline-Beecham, UK) was dissolved in 20% ethanol in saline at a concentration of 1 mg/ml, and further dilutions were made in normal saline. L-NAME, acetylcholine chloride, serotonin creatinine sulfate, nitroglycerin, and angiotensin II (Sigma

Chemical Co., St. Louis, MO) were dissolved in 0.9% NaCl. Nitric oxide solutions in methanol were prepared by first bubbling 10 ml of reagent-grade methanol (EM Industries, Cherry Hill, NJ) for 20 minutes with nitrogen in a gastight reaction vial to remove dissolved oxygen. The methanol was then bubbled with NO gas (Hydrocarbon Technologies, Sulfur, LA) for 20 minutes and stored in a freezer in a gas-tight vial. Working solutions of all agents were prepared in 0.9% NaCl frequently, stored in brown stoppered bottles, and kept on crushed ice during experiments. The isolated blood-perfused lung was allowed to stabilize for 30 minutes before experiments were begin. Partial tensions of arterial blood gas and pH were measured by a blood gas analyzer (Corning Model 178, Corning Instruments, Medfield, MA) at intervals during the experiment. Blood pH was maintained between 7.35 and 7.45 by adding small amounts of $NaHCO_3$ solution to the reservoir. To serve as a control for agonist injections, equivalent volumes of saline or vehicle controls were injected directly into the perfusion circuit. All injections were made in small volumes in a random sequence, and sufficient time was permitted between agonist injections for the pressure to return to baseline values. Since the pulmonary blood flow and outflow pressure were maintained constant, changes in perfusion pressure in this preparation reflect changes in pulmonary vascular resistance. All vascular pressures are expressed in absolute units (mm Hg) as means $\pm$ SE. The data were analyzed by the paired t-test or analysis of variance with post hoc Scheffé's F test (Stat View Co., Berkeley, CA) using a Quadra 700. A P value of less than 0.05 was used as the criterion for statistical significance.

In the first series of experiments, in order to ascertain if tonic release of NO and/or ATP-sensitive K^+ channel activation was involved in modulating pulmonary vascular tone of the rat, the effects of L-NAME and glybenclamide on baseline pulmonary perfusion pressure were studied under conditions of controlled pulmonary blood flow and ventilation. L-NAME and glybenclamide were injected into the perfusion circuit distal to the pump in a random sequence.

In the second series of experiments, in order to investigate the effect of inhibition of NO synthesis, inhibition of activation of soluble guanylate cyclase, inhibition of ATP-sensitive K^+ channel activity, and muscarinic receptor blockade on the pulmonary response to ventilatory hypoxia, the lung preparation was ventilated with a gas mixture of 3% O_2-5% CO_2-92% N_2 for 5–8 minutes until a stable maximal pressor response was obtained. This hypoxic trial was repeated three times with a 10- to 15-minute interval between each challenge to ensure reproducibility of the response to the hypoxic gas mixture. L-NAME, LY83583, atropine, and glybenclamide were then added to the perfusion reservoir and allowed to equilibrate for 10 minutes. The lung preparation was then rechallenged with the hypoxic gas mixture as described above. The effects of L-NAME on responses to angiotensin II and serotonin were also determined in these experiments.

In the third series of experiments, the selectivity of the NOS inhibitor or ATP-sensitive K^+-channel blocking agent was assessed by investigating the effects of L-NAME or glybenclamide on responses to agents that produce vasodilation by diverse mechanisms. The effects of L-NAME or glybenclamide on responses to acetylcholine or levcromakalim were investigated when pulmonary arterial tone was enhanced with hypoxic ventilation. In the elevated-tone experiments, injections of acetylcholine, nitroglycerin, levcromakalim, or NO were made into the pulmonary perfusion circuit before and after treatment with L-NAME or glybenclamide.

In the last series of experiments, the effects of the angiotensin II type 1 receptor antagonist L158809 on the response to hypoxia were investigated. Under baseline pulmonary vascular tone conditions, vasoconstrictor responses to angiotensin II were determined before and after the administration of L158809. The specificity of the effects of the angiotensin II receptor antagonist on responses to angiotensin II was assessed by studying the effects of L158809 on the pressor response to serotonin.

Influence of L-NAME and Glybenclamide on Baseline Tone

To determine if tonic release of NO or the activation of ATP-sensitive K^+ channels is involved in maintaining the pulmonary vascular bed in a dilated state, the effects of L-NAME, an inhibitor of NO synthase, and of glybenclamide, an ATP-sensitive K^+-channel blocking agent, were investigated under conditions of controlled pulmonary blood flow. These data are summarized in Figure 13.2. The administration of L-NAME in doses of $10-1,000$ µg into the pulmonary arterial perfusion circuit produced small increases in the baseline pulmonary arterial perfusion pressure that were significant at doses of 100, 300, and 1,000 µg (Figure 13.2, upper panel). In contrast, injections of glybenclamide in doses of $30-1,000$ µg caused significant increases in baseline pulmonary arterial perfusion pressure (Figure 13.2, lower panel). The Responses to glybenclamide were significantly greater than the responses to L-NAME when the pressor responses were compared at doses of 300 and 1,000 µg (Fig. 2).

Effects of L-NAME LY83583, Atropine, and Glybenclamide on the Response to Ventilatory Hypoxia

In addition to investigating the effects of L-NAME and glybenclamide on baseline tone, the effects of inhibition of NO synthesis, inhibition of activation of soluble guanylate cyclase, muscarinic receptor blockade, and inhibition of ATP-sensitive K^+ channels on the response to ventilatory hypoxia were investigated. These data are summarized in (Figure 13.3). During the control period, exposure of the lung to three trials of ventilation with the hypoxic gas mixture (3% O_2-5% CO_2-92% N_2) for 5-8 minutes increased pulmonary arterial perfusion pressure from 15 ± 0.6 to 33 ± 1 mm Hg ($\Delta 18 \pm 0.4$ mm Hg), and the response was reproducible after an interval of 10-15 minutes. After hypoxic ventilation was terminated, the, lung was ventilated with 95% O_2-5% CO_2 and the tone was allowed to return to baseline value. At this time, injections of L-NAME (100 nmol/ml) or LY83583 (100 nmol/ml) into the perfusion circuit increased pulmonary arterial perfusion pressure from a baseline value of 15 ± 0.6 mm Hg to peak values of 16

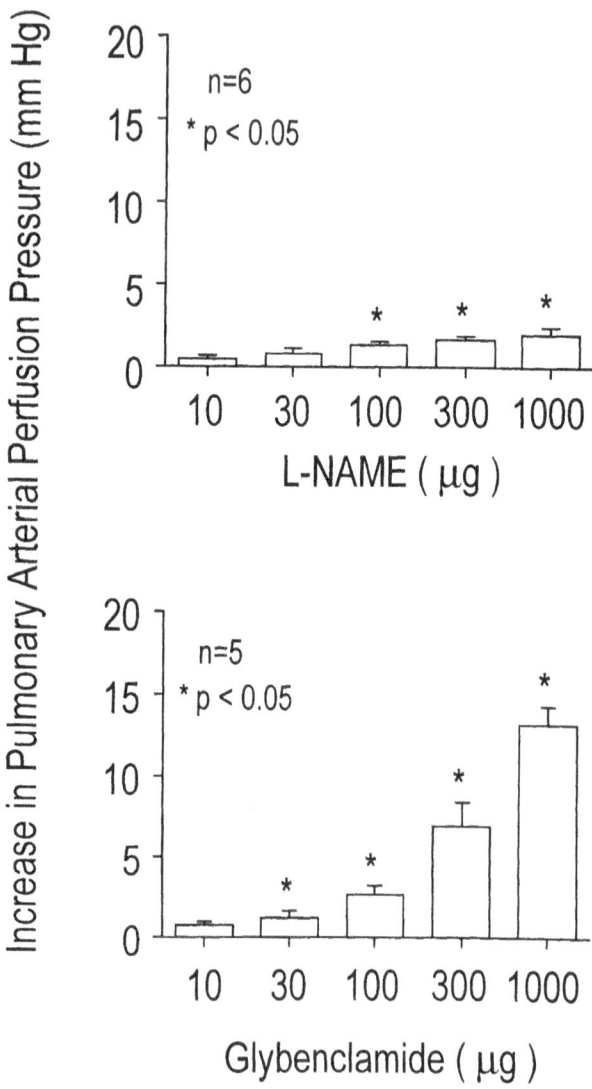

FIGURE 13.2. Influence of L-NAME and glybenclamide on baseline pulmonary arterial perfusion pressure. N L-NAME in doses 10 – 1,000 µg (top panel) and glybenclamide in doses of 10 – 1,000 µg (bottom panel) were injected into pulmonary artery perfusion circuit, n indicates number of animals, and * P < 0.05 indicates significantly different from control.

± 0.8 and 18 ± 1 mm Hg, respectively (Figure 13.3, upper panels). The muscarinic receptor antagonist atropine at a dose of 10 nmol/ml that blocked responses to acetylcholine in this preparation, had no significant effect on the pressor response to hypoxia (Figure 13.3, left lower panel). Injection of glybenclamide (100 nmol/ml) increased baseline pulmonary arterial perfusion pressure from 15 ± 1 to 24 ± 2 mm Hg (Figure 13.3, right lower panel). When the pulmonary arterial perfusion pressure attained a steady level after administra-

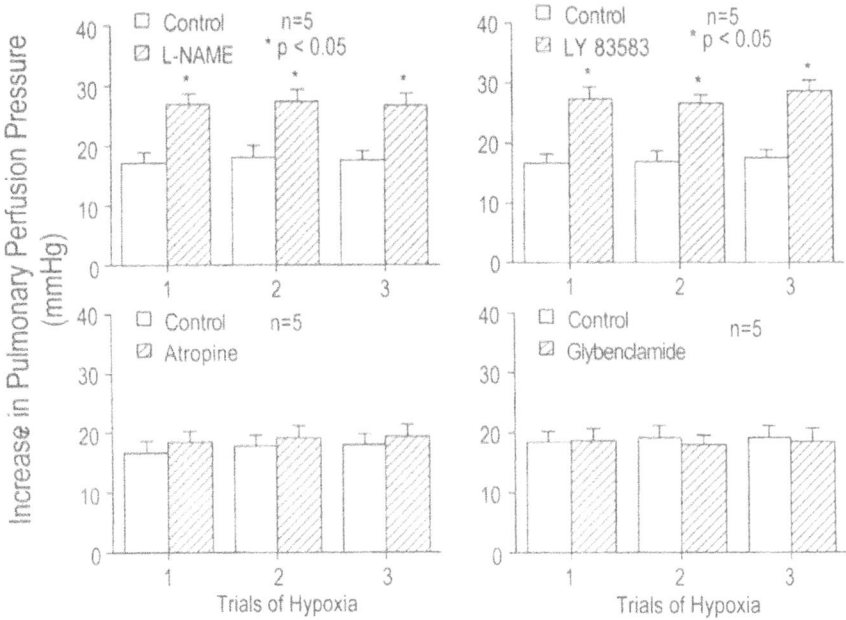

FIGURE 13.3. Influence of the nitric oxide synthase inhibitor, N L-NAME; the soluble guanylate cyclase inhibitor, LY83583; the muscarinic receptor antagonist, atropine; and the ATP-sensitive K^+ channel blocking agent, glybenclamide on the response to hypoxia. The response to 3 trials of hypoxia (3% O_2/5% CO_2/92% N_2 for 5–8 min) was compared before and after administration of L-NAME (100 nmol/ml; top left), LY 83583 (100 nmol/ml; top right), atropine (10 nmol/ml; bottom left) and glybenclamide (100 nmol/ml; bottom right). The blocking agents were added to the perfusion reservoir. n indicates number of animals. * P < 0.05 indicates significantly different from control.

tion of L-NAME, LY83583, or glybenclamide, the lung was rechallenged with the hypoxic gas mixture for 5–8 minutes. In these experiments, L-NAME and LY83583 significantly enhanced the pulmonary pressor response to ventilatory hypoxia (Figure 13.3 upper panels), whereas atropine and glybenclamide had no significant effect on the pressor response to hypoxic ventilation (Figure 13.3 lower panels).

Effects of L-NAME on Responses to Angiotensin II and Serotonin

In order to investigate the role of NO release in modulating pressor responses to angiotensin II and serotonin, the effects of L-NAME were investigated. During the control period, injections of angiotensin II produced dose-related increases in pulmonary arterial pressure (Figure 13.4, upper panel). After administration of L-NAME, pulmonary pressor responses to angiotensin II were significantly increased (Figure 13.4, upper panel). L-NAME shifted the dose–response curve for

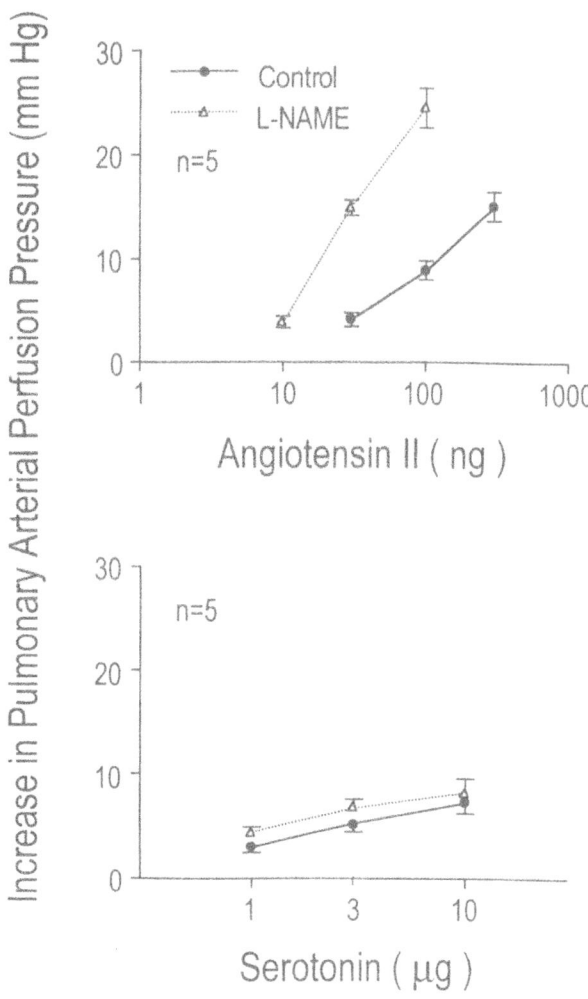

FIGURE 13.4. Influence of L-NAME on vasoconstrictor responses to angiotensin II (top panel) and serotonin (bottom panel). Responses to angiotensin II (10–100 ng) and serotonin (1–10 μg) were compared before and after administration of L-NAME (100 nmol/ml). n indicates number of animals. * P < 0.05 indicates significantly different from control.

angiotensin II to the left in a parallel manner (Figure 13.4, upper panel). Additionally, in these experiments, the effects of L-NAME on responses to serotonin were assessed, and injections of serotonin produced small increases in pulmonary arterial perfusion pressure (Figure 13.4, lower panel). After administration of L-NAME, pressor responses to serotonin were not significantly changed (Figure 13.4, lower panel).

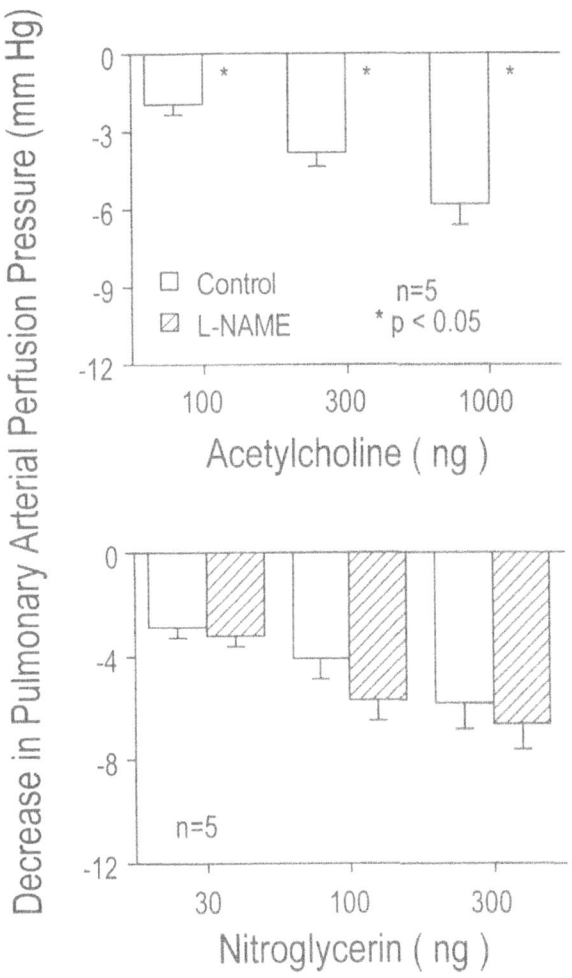

FIGURE 13.5. Influence of L-NAME on vasodilator responses to acetylcholine (top panel) and nitroglycerin (bottom panel) when tone was elevated by hypoxia. Responses to acetylcholine (100–1,000 ng; top panel) and nitroglycerin (30–300 ng; bottom panel) were compared before and after administration of L-NAME at a dose of 100 nmol/ml). n indicates number of animals. * $P < 0.05$ indicates significantly different from control.

Effects of L-NAME and Glybenclamide on Vasodilator Responses to Acetylcholine, Nitroglycerin, Levcromakalim, and NO

The effects of the NOS inhibitor and the ATP-sensitive K^+-channel blocking agent on responses to the vasodilator agents were also investigated. The data are summarized in (Figures 13.5 and 13.6). The pulmonary vasodilator responses to acetylcholine, nitroglycerin, levcromakalim, and NO were obtained when pulmonary arterial perfusion pressure was elevated by hypoxic ventilation. After the

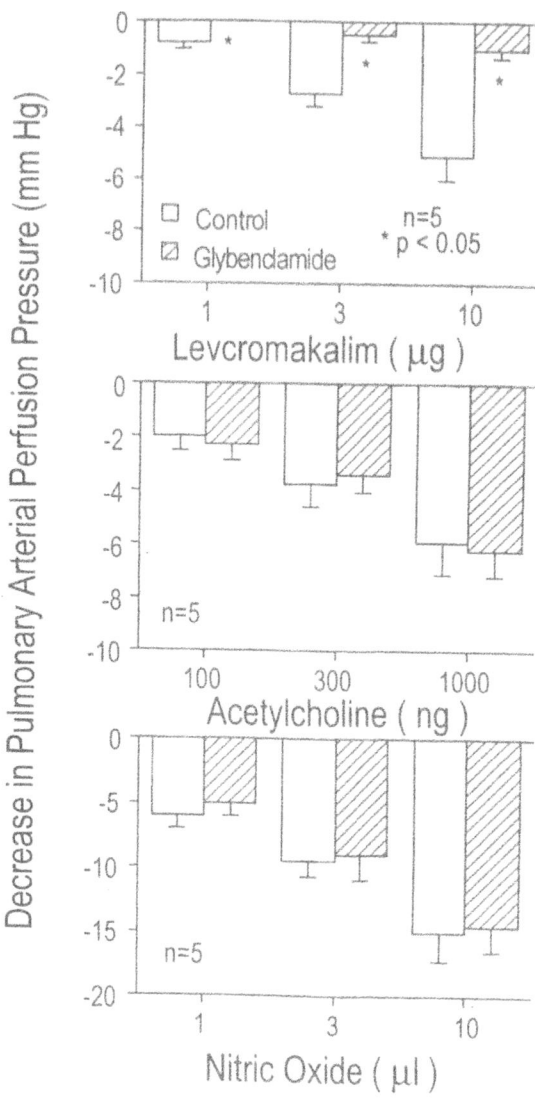

FIGURE 13.6. Effect of glybenclamide on responses to levcromakalim, acetylcholine, and nitric oxide. Vasodilator responses to levcromakalim (1–10 μg; top panel), acetylcholine (100–1,000 μg; middle panel) and nitric oxide (1–10 μl; bottom panel) were compared before and after administration of glybenclamide (100 nmol/ml). n indicates number of animals. * P < 0.05 indicates significantly different from control.

tone returned to baseline conditions, L-NAME (100 nmol/ml) or glybenclamide (100 nmol/ml) was added to the reservoir, and the tone was again increased by hypoxic ventilation. Injection of acetylcholine (100–1,000 ng) or nitroglycerin (30–300 ng) into the pulmonary arterial perfusion circuit reduced pulmonary arterial perfusion pressure in a dose-dependent manner. After the administration of 100 nmol/ml of L-NAME, a dose that increased the pulmonary pressor response to hypoxia, the decreases in pulmonary arterial perfusion pressure in response to acetylcholine were completely blocked (Figure 13.5, upper panel), whereas the vasodilator responses to nitroglycerin were not decreased (Figure 13.5, lower panel).

In experiments with the ATP-sensitive K^+-channel blocking agent glybenclamide, injections of levcromakalim (1–10 μg), acetylcholine (100–1,000 ng), or NO (1–10 μl) directly into the pulmonary arterial perfusion circuit significantly decreased pulmonary arterial perfusion pressure in a dose-dependent manner (Figure 13.6). After the administration of glybenclamide at a dose of 100 nmol/ml, the decreases in pulmonary arterial perfusion pressure in response to levcromakalim were significantly reduced (Figure 13.6, upper panel), whereas the vasodilator responses to acetylcholine and NO solutions were not altered (Figure 13.6, middle and lower panels).

Effects of L158809 on the Response to Hypoxia

Angiotensin II has been reported to modulate the pressor response to hypoxia in the pulmonary vascular bed of the rat (Liu et al. 1991). In the present experiments, the effect of L158809, an angiotensin II type 1 receptor antagonist, on the response to hypoxia was investigated. These data are presented in (Figure 13.7). When the lung was challenged with a hypoxic gas mixture (3% O_2/5% CO_2/92% N_2) for 5–8 minutes, pulmonary arterial perfusion pressure was increased from 15 ± 0.6 mm Hg to 33 ± 1 mm Hg. After the exposure to hypoxia has been terminated, the lung was ventilated with 95% O_2-5% CO_2 and the tone was permitted to return to baseline values, L158809 (10 nmol/ml) was then added to the reservoir and the lung was rechallenged with the hypoxic gas mixture. After administration of L158809 (10 nmol/ml), the increase in pulmonary arterial perfusion pressure in response to hypoxia was not altered (Figure 13.7, upper panel). Injections of angiotensin II and serotonin increased pulmonary arterial perfusion pressure in a dose-related manner (Figure 13.7 middle and lower panels). After the administration of L158809 (10 nmol/ml), the pressor response to angiotensin II was reduced significantly (Figure 13.7, middle panel), whereas the pressor response to serotonin was not changed (Figure 13.7, lower panel).

The results of these investigations demonstrate that the NOS inhibitor L-NAME enhances the pulmonary vasoconstrictor response to ventilatory hypoxia and, at higher doses, has a modest effect on baseline tone in the isolated blood-perfused rat lung. Since pulmonary blood flow and outflow pressure were constant, the increase in pulmonary arterial perfusion pressure in response to hypoxia reflects an

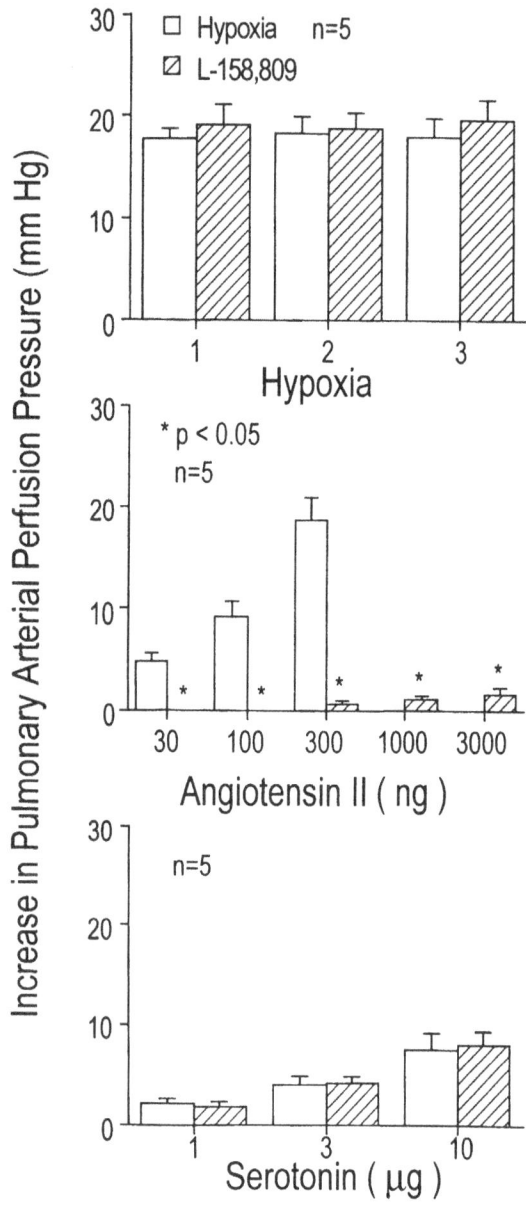

FIGURE 13.7. Influence of L158809 on pressor responses to hypoxia, serotonin, and angiotensin II. Responses to hypoxia (3% O_2-5% CO_2-92% N_2 for 5–8 min; top panel), angiotensin II (30–3,000 ng; middle panel) and serotonin (1–10 μg; bottom panel) were compared before and after administration of angiotensin II receptor type 1 antagonist, L158809 (10 nmol/ml). n indicates number of animals. * P < 0.05 indicates significantly different from control.

increase in pulmonary vascular resistance. L-NAME at a dose of 100 nmol/ml, which significantly increased the pressor response to hypoxia and selectively reduced the vasodilator response to acetylcholine, had only small effects on baseline pulmonary perfusion pressure, suggesting that tonic release of NO plays a minor role in maintaining the pulmonary vascular bed in a dilated state in the rat. These data are consistent with results from previous studies using the isolated rat lung preparation (Brashers et al. 1988; Robertson et al. 1990; Liu et al. 1991). This observation in the isolated perfused rat lung differs from the results of previous studies in the pulmonary vascular bed of the cat and dog, in which NOS inhibitors markedly increased systemic and pulmonary arterial pressures (Rees et al. 1989; Bellan et al. 1991; McMahon et al. 1991). The reason for the difference between the results in the pulmonary circulation of the rat and the cat is uncertain, but it may reflect important species differences in the role of basal endothelial cell NO release in regulating baseline tone in the pulmonary vascular bed. The observation that the pressor response to angiotensin II was greatly enhanced by L-NAME is in agreement with the results of previous studies in the rat lung and suggests that angiotensin II-induced EDRF release from endothelium plays a major role in modulating the angiotensin II-induced vasoconstriction in the pulmonary vascular bed of the rat (Liu et al. 1991)

When the pulmonary vascular tone was elevated with hypoxia, vasodilator responses to acetylcholine were completely inhibited by the NOS inhibitor L-NAME, suggesting that under increased pulmonary arterial tone conditions, the release of NO in the pulmonary circulation is important in the mediation of the vasodilator response to acetylcholine. In contrast, it has been reported that vasodilator responses to acetylcholine are only reduced by L-NAME in the pulmonary and hindlimb vascular beds of the cat, suggesting that a muscarinic receptor mechanism, in addition to NO release, they play a major role in mediating the response to acetylcholine in these species (Bellan et al. 1991; McMahon et al. 1991). It has also been reported that NOS inhibitors have no effect on the response to acetylcholine in the isolated perfused rat lung (Archer et al. 1989). The reason for the difference between the results of previous studies and the present study is uncertain but may reflect differences in the experimental methods used or in the perfusate composition employed. In some vascular beds, an endothelium-derived factor has been reported to hyperpolarize vascular smooth muscle through release of an unidentified endothelium-derived hyperpolarizing factor (Hasunuma et al. 1991). However, in the present study, glybenclamide, an ATP-sensitive K^+-channel blocking agent, had no effect on the response to acetylcholine, suggesting that ATP-sensitive K^+ channel activation and membrane hyperpolarization are not involved in the pulmonary vasodilator response to acetylcholine in the rat.

In the present study, glybenclamide increased baseline pulmonary perfusion pressure and, at a dose of 100 nmol/ml, selectively reduced the vasodilator response to levcromakalim. Glybenclamide, however, had no effect on the response to hypoxia. These data suggest that activation of ATP-sensitive K^+ channels may play a role in regulating the baseline tone in the pulmonary vascular bed, but that inactivation of ATP-sensitive K^+ channels does not modulate the pressor response to hypoxia. The results of the present study showing that glybenclamide had no effect on the hypoxic response are consistent with the results of previous studies (Hasunuma et al. 1991). However, the enhancement of baseline pulmonary arterial pressure by the ATP-sensitive K^+-channel blocking agent is not in agreement with the results of previous studies (Hasunuma et al. 1991). In the present study, glybenclamide, at a lower dose than that used in previous experiments, also caused an increase in the baseline pulmonary arterial pressure, and the glybenclamide vehicle had no significant effect on baseline tone. The reason for the difference between the results of the present study and those of previous studies in the isolated perfused rat lung is unknown. It may be related to differences in the perfusate composition used and requires further investigation.

In conclusion, the results of the present study demonstrate that baseline pulmonary arterial pressure is markedly increased by glybenclamide, but that L-NAME has only modest effects, suggesting that activation of ATP-sensitive K^+ channels, but not of tonic NO release, may play an important role in regulating baseline tone in the pulmonary vascular bed of the rat. The results of the present investigation show that L-NAME and LY83583 enhance the pressor response to hypoxia but glybenclamide and L158809 have no effect, suggesting that the pressor response to hypoxia is modulated by NO release and/or activation of soluble guanylate cyclase but not by activation of ATP-sensitive K^+ channels or angiotensin II receptors in the isolated perfused rat lung. These data suggest that NO release plays an important role in regulating the pulmonary vascular bed in the rat under conditions of elevated tone by mediating vasodilator responses to acetylcholine and under baseline tone conditions by restraining vasoconstrictor responses to ventilatory hypoxia and angiotensin II.

Nitric oxide-induced relaxation of arterial segments from the lung and peripheral organs is associated with increased cGMP levels in smooth muscle cells (Ignarro et al. 1986; Archer et al. 1989, 1994). However, the mechanism by which increased cGMP levels induce vasorelaxation is uncertain (Edwards et al. 1983). The hypothesis that activation of Ca^{2+}-sensitive K^+ channels plays a role in mediating vasorelaxant responses to NO was examined in isolated arteries from the rat lung and in vascular smooth muscle cells from rat conduit and resistance pulmonary arteries (Archer et al. 1994, 1996). The results of those studies suggest that NO and NO donor-induced relaxation is dependent on the activation of Ca^{2+}-sensitive K^+ channels (Archer et al. 1994, 1996). Although the effects of K^+ channel blockers on vascular tone in meclofenamate-treated, salt solution-perfused rat lungs have been examined, the effects of an inhibitor of Ca^{2+}-sensitive K^+ channels on responses to NO-donating agents have not been evaluated in the isolated blood-perfused rat lung (Hasunuma et al. 1991).

Inhibition of K^+ channels may also represent an important mechanism in mediating the pulmonary arterial pressor response to decreased PO_2 (Cornfield et al. 1992; Post et al. 1992; Feng et al. 1994; Nossaman et al. 1997). It has also been suggested that alterations in smooth muscle ion channel activity and Ca^{2+} homeostasis occur in response to stress in the pulmonary circulation (Post et al. 1992;

Rodman 1992; Yuan et al. 1993; Post et al. 1995). It has been suggested that there are four types of K^+ channels in vascular smooth muscle (Dumas et al. 1994; Tagaya et al. 1995; Yuan 1995). However, pulmonary artery vascular smooth muscle cells have been shown to have three types of K^+ channels: K_{Ca} (Post et al. 1992), delayed rectifier K_{DR} (Archer et al. 1993; Yuan et al. 1993; Post et al. 1995, Yuan 1995), and K_{ATP}-gated channels (Clapp and Gurney 1991, 1992; Clapp et al. 1993; Weiner et al. 1991). Moreover, the type of K^+ channel present in pulmonary artery vascular smooth muscle is important, since these channels may be involved in mediating the response to hypoxia (Yuan et al. 1993). The effects of various K^+ channel antagonists have been investigated in the airways and in the pulmonary vascular bed (Dumas et al. 1994; Yuan et al. 1993; Tagaya et al. 1995). However, the effects of charbydotoxin on responses to vasodilator agents and ventilatory hypoxia have not been investigated in the isolated blood-perfused rat lung. The purpose of these experiments was, therefore, to investigate the effects of charby-dotoxin, an inhibitor of large-conductance Ca^{2+}-sensitive K^+ channels, on va-sodilator responses to the NO donors nitroglycerin and nitroprusside, and on the response to ventilatory hypoxia in the isolated perfused rat lung. (Harvey et al. 1995).

In these 25 experiments (Nossaman et al. 1997), responses to the vasodilator agents were investigated when pulmonary arterial perfusion pressure was raised to a high steady level by addition of U 46619 to the perfusion reservoir. In some experiments, the tone was increased by ventilating the lung with a 3% O_2-5% CO_2-92% N_2 mixture, which resulted in a marked reduction in PO_2 but no change in PCO_2 or pH (pH: control 7.4 ± 0.05, hypoxia 7.4 ± 0.03; PaO_2: control 125 ± 10 mm Hg, hypoxia 42 ± 1 mm Hg; $PaCO_2$: control 28 ± 2 mm Hg, hypoxia 29 ± 1 mm Hg). Charybdotoxin (Bachem Bioscience, King of Prussia, PA), sodium nitroprusside, nitroglycerin, albuterol, and isoproterenol hydrochloride (Sigma Chemical Co., St. Louis, MO) were dissolved in 0.9% NaCl and injected directly into the perfusion circuit distal to the perfusion pump.

The effects of the Ca^{2+}-sensitive K^+-channel antagonist charybdotoxin were investigated on vasodilator responses to nitroprusside and nitroglycerin and to al-buterol and isoproterenol when tone was increased with U46619 to determine if Ca^{2+}-sensitive K^+ channels mediate or modulate the responses to the vasodilator agonists that act by releasing NO and increasing cGMP levels and by increasing cAMP levels (Edwards et al. 1983; Ignarro et al. 1986). In the second series of ex-periments, responses to charybdotoxin were investigated to determine the effects of the Ca^{2+}-sensitive K^+-channel antagonist on the pressor response to hypoxic ventilation.

Influence of Charybdotoxin on Responses to Nitroglycerin, Sodium Nitroprusside, Albuterol, and Isoproterenol

In order to determine the role of activation of Ca^{2+}-sensitive K^+ channels on va-sodilator responses in the rat lung, the effects of charybdotoxin on vasodilator

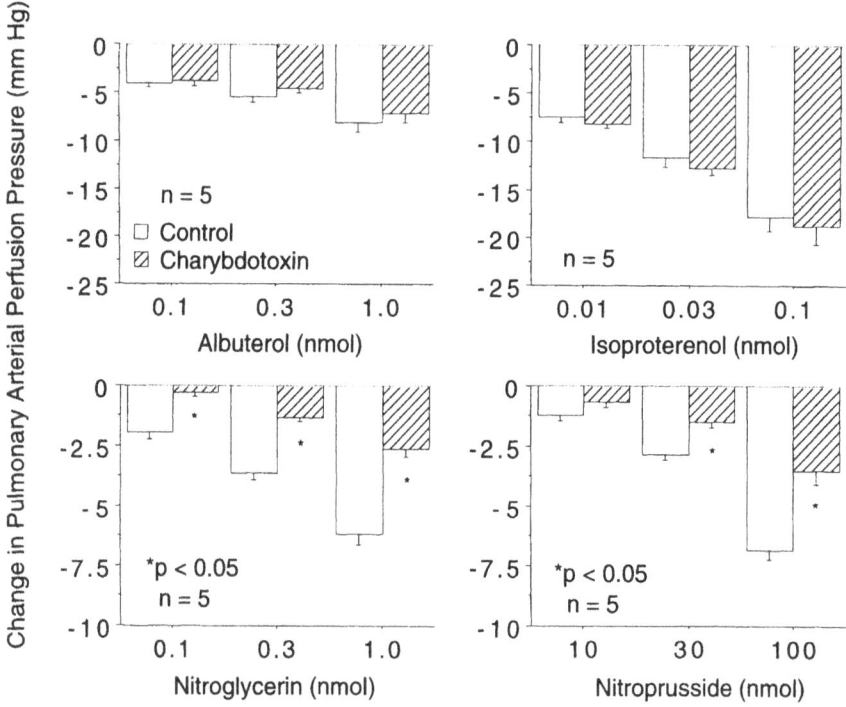

FIGURE 13.8. Dose-response relationships illustrating the effects of charybdotoxin in a final reservoir concentration of 50 ng/ml on vasodilator responses to albuterol (0.1–1.0 nmol), isoproterenol (0.01–0.1 nmol) (upper panels) and to nitroglycerin (0.1–1.0 nmol), and sodium nitroprusside (10–100 nmol) (lower panels) in the isolated blood perfused lung of the rat. n equals the number of experiments. The asterisks indicate that the responses are significantly different from control. The vertical bars indicate standard error of the mean.

responses to nitroglycerin, sodium nitroprusside, albuterol, and isoproterenol were investigated in the isolated perfused lung and the results of these experiments are summarized in figure 13.8. When the tone was increased with U46619, injections of albuterol (0.1–1.0 nmol), isoproterenol (0.01–0.1 nmol), nitroglycerin (0.1–1.0 nmol), and sodium nitroprusside (10–100 nmol) caused dose-related decreases in pulmonary arterial perfusion pressure (Fig. 13.8). Administration of charybdotoxin (final reservoir concentration, 50 ng/ml) had no significant effect on vasodilator responses to albuterol or isoproterenol in the pulmonary vascular bed of the rat (Fig. 13.8, upper panels), whereas pulmonary vasodilator responses to nitroglycerin and sodium nitroprusside were significantly decreased following treatment with the Ca^{2+}-sensitive K^+-channel antagonist (Fig. 13.8, lower panels). Injection of charybdotoxin into the perfusion reservoir had no significant effect on baseline pulmonary arterial perfusion pressure (control 13.5 ± 2.3 mm Hg, charybdotoxin 13.8 ± 2.7 mm Hg).

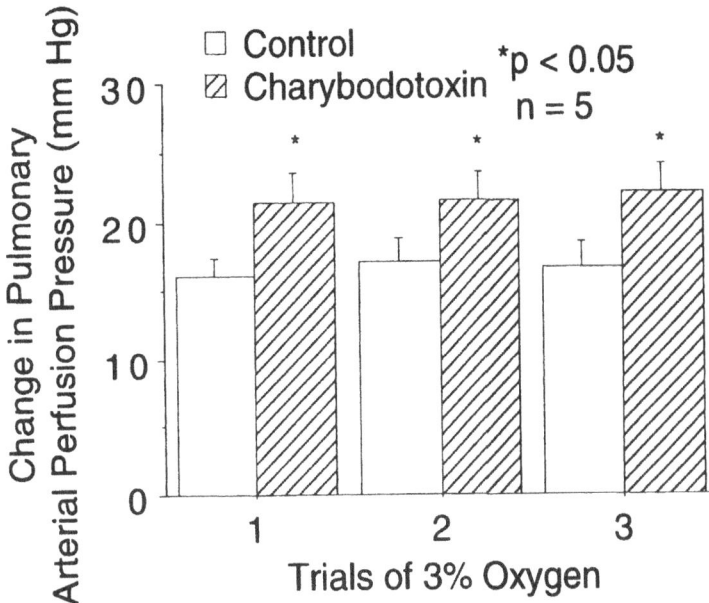

FIGURE 13.9. Influence of charybdotoxin on pressor responses to ventilatory hypoxia. Responses to three trials of hypoxia (3% O_2-5% CO_2-balance N_2 gas mixture for 5–8 min) were compared before and after administration of charybdotoxin in final reservoir concentration of 50 ng/ml. n equals the number of experiments. The asterisks indicate that the responses are significantly different from control. The vertical bars indicate standard error of the mean.

Influence of Charybdotoxin on Ventilatory Hypoxia

During the control period, exposure of the rat lung preparation to three trials of ventilation with a hypoxic gas mixture (3% O_2-5% CO_2-balance N_2) for 5–8 minutes increased pulmonary arterial perfusion pressure. The response was reproducible in the three trials after an interval of 10–15 minutes between trials (Fig. 13.9). After hypoxic ventilation had been terminated, the lung was again ventilated with 30% O_2-5% CO_2 balance N_2 gas mixture and the tone was allowed to return to baseline values. After administration of the Ca^{2+}-sensitive K^+-channel antagonist, the lung was again challenged with the hypoxic gas mixture for 5–8 minutes. After administration of charybdotoxin (final reservoir concentration, 50 ng/ml), a small but significant increase in the pulmonary pressor response to ventilatory hypoxia was observed (Fig. 13.9).

The results of these investigations show that vasodilator responses to nitroglycerin and sodium nitroprusside are reduced by charybdotoxin in the isolated blood-perfused rat lung. The inhibitory effects of charybdotoxin on responses to the NO-donating vasodilator agents were selective, in that vasodilator responses to the β-receptor agonists isoproterenol and albuterol were not altered. The results of the present investigation are consistent with results of studies in isolated arteries from the rat lung and provide support for the hypothesis that vasodilator responses to NO and NO donors are mediated, in part, by the activation of Ca^{2+}-sensitive K channels (Archer et al. 1994, 1996).

It has been reported that vasodilation induced by NO and NO-donating drugs is associated with an increase in smooth muscle cGMP levels (Edwards et al. 1983; Ignarro et al. 1986). However, the mechanism by which cGMP causes vasodilation is uncertain, although a decrease in intracellular Ca^{2+} concentration is involved (Archer et al. 1994). It has been suggested that cGMP-mediated vasodilation is associated with changes in smooth muscle membrane potential (Furchgott 1984). Nitric oxide has been shown to hyperpolarize vascular smooth muscle cells, and it has been shown that agonists that increase cGMP levels activate K^+ channels (Fujino et al. 1991; Robertson et al. 1993). K^+ channel activation causes membrane hyperpolarization, which inhibits voltage-dependent calcium entry and promotes vasorelaxation. The results of the present study with charybdotoxin, which inhibits the large-conductance calcium-activated K^+ channel, are consistent with the results of studies in isolated arterial segments from the rat lung and provide support for the hypothesis that NO-donating drugs induce vasodilation by increasing K^+ conductance through the calcium-sensitive K^+ channel and hyperpolarize resistance vessel channels in the rat lung (Archer et al. 1994).

The results of the present investigation also demonstrate that the Ca^{2+}-sensitive K^+ channel antagonist charybdotoxin causes a small but significant increase in the pulmonary vasoconstrictor response to ventilatory hypoxia in the isolated blood-perfused rat lung. The observation that a number of K^+-channel inhibitors simulate hypoxic pulmonary vasoconstriction by increasing tension in intact pulmonary arterial rings and elevate pulmonary arterial pressure in the isolated lung suggests that K^+ channel inhibition may be an important event in hypoxic pulmonary vasoconstriction (McMurty 1984; Hasunuma et al. 1991; Post et al. 1992; Yuan 1995; Archer et al. 1996). This hypothesis has received support from recent studies demonstrating that hypoxia inhibits macroscopic whole-cell K^+ currents, causing depolarization of the resting membrane potential in both acutely isolated and cultured pulmonary arterial smooth muscle cells (Post et al. 1992; Yuan et al. 1993). Our results with charybdotoxin provide support for the hypothesis that once hypoxia-induced pulmonary arterial vasoconstriction is initiated, the Ca^{2+}-sensitive K^+ channels modulate or limit the pulmonary arterial pressor response to hypoxic ventilation in the isolated blood-perfused rat lung.

Charybdotoxin at a concentration of 50 ng/ml, which significantly increased the pressor response to hypoxia and selectively reduced vasodilator responses to nitroglycerin, and sodium nitroprusside had only small effects on baseline pulmonary perfusion pressure, suggesting that the Ca^{2+}-sensitive K^+ channels play a minor role in maintaining the pulmonary vascular bed in a dilated state in the isolated blood-perfused rat lung.

In conclusion, the results of the present study demonstrate that charybdotoxin inhibits vasodilator responses to the NO-donating agents nitroglycerin and nitroprusside, without altering responses to albuterol or albuterol. Since baseline pulmonary arterial pressure is not markedly increased by charybdotoxin, these results suggest that activation of Ca^{2+}-sensitive K^+ channels does not play an important role in regulating baseline tone in the pulmonary vascular bed of the rat. The results of the present investigation showing that charybdotoxin enhances the pressor response to ventilatory hypoxia suggest that the pressor response is modulated by the activation of Ca^{2+}-sensitive K^+ channels in the isolated blood-perfused rat lung. These data suggest that Ca^{2+}-sensitive K^+ channel activation plays an important role in mediating vasodilator responses to the NO donors, nitroglycerin and sodium nitroprusside, and a smaller role in modulating or limiting the pressor response to ventilatory hypoxia.

References

Adatia I, Thomson J, Landzberg M, et al. 1993. Inhaled nitric oxide in chronic obstructive lung disease. *Lancet 341:*307–308.

Anbar M. 1996. The role of nitric oxide in the evolution of living systems from primordial prokaryotes to eukaryotes: a hypothesis. In: Moncada S, Stamler J, Gross S, and Higgs EA, eds. *The Biology of Nitric Oxide, Part 5.* London: Portland Press, p 226.

Archer SL, Tolins JP, Raij L, & Weir EK. 1989. Hypoxic pulmonary vasoconstriction is enhanced by inhibition of the synthesis of an endothelium-dependent derived relaxing factor. *Biochem. Biophys. Res. Commun. 164:*1198–1205.

Archer SL, Weir EK, & McMurty IF. 1989. Mechanism of acute hypoxic and hyperoxic changes in pulmonary vascular reactivity. In: Weir EK, Reeves JT, eds. *Pulmonary Vascular Physiology and Pathophysiology.* New York: Marcel Dekker, pp 241–290.

Archer SL, Huang J, Henry T, Peterson D, & Weir EK. 1993. A redox based oxygen sensor in rat pulmonary vasculature. *Circ. Res. 73:*1100–1112.

Archer SL, Huang V. Hampl, Nelson DP, Shultz PJ, & Weir EK. 1994. Nitric oxide and cGMP cause vasorelaxation by activation of a charybdotoxin-sensitive K channel by cGMP-dependent protein kinase. *Proc. Natl. Acad. Sci. USA 91:*7583–7587.

Archer SL, Huang JMC, Reeve HL, Hampl V, Tolarová S, Michelakis E, & Weir EK. 1996. Differential distribution of electrophysiologically distinct myocytes in conduit and resistance arteries determines their response to nitric oxide and hypoxia. *Circ. Res. 78:*431–442.

Atz AM, Adatia I, Jonas RA, & Wessel DL. 1996. Inhaled nitric oxide in children with pulmonary hypertension and congenital mitral stenosis. *Am. J. Cardiol. 77:*316–319.

Bansal V, Toga H, & Usha Raj J. 1993. Tone dependent nitric oxide production in ovine vessels in vitro. *Respir. Physiol. 93:*249–260.

Barbera JA, Roger N, Roca J, Rovira I, Higenbottam TW, & Rodriguez-Roisin R. 1996. Worsening of pulmonary gas exchange with nitric oxide inhalation in chronic obstructive pulmonary disease. *Lancet. 347(8999):*436–440.

Bellan JA, Minkes RK, McNamara DB, & Kadowitz PJ. 1991. N^ω-nitro-L-arginine selectively inhibits vasodilator responses to acetylcholine and bradykinin in cats. *Am. J. Physiol. 260:*H1025–H1033.

Berner M, Beghetti M, Spahr-Schopfer I, Oberhansli I, & Friedli B. 1996. Inhaled nitric oxide to test the vasodilator capacity of the pulmonary vascular bed in children with long-standing pulmonary hypertension and congenital heart disease. *Am. J. Cardiol.* *77:*532–535.

Blomqvist H, Wickerts CJ, Andreen M, Ullberg U, Ortqvist A, & Frostell C. 1993. Enhanced pneumonia resolution by inhalation of nitric oxide? *Acta Anaesthesiol Scand 37:*110–114.

Brashers VJ, Pearch MJ, & Rose CE Jr. 1988. Augmentation of hypoxic pulmonary vasoconstriction in the isolated perfused rat lung by in vitro antagonists of endothelium-dependent relaxation. *J. Clin. Invest. 82:*1495–1502.

Chiuodi H, & Mohler JG. 1985. Effects of exposure of blood hemoglobin to nitric oxide. *Environ. Res. 37:*355–363.

Clapp L, & Gurney A. 1991. Outward currents in rabbit pulmonary artery cells disassociated with a new technique. *Exp. Physiol. 76:*667–693.

Clapp LH, & Gurney AM. 1992. ATP-sensitive K^+ channels regulate resting potential of pulmonary arterial smooth muscle cells. *Am. J. Physiol. 262:*H916–H920.

Cornfield DN, McQueston JA, McMurtry IF, Rodman DM, & Abman SH. 1992. Role of ATP-sensitive potassium channels in ovine fetal pulmonary vascular tone. *Am. J. Physiol. 263:*H1363–H1368.

Culotta E, & Koshland DE Jr. 1992. Molecule of the year: NO news is good news. *Science 258:*1862–1865.

Doyle MP, & Hoekstra JW. 1981. Oxidation of nitrogen oxides by bound dioxygen in hemoproteins. *J. Inorg. Biochem. 14:*351–358.

Dumas JP, Dumas M, Sgro C, Advenier C, & Giudicelli JF. 1994. Effects of two K^+ channel openers, aprikalim and pinacidil, on hypoxic pulmonary vasoconstriction. *Eur. J. Pharmacol. 263:*17–23.

Edwards JC, Ignarro LJ, Hyman AL, & Kadowitz PJ. 1983. Relaxation of intrapulmonary artery and vein by nitrogen oxide-containing vasodilators and cyclic GMP. *J. Pharmacol. Ther. 228:*33–42.

Feng CJ, Cheng DY, Kaye AD, Kadowitz PJ, & Nossaman BD. 1994. Influence of N^{ω}-L-nitro-*L*-arginine methyl ester, LY83583, glybenclamide and L158809 in the perfused rat lung. *Eur. J. Pharm. 263:*133–140.

Fierobe L, Brunet F, Dhainaut JF, Monchi M, Belghith M, Mira JP, Dall'ava-Santucci J, & Dinh-Xuan AT. 1995. Effect of inhaled nitric oxide on right ventricular function in adult respiratory distress syndrome. *Am. J. Respir. Crit. Care. Med. 151:* 1414–1419.

Fishman AP. The enigma of hypoxic pulmonary vasoconstriction. 1990. In: Fishman AP, ed. *The Pulmonary Circulation: Normal and Abnormal.* Philadelphia: University of Pennsylvania Press, pp 109–129.

Frostell CG, Fratacci MD, Wain JC, Jones R, & Zapol WM. 1991. Inhaled nitric oxide. A selective pulmonary vasodilator reserving hypoxic pulmonary vasoconstriction. *Circulation 83:*2038–2047.

Frostell CG, Bloomqvist H, Hedenstierna G, Lundberg J, & Zapol WM. 1993. Inhaled nitric oxide selectively reverses human pulmonary vasoconstrictions without causing system vasodilation. *Anesthesiology 78:*427–435.

Fujino K, Nakaya S, Wakatsuki T, Miyoshi Y, Nakaya Y, Mori H, & Inoue I. 1991. Effects of nitroglycerin on ATP-induced Ca(++)-mobilization, Ca(++)-activated K channels and contraction of cultured smooth muscle cells of porcine coronary artery. *J. Pharmacol. Exp. Ther. 256:*371–377.

Furchgott RF, & Bhadrakom SJ. 1953. *Pharmacol. Exp. Ther. 108:*129–143.

Furchgott RF. The pharmacology of vascular smooth muscle. 1955. *Pharmacol. Rev.* 7:183–265.

Furchgott RF, & Zawadzki JV. 1980. The obligatory role of endothelial cells in the relaxation of arterial smooth muscle with acetylcholine. *Nature 288:*373–376.

Furchgott RF. 1984. The role of endothelium in the responses of vascular smooth muscle to drugs. *Annu. Rev. Pharmacol. 24:*175–197.

Furchgott RF. 1988. Studies on relaxation of rabbit aorta by sodium nitrite: the basis for the proposal that the acid-activatable inhibitory factor from retractor penis is inorganic nitrite and the endothelium-derived relaxing factor is nitric oxide. In: Vanhoutte PM, ed. *Vasodilation: Vascular Smooth Muscle, Peptides, Autonomic Nerves and Endothelium.* New York: Raven Press, pp 401–414.

Gerlach H, Rossaint R, Pappert D, Knorr M, & Falke KJ. 1994. Autoinhalation of nitric oxide after endogenous synthesis in nasopharynx. *Lancet 343:*518–519.

Gibson QH, & Roughton FJW. 1957. The kinetics of equilibria of the reactions of nitric oxide with sheep hemoglobin. *J. Physiol. (Lond) 136:*507–526.

Griffith TM, Edwards DH, Lewis MJ, Newby AC, & Henderson AH. 1984. The nature of endothelium-derived vascular relaxant factor. *Nature 308:*645–647.

Griffith TM, Edwards DH, Lewis MJ, et al. 1985. Evidence that cyclic guanosine monophosphate (cGMP) mediates endothelium-dependent relaxation. *Eur. J. Pharmacol. 112:*195–202.

Grover RF, Wagner WW, McMurtry IF, & Reeves JT. 1983. Pulmonary circulation. In: Shepherd JT, Abboud FM, eds. *Handbook of Physiology. Section 2: The Cardiovascular System, Vol III: Peripheral Circulation and Organ Blood Flow, Part I.* Baltimore: Williams and Wilkins, pp 103, 137.

Grover R, Murdoch I, Smithes M, Mitchell I, & Bihari D. 1992. Nitric oxide during hand ventilation in a patient with acute respiratory failure. *Lancet 340:*1038–1039.

Hales CA, Rouse ET, & Slate JL. 1978. Influence of aspirin and indomethacin on variability of alveolar-hypoxic vasoconstriction. *J. Appl. Physiol. 45:*33–39.

Hales CA, & Westphal DM. 1979. Pulmonary hypoxic vasoconstriction: not affected by chemical sympathectomy. *J. Appl. Physiol. 46:*529–533.

Harvey AL, Vatanpour H, Rowan EG, Pinkasfeld S, Vita C, Menez A, & Martin-Eauclaire MF. 1995. Structure-activity studies on scorpion toxins that block potassium channels. *Toxicon. 33:*425–436.

Hasunuma K, Rodman DM, & McMurtry IF. 1991. Effects of K$^+$ channel blockers on vascular tone in the perfused rat lung. *Am. Rev. Respir. Dis. 144:*884–887.

Hasunuma K, Yamaguchi T, Rodman DM, O'Brien RF, & McMurtry IF. 1991. Effects of inhibitors of EDRF and EDHF on vasoreactivity of perfused rat lungs. *Am. J. Physiol. 260:*L97–xx.

Hauge A. 1968. Conditions governing the pressor response to ventilation hypoxia in isolated perfused rat lungs. *Acta Physiol. Scand. 72:*33–44.

Holzmann S. 1982. Endothelium-induced relaxation by acetylcholine associated with larger rises in cGMP in coronary arterial strips. *J. Cyclic Nucl. Res. 8:*409–419

Hyman AL, Lippton HL, & Kadowitz PJ. 1991. Methylene blue prevents hypoxic pulmonary vasoconstriction in cats. *Am. J. Physiol. 260:*H586–H592.

Ignarro LJ. 1989. Endothelium-derived nitric oxide: actions and properties. *FASEB J. 3:*31–36.

Ignarro LJ, Burke TM, Wood KS, et al. 1983. Association between cyclic GMP accumulation and acetylcholine-elicited relaxation of bovine intrapulmonary artery. *J. Pharmacol. Exp. Ther. 338:*682–690.

Ignarro LJ, Harbison RG, Wood KS, & Kadowitz PJ. 1986. Activation of purified soluble guanylate cyclase by endothelium-derived relaxing factor from intrapulmonary artery and vein: stimulation by acetylcholine, bradykinin and arachidonic acid. *J. Pharmacol. Exp. Ther.* 237:893–897.

Ignarro LJ, Buga GM, Wood KS, Byrns RE, & Chaudhuri G. 1987a. Endothelium-derived relaxing factor produced and released from artery and vein is nitric oxide. *Proc. Natl. Acad. Sci. USA* 86:9265–9269.

Ignarro LJ, Buga RM, Byrns RE, Wood KS, & Chaudhuri G. 1987b. Endothelium-derived relaxing factor and nitric oxide possess identical pharmacological properties as relaxants of smooth muscle. *J. Pharmacol. Exp. Ther.* 228:682–690.

Ignarro LJ, Byrns RE, & Wood KS. 1988. Studies on relaxation of rabbit aorta by sodium nitrite; the basis for the proposal that the acid-activatable inhibitory factor from retractor penis is inorganic nitrite and the endothelium-derived relaxing factor is nitric oxide. In: Vanhoutte PM, ed. *Vasodilation: Vascular Smooth Muscle, Peptides, Autonomic Nerves and Endothelium.* New York: Raven Press, pp 427–436.

Janssens SP, Shimouchi A, Quertermous T, et al. 1992. Cloning and expression of a cDNA encoding human endothelium-derived relaxing factor/nitric oxide synthase. *J. Biol. Chem.* 26:14519–14522.

Johns RA, Linden JM, & Peach MJ. 1989. Endothelium-dependent relaxation and cyclic GMP accumulation in rabbit pulmonary artery are selectively impaired by moderate hypoxia. *Circ. Res.* 65:1508–1515.

Kazemi H, Bruecke PE, & Parsons EF. 1972. Role of autonomic nervous system in the hypoxic response of the pulmonary vascular bed. *Respir. Physiol.* 15:245–254.

Liu SF, Crawley DE, Barnes PJ, & Evans TW. 1991. Endothelium-derived relaxing factor inhibits hypoxic pulmonary vasoconstriction in rats. *Am. Rev. Resp. Dis.* 143:32–37.

Lonigro AJ, Sprague RS, Stephenson AH, & Dahms TE. 1988. Relationship of leukotriene C_4 and D_4 to hypoxic vasoconstriction in dogs. *J. Appl. Physiol.* 64:2538–2543.

Lundberg JON, Rinder J, Weitzberg E, Lundberg JM, & Alving K. 1994. Nasally exhaled nitric oxide in humans originates mainly in the paranasal sinuses. *Acta Physiol. Scand.* 152:431–432.

Lundberg JON, Lundberg JM, Settergren G, Alving K, & Weitzberg E. 1995. Nitric oxide, produced in the upper airways, may act in an 'aerocrine' fashion to enhance pulmonary uptake in human. *Acta Physiol. Scand.* 155:467–468.

MacLeod KM, Ng DDW, Harris KH, et al. 1987. Evidence that cGMP is the mediator of endothelium-dependent inhibition of contractile responses of rat arteries to α-adrenoceptor stimulation. *Mol. Pharmacol.* 32:59–64.

Madden JA, Vadula MS, & Kurup VP. 1992. Effects of hypoxia and other vasoactive agents on pulmonary and cerebral artery smooth muscle cells. *Am. J. Physiol.* 263:L384–L393.

Malik AB, & Kidd BS. 1973. Time course of pulmonary vascular response to hypoxia in dogs. *Am. J. Physiol.* 224:1–6.

Marshall BE, Marshall C, Benunoff J, & Saidman LJ. 1981. Hypoxic pulmonary vasoconstriction in dogs: effects of lung segment size and oxygen tension. *J. Appl. Physiol.* 51:1543–1551.

Marshall C, & Marshall BE. 1983. Influence of perfusate PO_2 on hypoxic pulmonary vasoconstriction in rats. *Circ. Res.* 52:691–696.

Martin W, Furchgott RF, Villani GM, & Jothianandan D. 1986. Phosphodiesterase inhibitors induce endothelium-dependent relaxation of rat and rabbit aorta by potentiating the effects of spontaneously released endothelium-derived relaxing factor. *J. Pharmacol. Exp. Ther.* 237:539–547.

Mazmanian G-M, Baudet B, Brink C, et al. 1989. Methylene blue potentiates vascular reactivity in isolated rat lungs. *J. Appl. Physiol. 66:*1040–1045.

McCormick DG, & Paterson NAM. 1993. Loss of hypoxic pulmonary vasoconstriction in
chronic pneumonia is not mediated by nitric oxide. *Am. J. Physiol. 265:*H1523–H1528.

McMahon TJ, Hood JS, Bellan JA, & Kadowitz PJ. 1991. N^{ω}-nitro-L-arginine methyl ester
selectively inhibits pulmonary vasodilator responses to acetylcholine and bradykinin.
*J. Appl. Physiol. 71:*2026–2031.

McMurtry IF. 1984. Angiotensin is not required for hypoxic constriction in salt solution-
perfused lungs. *J. Appl. Physiol. 56:*375–380.

McMurtry IF, Rodman DM, Yamaguchi T, & O'Brien RF. 1988. Pulmonary vascular reactivity. *Chest 93:*88S–93S.

Moinard J, Manier G, Pillet O, & Castaing Y. 1994. Effect of inhaled nitric oxide on hemodynamics and VA/Q inequalities in patients with chronic obstructive pulmonary disease. *Am. J. Respir. Crit. Care Med. 149:*1482–1487.

Moncada S, Palmer RM, & Higgs EA. 1991. Nitric oxide: physiology, pathophysiology,
and pharmacology. *Pharmacol. Rev. 43:*109–142.

Morganroth ML, Stenmark KR, Morris KG, Murphy RC, Mathias M, Reeves JT, et al.
1985. Diethylcarbamizine inhibits acute and chronic hypoxic pulmonary hypertension
in awake rats. *Am. Rev. Respir. Dis. 131:*488–492.

Murray TR, Chen L, Marshall BE, et al. 1990. Hypoxic contraction of cultured pulmonary
vascular smooth muscle cells. *Am. J. Respir. Cell. Mol. Biol. 3:*457–465.

Myers PR, Minor RLJ, Guerra RJ, Bates JN, & Harrision DG. 1990. Vasorelaxant properties of the endothelium-derived relaxing factor more closely resemble *S*-nitrosocystein
than nitric oxide. *Nature 345:*161–165.

Nandiwada PA, Hyman AL, & Kadowitz PJ. 1983. Pulmonary vasodilator response to
vagal stimulation and acetylcholine in the cat. *Circ. Res. 53:*86–95.

Nathan C. 1992. Nitric oxide as a secretory product of mammalian cells. *FASEB J.
6:*3051–3064.

Nossaman BD, Kaye AD, Feng CJ, & Kadowitz PJ. 1997. Effects of charybdotoxin on
responses to nitrosovasodilators and hypoxia in the rat lung. *Am. J. Physiol. 272:*
L573–L579.

Oda H, Kusumoto S, & Nakajimia T. 1975. Nitrosyl-hemoglobin formation in the blood of
animals exposed to nitric oxide. *Arch. Environ. Health 30:*453–465.

Ogata M, Ohe M, Katayose D, et al. 1992. Modulatory role of EDRF in hypoxic contraction of isolated porcine pulmonary arteries. *Am. J. Physiol. 262:*H691–H697.

Ohe M, Ogata M, Katayose D, et al. 1992. Hypoxic contraction of pre-stretched human
pulmonary artery. *Resp. Physiol. 87:*105–114.

Oka M, Hasunuma K, Webb JSA, et al. 1993. EDRF suppresses an unidentified vasoconstriction mechanism in hypertensive rat lungs. *Am. J. Physiol. 264:*L587–L597.

Palmer RMJ, Ashton DS, & Moncada S. 1988a. Vascular endothelial cells synthesize nitric
oxide from L-arginine. *Nature 333:*664–666.

Palmer RM, Rees DD, Ashton DS, & Moncada S. 1988b. L-Arginine is the physiological
precursor for the formation of nitric oxide endothelium-dependent relaxation. *Biochem.
Biophys. Res. Commun. 153:*1251–1258.

Pepke-Zaba J, Higgenbottam TW, Tuan Dinh-Xuan AT, & Stone D. 1991. Inhaled nitric
oxide as a cause of selective pulmonary vasodilation in pulmonary hypertension. *Lancet
338:*1173–1174.

Persson MG, Gustafsson LE, Wiklund NP, et al. 1990. Endogenous nitric oxide as a modulator of rabbit skeletal muscle microcirculation in vivo. *Br. J. Pharmacol. 100:*463–466.

Pollock JS, Forstermann U, Mitchell JA, Warner TD, Schmidt HH, Nakane M, & Murad F. 1991. Purification and characterization of particulate endothelium-derived relaxing factor synthase from cultured and native ovine aortic endothelial cells. *Proc. Natl. Acad. Sci. USA 88:*10480–10484.

Porcelli RJ, Viau AT, Naftchi NE, & Bergofsky EH. 1977. β-Receptor influence on lung vasoconstrictor responses to hypoxia and humoral agents. *J. Appl. Physiol. 43:*612–616.

Post JM, Hume JR, Archer SL, & Weir EK. 1992. Direct role for potassium channel inhibition in hypoxic pulmonary vasoconstriction. *Am. J. Physiol. 262:*C882–C890.

Post JM, Gelband CH, & Hume JR. 1995. [Ca^2]$_i$ Inhibition of K$^+$ channels in canine pulmonary artery: novel mechanism for hypoxia-induced membrane depolarization. *Circ. Res. 77:*131–139.

Radermacher P, & Rammos S. 1994. Low dose inhaled nitric oxide causing selective pulmonary vasodilation in child with idiopathic pulmonary hypertension. *Eur. J. Pediatr. 153:*691–693.

Rapoport RM, & Murad F. 1983. Agonist-induced endothelium-dependent relaxation in rat thoracic aorta may be mediated through cGMP. *Circ. Res. 52:*352–357.

Rees DD, Palmer RMJ, & Moncada S. 1989. Role of endothelium-derived nitric oxide in the regulation of blood pressure. *Proc. Natl. Acad. Sci. USA 86:*3375–3378.

Roberts JD Jr, Lang P, Bigatello LM, et al. 1993. Inhaled nitric oxide in congenital heart disease. *Circulation 87:*447–453.

Robertson BE, Warren JB, & Nye PGG. 1990. Inhibition of nitric oxide synthesis potentiates hypoxic vasoconstriction in isolated rat lungs. *Exp. Physiol. 75:*255–257.

Robertson BE, Schubert R, Hescheler J, & Nelson MT. 1993. cGMP-dependent protein kinase activates Ca-activated K channels in cerebral artery smooth muscle cells. *Am. J. Physiol. 265:*C299–C303.

Rodman DM. 1992. Chronic hypoxia selectively augments rat pulmonary artery Ca^{2+} and K$^+$ channel-mediated relaxation. *Am. J. Physiol. 263:*L88–L94.

Rodman DM, Yamaguchi T, Hasunuma K, et al. 1990. Effects of hypoxia on endothelium-dependent relaxation of rat pulmonary artery. *Am. J. Physiol. 258:*L207–L214.

Rossetti M, Guenard H, & Gabinski C. 1996. Effects of nitric oxide inhalation on pulmonary serial vascular resistances in ARDS. *Am. J. Respir. Crit. Care Med. 154:*1375–1381.

Rubin LJ. 1995. Pulmonary hypertension and cor pulmonale. In: Bone RC, ed. *Pulmonary and Critical Care Medicine.* Vol. 2, L2, pp 1–9.

Schuster DP, & Dennis DR. 1987. Leukotriene inhibitors do not block hypoxic pulmonary vasoconstriction in dogs. *J. Appl. Physiol. 62:*1808–1813.

Silove ED, Inoue T, & Grover RF. 1968. Comparison of hypoxia, pH, and sympathomimetic drugs on bovine pulmonary muscle vasculature. *J. Appl. Physiol. 24:*355–365.

Silove ED, Inoue T, & Grover RF. 1968. Effects of alpha-adrenergic blockage and tissue catecholamine depletion on pulmonary vascular response to hypoxia. *J. Clin. Invest. 47:*274–285.

Stamler JS, Loh E, Roddy M-A, Currie KE, & Creager MA. 1994. Nitric oxide regulated basal systemic and pulmonary vascular resistance in healthy humans. *Circulation 89:*2035–2040.

Szidon JP, & Flint JF. 1977. Significance of sympathetic innervation of pulmonary vessels in response to acute hypoxia. *J. Appl. Physiol. 43:*65–71.

Tagaya E, Tamaoki J, Chiyotani A, Yamawaki I, Takemura H, & Konno K. 1995. Regulation of airway cholinergic neurotransmission by Ca(2+)-activated K$^+$ channel and Na(+)-K$^+$ adenosinetriphosphatase. *Exp. Lung Res. 21:*683–694.

Voelkel NF. 1986. Mechanism of hypoxic pulmonary vasoconstriction. *Am. Rev. Respir. Dis. 133*:1186–1195.

Voekel NF, Gerber JG, McMurtry IF, Nies AS, & Reeves JT. 1981. Release of vasodilator prostaglandin, PGI₂, from isolated rat lung during vasoconstriction. *Circ. Res. 48*:207–213.

Von Euler US, & Liljestrand G. 1946. Observation on the pulmonary arterial blood pressure in the cat. *Acta Physiol. Scand. 12*:301–320.

Walmrath D, Schneider T, Schermuly R, Olschewski H, Grimminger F, and Seeger W. 1996. Direct comparison of inhaled nitric oxide and aerosolized prostacyclin in acute respiratory distress syndrome. *Am. J. Respir. Crit. Care Med. 153*:991–996.

Weiner CM, Dunn A, & Sylvester JT. 1991. ATP-dependent K^+ channels modulate vasoconstrictor responses to severe hypoxia in isolated ferret lung. *J. Clin. Invest. 88*:500–504.

Weir EK. 1978. Does normoxic pulmonary vasodilatation rather than hypoxic vasoconstriction account for the pulmonary pressor response to hypoxia? *Lancet i*: 476–477.

Weir EK, McMurtry IF, Tucker A, Reeves JT, & Grover RF. 1976. Prostaglandin synthetase inhibitors do not decrease hypoxic pulmonary vasoconstriction. *J. Appl. Physiol. 41*:714–718.

Yaghi A, Paterson NA, & McCormack , DG. 1993. Nitric oxide does not mediate the attenuated pulmonary vascular reactivity of chronic pneumonia.*Am. J. Physiol. 265(3 Pt 2)*: H943–H948.

Yuan X-J, Tod ML, Rubin LJ, et al. 1990. Contrasting effects of hypoxia on tension in rat pulmonary and mesenteric arteries. *Am. J. Physiol. 259*:H281–H289.

Yuan X-J, Goldman W, Tod M, Rubin L, & Blaustein M. 1993. Ionic currents in rat pulmonary and mesenteric arterial myocytes in primary culture and subculture. *Am. J. Physiol. 264*:L107–L115.

Yuan X-J, Goldman W, Tod M, Rubin L, & Blaustein M. 1993. Hypoxia reduces potassium currents in cultured rat pulmonary but not mesenteric arterial myocytes. *Am. J. Physiol. 264*:L116–L123.

Zapol W, & Hurford W. 1993. Inhaled nitric oxide in adult respiratory distress syndrome and other lung diseases. *New Horizons 1*:638–650.

Zwissler R, Kemming G, Habler O, Kleen M, Merkel M, Haller M, Briegel J, Welte M, & Peter K. 1996Inhaled prostacyclin (PGI₂) versus inhaled nitric oxide in adult respiratory distress syndrome. *Am. J. Respir. Crit. Care. Med. 154(6 Pt 1)*: 1671–1677.

C. Skeletal

14
Role of Nitric Oxide in the Skeletal Muscle Vascular Bed

Hunter C. Champion and Philip J. Kadowitz

Nitric oxide (NO) is a free radical synthesized by NO synthase (NOS), an enzyme that exists in three identified isoforms in the body (Kerwin et al. 1995). The isoforms of NOS have been named on the basis of location: neural constitutive NOS (type I or nNOS), inducible NOS (type II or iNOS), and endothelial constitutive NOS (type III or eNOS) (Kerwin et al. 1995). Although all forms are distinct based on their location, each form uses L-arginine in the formation of NO in a reaction that requires molecular oxygen and NADPH as cosubstrates with L-citrulline produced as a by-product of the reaction. NOS types I, II, and III are found in the skeletal muscle vascular bed, with types I and III being found under physiological conditions, and the mRNA for type II NOS being up-regulated under pathophysiological conditions, such as in gram-negative sepsis (Kerwin et al. 1995; Reid 1998).

Role of NO in Vascular Tone

The tone of vascular smooth muscle is regulated by a number of vasoactive agents that reach the vessel wall through the bloodstream or are released by adrenergic, cholinergic, or other nerve terminals in the vessel wall, and by locally produced tissue metabolites, such as adenosine lactate and carbon dioxide. All of these agents interact to maintain the blood pressure within predictable values, and failure to do so leads to abnormal changes, which can be observed in hypertension and vasospastic disorders. At the local tissue level, it is generally believed that NO plays an important role in maintaining the tone of skeletal muscle resistance vessels (Monoada et al. 1976; Furchgott and Zawadzki 1980; Furchgott 1983; Ignarro 1990; Kerwin et al. 1995; Joyner and Dietz 1997).

The study of the role of NO in maintaining baseline vascular tone in the skeletal muscle vascular bed has been aided greatly by the development of L-arginine analogues that act as selective inhibitors of NOS. Such analogues include N^G-nitro-L-arginine (L-NA), N^G-nitro-L-arginine methylester (L-NAME), and N^G-monomethyl-L-arginine (L-NMMA). When anesthetized rabbits were treated with

L-NMMA, L-NA, or L-NAME, blood pressure increases. This hypertension was reversed by the administration of the NO substrate L-arginine but not D-arginine. The hypertension observed in these rabbits was associated with reduced release of NO from a perfused aortic segment obtained from the treated animals, and the reduced NO release was also reversed by the administration of L-arginine but not D-arginine (Moncada et al. 1976; Bellan et al. 1991, 1993; Majid and Navar 1992; Majid et al. 1992, 1996; Pucci et al. 1992; McMahon and Kadowitz 1993; DeWitt et al. 1994, 1996, 1997; Feng et al. 1994; Nossaman et al. 1994, 1996; Santiago et al. 1994, 1995, 1997; Kerwin et al. 1995; Minkes et al. 1995; Champion and Kadowitz 1997a,b, 1998; Champion et al. 1997a,b; Joyner and Dietz 1997). Similarly, in awake, chronically instrumented dogs, L-NMMA produced dose-related vasoconstriction in the coronary circulation and reductions in resting phasic coronary blood flow, and these changes were reversed by L-arginine (Reid 1998). These data provide evidence in support of the hypothesis that the basal release of NO plays an important role in regulating vascular tone, blood flow, and blood pressure. The magnitude of the basal release of NO appears to correlate best with shear stress, and basal NO release may play a role in opposing or limiting the myogenic vasoconstrictor response induced by increases in blood flow (Reid 1998).

The pressor response to arginine analogues such as L-NMA appears to be mediated by the removal of a tonically active vasodilator tone that is due to the basal release of NO. It is uncertain, however, whether this basally released NO influences blood pressure solely by suppressing the release of NO, or by opposing the actions of endogenous vasoconstrictor hormones or some intrinsic vasoconstrictor mechanism. The pressor response and decreases in regional flow in response to L-NA in anesthetized rats were not altered by the α_1-adrenergic receptor antagonist prazosin or the vasopressinergic teceptor antagonist $d(CH_2)_5Tyr(Me)AVP$, or by treatment with the angiotensin-converting enzyme inhibitor captopril or the cyclooxygenase inhibitor indomethacin (Pucci et al. 1992). These results argue against significant contributions of the sympathetic nervous system, the renin–angiotensin system, vasopressor prostanoids, or vasopressin to the L-NA-induced elevation of mean arterial pressure or to regional vasoconstriction in anesthetized rats (Pucci et al. 1992). Similarly, in anesthetized dogs pretreated with DuP 753, L-NA infusion increases systemic arterial pressure and renal vascular resistance and decreases renal blood flow, and the changes in urine flow and sodium excretion in response to changes in renal arterial pressure were similar to those produced in response to L-NA infusion in control animals (Majid and Navar 1992; Majid et al. 1992). However, the decreases in afferent and efferent arteriolar diameter in response to L-NA in isolated, blood-perfused rat kidneys were markedly reduced by pretreatment with the angiotensin-receptor antagonist DuP 753, suggesting that an interaction between the peptide angiotensin II and NO may influence both afferent and efferent arteriolar resistance. Recent data from the hindlimb vascular bed of the cat suggest that the renin–angiotensin system and the adrenergic system may play a role in mediating a part of the hypertensive response to L-NAME and L-NA, while vasopressin and endothelin receptor activation does not appear to influence the hypertensive response to the NOS inhibitors.

Role of NO in Mediating Responses to Vasodilator Agonists

Acetylcholine, a potent hypotensive agent, was known on occasion either to have no effect or to produce contraction of isolated blood vessels in vitro. The reason for this discrepancy was revealed in 1980 when Furchgott and Zawadzki showed that vasorelaxation induced by acetylcholine was dependent on the presence of the endothelium, and these investigators provided evidence for the release of a labile humoral factor that causes relaxation (Furchgott and Zawadzki 1980; Furchgott 1983). This substance, which was later called endothelium-derived relaxing factor (EDRF), was shown to be distinct from prostacyclin, since endothelium-dependent relaxation was observed in vessels treated with indomethacin (Furchgott and Zawadzki 1980; Furchgott 1983). This demonstration was important, since it was known that the arachidonic acid metabolite prostacyclin, a potent vasodilator, could also be generated by the vascular endothelium (Moncada et al. 1976). Although other factors, such as endothelium-derived hyperpolarizing factor (EDHF), may be involved in the response to endothelium-dependent vasodilators, it is generally believed that NO or an S-nitrosylated intermediate is in fact EDRF (Kerwin et al. 1995; Joyner and Dietz 1997; Reid 1998).

Bradykinin is an endothelium-dependent vasodilator agent. This nine-amino-acid-containing peptide and kallidin are formed by the action of kallikrein on high- and low-molecular weight kininogen, respectively (Furchgott 1983; Bellan et al. 1991). Formation of the kinins results in an inflammatory response characterized by vascular permeability changes, vasodilation, and pain production. However, there is increasing evidence for a role of locally formed kinins in improving cardiovascular function following inhibition of the angiotensin-converting enzyme with angiotensin-converting enzyme (ACE) inhibitors. Kinin receptors are classified as B_1 and B_2 based on the pharmacology of selective kinin receptor agonists and antagonists. The majority of the biological activity of the kinins is mediated by B_2 receptors. In contrast, the B_1 receptor is thought to be up-regulated in pathological conditions and is not normally expressed in tissues. Moreover, following up-regulation of the kinin B_1 receptor by *Escherichie* coli endotoxin, both vasodilator and vasoconstrictor responses have been reported. Recent studies have demonstrated hypotensive activity in normal dogs in response to intravenously or intrarenally administered des-arg[9]-bradykinin (DABK), a B_1 agonist, suggesting that kinin B_1 receptors may be normally expressed in some vascular beds in some species (Regoli and Barabe 1980; Regoli et al. 1990).

The mechanism by which des-arg[9]-kallidin (DAK) and kallidin decrease hindlimb vascular resistance has been investigated in the hindlimb circulation of the cat (Santiago et al. 1995). Inhibition of NO synthesis with L-NAME significantly decreased vasodilator responses to DAK and kallidin, as well as to acetylcholine, indicating that kinin B_1 and B_2 receptor activation releases NO from the endothelium (Santiago et al. 1995). The kinin B_2 receptor has been shown to be functionally linked to the release of NO from the vascular endothelium. Moreover, L-NAME decreased responses to DABK and DAK, which suggests that kinin B_1 receptors are functionally coupled to the release of NO in the hindlimb vascular bed of the cat. NO release mediated by B_1 receptor activation has also been studied in other vascular beds. (Bellan et al. 1991; 1993; McMahon and Kadowitz 1993; DeWitt et al. 1994, 1996, 1997; Nossaman et al. 1994; Santiago et al. 1994, 1995, 1997).

Vasodilator responses to DAK and kallidin were not altered after the administration of meclofenamate, atropine, or glibenclamide. These data indicate that vasodilator prostaglandin release, muscarinic receptor activation, or K^+-ATP channel activation does not play a role in mediating hindlimb vasodilator responses to B_1 and B_2 receptor activation by DAK and kallidin. In the rat cremaster skeletal muscle, and in the canine coronary vascular bed, responses to bradykinin are mediated by kinin B_2 receptors coupled to the release of vasodilator prostaglandins (Santiago et al. 1994, 1995, 1997). However, it has also been shown in the blood-perfused hindlimb vascular bed of the rat that vasodilator responses to bradykinin are not altered after administration of the NOS inhibitor L-NAME or the cyclooxygenase inhibitor meclofenamate. The reason for the difference in results in studies in the hindlimb vascular bed of the cat and rat and the coronary vascular bed of the dog is uncertain, but it may reflect differences in the species and/or the vascular bed studied (Santiago et al. 1994, 1995, 1997; Nossaman et al. 1996).

Histamine is released from a variety of cell types in inflammatory and allergic reactions, and this mediator has marked activity on vascular smooth muscle. The vascular actions of histamine have been attributed to a direct interaction with the three histamine receptor subtypes designated H_1, H_2, and H_3. Although histamine has been shown to produce vasodilation in a number of vascular beds, vasoconstriction in response to histamine has also been reported (Neely et al. 1995). In the pulmonary vascular bed of the cat, histamine produces tone-dependent responses, with H_1 receptors mediating vasoconstriction at low tone and H_2 receptors mediating vasodilation during conditions of elevated pulmonary vascular tone (Neely et al. 1995). H_3 receptor-mediated vasopressor activity has been reported in a number of experimental models, but the physiological role of this receptor is not yet known (Furohgott and Zawadzki 1980).

The mechanism by which vasodilator responses to histamine are mediated appears to vary with the species and vascular bed studied. Although histamine has been shown to induce vasodilation through the H_1-receptor-mediated release of NO and prostacyclin (prostagladin I_2, PGI_2), other reports demonstrate that histamine induces vasoconstriction as a result of activation of the H_1 receptor. H_2 receptor-mediated vasodilation has been reported to involve increased accumulation of cAMP in vascular smooth muscle cells. However, other studies report the involvement of NO and cyclooxygenase products when H_2 receptors are stimulated. It has been reported that vasodilation in response to H_3 receptor activation results from inhibition of sympathetic neurotransmitter release, and studies in the isolated cerebral artery of the dog suggest that relaxation induced by selective H_3 receptor agonists is mediated by the release of NO and PGI_2, suggesting that the mechanism may be dependent on the species or the vascular preparation studied (Champion and Kadowitz 1997b).

In the hindlimb vascular bed of the cat, administration of the NO synthase inhibitor L-NAME at a dose that attenuated vasodilator responses to acetylcholine caused a significant reduction in responses to histamine, the H_1 receptor agonist HTMT, the H_2 receptor agonist dimaprit, and the H_3 receptor agonist $R(-)$-α-methyl-histamine. These data suggest that vasodilator responses to histamine and to the H_1, H_2, and H_3 receptor agonists are mediated, at least in part, by the release of NO or a labile nitroso derivative formed from L-arginine (Champion and Kadowitz 1997b). These data are in agreement with previous studies in which NO has been shown to mediate vasodilator responses to histamine (Champion and Kadowitz 1997b). These data further suggest the presence of H_1, H_2, and H_3 receptors coupled to NO release from the endothelium in the hindlimb vascular bed of the cat. In addition to being attenuated by L-{6-[2-(4-imidazolyl)ethylamino]-N-(4-trifuormethylphenyl)-heptanecardoxamide dimaleate} (Tocris-Cookson, St. Louis, MO) NAME, hindlimb vasodilator responses to histamine and to HTMT were reduced by the K^+-ATP channel antagonist U-37883A at a dose that attenuated vasodilator responses to the K^+-ATP channel opener levcromakalim (Champion and Kadowitz 1997b). These data suggest that in addition to releasing NO, vasodilator responses to histamine and the H_1 receptor agonist HTMT are mediated in part by the opening of a K^+-ATP channel in the hindlimb vascular bed of the cat (Champion and Kadowitz 1997b). The observation that responses to dimaprit and to $R(-)$-α-methyl-histamine were not altered by U-37883A suggests that opening of a K^+-ATP channel does not play a role in mediating vasodilator responses to the H_2 and H_3 receptor agonists (Champion and Kadowitz 1997b). Although responses to histamine and the H_1, H_2, and H_3 agonists appear to be mediated by the release of NO, only responses to histamine and HTMT were attenuated by U-37883A, suggesting that only histamine, through an H_1 receptor mechanism, possesses the ability to open a K^+-ATP channel, thereby decreasing hindlimb vascular resistance by two separate but complementary mechanisms (Champion and Kadowitz 1997b). The opening of a K^+-ATP channel by histamine is not dependent on the release of NO, since hindlimb vasodilator responses to the H_2 and H_3 receptor agonists and to the NO donor, DEA/NO (diethylamine/nitric oxide complex) were not altered by the K^+-ATP channel antagonist U-37883A (Champion and Kadowitz 1997b). The data with the NO synthase inhibitor and with the K^+-ATP channel antagonist suggest that histamine acts by two complementary yet independent mechanisms to release NO from the endothelium and to open a K^+-ATP channel in vascular smooth muscle cells to induce vasodilation in the hindlimb vascular bed of the cat (Champion and Kadowitz 1997b).

The mechanism by which histamine and the H_1, H_2, and H_3 receptor agonists decrease vascular resistance in the hindlimb vascular bed was further studied using phosphodiesterase inhibitors. Following treatment with the type 5, cGMP-selective phosphodiesterase inhibitor zaprinast, responses to histamine and the H_1, H_2, and H_3 receptor agonists were increased in duration. Zaprinast enhanced the duration of responses to the NO donor DEA/NO but did not alter the duration of vasodilator responses to the adrenergic β_2-agonist albuterol. These data, along with the results from experiments with L-NAME, provide support for the hypothesis that hindlimb vasodilator responses to histamine and to the H_1, H_2, and H_3 receptor agonists are mediated at least in part by the release of NO and the activation of soluble guanylate cyclase, leading to an increase in smooth muscle cGMP levels (Lincoln 1989; Champion and Kadowitz 1997b). A diagram illustrating the putative actions of histamine in the hindlimb vascular bed of the cat is shown in Figure 14.1.

Adrenomedullin (ADM) is a novel hypotensive peptide first isolated from human pheochromocytoma cells. Human adrenomedullin (hADM) consists of 52 amino acids and a disulfide bond that forms a six-membered ring structure similar to the ring structure found in calcitonin gene-related peptide (CGRP) and pancreatic amylin. Although ADM shares slight (21%) sequence homology with CGRP, it is considered to be a member of the CGRP/amylin superfamily of peptides, based largely on the presence of the six-membered ring structure and C-terminal amidation that is highly conserved in this peptide family (Champion et al. 1997a,b).

Although ADM was first isolated from pheochromocytoma cells, the peptide is localized in a variety of tissues. ADM is expressed in the adrenal medulla, lung, kidney, ventricle, spinal cord, stomach, anterior pituitary, thalamus, and hypothalamus. It has also been reported that ADM is actively synthesized and secreted by vascular endothelial cells and that this secretion rate is similar to the rate for endothelin-1. The observation that ADM is synthesized and released by the endothelium may be interpreted to suggest that ADM may act as an EDRF. In addition to the presence of ADM in tissue and plasma under physiologic conditions, plasma levels of ADM have been found to be increased in disease states, such as congestive heart failure, hypertension, renal failure, septic shock, primary aldosteronism, diabetes, myocardial infarction, and pulmonary hypertension. This peptide has been shown to have actions in the central nervous system, gut, and kidney in addition to having vasodilator activity in many species and vascular beds, including the hindlimb, mesenteric, and pulmonary vascular beds. Although ADM has pronounced effects on renal vascular resistance and electrolyte excretion in the dog, it has been shown that constant infusion of ADM at a dose that results in a minimal hypotensive effect increases renal plasma flow and urinary sodium excretion in the rat (Majid et al. 1996; Champion et al. 1997a,b). These data suggest that ADM may act as a natriuretic peptide.

The DNA sequence encoding the ADM precursor proadrenomedullin has been determined in rat, porcine, and human tissue. Proadrenomedullin contains 185 amino acids, and enzymatic cleavage of the signal peptide between amino acids

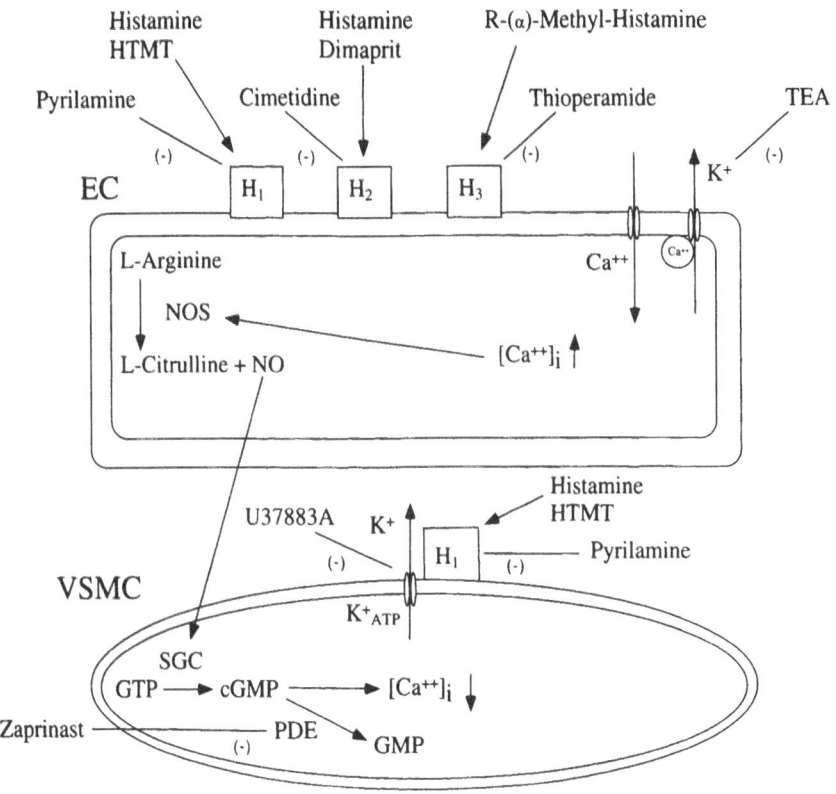

FIGURE 14.1. Drawing showing the postulated sites of action of histamine, H_1, H_2, and H_3 receptor agonists and antagonists the K^+ channel antagonist TEA (tetraethylammonium) and HTMT, the K^+_{ATP} channel antagonist U-37883A, and the type V cGMP phosphodiesterase (PDE) inhibitor zaprinast on endothelial cell (EC) and vascular smooth muscle cell (VSMC) in the hindlimb vascular bed of the cat. NOS, Nitric oxide synthase; $[Ca^{2+}]_i$, intracellular Ca^{2+} concentration; SGC, soluble guanylate cyclase. In the postulated sequence of biologic events, histamine and HTMT binds to H_1 receptors on the EC and VSMC in resistance vessel elements in the hindlimb circulation. In the EC, NOS is activated when intracellular $[Ca^{++}]$ is increased and NO is formed from L-arginine. The NO diffuses into the overlying VSMC and activates SGC, increasing cGMP levels, which leads to a decrease in $[Ca^{++}]$ in VSMC and induces relaxation. Histamine, by binding to H_1 receptors on VSMC membrane, opens K^+_{ATP} channels and hyperpolarizes the VSMC, decreasing Ca^{++} entry. From Champion and Kadowitz (1997b). Copyright 1997 by the American Physiological Society. Adopted with permission.

Thr21 and Ala22 yields a shortened propeptide composed of 164 amino acids, which contains ADM. Three usually paired basic amino acids that can be sites for enzymatic processing are found in the proadrenomedullin molecule. The group Gly42 Lys43 Arg44, which is the first group of basic amino acids, is a typical site for enzymatic cleavage releasing proadrenomedullin NH_2-terminal 20 peptide (PAMP). PAMP and ADM are distinct products of the ADM gene, and both possess hypotensive activity. ADM was isolated on the basis of its ability to increase cAMP levels in rat platelets and has been shown to decrease vascular resistance by a direct action on vascular smooth muscle. However, it has also been shown that ADM can release NO from the endothelium in the pulmonary, mesenteric, and hindlimb vascular beds of the rat. PAMP has been reported to induce vasodilation by inhibiting the release of norepinephrine from adrenergic nerve terminals in the mesenteric vascular bed of the rat, but has also been shown to have direct cAMP-mediated vasodilator activity in the hindlimb vascular bed of the cat and rat. Recent data suggest that ADM and PAMP possess significant vasodilator activity in human resistance vessels that is mediated by the release of NO and the activation of soluble guanylate cyclase (Majid et al. 1996; Nossaman et al. 1996; Champion et al. 1997a,b).

Role of K^+ Channels in Agonist-Induced NO Release

Agonist-induced release of EDRF from the endothelium has been shown to depend on the presence of extracellular Ca^{2+}, and the activation of constitutive NOS is dependent on an increase in intracellular Ca^{2+} levels, which results from Ca^{2+} release from intracellular stores and/or the influx of extracellular Ca^{2+}. It has been postulated that the increase in intracellular Ca^{2+} is related to agonist-induced endothelial cell hyperpolarization, which is associated with an increase in membrane permeability to K^+. Furthermore, it has been suggested that endothelium-dependent vasodilator agents cause an increase in cytoplasmic Ca^{2+} concentration, leading to the opening of Ca^{2+}-dependent K^+ channels in the endothelium (Champion and Kadowitz 1997a). The opening of the Ca^{2+}-dependent K^+ channels increases K^+ permeability and hyperpolarizes endothelial cells, increasing the electrochemical gradient for Ca^{2+} entry. In some vascular preparations, activation of muscarinic receptors on the endothelium by acetylcholine leads to the release of NO, as well as an EDHF, resulting in the relaxation of vascular smooth muscle by increasing K^+ permeability and hyperpolarizing smooth muscle cells. Acetylcholine-induced NO release has been shown to be antagonized by inhibitors of Ca^{2+}-dependent K^+ channels, and it has been postulated that acetylcholine, either directly or through an elevation of intracellular Ca^{2+} concentration, increases K^+ permeability of endothelial cells. The resulting hyperpolarization may maintain the Ca^{2+} influx required for EDRF and/or EDHF production, which leads to the activation of soluble guanylate cyclase and the opening of K^+ channels and hyperpolarization in vascular smooth muscle cells (Kerwin et al. 1995; Champion and Kadowitz 1997a).

In recent studies TEA (Tetraethylammonium) has been shown to inhibit large-conductance, Ca^{2+}-dependent K^+ channels and weakly inhibit other types of K^+ channels. It has been reported that TEA inhibits acetylcholine-induced release of EDRF in rabbit aortic endothelial cells by, blocking the large-conductance, Ca^{2+}-dependent K^+ channels. This inhibition results in the reduced release of EDRF and the failure of vascular smooth muscle cells to relax to acetylcholine. TEA has also been shown to prevent the effects of EDHF (Kerwin et al. 1995; Champion and Kadowitz 1997a).

The inhibition of vasodilator responses to acetylcholine, bradykinin, histamine, and substance P in the hindlimb vascular bed of the cat is consistent with in vitro studies in which TEA reduced relaxant responses to acetylcholine in isolated rabbit aorta. The results of recent studies provide additional support for the hypothesis that vasodilator responses to acetylcholine, bradykinin, histamine, and substance P in the hindlimb vascular bed in vivo are endothelium- or NO-dependent (Champion and Kadowitz 1997a). This TEA-sensitive mechanism may be associated with the blockade of K^+ channels in endothelial cells and the subsequent inhibition of the release of EDRF from the endothelium in the hindquarters vascular bed. However, responses to acetylcholine and bradykinin are reduced approximately 50% by TEA. These data suggest that perhaps the concentration of TEA used is not high enough to afford complete blockade or, alternatively, these data may be interpreted to suggest that responses to the endothelium-dependent vasodilators are only partially dependent on a TEA-sensitive mechanism (Champion and Kadowitz 1997a).

An antagonistic action of TEA on muscarinic and nicotinic receptors has been reported in autonomic ganglia and at the neuromuscular junction. However, the results of a recent study suggest that the inhibitory effects of TEA on the endothelium are more likely due to an inhibition of the release of EDRF by virtue of the blockade of K^+ channels in the endothelium and are not due to the inhibition of muscarinic receptors on the endothelium (Champion and Kadowitz 1997a). Moreover, the results of recent studies demonstrate an inhibitory effect of TEA on the response to bradykinin, histamine, and substance P, agonists that do not utilize muscarinic receptors to elicit a vasodilator response; studies in the hindlimb vascular bed of the cat with hexamethonium, a nicotinic receptor antagonist, suggest that this agent had no significant effect on responses to acetylcholine or bradykinin when administered in doses up to 20 mg/kg iv (Champion and Kadowitz 1997a).

Role of NO in Reactive Hyperemia

After a period of ischemia, reactive hyperemia occurs and involves a significant rise in hindlimb blood flow that peaks quickly and decays exponentially (Minkes et al. 1995). The degree of reactive hyperemia is dependent upon the duration of the arterial occlusion. It has been postulated that reactive hyperemia does not occur solely in an effort to repay the metabolic debt in the ischemic tissue, since the resulting increase in blood flow exceeds any metabolic debt incurred during the ischemic insult. A number of mediators and mechanisms have been postulated to be involved in the hyperemic response. However, no single mediator has been consistently identified in the many species and experimental preparations in which the hyperemic response has been studied. In human studies, inhibition of the cyclooxygenase pathway with nonsteroidal antiinflammatory agents did not alter resting blood flow but decreased the reactive hyperemic response by

15–30%. In human studies using L-NMMA, resting blood flow was significantly reduced following administration of the NOS inhibitor. The reactive hyperemic response was significantly reduced, although not abolished, after administration of L-NMMA. These data provide evidence in support of the hypothesis that NO is involved in mediating a portion of the hyperemic response. Moreover, these data provide evidence to support the hypothesis that additional mechanisms are involved, since the response was not abolished by NOS inhibitors (Minkes et al. 1995).

In animal models, the response to arterial inflow occlusion has been studied under constant-flow conditions in the denervated hindquarters vascular bed of the cat. The response to brief periods of ischemia was characterized by decreases in hindquarters vascular resistance in which the magnitude and duration of the response were dependent on the period of inflow occlusion (Minkes et al. 1995). The reactive hyperemic response was reproducible with respect to time and was not dependent on the presence of an intact sympathetic innervation. Under constant-flow conditions, L-NAME decreased the duration of the reactive hyperemic response, the maximal decrease in vascular resistance, and the area under the perfusion pressure–time curve (Minkes et al. 1995). Moreover, the K^+-ATP channel antagonists glibenclamide and U-37883A markedly reduced the reactive hyperemia and attenuated the vasodilator response to the K^+-ATP channel opener levcromakalim, but had no effect on responses to acetylcholine, bradykinin, or ATP (Minkes et al. 1995). The reduction in response to brief periods of ischemia following treatment with glibenclamide or U-37883A in doses that markedly attenuated vasodilator responses to the K^+-ATP channel opener levcromakalim suggests that reactive hyperemia in the hindlimb circulation of the cat is mediated in part by the opening of K^+-ATP channels and the release of NO in the hindquarters vascular bed of the cat. These data provide additional evidence in support of the hypothesis that multiple complementary, yet independent, pathways are involved in mediating the reactive hyperemic response (Minkes et al. 1995).

Role of NO in Muscle Metabolism

In addition to the effects of NO in mediating or modulating vasodilation in the skeletal muscle vascular bed, it also plays an important role in muscle metabolism. In general, NO functions to improve intracellular energy stores by acting upon muscle metabolism, promoting glucose uptake, and inhibiting mitochondrial respiration, glycolysis, and phosphocreatine breakdown. Nitric oxide has been shown to modulate glucose metabolism through two separate mechanisms. Experiments using NOS inhibitors and NO donor agents have been carried out to demonstrate that NO increases-2-deoxyglucose uptake in rat hindlimb skeletal muscle. Nitric oxide has also been shown to inhibit glyceraldehyde-3-phosphate dehydrogenase (GADPH) through S-nitrosylation of cysteine 149, thus regulating glycolysis in skeletal muscle (Kerwin et al. 1995, Reid 1998). Nitric oxide has been shown to reversibly inhibit mitochondrial respiration by competing with molecular oxygen for the cytochrome oxidase enzyme. The inhibition of oxidative phosphorylation by NO has been observed in a number of cell types (Reid 1998). Although NO may have a metabolic function in skeletal muscle of the hindlimb, it is still uncertain whether such actions can be demonstrated in other organ systems (Kerwin et al. 1995). The constitutive form of NOS (type III, cNOS) colocalizes with mitochondria in muscle fibers, and blockade of NOS in the hindlimb vascular bed of the dog results in increased oxygen consumption. The inhibition by NO of phosphocreatine breakdown by creatine kinase depresses calcium ATPase activity in the sarcoplasmic reticulum of striated muscle. The resulting decrease in ATP levels limits skeletal muscle function (Reid 1998). Therefore, in the hindlimb circulation of the dog, NO has important metabolic actions.

Role of NO in Muscle Contractile Function

It is well known that NO dilates vascular smooth muscle, inhibits vascular smooth muscle contraction, inhibits contraction of the smooth muscle of the uterus and gut, and depresses myocardial contractility. Nitric oxide also inhibits contractility in skeletal muscle. The mechanism of action of this inhibitory effect on muscle contraction has been studied extensively in rat diaphragm and other skeletal muscles. It has been shown that NO inhibits contractile force by two complementary pathways. As observed in vascular smooth muscle, skeletal muscle expresses soluble guanylate cyclase, which converts GTP to cGMP. Although pharmacologic probes that block NOS, guanylate cyclase, and phosphodiesterase have demonstrated that the NO/cGMP pathway plays a role in contractile function, the actions of cGMP alone do not account for all of the inhibitory effects observed. Moreover, the subcellular targets for cGMP in the skeletal muscle vascular bed are not well understood at the present time. On the basis of these findings, it has been suggested that NO may act through a second independent, yet complementary, mechanism to depress contractile function. It has been suggested that the opening of calcium channels of the sarcoplasmic reticulum could be inhibited by NO. Recent data provide support for the hypothesis and have demonstrated that NO has a concentration-dependent effect on calcium channels whereby low concentrations of NO may inhibit oxidative modification of sulfhydryl groups, resulting in the inhibition of calcium channel opening. At higher concentrations, NO may induce oxidative crosslinking between calcium channel subunits, resulting in an increase in the probability of channel opening. In addition, it has been hypothesized that NO may have a function in modulating sarcoplasmic reticulum ATPase and myosin heavy chains, both of which show reversible redox sensitivity (Reid 1998).

In summary, the results of studies reviewed in this chapter provide a good deal of evidence in support of the concept that NO or a labile nitrosyl product derived from L-arginine plays an important role in the regulation of vascular resistance, the mediation of the reactive hyperemic response, and the mediation of vasodilator responses in the hindlimb vascular bed, which is comprised mainly of skeletal muscle blood vessels. The mechanism by which NO induces vasodilation in the skeletal muscle vascular bed is not completely understood, although a great deal of experimental evidence suggests that responses are mediated by the activation of soluble guanylate cyclase, increased cGMP levels, and decreased intracellular Ca^{2+} concentration. The role of activation of K^+ channels and membrane hyperpolarization and other mechanisms in the regulation of the hindlimb circulation remains to be determined in future experiments.

In regard to the role of K^+-ATP channels, there is evidence that the reactive hyperemic response in the hindlimb vascular bed is mediated in part by the opening of K^+-ATP channels and by the release of NO from the endothelium (Minkes et al. 1995). The observation that two independent but complementary mechanisms may be involved in mediating vasodilator responses has also been made in recent studies with histamine in the hindlimb circulation (Champion and Kadowitz 1997b). The mechanism by which histamine and H_1, H_2, and H_3 receptor agonists induce vasodilation in the hindlimb circulation of the cat is illustrated in Figure 14.1. Injections of histamine and of H_1, H_2, and H_3 receptor agonists decrease vascular resistance. Pyrilamine reduced responses to histamine by approximately 80%, whereas cimetidine had only a small effect, and the H_3 receptor antagonist thioperamide had no effect. These data suggest that vasodilator responses to histamine are mediated mainly by the activation of H_1 receptors. Vasodilator responses to histamine and the H_1, H_2, and H_3 receptor agonists are reduced by the NOS inhibitor L-NAME and are enhanced in duration by the cGMP phosphodiesterase inhibitor zaprinast, suggesting that responses are mediated in part by the release of NO and an increase in cGMP levels. Vasodilator responses to histamine and an H_1 receptor agonist are reduced by K^+-ATP channel inhibitors. These data suggest that vasodilator responses to histamine are mediated by the activation of H_1 receptors, which leads to the release of NO from the endothelium and opening of K^+-ATP channels on vascular smooth muscle cells, which leads to a complementary, yet independent, vasodilator response, as illustrated in Figure 14.1.

References

Bellan JA, Minkes RK, McNamara DB, & Kadowitz PJ, 1991. N^M-Nitro-L-arginine selectively inhibits vasodilator responses to acetylcholine and bradykinin in cats. *Am J Physiol 260:*H1125–H1129.

Bellan JA, McNamara DB, & Kadowitz PJ, 1993. Differential effects of nitric oxide synthesis inhibitors on vascular resistance and responses on cats. *Am J Physiol 264:*H45–H52.

Champion HC, & Kadowitz PJ, 1997. Vasodilator responses to acetylcholine, bradykinin, and substance P are mediated by a TEA-sensitive mechanism. *Am J Physiol 272:*R414–R422.

Champion HC, & Kadowitz PJ, 1997. Nitric oxide release and the opening of K^+_{ATP} channels mediate vasodilator responses to histamine in the hindlimb vascular bed of the cat. *Am J Physiol 273:*H928–H937.

Champion HC, & Kadowitz PJ, 1998. [D-Ala^2]-Endomorphin 2(TAPP) has novel endothelium-dependent vasodilator activity in the rat. *Am J Physiol 274:*H1690–H1697.

Champion HC, Murphy WA, Coy DH, & Kadowitz PJ, 1997a. Proadrenomedullin NH_2-terminal peptide, a product of the adrenomedullin gene, has direct vasodilator activity in the cat. *Am J Physiol 272:*R1047–R1054.

Champion HC, Santiago JA, Murphy WA, Coy DH, & Kadowitz PJ, 1997b. Adrenomedullin(22–52) antagonizes vasodilator responses to calcitonin gene-related peptide but not adrenomedullin in the cat. *Am J Physiol 272*:R234–R243.

DeWitt BJ, Cheng DY, & Kadowitz PJ, 1994. Des-arg[9]-bradykinin produces tone-dependent kinin β1-receptor-mediated responses in the pulmonary vascular bed. *Circ Res 75*:1064–1072.

DeWitt BJ, Cheng DY, McMahon TJ, Marrone JR, Champion HC, & Kadowitz PJ, 1996. Differential effects of U-37883A, a novel K^+_{ATP} channel antagonist, in the pulmonary and hindlimb vascular beds of the cat. *Am J Physiol 271*:L924–L931.

DeWitt BJ, Champion HC, Marrone JR, McNamara DB, Giles TD, Greenberg SS, & Kadowitz PJ, 1997. Differential effects of L-N^5-(1-iminoethyl) ornithine on tone and endothelium-dependent vasodilator responses. *Am J Physiol 272*:L588–L594.

Feng CJ, Kang B, Kaye AD, Kadowitz PJ, & Nossaman BD, 1994. L-NAME modulates responses to adrenomedullin in the hindquarters vascular bed of the cat. *Life Sci 55*:PL433–538.

Furchgott RF, 1983. Role of endothelium in responses of vascular smooth muscle. *Circ. Res. 53*:557–573.

Furchgott RF, & Zawadzki JV, 1980. The obligatory role of endothelial cells in the relaxation of arterial smooth muscle by acetylcholine. *Nature 288*:373–376.

Ignarro LJ, 1990. Biosynthesis and metabolism of endothelium-derived nitric oxide. *Annu Rev Pharmacol Toxicol 30*:535–560.

Joyner MJ, & Dietz NM, 1997. Nitric oxide and vasodilation in human limbs. *J Appl Physiol 83*:1785–1796.

Kerwin JF, Lancaster JR, & Feldman PL, 1995. Nitric oxide: a new paradigm for second messengers. *J Med Chem 38*:4343–4360.

Lincoln TM, 1989. Cyclic GMP and mechanisms of vasodilation. *Pharmacol Ther 41*:479–502.

Majid DSA, & Navar LG, 1992. Suppression of blood flow autoregulation plateau during nitric oxide blockade in canine kidney. *Am J Physiol 262*:F40–F46.

Majid DSA, Williams A, & Navar LG. 1992. Effect of DuP 753 on renal responses to endothelium-derived nitric oxide (EDNO) inhibition in dogs. *FASEB J. 6*:A1511.

Majid DSA, Kadowitz PJ, Coy DH, & Navar LG. 1996. Renal responses to intra-arterial administration of adrenomedullin in dogs. *Am J Physiol 270*:F200–F205.

McMahon TJ, & Kadowitz PJ. 1993. Analysis of pulmonary vascular responses to substance P in the cat. *Am J Physiol 267*:H394–H402.

Minkes RK, McMahon TJ, & Kadowitz PJ. 1995. Role of K^+_{ATP} channels and endothelium-derived relaxing factor in reactive hyperemia in the hindquarters vascular bed of the cat. *Am J Physiol 269*:1704–1712.

Moncada S, Gryglewski RJ, Bunting S, & Vane JR. 1976. An enzyme isolated from arteries transforms prostaglandin endoperoxides to an unstable substance that inhibits platelet aggregation. *Nature (Lond.) 263*:663–665.

Neely CF, Matot I, Haile D, Nguyen J, & Batra V. 1995. Tone-dependent responses to histamine in feline pulmonary vascular bed. *Am J Physiol 268*:H653–H661.

Nossaman BD, Feng CJ, & Kadowitz PJ. 1994. Analysis of responses to bradykinin and influence of HOE 140 in the isolated perfused rat lung. *Am J Physiol 266*:H2452–H2461.

Nossaman BD, Feng CJ, Kaye AD, DeWitt BJ, Coy DH, Murphy WA, & Kadowitz PJ. 1996. Pulmonary vasodilator responses to adrenomedullin are reduced by nitric oxide synthase inhibitors in rats but not in cats. *Am J Physiol 270*:L782–L789.

Pucci ML, Lin L, & Nasjletti A. 1992. Pressor and renal vasoconstrictor effects of N^G-nitro-L-arginine as affected by blockade of pressor mechanisms mediated by the sympathetic nervous system, angiotensin, prostanoids, and vasopressin. *J Pharmacol Exp Ther 261*:240–245.

Regoli D, & Barabe J. 1980. Pharmacology of bradykinin and related kinins. *Pharmacol Rev 32*:1–46.

Regoli D, Rahaleb N-E, Dion S, & Drapeau G. 1990. New selective bradykinin receptor antagonists and bradykinin B_2 receptor characterization. *Trends Pharmacol Sci 11*:156–161.

Reid MB. 1998. Role of nitric oxide in skeletal muscle: synthesis, distribution and functional importance. *Acta Physiol Scand. 162*:401–409.

Santiago JA, Garrison EA, & Kadowitz PJ. 1994. Analysis of responses to bradykinin: effect of Hoe 140 in the hindquarters vascular bed of the cat. *Am J Physiol 267*:H828–H836.

Santiago JA, Garrison EA, Champion HC, Smith RE, Del Rio O, & Kadowitz PJ. 1995. Analysis of responses to kallidin, DABK, and DAK in feline hindlimb vascular bed. *Am J Physiol 269*:H2057–H2064.

Santiago JA, Champion HC, & Kadowitz PJ. 1997. T-kinin has endothelial-dependent vasodilator activity in the cat. *Am J Physiol 272*:H1491–H1498.

D. GI/GU

15
Nitric Oxide and the Hepatic Circulation

W. WAYNE LAUTT AND M. PAULA MACEDO

Introduction

More is known about nitric oxide (NO) than is true. The research on NO is proceeding at such a rate that virtually all areas of biological science are being swept into the enthusiastic vortex. Principles that we will espouse in this chapter include the following. In vivo and in vitro observations are often at odds: life is in vivo. Knowledge, or even information, related to NO derived from one tissue cannot necessarily be extrapolated to other systems: NO shows remarkable heterogeneity from tissue to tissue, organ to organ, and cell to cell. Given these philosophical biases, we will describe what we know of the effects of NO in the hepatic circulation.

Approximately 25% of the cardiac output passes through the liver, with two-thirds of this flow being accounted for by the portal vein. Hepatic blood flow is important not only for the supply of nutrients to the liver but also for the processing of this blood that the liver performs. Hepatic blood flow tends to be regulated in a way that maintains the blood flow at a fairly constant level. This is important for several homeostatic reasons. First, hepatic clearance of many compounds, including hormones (Lautt 1980), is blood flow-dependent. In order for an endocrine gland to be the primary regulator of the blood levels of its secreted hormones, it is essential that the hormonal clearance be maintained at a relatively constant rate. Because hormonal clearance is primarily regulated by hepatic metabolism in most cases, it is important for homeostatic integrity that the hepatic blood flow be maintained as constant as possible.

Although we currently have some understanding of the normal regulation of the hepatic circulation, the disruption of hepatic circulation in many disease states is well recognized, but poorly understood. In this chapter, we will describe the role of NO in the healthy hepatic circulation and briefly review the role of NO in pathological conditions of the liver. We will also provide a background to deal with some of the rapidly developing areas related to unique cellular involvement with NO and local blood flow. As a background, one must have a basic grasp of some of the unique features of the hepatic circulation. A conceptual review of the hepatic circulation can be consulted for details (Lautt and Greenway 1987).

Basic Hepatic Vascular Physiology

The hepatic artery provides one-quarter to one-third of the blood flow to the liver under normal arterial blood pressure of approximately 100 mmHg and normal arterial oxygen content. The portal vein provides the majority of the blood flow to the liver, but the hydrostatic pressure is only about 5–10 mmHg and the oxygen content is substantially reduced because the arterial blood has passed through the splanchnic organs prior to entering the portal vein. The portal blood flow is rich in nutrients absorbed from the gastrointestinal tract and in hormones secreted from these organs. The ratio of hepatic arterial to portal venous flow can vary dramatically, depending upon the physiological status of the organism. These two supplying vessels undergo progressive brachiation, with the final distribution within the liver occurring through regions referred to as the portal triads, which are seen on histological examination as a small zone circled by a limiting plate of hepatocytes and containing terminal branches of the hepatic artery, portal vein, and bile ducts. Through this region of the portal triad also pass the hepatic lymphatics and hepatic nerves. The fluid space surrounding these structures is referred to as the space of Mall and is an important area for hepatic vascular regulation.

The hepatic artery and portal vein supply minute branches to approximately 100,000 microvascular units of the liver, referred to as the hepatic acini, which can be conceived of as a cluster of hepatic parenchymal cells grouped around the terminal branches of the hepatic artery and portal vein like a 2-mm-diameter berry on a vascular stalk. Blood rich in oxygen and nutrients flows into the center of the acinus, which is referred to as zone 1. The blood flows through hepatic sinusoids, passing approximately 16 hepatocytes on average, and enters the hepatic venules, which drain into progressively larger veins, finally exiting the liver and draining into the inferior vena cava.

The hepatic sinusoids represent a unique equivalent of the capillary bed in other tissues. The endothelial cells within the sinusoids have extremely large fenestrations that allow albumin to pass relatively freely from the plasma into the space of Disse, which is the equivalent of the interstitial space of other tissues. Compounds entering the space of Disse are in direct contact with the hepatocytes, which have microvilli that serve to increase the hepatocyte contact with this fluid compartment. Circulating substances as large as lipoproteins can pass through the hepatic fenestrae from the plasma and make direct contact with the hepatocytes. Hepatocytes are connected as interconnected sheets of hepatocytes, with sinusoids passing, on the average, on both sides of the sheet, so that most hepatocytes are in relatively direct contact with the hepatic circulation on two sides. In chronic liver disease, the endothelial cells of the liver undergo a process that is referred to as capillarization in which the hepatic fenestrae decrease in size and the endothelial cells take on an appearance more similar to what is seen in other vascular beds.

Given the unusual nature of the hepatic circulation, it should not be surprising to discover that the role of NO in the liver is also substantially different from its role in other vascular beds. In several instances, we will contrast the role of NO that is understood to occur in the liver with its role in other vascular beds.

Nitric Oxide in the Normal Hepatic Circulation

Blockade of nitric oxide synthase (NOS) by 2.5 mg/kg of intravenously administered N^G-nitro-L-arginine methylester (L-NAME) in anesthetized cats resulted in an elevation in systemic arterial blood pressure and vascular tone in the superior mesenteric artery. In dramatic contrast, the basal vascular tone in the hepatic artery and the portal vein was not increased. In order to ensure that these responses were not modified by secondary responses to alterations in systemic arterial blood pressure, the hepatic circulation was perfused using a vascular circuit that shunted blood from the femoral arteries to the hepatic artery or superior mesenteric artery. This is one example of organ specificity, in which the blood supplies of the intestines and liver are regulated by NO in substantially different manners (Macedo and Lautt 1997a).

A similar organ-specific difference was noted in which endogenous NO production appears to antagonize non-endothelial-dependent vasodilation induced by adenosine or isoproterenol in the intestine. L-NAME resulted in a significantly potentiated vasodilator response to these agents in the intestine, but this effect was absent in the liver (Macedo and Lautt, 1997a). The lack of effect of L-NAME on hepatic arterial basal tone (Mathie et al. 1991; Greenblatt et al. 1993; Browser et al. 1994; Macedo and Lautt, 1997a) does not guarantee that hepatic arterial blood flow will not change after the administration of L-NAME. The hepatic arterial blood flow is affected by a number of stimuli external to the liver, including hormones and hepatic sympathetic nerves (reviewed by Lautt 1996a,b) and the hepatic arterial buffer response to changes in portal flow (reviewed by Lautt 1996c).

The principal intrinsic regulator of the hepatic artery is the hepatic arterial buffer response, which is a mechanism that is dependent upon local concentrations of adenosine that are produced at a constant rate, independently of hepatic parenchymal cell metabolism, whereby the concentration of adenosine in the region of the hepatic arterial resistance vessels is regulated by washout into either the portal vein or the hepatic artery. By this mechanism, a decrease in portal flow will result in less adenosine washout and the constant rate of production will lead to an accumulation of adenosine in the fluid space (space of Mall) that surrounds the hepatic artery, thus leading to arterial vasodilation. This is the proposed mechanism of the hepatic arterial buffer response (Lautt et al. 1985; Lautt 1996c). The buffer capacity is not altered by L-NAME (Macedo, Han, and Lautt, unpublished observation).

The other form of intrinsic regulation of the hepatic artery is referred to as arterial autoregulation, which was previously thought to be a myogenic response to changes in arterial pressure. It appears rather that hepatic arterial autoregulation is explained by the same mechanism as the hepatic arterial buffer response, in which an increase in systemic arterial blood pressure results in an increase in hepatic arterial flow that washes away adenosine and leads to arterial constriction (Ezzat and Lautt 1987). L-NAME also did not affect autoregulation in the liver (Macedo and Lautt, unpublished observation). The hepatic circulation thus appears well designed to maintain hepatic blood flow at relatively constant levels. The lack of control of basal vascular tone by NO and the lack of interaction between NO and adenosine ensures that the hepatic circulation is primarily regulated according to these two intrinsic regulatory systems, with the mechanisms serving to regulate total hepatic blood flow.

This is in contrast to the intestine, in which NO normally contributes to vascular tone and in which elimination of this endothelial-dependent vasodilator tone results in a compensatory potentiation of other vasodilators (Macedo and Lautt 1997a). Furthermore, blockade of NOS leads to potentiation of autoregulation in the intestine, suggesting that NO normally opposes the adenosine-mediated autoregulatory mechanism in the gut (Macedo and Lautt 1996a).

Despite the fact that NO does not directly interact with adenosine and the intrinsic regulatory mechanisms of the hepatic artery, the hepatic arterial flow can undergo indirect vascular responses to NOS antagonism, depending upon what the net effect is on portal blood flow. If the vascular resistance in the superior mesenteric artery increases to a greater proportional extent than the increase in arterial blood pressure, the superior mesenteric arterial flow and, therefore, the portal venous flow will decrease and thus activate the buffer response, thereby leading to a dilation of the hepatic artery. This combination of responses is demonstrated in the conscious rat (Greenblatt et al. 1993). This observation demonstrates the precaution that must be applied in interpreting the effect of stimuli on the hepatic artery.

Nitric Oxide as a Neuromodulator

The literature is rife with contradictory data regarding the ability of NO to inhibit the vascular reaction to sympathetic nerves, with some reports suggesting a modulatory effect (Cohen and Weisbrod 1988; Tesfamariam and Cohen 1988; Greenberg et al. 1989; Najafipour and Ferrell 1993; Macedo and Lautt 1994, 1996b; Thorin and Atkinson 1994; Yasuhiro et al. 1994), and other reports being inconsistent with such an effect (Hynes et al. 1988; Wennmalm et al. 1991; Thatikunta et al. 1993; Thorin and Atkinson 1994). We have suggested that this controversy in many cases can be readily explained by the methodological differences used in the different reports. We have suggested that NO serves as a modulator of sympathetic nerve-induced vasoconstriction only when the vasoconstriction results in an increase in shear stress, leading to a release of NO. Although a full discussion is beyond the scope of this chapter, we provided calculations that predicted, and data that confirm, that vasoconstriction will not result in an increase in shear stress at the site of the constriction if blood flow is allowed to decline and the perfusion pressure is not allowed to rise (Macedo and Lautt 1996b, 1997b). If perfusion pressure rises during the vasoconstriction, shear stress will be increased. An increase in shear stress is obtained when a vascular perfusion circuit is used that holds blood flow constant during nerve stimulation (Macedo and Lautt 1996b).

This effect is again dramatically organ-specific, with shear stress resulting in an NO-dependent suppression of vasoconstriction in response to sympathetic nerves but not to norepinephrine infusion in the intestine. This suggests that the shear stress-induced release of NO suppresses norepinephrine release through a presynaptic mechanism in the intestine (Macedo and Lautt 1996b). In dramatic contrast, in the liver, vasoconstriction in both the hepatic artery and the portal vein is also shown to result in shear stress-dependent release of NO. However, the mechanism of suppression of the vasoconstriction appears to be postsynaptic, since the vasoconstriction induced by sympathetic nerve stimulation and exogenous norepinephrine infusion is inhibited to a similar degree (Macedo and Lautt 1997b). The unique role of hepatic sympathetic nerves in causing glycogenolysis to supply glucose to the extrahepatic tissues (Lautt et al, 1982) is thus preserved and, in fact, is potentiated by NO (Ming and Lautt, unpublished observation).

In studies in which vascular flow and pressure are not carefully monitored and reported, it is virtually impossible to determine if shear stress was altered during vascular stimulation. In such cases, therefore, it is also impossible to predict a neuromodulatory role for NO.

The liver represents a unique biological situation, in that the normal physiological response to isolated vasoconstrictor stimuli delivered to the liver, such as sympathetic nerve stimulation, results in a dual effect. The hepatic artery responds in the way anticipated from studies in other arteries. That is, the arterial blood flow decreases and arterial pressure may rise only modestly. If arterial pressure does not rise significantly, the decrease in hepatic arterial flow does not result in an increase in shear stress and would, therefore, not be expected to result in the release of NO. In contrast, the normal physiological response to vasoconstriction of the vessels controlling portal flow through the liver is an elevation of portal pressure with no change in portal flow. Recall that portal blood flow is not directly regulated by the liver but is controlled by the total venous outflow from the extrahepatic splanchnic organs (stomach, spleen, pancreas, intestines, and omentum). Therefore, in the face of a constant portal flow, local constriction of the portal vessels will lead to an increase in shear stress, with a resultant release of NO and suppression of the constrictor response.

The shear stress-dependent modulation of vasoconstrictors within the liver would appear to be a mechanism well suited to protect the hepatic endothelial cells from increased shear stress without decreasing norepinephrine release from sympathetic nerves. This postsynaptic suppression of vasoconstriction is important, in that it allows a normal metabolic response to stress to occur, including dramatic glycogenolysis and glucose release from the liver, while affording protection of the endothelial cells against shear stress.

Role of Vascular NO in Liver Regeneration

We have proposed the hypothesis that the hemodynamic events that occur subsequent to partial hepatectomy constitute the trigger that initiates the onset of the liver regeneration cascade. According to this hypothesis, surgical removal of two-thirds of the liver results in all of the portal blood flow being forced through the small remaining portion of the liver, with the result that the increase in shear stress generates NO, which serves as the trigger leading to production of a wide range of growth factors, leading eventually to restoration of the liver mass. For these studies, we used the appearance of proliferating factors in the plasma of rats as an index that the regeneration cascade had been triggered. Blockade of NOS by L-NAME completely prevented the appearance of proliferating factors in the plasma following partial hepatectomy and similarly led to an inhibition of liver regeneration. L-Arginine restored the ability of the partial hepatectomy to produce proliferating factors (Wang and Lautt 1997, Wang and Lautt, 1998). A recent study (Yamakado et al. 1997) reported that portal vein embolization induced atrophy of the embolized hepatic parenchyma and hypertrophy of the unembolized liver, but only if the extent of portal vein embolization was sufficient to cause an elevation in portal venous pressure. This is consistent with the conclusion that an elevated portal pressure is an index of increased shear stress and that the shear stress leads to NO release and triggering of the regeneration cascade.

A similar hemodynamic perturbation of the liver is produced by selective ligation of the left branch of the portal vein, which also deprives two-thirds of the liver of portal blood flow. Portal blood flow was shown not to decrease, so that the total portal flow was forced through the unligated portal branches and led to an increase in portal venous pressure. The unligated lobes underwent hypertrophy, which became stabilized after 7 days, when the portal pressure also stabilized (Um et al. 1994). It should be noted that the generation of shear stress involved with the control of regeneration appears to be dependent only upon the portal flow, since the increase in flow through the unligated liver lobes resulted in activation of the hepatic arterial buffer response and a subsequent decrease in hepatic arterial blood flow. We have demonstrated that this model also results in a similar elevation of proliferating factors in the plasma of rats 4 hours after selective portal vein ligation, with the proliferating factors reaching identical levels to that seen 4 hours after a two-thirds partial hepatectomy (Wang and Lautt, unpublished observations). This model is also, therefore, consistent with the hypothesis that shear stress release of NO in the liver serves as a trigger to initiate the complex cascade leading to hepatic hyperplasia.

Although NO appears to be a pivotal dilator in conditions of chronic liver disease associated with portal hypertension and portocaval shunts, the effects of NO in these conditions are primarily on nonhepatic tissue, and the source of the NO is not clear. For these reasons, we have declined to include this vast literature as part of the present review.

In all of the examples discussed above, a role for NO was studied in a way that could not implicate specific cell types as regulators of NO production. The following section focuses on the specific cells.

Nitric Oxide and Liver Cell Types

The cells that constitute the hepatic sinusoidal wall are the sinusoidal endothelial cells, stellate cells, and Kupffer cells. These cells are now recognized to be very important in the physiology and pathology of the liver. They function individually or in coordination to regulate the functions of other intrahepatic cells, including the vastly more numerous hepatocytes.

Stellate Cells

Stellate cells (fat-storing cells, Ito cells, or hepatic perisinusoidal lipocytes) are mesenchymal cells that reside in the perisinusoidal space of Disse and can undergo reversible contraction (Kawada et al. 1993; Rockey and Chung, 1995). Recently it has been suggested that intrahepatic resistance and portal pressure are affected by stellate cells (Kawada et al. 1993, 1996). This theory is consistent with their histological and anatomic features, since they express smooth muscle-specific intermediate filament desmin, which allows these cells to contract (Rockey 1997). Studies on isolated stellate cells showed that vasoconstrictors, such as angiotensin II, thrombin, and endothelin-1, increase intracellular free calcium, which is coupled with effective cell contraction (Pinzani et al. 1992): Kawada et al. (1992) showed an increase or decrease in traction forces exerted by stimulated stellate cells by culturing the cells on a flexible silicone rubber membrane. These effects on the stellate cells were elicited by a thromboxane A_2 analogue, prostaglandin $F_{2\alpha}$ ($PGF_{2\alpha}$), and endothelin-1, whereas prostacyclin (prostaglandin I_2, PGI_2) analogues and prostaglandin E_2 (PGE_2) induced cell relaxation (Kawada et al. 1992, 1993). Moreover, it has been suggested that stellate cells act as pericytes, which was demonstrated by using isolated rat livers under high-power intravital microscopy (Zhang et al. 1994). These authors suggested that the sinusoidal constriction induced by endothelin-1 may be mediated by stellate cells, and this effect was inhibited by NO donors (Zhang et al. 1994). All these data provide strong evidence that stellate cells are contractile and function as a regulator of intrahepatic resistance. It is suggested that a balance between dilators and constrictors becomes disrupted in conditions of reperfusion injury or endotoxin shock, and that the balance favors stellate cell constriction, thus leading to sinusoidal regional blood flow heterogeneity (Clemens et al. 1997).

In the liver, NO can exert paracrine (generated from hepatocytes, Kupffer cells, or endothelial cells) or autocrine (intracrine) effects, and its release is elicited by interferon-γ (IFN-γ) and other cytokines with or without lipopolysaccharide (LPS, endotoxin) (Lyons et al. 1992; Geller et al. 1993; Feder and Laskin 1994; Helyar et al. 1994; Spitzer 1994; Rockey and Chung 1995). Recently a connection has been suggested between NO and the regulation of the sinusoidal tonus via relaxation of stellate cells, based on experiments in which sodium nitroprusside induced stellate cell relaxation and caused partial degradation of actin stress fibers in these cells (Kawada et al. 1993). The suggestion that NO reduces the traction forces of the stellate cells is corroborated by the demonstration that NO suppressed the expression of smooth muscle actin in these stellate cells (Kawada et al. 1996). The consequent relaxation may be mediated by the stimulation of soluble guanylate cyclase, leading to the generation of cGMP, since sodium nitroprusside and 3-morpholinosydnonimine (SIN-1) induced the accumulation of this nucleotide in these cells (Kawada et al. 1993). The type of NOS that is present in the stellate cells seems to be the inducible form, as supported by the results of Helyar et al. (1994). This group has shown that IFN-γ and endotoxin acted synergistically in vitro to induce expression of NOS protein. Rockey and Chung (1995) also showed that an in vivo single exposure to IFN-γ tumor necrosis factor-α (INFα), and endotoxin rapidly increased inducible NOS mRNA (Heylar et al. 1994; Rockey and Chung 1995). Even though the in vivo microscopy data showed variation of contractility, it would be difficult to distinguish whether the site of the effect is on terminal smooth muscle cells or on perisinusoidal cells. The development and use of specific pharmacological tools for stellate cell contraction might solve this issue. However, experiments in cultured cells seem to be relatively insensitive to contraction, and these results should be extrapolated to the intact liver cautiously.

Some groups claim a cytoprotective action by NO in the liver when, for instance, damage is induced by lipopolysaccharide plus *Corynebacterium parvum*, which may be due to the maintenance of sinusoidal blood flow by NO (Billiar et al. 1990). The trigger for NOS induction seems to be cytokines, which are abundant in patients with inflammatory or chronic liver diseases (Andus et al. 1991). Cytokines, tumor necrosis factor, and interleukin-1 (IL-1) are released by endotoxin-stimulated Kupffer cells (Decker 1990). When liver injury occurs, stellate cells show morphological and functional alterations that are characterized by the development of stress fibers, production of extracellular matrix, and de novo expression of actin (Rockey et al. 1992; Rockey 1997). The modified stellate cells resemble myofibroblasts, which, under the influence of endothelin and other stellate cell constrictors, could alter hepatic microcirculation (Rockey 1997). During liver injury produced by bile duct ligation, an increase in NOS expression is observed, suggesting that NO may serve as a counterregulator of "myofibroblast" contraction (Rockey and Chung 1995). Nitric oxide may thus play an important role in the modulation of stellate cell contractility not only by promoting relaxation but also by counteracting the effect of contractile endogenous substances, resulting in the regulation of local sinusoidal blood flow, especially in the injured liver.

Kupffer Cells

Kupffer cells are liver-specific macrophages that reside in the sinusoidal lumen, where they are exposed to the systemic circulation via the hepatic artery and to the splanchnic circulation via the portal vein. They constitute approximately 80% of the total population of macrophages in the body (Obolenskaya et al. 1994). Because of their phagocytic capacity, they participate in the host defense system to clear circulating endotoxin from the blood (Billiar and Curran 1990; Winwood and Arthur 1993). Kupffer cells are also capable of secreting mediators involved in host responses to inflammation, such as cytokines, endothelins, prostanoids, and NO (Clemens et al. 1994; Roland et al. 1994).

Kupffer cells synthesize inducible NOS, producing large amounts of NO (Spitzer 1994). This enzyme is not present in resting cells and, after stimulation, requires a period of mRNA expression and new protein synthesis to detect enzyme activity (Billiar et al. 1989; Spitzer 1994). Different stimuli, such as cytokines, interferon-γ, LPS, tumor necrosis factor, IL-1 are able to induce the synthesis and to release NO from Kupffer cells (Decker 1990; Kurose et al. 1996).

It has been suggested that Kupffer cells not only play an important role in the body's defense system but may also contribute to liver damage (Winwood and Arthur 1993). Extensive NO production by Kupffer cells might cause significant cytotoxicity. The damaging effects of NO might be due to a cooperative action with superoxide, yielding the peroxynitrite anion ($ONNO^-$). Peroxynitrite, known to oxidize sulfhydryls and to generate products indicative of hydroxyl radical reaction with deoxyribose and dimethyl sulfoxide, induces lipid peroxidation (Beckman et al. 1990; Radi et al. 1991). However, one should keep in mind that NO cytotoxic effects are dependent on the concentration of superoxide radical (Rubbo et al. 1994). In chronic alcoholic rats, it has been observed that the production of both superoxide radical and NO in isolated Kupffer cells seems to be at the same rate (Bautista et al., 1992; Wang et al. 1995). Similarly, the superoxide radical from mitochrondria and microsomal enzymes increases as NO concentrations increase in hepatocytes from alcoholic (LPS)-treated rats, contributing to the enhancement of hepatic injury (Boveris et al. 1983; Kukielka et al. 1994; Wang et al. 1995). On the other hand, NO by itself might have a protective effect by regulating the production of specific inflammatory mediators by Kupffer cells during sepsis (Stadler et al. 1993). In the septic liver, NO has a profound inhibitory effect on the production of prostaglandin E_2 (PGE_2), thromboxane B_2 (TXB_2), and interleukin-6 (IL-6) (Stadler et al. 1993). In this case, NO might be an autoregulator of inflammatory reactions.

Kupffer cells could play a role in the regulation of the microvascular flow by protruding their pseudopods towards the vascular space in response to endotoxin administration in vivo (McCuskey et al. 1982; Nishida et al. 1994). Moreover, the NO produced by Kupffer cells could affect microvascular flow because of the strong vasorelaxant properties acting on other cells. Indirectly, NO, by suppressing Kupffer cell eicosanoid synthesis, could also have an impact on vascular tone.

Sinusoidal Endothelial Cells

Endothelial cells have classically been associated with constitutive production of NO, facilitating the relaxation of the underlying smooth muscle cells by activating guanylate cyclase. Recently, it has been reported that hepatic endothelial cells are capable of expressing NOS after an inflammatory stimulus, which might become important in physiological processes in the liver (Spolarics et al. 1993; Rockey and Chung 1996). Both isoforms are now suggested to alter vascular tone or reactivity under normal (constitutive isoform) or pathological (constitutive and inducible isoforms) conditions.

In contrast to other liver cell types, the expression of inducible NOS (iNOS) mRNA in sinusoidal endothelial cells is prominently caused by IFN-γ (Rockey and Chung 1996). Even though other cytokines, such as IL-1, TNFα, or endotoxin, have little or no effect on the induction of iNOS mRNA, they can cause a synergistic increase in iNOS mRNA with IFN-γ. Nitrite formation parallels iNOS mRNA levels when IFN-γ is the stimulus, and a combination of IFN-γ with IL-1, LPS, or TNFα potentiates nitrite formation (Rockey and Chung 1996). These results are divergent from the ones obtained for other hepatic cell types, suggesting some selectivity for the stimuli responsible for iNOS expression and NO production (Rockey and Chung, 1996). Cultured sinsuoidal endothelial cells produce as much nitrite after exposure to IFN-γ plus TNFα as do hepatocytes, which is equivalent to 70% of that generated by Kupffer cells, suggesting that endothelial cells have a prominent role in NO production in pathological states where inflammatory cytokines are present in the environment (Rockey Chung, 1996).

Injury of the sinusoidal microvasculature is one of the first events in the sequel of developing hepatic failure during severe sepsis, endotoxemia, or reperfusion injury (Hirata et al. 1989). The effect on sinusoidal endothelial cells might lead to alterations in NO levels. Increased release of NO by Kupffer cells could have an impact on sinusoidal blood flow by directly relaxing stellate cells.

Hepatocytes

Some cytokines are capable of inducing the expression of iNOS in hepatocytes, resulting in their serving as a major source of NO (Curran et al. 1990; Nussler et al. 1992). However, there is now evidence that the direct action of endotoxin (LPS) on hepatocytes and endothelial cells does not alter either constitutive NOS (cNOS) or iNOS activity or nitrite and nitrate levels (Kurose et al. 1996). In contrast, iNOS activity, but not cNOS activity, as well as nitrite and nitrate levels are increased in LPS-activated cocultured Kupffer cells and hepatocytes, resulting in an additive net effect on NO production. These results are in accordance with a previous report demonstrating that hepatocytes produce NO from L-arginine in response to inflammatory products from Kupffer cells (Curran et al. 1989; Kurose et al. 1996). Moreover, TNFα per se does not increase the NO production of cultured hepatocytes (Kurose et al. 1996). From these observations, one can hypothesize that multiple stimulators and cytokines released by Kupffer cells are required for induction of iNOS in the hepatocytes and consequently NO production.

Even though the mechanisms of NO production in hepatocytes are becoming clear, the functional role of hepatocyte NO production is not. It was recently suggested that hepatocytes might serve as immunomodulator cells by synthesizing and secreting immunomodulatory factors involved in the acute phase of inflammatory reactions (Fey and Gauldie 1990; Volpes et al. 1992). It was never assessed whether the NO released by hepatocytes was able to have an impact on the regulation of sinusoidal blood flow. However, in pathological conditions that lead to excess NO production, Kupffer cells, endothelial cells, hepatocytes, and stellate cells are capable of producing NO in amounts that will affect liver hemodynamics. Whether the amount of NO produced by hepatocytes has a significant role in hepatic blood flow regulation is a question that remains to be answered.

Summary and Conclusions

The liver has a large and unique vascular bed in terms of both intrinsic regulatory mechanisms and the role of NO interactions with the vascular bed. NO does not influence basal resting vascular tone, nor does it interfere with the adenosine-mediated intrinsic vascular regulatory mechanisms. NO does not interact with other nonendothelial-dependent vasodilators. This is in contrast to the intestine, where NO does contribute to the basal vascular tone by antagonizing adenosine-induced vasodilation and antagonizing autoregulation. Nitric oxide serves as a neuromodulator and is released by shear stress that is induced by vasoconstriction. Vasoconstriction causes an increase in shear stress only in situations where the perfusion pressure rises during the constriction. This is seen during the normal constrictor response in the portal vein, where portal blood flow is not decreased but portal pressure and shear stress are elevated by sympathetic nerve stimulation or blood-borne vasoconstrictors. Shear stress does not normally increase in the hepatic artery, where arterial perfusion pressure remains stable and hepatic arterial flow decreases. If generalized sympathetic activation throughout the body occurs such that perfusion pressure rises, the hepatic artery also releases NO. Shear-induced NO release results in postsynaptic inhibition of vasoconstriction, in contrast to the intestine, where the shear stress-induced vasoconstriction is presynaptic and decreases sympathetic nerve-induced constriction, but not norepinephrine-induced constriction.

Shear stress induced as a result of the hemodynamic consequences of partial hepatectomy also appears to result in a release of NO that serves as a trigger to initiate liver regeneration. The hypertrophy of the liver continues until the mass of the liver returns to an equilibrium with blood flow such that portal pressure and shear stress are returned to normal levels.

The relationship of NO production in specific hepatic cell types and vascular regulation is not clearly understood. However, there is substantial evidence implicating stellate cells as regulators of intrahepatic blood flow distribution. The stellate cells respond to a number of contractile stimuli, including endothelin, and relax in response to NO. The balance between endogenous dilators and constrictors appears to become disturbed in a number of situations, including exposure to endotoxin, reperfusion injury, and shock. NO released by other cell types may also affect stellate cells. Kupffer cells are macrophages that use NO as a form of defense against bacterial invasion, but when excessively stimulated can lead to detrimental effects within the liver. Endothelial cells in the liver also release NO. It is not clear whether NO is able to affect sinusoidal endothelial permeability and whether endothelial cells are capable of releasing NO to interact with other cell types. The hepatocytes are also able to release NO in response to specific stimuli; the response to cytokines released from Kupffer cells is especially notable. Whether NO released from hepatocytes has any action on the hepatic circulation is not known. In virtually all cases involving in vivo responses to complex stimuli, it is unknown which cell type and often which form of NOS is involved in the production of NO. In complex disease states, it is highly likely that both the constitutive and the inducible forms of NO are involved, possibly at different times in different cells. These considerations and the unusual nature of the hepatic circulation make studies of endogenous regulators of this large circulation a challenge that requires a full range of experimental approaches in order to convert a vast amount of information into usable forms of knowledge.

Acknowledgments. Manuscript preparation was by Karen Sanders. The Medical Research Council of Canada and the Heart and Stroke Foundation of Manitoba funded the work of the authors cited in this chapter. Dallas Legare provided technical support and collaboration with these studies.

References

Andus T, Bauer J, & Gerok W. 1991. Effects of cytokines on the liver. *Hepatology 13:*364–375.

Bautista AP, D'Souza NB, Lang CH, & Spitzer JJ. 1992. Modulation of f-met-leu-phe-induced chemotactic activity and superoxide production by neutrophils during chronic ethanol intoxication. *Alcohol Clin Exp Res 16:*788–794.

Beckman JS, Beckman TW, Chen J, Marshall PA, & Freeman BA. 1990. Apparent hydoxyl radical production by peroxynitrite: implications for endothelial injury from nitric oxide and superoxide. *Proc Natl Acad Sci USA 87:*1620–1624.

Billiar TR, & Curran RD. 1990. Kupffer cell and hepatocyte interactions: a brief overview. *J Parenter Enteral Nutr 14:*175S–180S.

Billiar TR, Curran RD, Stuehr DJ, West MA, Bentz BG, & Simmons RL. 1989. An L-arginine-dependent mechanism mediates Kupffer cell inhibition of hepatocyte protein synthesis in vitro. *J Exp Med 169:*1467–1472.

Billiar TR, Curran RD, Harbercht BG, Stuehr DJ, Demetris AJ, & Simmons RL. 1990. Modulation of nitrogen oxide synthesis in vivo: N^G-monomethyl-L-arginine inhibits endotoxin-induced nitrite/nitrate biosynthesis while promoting hepatic damage. *J Leukocyte Biol 48:*565–569.

Boveris A, Fraga CG, Varsavsky AI, & Koch OR. 1983. Increased chemiluminescence and superoxide production in the liver of chronically ethanol-treated rats. *Arch Biochem Biophys 227:*534–541.

Browser DJ, Mathie RT, Benjamin IS, & Alexander B. 1994. The transhepatic action of ATP on the hepatic arterial and portal venous vascular beds of the rabbit: the role of nitric oxide. *Br J Pharmacol 113:*987–993.

Clemens MG, Bauer M, Gingalewski C, Miescher E, & Zhang J. 1994. Hepatic intercellular communication in shock and inflammation. *Shock 2:*1–9.

Clemens MG, Bauer M, Pannen BHJ, Bauer I, & Zhang JX. 1997. Remodeling of hepatic microvascular responsiveness after ischemia/reperfusion. *Shock 8:*80–85.

Cohen RA, & Weisbrod RM. 1988. Endothelium inhibits norepinephrine release from adrenergic nerves of rabbit carotid artery. *Am J Physiol 254:*H871–H878.

Curran RD, Billiar TR, Stuehr DJ, Hofmann K, & Simmons RL. 1989. Hepatocytes produce nitrogen oxides from L-arginine in response to inflammatory products of Kupffer cells. *J Exp Med 170:*1769–1774.

Curran RD, Billiar TR, Stuehr DJ, Ochoa JB, Harbrecht BG, Flint SG, & Simmons RL. 1990. Multiple cytokines are required to induce hepatocyte nitric oxide production and inhibit total protein synthesis. *Ann Surg 212:*462–469.

Decker K. 1990. Biologically active products of stimulated liver macrophages (Kupffer cells). *Eur J Biochem 192:*245–261.

Ezzat WR, & Lautt WW. 1987. Hepatic arterial pressure-flow autoregulation is adenosine mediated. *Am J Physiol 252:*H836–H845.

Feder IS, & Laskin DL. 1994. Regulation of hepatic endothelial and macrophage proliferation and nitric oxide production by GM-CSF, M-CSF, and IL-1β following acute endotoxemia. *J Leukocyte Biol 55:*507–513.

Fey GH, & Gauldie J. 1990. The acute phase response of the liver in inflammation. *Prog Liver Dis 9:*89–116.

Geller DA, Lowenstein CJ, Shapiro RA, Nussler AK, Di Silvio M, Wang SC, Nakayama DK, Simmons RL, Snyder SH, & Billiar TR. 1993. Molecular cloning and expression of inducible nitric oxide synthase from human hepatocytes. *Proc Natl Acad Sci USA 90:*3491–3495.

Greenberg SS, Diecke FPJ, Peevy K, & Tanaka TP. 1989. The endothelium modulates adrenergic neurotransmission to canine pulmonary arteries and veins. *Eur J Pharmacol 162:*67–80.

Greenblatt EP, Loeb AL, & Longnecker DE. 1993. Marked regional heterogeneity in the magnitude of EDRF/NO-mediated vascular tone in awake rats. *J Cardiovasc Pharmacol 21:* 235–240.

Helyar L, Bundschuh DS, Laskin JD, & Laskin DL. 1994. Induction of hepatic Ito cell nitric oxide production after acute endotoxemia. *Hepatology 20:*1509–1515.

Hirata K, Ogata I, Ohta Y, & Fujiwara K. 1989. Hepatic sinusoidal cell destruction in the development of intravascular coagulation in acute liver failure of rats. *J Pathol 158:*157–165.

Hynes MR, Dang H, & Duckles SP. 1988. Contractile responses to adrenergic nerve stimulation are enhanced with removal of endothelium in rat caudal artery. *Life Sci 42:*357–365.

Kawada N, Klein H, & Decker K. 1992. Eicosanoid-mediated contractility of hepatic stellate cells. *Biochem J 285:*367–371.

Kawada N, Tran-Thi T, Klein H, & Decker K. 1993. The contraction of hepatic stellate (Ito) cells stimulated with vasoactive substances. Possible involvement of endothelin 1 and nitric oxide in the regulation of the sinusoidal tonus. *Eur J Biochem 213:*815–823.

Kawada N, Kuroki T, Uoya M, Inoue M, & Kobayashi K. 1996. Smooth muscle α-actin expression in rat hepatic stellate cell is regulated by nitric oxide and cGMP production. *Biochem Biophys Res Commun 229:*238–242.

Kukielka E, Dicker E, & Cederbaum AI. 1994. Increased production of reactive oxygen species by rat liver mitochondria after chronic ethanol treatment. *Arch Biochem Biophys, 309:*377–386.

Kurose I, Miura S, Higuchi H, Watanabe N, Kamegaya Y, Takaishi M, Tomita K, Fukumura D, Kato S, & Ishii H. 1996. Increased nitric oxide synthase activity as a cause of mitochondrial dysfunction in rat hepatocytes: roles for tumor necrosis factor alpha. *Hepatology 24:*1185–1192.

Lautt WW. 1980. Control of hepatic arterial blood flow: independence from liver metabolic activity. *Am J Physiol 239:*H559–H564.

Lautt WW. 1996a. Hepatic circulation. In: Bennet T, Gardiner S, ed. *Nervous Control of Blood Vessels.* London: Harwood Academic Publishers, pp 465–503.

Lautt WW. 1996b. Hepatic vascular control by sympathetic nerves in vivo: In: Shimazu T, ed. *Liver Innervation.* London: John Libbey & Company, pp 239–245.

Lautt WW. 1996c. The 1995 Ciba-Geigy Award Lecture. Intrinsic regulation of hepatic blood flow. *Can J Physiol Pharmacol 74:*223–233.

Lautt WW, & Greenway CV. 1987. Conceptual review of the hepatic vascular bed. *Hepatology 7:*952–963.

Lautt WW, Dwan PD, & Singh RR. 1982. Control of the hyperglycemic response to hemorrhage in cats. *Can J Physiol Pharmacol 60:*1618–1623.

Lautt WW, Legare DJ, & d'Almeida MS. 1985. Adenosine as putative regulator of hepatic arterial flow (the buffer response). *Am J Physiol 248:*H331–H338.

Lyons CR, Orloff GJ, & Cunningham JM. 1992. Molecular cloning and functional expression of Na inducible nitric oxide synthase from a murine macrophage cell line. *J Biol Chem 267:*6370–6374.

Macedo MP, & Lautt WW. 1994. Nitric oxide suppression of norepinephrine release from nerves in the superior mesenteric artery. *Proc West Pharmacol Soc 37:*103–104.

Macedo MP, & Lautt WW. 1996a. Autoregulatory capacity in the superior mesenteric artery is attenuated by nitric oxide. *Am J Physiol 271:*G400–G404.

Macedo MP, & Lautt WW. 1996b. Shear-induced modulation by nitric oxide of sympathetic nerves in the superior mesenteric artery. *Can J Physiol Pharmacol 74:*692–700.

Macedo MP, & Lautt WW. 1997a. Potentiation to vasodilators by nitric oxide synthase blockade in superior mesenteric but not hepatic artery. *Am J Physiol 270:*G507–G514.

Macedo MP, & Lautt WW. 1998. Shear-induced modulation of vasoconstriction in the hepatic artery and portal vein by nitric oxide. *Am J Physiol 274:*G253–G260.

Mathie RT, Ralevic V, Alexander B, & Burnstock G. 1991. Nitric oxide is the mediator of ATP-induced dilation of the rabbit hepatic arterial vascular bed. *Br J Pharmacol 103:*1103–1107.

McCuskey RS, Urbaschek R, McCuskey PA, & Urbaschek B. 1982. In vivo microscopic studies of the responses of the liver to endotoxin. *Klin Wochenschr 60:*749–751.

Najafipour H, & Ferrell WR. 1993. Nitric oxide modulates sympathetic vasoconstriction and basal blood flow in normal and acutely inflamed rabbit knee joints. *Exp Physiol 78:*615–624.

Nishida J, McCuskey RS, McDonnell D, & Fox ES. 1994. Protective role of NO in hepatic microcirculatory dysfunction during endotoxemia. *Am J Physiol 267:*G1135–
G1141.

Nussler AK, Di Silvio M, Billiar TR, Hoffman RA, Geller DA; Selby R, Madariaga J, &
Simmons RL. 1992. Stimulation of the nitric oxide synthase pathway in human hepatocytes by cytokines and endotoxin. *J Exp Med 176:*261–264.

Obolenskaya MY, Vanin AF, Mordvintcev PI, Mulsch A, & Decker K. 1994. EPR evidence
of nitric oxide production by the regenerating rat liver. *Biochem Biophys Res Commun
202:*571–576.

Pinzani M, Failli P, Ruocco C, Casini A, Milani S, Baldi E, Giotti A, & Gentilini PJ. 1992.
Fat-storing cells as liver-specific pericytes. Spatial dynamics of agonist-stimulated intracellular calcium transients. *Clin Invest 90:*642–646.

Radi R, Beckman JS, Bush KM, & Freeman BA. 1991. Peroxynitrite-induced membrane
lipid peroxidation: the cytotoxic potential of superoxide and nitric oxide. *Arch Biochem
Biophys 288:*481–487.

Rockey DC. 1997. The cellular pathogenesis of portal hypertension: stellate cell contractility, endothelin, and nitric oxide. *Hepatology 25:*2–5.

Rockey DC, & Chung JJ. 1995. Inducible nitric oxide synthase in rat hepatic lipocytes and
the effect of nitric oxide on lipocyte contractility. *J Clin Invest 95:*1199–1206.

Rockey DC, & Chung JJ. 1996. Regulation of inducible nitric oxide synthase in hepatic sinusoidal endothelial cells. *Am J Physiol 271:*G260–G267.

Rockey DC, Boyles JK, Gabbiani G, & Friedman SL. 1992. Rat hepatic lipocytes express
smooth muscle actin upon activation in vivo and in culture. *J Submicrosc Cytol Pathol
24:*193–203.

Roland CR, Goss JA, Mangino MJ, Hafenichter D, & Flye MW. 1994. Autoregulation by
eicosanoids of human Kupffer cell secretory products: a study of interleukin-1,
interleukin-6, tumor necrosis factor-alpha, transforming growth factor-beta and nitric
oxide. *Ann Surg 219:*389–399.

Rubbo H, Radi R, Trujillo M, Telleri R, Kalyanaraman B, Barnes S, Kirk M, & Freeman,
BA. 1994. Nitric oxide regulation of superoxide and peroxynitrite-dependent lipid per-
oxidation. Formation of novel nitrogen-containing oxidized lipid derivatives. *J Biol
Chem 269:*26066–26075.

Spitzer JA. 1994. Cytokine stimulation of nitric oxide formation and differential regulation in hepatocytes and nonparenchymal cells of endotoxemic rats. *Hepatology 19:*
217–228.

Spolarics Z, Spitzer JJ, Wang JF, Xie J, Kolls J, & Greenberg S. 1993. Alcohol administration attenuates LPS-induced expression of inducible nitric oxide synthase in Kupffer
and hepatic endothelial cells. *Biochem Biophys Res Commun 197:*606–611.

Stadler J, Harbrecht BG, Di Silvio M, Curran RD, Jordan ML, Simmons RL, & Billiar, TR.
1993. Endogenous nitric oxide inhibits the synthesis of cyclooxygenase products and
interleukin-6 by rat Kupffer cells. *J Leukocyte Biol 53:*165–172.

Tesfamariam B, & Cohen RA. 1988. Inhibition of adrenergic vasoconstriction by endothelial cell shear stress. *Circ Res 63:*720–725.

Thatikunta P, Chakder S, & Rattan S. 1993. Nitric oxide synthase inhibitor inhibits catecholamine release caused by hypogastric sympathetic nerve stimulation. *J Pharmacol
Exp Ther 267:*1363–1368.

Thorin E, & Atkinson J. 1994. Modulation by the endothelium of sympathetic vasoconstriction in an in vitro preparation of the rat tail artery. *Br J Pharmacol 111:*351–357.

Um S, Nishida O, Tokubayashi M, Kimura F, Takimoto Y, Yoshioka H, Inque R, & Kita T. 1994. Hemodynamic changes after ligation of a major branch of the portal vein in rats: comparison with rats with portal vein constriction. *Hepatology 19:*202–209.

Volpes R, van den Oord JJ, & Desmet VJ. 1992. Can hepatocytes serve as "activated" immunomodulating cells in the immune response? *J Hepatol 16:*228–240.

Wang JF, Greenberg SS, & Spitzer JJ. 1995. Chronic alcohol administration stimulates nitric oxide formation in the rat liver with or without pretreatment by lipopolysaccharide. *Alcohol Clin Exp Res 19:*387–393.

Wang H, & Lautt WW. 1997. Evidence of nitric oxide (NO) being a trigger of liver regeneration. *Can J Gastroenterol 11(Suppl A):*63A.

Wang H, & Lautt WW. 1998. Evidence of nitric oxide, a flow-dependent factor, being a trigger of liver regeneration in rats. *Can J Physiol Pharmacol. 76:*1072–1079.

Wennmalm A, Benthin G, Karwatowska-Prokopczuk E, Lundberg J, & Peterson A. 1991. Release of endothelial mediators and sympathetic transmitters at different coronary flow rates in rabbit hearts. *J Physiol 435:*163–173.

Winwood PJ, & Arthur MJP. 1993. Kupffer cells: their activation and role in animal models of liver injury and human liver disease. *Semin Liver Dis 13:*50–59.

Yamakado K, Takeda K, Matsumura K, Nakatsuka A, Hirano T, Kato N, Sakuma H, Nakagawa T, & Kawarada Y. 1997. Regeneration of the un-embolized liver parenchyma following portal vein embolization. *J Hepatol 27:*871–880.

Yasuhiro E, Matsumara Y, Murata S, Umekawa T, Hisaki K, Takaoka M, & Morimoto S. 1994. The effect of *N*-nitro-L-arginine, a nitric oxide synthetase inhibitor, on norepinephrine overflow and antidiuresis induced by stimulation of renal nerves in anesthetized dog. *J Pharmacol Exp Ther 269:*529–535.

Zhang JX, Pegoli WJ, & Clemens MG. 1994. Endothelin-1 induces direct constriction of hepatic sinusoids. *Am J Physiol 266:*G624–G632.

16
Nitric Oxide and the Gastrointestinal Circulation

PHILIPPE BAUER, ZSUZSANNA ROZSA, AND D. NEIL GRANGER

Introduction

Nitric oxide (NO), a small, relatively unstable free radical, is now widely recognized to be an important modulator of different physiological processes in a variety of organ systems. The blood circulation appears to be particularly sensitive to the modulating influence of NO, which may be explained by the capacity of endothelial cells to rapidly generate large quantities of this inorganic gas. In the gastrointestinal tract, alterations in NO production or bioavailability has been implicated in microvascular responses to ischemia, inflammation, sepsis, and chronic portal hypertension (Crissinger and Granger 1995). This chapter discusses the potential role of NO as a modulator of gastrointestinal blood flow as well as the contribution of an altered tissue NO level to mediating the microvascular alterations associated with acute inflammation.

Nitric Oxide and Blood Flow Regulation

Vascular endothelial cells are known to produce agents that act on the underlying smooth muscle to promote protein phosphorylation, leading in turn to a decrease in smooth muscle tone (Ignarro 1989). Nitric oxide produced by endothelial cells diffuses to adjacent vascular smooth muscle, where it binds to soluble guanylate cyclase, which in turn increases the production of cGMP, with subsequent relaxation of smooth muscle (Wolin et al. 1982). Endothelial cells contain both a constitutive and an inducible form of nitric oxide synthase (NOS). The constitutive form of the enzyme is Ca^{2+}/calmodulin- and NADPH-dependent (Grisham 1992), and it produces NO in response to receptor (e.g., acetylcholine) or physical (e.g., shear stress) stimulation (Rubyani et al. 1986; Hishikawa et al. 1992). The inducible form of NOS, which can produce larger quantities and a more sustained release of NO than the constitutive isoform, is increased in response to certain stimuli that act directly on endothelial cells (e.g., cytokines) (Grisham 1992).

The role of NO in the intrinsic regulation of gastrointestinal blood flow has not been well defined. In vitro studies using monolayers of cultured endothelial cells indicate that the rate of production of NO is influenced by physical forces, such as shear rate and transmural pressure (Rubyani et al. 1986; Hishikawa et al. 1992). The production of NO by endothelial cells appears to be directly coupled to shear rate, yet it is inversely related to transmural pressure. The relation between NO production and shear rate does not favor a role for NO in the gastrointestinal vasodilation associated with reductions in local arterial pressure (i.e., pressure–flow autoregulation); however, the increased NO production that would be elicited by a decline in endothelial cell transmural pressure may provide a chemical basis for the phenomenon of pressure–flow autoregulation. The latter possibility is supported by the observation that pressure–flow autoregulation in the mesenteric circulation is an endothelium-dependent phenomenon (Randall and Hiley 1988).

The contribution of NO to basal vascular tone in the gastrointestinal tract has been demonstrated using analogues of L-arginine. Inhibition of NOS reduces resting blood flow in the gastrointestinal tract (Pique et al. 1989; Stark and Szurszewski 1992). The endothelium-dependent relaxation of gastric vascular smooth muscle by agents such as acetylcholine and bradykinin is also attenuated by inhibitors of NO synthesis (Stark and Szurszewski 1992). There is also a growing body of evidence that implicates NO in the gastrointestinal hyperemia associated with conditions as diverse as portal hypertension and central vagal stimulation (Stark and Szurszewski 1992; Tanaka et al. 1993). Inactivation of the NO that is basally produced by endothelial cells may also explain the increased vascular resistance that is observed following reperfusion of ischemic tissues. Superoxide rapidly and efficiently inactivates NO (Gryglewski et al. 1986); consequently, conditions such as ischemia and reperfusion that are associated with an accelerated production of superoxide (Halliwell and Gutteridge 1989) may abolish endothelium-dependent vascular tone and reduce blood flow.

Interpretation of the blood flow responses to NOS inhibitors is complicated by the actions of these agents on circulating blood cells. The NO produced by endothelial cells appears to play an important role in preventing intravascular adhesion and aggregation of platelets and leukocytes. As discussed below, inhibitors of NO synthesis promote the adhesion and aggregation of platelets and leukocytes within the microcirculation (Kubes et al. 1991; Granger et al. 1995). Hence, a potential consequence of these events is increased vascular resistance secondary to obstruction of microvessels with platelet–leukocyte aggregates.

Nitric Oxide Modulates Inflammatory Cell–Endothelial Cell Interactions

The ability of endothelial cell-derived NO to influence the adhesive interactions between circulating inflammatory cells and vascular endothelium is now well established. Nitric oxide has been reported to decrease the binding of platelets, monocytes, macrophages, and neutrophils to endothelial cells. Nitric oxide plays an important role in neutrophil adhesion at the inflammatory site, as well as in remote organs (Fukatsu et al. 1997). Nitric oxide also appears to inhibit both the homotypic and heterotypic aggregation of platelets and leukocytes. In vascular diseases, such as ischemia–reperfusion, an enhanced formation of reactive oxygen metabolites, as well as a decreased bioavailability of NO, appears to play an important role in initiating leukocyte–endothelial cell adhesion and the formation of platelet–leukocyte aggregates.

Intravital Studies (NOS Inhibitors and NO Donors)

Intravital microscopic techniques have allowed investigators to focus on inflammatory responses in postcapillary venules, which is the primary site of leukocyte adhesion and vascular protein leakage. This technology has also been employed to delineate the modulatory role of NO in ischemia–reperfusion-induced leukocyte rolling, adhesion, and emigration in postcapillary venules. In vivo studies have demonstrated that inhibition of NOS with the L-arginine analogues N^G-monomethyl-L-arginine (L-NMMA) and N^G-nitro-L-arginine methylester (L-NAME) increases the number of adherent and emigrated leukocytes and reduces the wall shear rate in mesenteric postcapillary venules (Kubes et al. 1991), whereas NO-generating agents attenuate (Kurose et al. 1994) the recruitment of adherent and emigrated leukocytes in different experimental models of inflammation.

In Vitro Studies

Similar responses to agents that alter NO levels have been observed in vitro in models that examine the adhesion of leukocytes or platelets to monolayers of cultured endothelial cells. However, the intensity of these responses is blunted relative to those observed in vivo, possibly reflecting the lower level of NO production by cultured endothelial cells (Niu et al. 1994). As expected, superoxide dismutase potentiates the effects of NO on neutrophil adhesion and platelet aggregation in vitro.

Role of Shear Rate

Inhibitors of NO synthesis (L-NMMA and L-NAME) reduce, and NO-donating agents increase, wall shear rate in postcapillary venules (Kurose et al. 1993, 1995). These responses are explained by the actions of the NO-modulating compounds on the tone of upstream arterioles. Although alterations in venular shear rate have been shown to exert a profound influence on leukocyte–endothelial cell adhesion in vivo (increases in shear rate blunt, and reductions in shear rate promote, leukocyte adhesion) (Perry and Granger 1991), the magnitude of the changes in shear rate elicited by NOS inhibitors and NO donors can account for only a small proportion of the changes in leukocyte–endothelial cell adhesion (Kubes et al. 1991; Kurose et al. 1995).

Role of Adhesion Molecules

Adhesion molecules expressed on the surface of leukocytes (CD11/CD18, L-selectin) or endothelial cells (ICAM-1, P- and E-selectin) mediate the leukocyte–endothelial cell interactions observed in inflamed postcapillary venules (Granger and Kubes 1994). Inhibition of NO synthesis leads to adhesive interactions between leukocytes and venular endothelial cells that are mediated by CD11/CD18 on leukocytes and intracellular adhesion molecule-1 (ICAM-1) on endothelial cells, with P-selectin modulating leukocyte rolling (Kurose et al. 1993; Kubes 1997). Nitric oxide attenuates adherence of blood cells to the endothelium, at least in part, by blunting the transcription-dependent expression of the endothelial cell adhesion molecules vascular cell adhesion molecule-1 (VCAM-1) and ICAM-1 (De Caterina et al. 1995; Takahashi et al. 1996). There is also evidence for a role of NO in regulating the rapid mobilization of P-selectin (Davenpeck et al. 1994; Armstead et al. 1997) from preformed pools in endothelial cells of mesenteric venules, that is, L-NAME promotes the rapid expression of P-selectin, whereas NO donors blunt the rapid P-selectin expression elicited by ischemia–reperfusion injury (Eppihimer et al. 1997). These changes in P-selectin expression are accompanied by corresponding changes in endothelial cell adhesiveness to circulating leukocytes. It has been shown using monolayers of cultured human umbilical vein endothelial cells that exogenous NO inhibits P-selectin mRNA expression and protein synthesis, whereas a reduction in NO production results in increased synthesis of P-selectin RNA and protein (Armstead et al. 1997).

Role of Mast Cells

Mast cell degranulation causes histamine and P-selectin-dependent leukocyte rolling and platelet activating factor (PAF)- and CD18-associated leukocyte adhesion (Gaboury et al. 1996). It has been demonstrated that NO-donating agents (e.g., spermine-NO) attenuate the mast cell-dependent leukocyte–endothelial cell adhesion elicited by a variety of stimuli (Kubes and Granger 1996). Part of the protective effect of NO donors in mast cell-dependent models of inflammation can be attributed, at least in part, to a mast cell-stabilizing action of NO. The view that NO contributes to mast cell stability and mast cell-mediated inflammation is supported by the observation that NOS inhibitors (e.g., L-NAME) elicit the degranulation of mesenteric mast cells and that mast cell-stabilizing agents attenuate the leukocyte–endothelial cell adhesion induced by L-NAME (Kurose et al. 1993; Kubes and Granger 1996). It has been shown that mast cells enhance L-NAME-induced neutrophil adhesion to monolayers of cultured endothelial cells via a mechanism that involves PAF and the adhesion molecules CD11/CD18 and ICAM-1 (Niu et al. 1996).

Lipid mediators (PAF and Leukotriene B_4)

The recruitment of adherent and emigrated leukocytes in rat mesenteric venules induced by L-NAME is significantly attenuated by antagonists to leukotriene B_4

(LTB$_4$) or PAF as well as an inhibitor of phospholipase A$_2$ (Arndt et al. 1993; Kurose et al. 1993). Restoring cellular cGMP levels with 8-bromo-cGMP also attenuates the leukocyte adhesion responses elicited by L-NAME (Kurose et al. 1993). These findings indicate that inhibition of NOS and the subsequent fall in cellular cGMP leads to the activation of phospholipase A$_2$, which results in the enhanced production of LTB$_4$ and PAF, with a consequent increase in the adherence and emigration of leukocytes. Since both PAF and LTB$_4$ are known to increase the expression of β_2-integrins (CD11/CD18) on leukocytes (Granger and Kubes 1994), this probably explains how L-NAME causes CD11/CD18-dependent leukocyte adhesion in postcapillary venules.

Role of Reactive Oxygen Metabolites

There are several lines of evidence that implicate reactive oxygen metabolites (ROMs) in the increased leukocyte–endothelial cell adhesion associated with NOS inhibition. It has been shown that L-NAME and L-NMMA elicit an oxidant stress in postcapillary venules (Kurose et al. 1995) and in monolayers of cultured endothelial cells (Niu et al. 1994). Second, superoxide dismutase and catalase effectively blunt the inflammatory responses to L-NAME whereas inhibitors of endogenous glutathione peroxidase exacerbate the inflammatory responses to L-NAME (Kurose et al. 1995). These findings are consistent with the view that NO avidly reacts with superoxide, thereby imparting a radical scavenging role to NO. Indeed, when rat mesenteric venules are exposed to the ROM-generating system hypoxanthine-xanthine oxidase, there is an intense recruitment of rolling and adherent leukocytes that can be prevented by administration of either superoxide dismutase or the NO donor 3-morpholinosydnonimine (SIN-1) (Gaboury et al. 1993).

Inhibition of NO synthase with L-NAME results in two phases of oxidant production: an initial increase that is leukocyte-independent and a later phase that is leukocyte-dependent. The initial phase probably reflects the sudden elimination of the superoxide scavenger NO. This results in the accumulation of hydrogen peroxide and the consequent activation of phospholipase A$_2$ and degranulation of mast cells, both of which increase tissue levels of PAF and LTB$_4$, and increases the expression of P-selectin and ICAM-1 on endothelial cells. The later phase of L-NAME-induced oxidant stress can be prevented by monoclonal antibodies that prevent leukocyte–endothelial cell adhesion (Kurose et al. 1994). This leukocyte-dependent phase of oxidant production probably reflects the generation of ROM by the adherent and emigrated leukocytes.

Nitric Oxide Regulates Endothelial Barrier Function

There is a growing body of evidence that implicates NO in the vascular permeability changes associated with different pathologic conditions, including ischemia-reperfusion injury, sepsis, and nonsteroidal antiinflammatory drug (NSAID)-induced gastropathy. The possibility that NO modulates the barrier function of vascular endothelial cells was first suggested by studies employing NOS inhibitors, which tend to invoke a protective role of NO derived from the constitutive form of endothelial NOS (Kubes and Granger 1992). However, there is also a substantial body of evidence that implicates NO as a mediator of the increased vascular permeability that is associated with the induction of the calcium-independent form of NO synthase (iNOS). These seemingly inconsistent results suggest that the action of NO on endothelial barrier function is influenced by the amount, the chemical products, and/or the source of NO.

Protective Effects of Endothelial Cell (cNOS)-Derived NO

Several experimental approaches have been used to define the acute effects of NOS inhibition on microvascular permeability. These include whole organ strategies, such as measurement of lymphatic protein flux or tissue accumulation of radiolabeled albumin, as well as intravital microscopic quantification of the extravasation of fluorescein isothiocyanate (FITC)-labeled albumin across single postcapillary venules (Kubes and Granger 1992; Kurose et al. 1993; Laszlo et al. 1994). Local intraarterial infusion of L-NAME in the cat intestine does not alter capillary hydrostatic pressure but results in a five-fold increase in lymphatic fluid and protein fluxes. Estimates of the capillary osmotic reflection coefficient from the lymphatic protein flux data suggest that L-NAME nearly doubles vascular permeability to plasma proteins in feline small intestine (Kubes and Granger 1992). The NO-liberating compound nitroprusside blunted the L-NAME-induced increase in vascular permeability as well as the corresponding recruitment of adherent leukocytes in postcapillary venules.

The findings from studies that examine the tissue accumulation of radiolabeled albumin as a measure of vascular permeability do not predict an altered endothelial barrier function after NOS inhibition (Laszlo et al. 1994; Filep et al. 1997). Although this inability to detect increases in albumin extravasation may reflect an insensitivity of the technique, large increments in radiolabeled albumin accumulation in gastrointestinal organs have been noted following iNOS activation (discussed in the following section).

Data derived from intravital microscopic studies of FITC-albumin leakage from intact, blood-perfused mesenteric venules support the view that acute inhibition of endothelial NOS results in increased vascular permeability (Kurose et al. 1993). L-NAME appears to elicit an initial rapid increase, followed by a slower rate of albumin accumulation in the perivenular compartment. The initial phase of albumin leakage precedes the L-NAME-induced leukocyte adherence and emigration, suggesting that there is clearly a leukocyte-independent component to the albumin leakage response. The magnitude of the albumin leakage observed in the later phase of L-NAME exposure is highly correlated with the number of adherent and emigrated leukocytes in the same segment of venule, suggesting that there is also a leukocyte-dependent component to the L-NAME-induced albumin leakage response. Support for the existence of a leukocyte-dependent component of

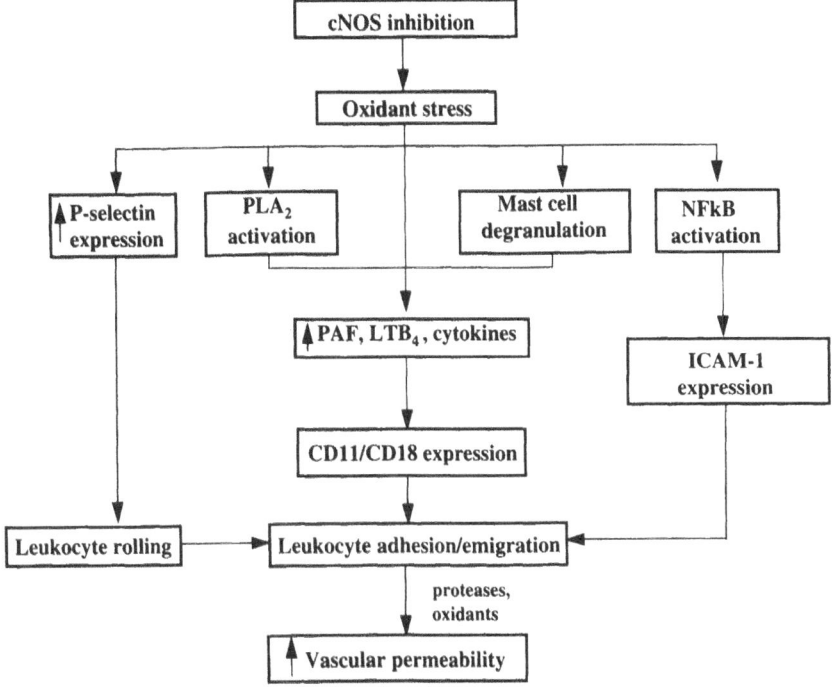

FIGURE 16.1. Mechanism proposed to explain the leukocyte–endothelial cell adhesion and increased vascular permeability associated with inhibition of endothelial nitric oxide synthase (NOS). PLA_2, Phospholipase A_2; PAF, platelet activating factor; LTB_4, leukotriene B_4; ICAM-1, intercellular adhesion molecule-1.

albumin extravasation is provided by reports describing a blunted leakage response to L-NAME in animals receiving either monoclonal antibodies (against CD11/CD18, ICAM-1, or P-selectin) that inhibit leukocyte–endothelial cell adhesion (Kubes and Granger 1992; Kurose et al. 1993) or NO-donating agents (Kurose et al. 1994; Kubes 1997). Furthermore, mesenteric venules perfused with buffer not containing leukocytes do not respond to NOS inhibitors with a rise in albumin extravasation. Indeed, the response of these artificially perfused vessels suggests that NO per se acts to diminish endothelial barrier integrity (Yuan et al. 1992).

Overall, the pharmacological intervention studies that have focused on the increased vascular permeability elicited by acute inhibition of endothelial NOS in blood-perfused venules are consistent with the mechanism outlined in Figure 16.1. The evidence indicates that reducing NO availability results in an oxidant stress, which in turn leads to increased P-selectin expression, phospholipase A_2 activation, mast cell degranulation, and activation of the nuclear transcription factor NFkB. Phospholipase A_2 activation and mast cell degranulation result in the production, release, and accumulation of PAF, LTB_4, and cytokines (e.g., tumor necrosis factor). The lipid mediators PAF and LTB$_4$ promote the increased expression and activation of β$_2$-integrins (CD11/CD18) on leukocytes. The counter-receptor for CD11/CD18 on the endothelial cell is ICAM-1, whose expression is increased as a result of cytokine release and NFkB activation. Consequences of the up-regulation of P-selectin, E-selectin, and ICAM-1 on endothelial cells are leukocyte rolling, adhesion, and emigration in postcapillary venules. These activated and adherent leukocytes release oxidants and proteases, which can then mediate the increased vascular permeability that is associated with cNOS inhibition.

Deleterious Effects of iNOS-Derived NO

There are several reports of a time-dependent role of NO in mediating the increased vascular permeability that results from the administration of bacterial endotoxin (lipopolysaccharide, LPS) or indomethacin. For example, in rat intestine LPS caused a time-dependent increase in albumin extravasation, which was exacerbated when L-NAME was administered concurrently with LPS. However, when L-NAME (or L-NMMA) was administered after the initial induction of iNOS (occurring at about 3 hours), a reduction in albumin leakage was observed (Laszlo et al. 1994). Another study (Filep et al. 1997) showed that inhibition of cNOS potentiated, whereas inhibition of iNOS markedly attenuated, the loss of plasma volume and multiorgan albumin extravasation associated with endotoxin shock. It was suggested from these findings that selective inhibitors of iNOS may be more effective that nonselective inhibitors of all forms of NOS (e.g., L-NMMA or L-NAME) in the treatment of septic shock. Similar responses were noted in rats that manifested a time-dependent increase in albumin extravasation after indomethacin treatment (Whittle et al. 1995).

The mechanisms that underlie the ability of NO to mediate increased vascular permeability under conditions of iNOS induction have not been clearly defined. However, it has been proposed that the deleterious effects of iNOS-derived NO may be related to the much higher levels of the unstable gas that is produced by this isoform. Another proposal is that iNOS-derived NO reacts with superoxide (generated from leukocytes and/or other cells) to form peroxynitrite, which is a highly reactive species that could damage endothelial cells and result in increased vascular permeability (Grisham 1992). It is also conceivable that iNOS-derived NO binds to and inactivates key metal-containing enzymes that contribute to the maintenance of endothelial barrier function.

Conclusions

Evidence has been provided of a role for NO as both a protective and a injurious agent in the microcirculation. The prevailing view in the literature is that the role of NO in different pathological conditions is dictated by its source. With the advent of mutant mice that have been rendered genetically deficient in either the constitutive or the inducible (or both) isoforms of NOS, it is likely that this view can be more rigorously tested in the future. These studies should provide more insight into the feasibility of using NOS isoform-specific inhibitors or NO-donating agents in the clinical management of disease states associated with ischemia–reperfusion, sepsis, or inflammation.

References

Armstead VE, Minchenko AG, Schuhl RA, Hayward R, Nossuli TO, & Lefer AM. 1997. Regulation of P-selectin expression in human endothelial cells by nitric oxide. *Am J Physiol 273:* H740–H746.

Arndt H, Russell JB, Kurose I, Kubes P, & Granger DN. 1993. Mediators of leukocyte adhesion in rat mesenteric venules elicited by inhibition of nitric oxide synthesis. *Gastroenterology 105:* 675–680.

Crissinger KD, & Granger DN. 1995. Gastrointestinal blood flow. In: Yamada T, ed. *Yearbook of Gastroenterology.* 2nd edition. Philadelphia: JB Lippincott, pp 518–545.

Davenpeck KL, Gauthier TW, & Lefer AM. 1994. Inhibition of endothelial-derived nitric oxide promotes P-selectin expressions and actions in the rat microcirculation. *Gastroenterology 107:*1050–1058.

De Caterina R, Libby P, Peng HB, Thannickal VJ, Rajavashisth TB, Gimbrone MA Jr, Shin WS, & Liao JK. 1995. Nitric oxide decreases cytokine-induced endothelial activation. Nitric oxide selectively reduces endothelial expression of adhesion molecules and proinflammatory cytokines. *J Clin Invest 96:*60–68.

Eppihimer MJ, Russell J, Anderson DC, Epstein CJ, Laroux S, & Granger DN. 1997. Modulation of P-selectin expression in the postischemic intestinal microvasculature. *Am J Physiol 273:*G1326–G1332.

Filep JG, Delalandre A, & Beauchamp M. 1997. Dual role of nitric oxide in the regulation of plasma volume and albumin escape during endotoxin shock in conscious rats. *Circ Res 81:*840–847.

Fukatsu K, Saito H, Han I, Furukawa S, Hashiguchi Y, Lin MT, Matsuda T, Inaba T, Inoue T, Ikeda S, Yasuhara H, & Muto T. 1997. Nitric oxide inhibition decreases neutrophil adhesion at the inflammatory site, while increasing adhesion in remote organs in peritonitis. *J Surg Res 68:*79–86.

Gaboury J, Woodman RC, Granger DN, Reinhardt P, & Kubes P. 1993. Nitric oxide prevents leukocyte adherence: role of superoxide. *Am J Physiol 265:*H862–H867.

Gaboury JP, Niu XF, & Kubes P. 1996. Nitric oxide inhibits numerous features of mast cell-induced inflammation. *Circulation 93:*318–326.

Granger DN, & Kubes P. 1994. The microcirculation and inflammation: modulation of leukocyte-endothelial cell adhesion. *J Leukoc Biol 55:*662–675.

Granger DN, Kurose I, & Kubes P. 1995. Nitric oxide: a modulator of cell-cell adhesion and protein exchange in postcapillary venules. In: Schlag G, Redl H, eds. *Shock, Sepsis and Organ Failure—nitric oxide.* Berlin: Springer-Verlag, pp 121–131.

Grisham MB. 1992. *Reactive Metabolites of Oxygen and Nitrogen in Biology and Medicine. Austin, Tex.:* RG Landes Co.

Gryglewski RJ, Palmer RM, & Moncada S. 1986. Superoxide anion is involved in the breakdown of endothelium-derived vascular relaxing factor. *Nature 320:*454–456.

Halliwell B, & Gutteridge JMC. 1989. *Free Radicals in Biology and Medicine.* Oxford: Clarendon Press.

Hishikawa K, Nakaki T, Suzuki H, Saruta T, & Kato R. 1992. Transmural pressure inhibits nitric oxide release from human endothelial cells. *Eur J Pharmacol 215:*329–331.

Ignarro LJ. 1989. Biological actions and properties of endothelium-derived nitric oxide formed and released from artery and vein. *Circ Res 65:*1–21.

Kubes P. 1997. Nitric oxide: a modulator of the inflammatory response. In: Vincent JL, ed. *Yearbook of Intensive and Critical Care Medicine.* Berlin: Springer-Verlag, pp 197–210.

Kubes P, & Granger DN. 1992. Nitric oxide modulates microvascular permeability. *Am J Physiol 262:*H611–H615.

Kubes P, & Granger DN. 1996. Leukocyte-endothelial cell interactions evoked by mast cells. *Cardiovasc Res 32:*699–708.

Kubes P, Suzuki M, & Granger DN. 1991. Nitric oxide: an endogenous modulator of leukocyte adhesion. *Proc Natl Acad Sci USA 88:*4651–4655.

Kurose I, Kubes P, Wolf R, Anderson DC, Paulson J, Miyasaka M, & Granger DN. 1993. Inhibition of nitric oxide production: mechanisms of vascular albumin leakage. *Circ Res 73:*164–171.

Kurose I, Wolf R, Grisham MB, Aw TY, Specian RD, & Granger DN. 1995. Microvascular responses to inhibition of nitric oxide production. Role of active oxidants. *Circ Res 76:*30–39.

Kurose I, Wolf R, Grisham MB, & Granger DN. 1994. Modulation of Ischemia/-reperfusion-induced microvascular dysfunction by nitric oxide. *Circ Res 74:*376–382.

Laszlo F, Whittle BJ, & Moncada S. 1994. Time-dependent enhancement or inhibition of endotoxin-induced vascular injury in rat intestine by nitric oxide synthase inhibitors. *Br J Pharmacol 111:*1309–1315.

Niu XF, Smith CW, & Kubes P. 1994. Intracellular oxidative stress induced by nitric oxide synthesis inhibition increases endothelial cell adhesion to neutrophils. *Circ Res 74:* 1133–1140.

Niu XF, Ibbotson G, & Kubes P. 1986. A balance between nitric oxide and oxidants regulates mast cell-dependent neutrophil-endothelial cell interactions. *Circ Res 79:* 992–999.

Perry MA, & Granger DN. 1991. Role of CD11/CD18 in shear rate -dependent leukocyte-endothelial cell interactions in cat mesenteric venules. *J Clin Invest 87:*1798–1804.

Pique JM, Whittle BJ, & Esplugues JV. 1989. The vasodilator role of endogenous nitric oxide in the rat gastric microcirculation. *Eur J Pharmacol 174:*293–296.

Randall MD, & Hiley CR. 1988. Detergent and methylene blue affect endothelium-dependent vasorelaxation and pressure-flow relations in rat blood perfused mesenteric arterial bed. *Br J Pharmacol 95:*1081–1088.

Rubyani GM, Romero JC, Vanhoutte PM. 1986. Flow-induced release of endothelium-derived relaxing factor. *Am J Physiol 250:*H1145–H1149.

Stark ME, & Szurszewski JH. 1992. Role of nitric oxide in gastrointestinal and hepatic function and disease. *Gastroenterology 103:*1928–1949.

Takahashi M, Ikeda U, Masuyama J, Funayama H, Kano S, & Shimada K. 1996. Nitric oxide attenuates adhesion molecule expression in human endothelial cells. *Cytokine 8:*817–821.

Tanaka T, Guth P, & Tache Y. 1993. Role of nitric oxide in gastric hyperemia induced by central vagal stimulation. *Am J Physiol 264:*G280–G284.

Whittle BJ, Laszlo F, Evans SM, & Moncada S. 1995. Induction of nitric oxide synthase and microvascular injury in the rat jejunum provoked by indomethacin. *Br J Pharmacol 116:*2286–2290.

Wolin MS, Wood KS, & Ignarro LJ. 1982. Guanylate cyclase from bovine lung. A kinetic analysis of the regulation of the purified soluble enzyme by protoporphyrin IX, heme, and nitrosyl-heme. *J Biol Chem 257:*13312–13320.

Yuan Y, Granger HJ, Zawieja DC, & Chilian WM. 1992. Flow modulates coronary venular permeability by a nitric oxide-related mechanism. *Am J Physiol 263:*H641–H646.

17
Uterine Effects of Nitric Oxide

KENNETH E. CLARK AND LESLIE MYATT

Introduction

In contrast to the vasculature of many other organs in the body, the uterine vasculature is very dynamic and is continuously undergoing significant vascular changes and remodeling. The uterus is exposed to a constantly changing hormonal milieu associated with fluctuations in estrogen and progesterone levels. The observed hemodynamic changes are mediated by alterations in numerous endogenous hormonal substances that are produced locally in the endothelium, vascular smooth muscle, and surrounding uterine tissues. These modulators include both vasodilators, such as nitric oxide (NO) and prostaglandins, and vasoconstrictors, such as endothelin-1, throboxane A_2, and prostaglandin $F_{2\alpha}$ ($PGF_{2\alpha}$). In females, this modulation occurs throughout, life including the menstrual cycle, pregnancy, and finally during menopause. In the present chapter we will concentrate mainly on NO as a regulator of uterine and uteroplacental hemodynamics in the pregnant and nonpregnant states. Additionally we will discuss the role of NO in regulating umbilical blood flow, myometrial activity of the uterus, and hemodynamic changes during menopause.

Nonpregnant Uterine Vasculature

The uterine vasculature is lined with endothelial cells that have been shown to produce many vasoactive substances, including NO (Ignarro 1989) and prostacyclin. Endothelial cells are in an ideal location for modulation of vascular resistance and vascular responsiveness, since vasoactive substances must come in contact with the endothelium prior to reaching the end organ, vascular smooth muscle. In recent years, a significant amount of research has been conducted into the role of NO and its ability to dilate uterine vessels directly and act as a transducer for other substances that bind to the endothelium. As described in much greater detail elsewhere, NO is produced in the endothelial cells during the conversion of L-arginine to L-citrulline by the enzyme nitric oxide synthase (NOS) (Palmer et al. 1988) and can be antagonized competitively by arginine analogues such as N^G-nitro-L-arginine methylester (L-NAME) and N^G-monomethyl-L-arginine (L-NMMA) (Moore et al. 1990; Rees et al. 1990). These antagonists are used routinely to verify that NO is important in producing the effect being studied. Once administered, the inhibitor should block the normal response mediated by synthesis of NO and these effects should be stereospecific (i.e., D-NAME has no effect) and reversed by L-arginine.

Vasoactive substances bind to their specific receptors and have direct vascular effects, but they also can stimulate the release or synthesis of NO and prostacyclin from the endothelial cell, thus modulating their vascular responses. These vascular responses are dependent on both the basal level of NO in the endothelium and the underlying vascular smooth muscle and the amount of NO synthesized following stimulation of their receptors. Acetylcholine, adenosine, bradykinin, histamine, and serotonin (De Nucci et al. 1988; Dainty et al. 1990; Carpenter et al. 1997) all work at least in part via this mechanism. In addition to modulation of vascular responses, it is clear that the higher the basal levels of NO in the endothelial cells, the greater the effect it will have on the basal vascular resistance of the organ, i.e., increases in endothelial cell NO lead to relaxation of the underlying vascular smooth muscle.

Probably the most dramatic hemodynamic change that occurs in the uterine vasculature is the increase in uterine blood flow that occurs in response to endogenous and exogenous estrogens. It is clear from the work of Rosenfeld et al. (1973) that the magnitude of this vasodilation is greatest in reproductive tissues (uterus, vagina, and mammary gland), but it also occurs in many other tissues, including the skin, myocardium, and thyroid gland, and most recently has been shown to occur in the coronary circulation (Lang et al. 1997). In the sheep, the response to estrogen in the uterine circulation undergoes a classical genomic response in which uterine blood flow remains constant for the first 30 minutes. By approximately 35 to 40 minutes after estrogen administration uterine vascular resistance begins to fall and uterine blood flow begins to rise, reaching a maximum at approximately 105 to 120 minutes. Over 8 to 12 hours, it slowly returns toward baseline blood flow. Similar time-related changes are observed in heart rate, cardiac output (Rosenfeld et al. 1973), and coronary blood flow (Lang et al. 1997). Although the uterine vasodilator response to estrogen reaches a maximum at approximately 120 minutes in rats (Brody et al. 1974) and sheep (Lang et al. 1997) and 60 minutes in mice (Brody et al. 1974), the timing of uterine vascular response in humans is unknown. These estrogenic vascular effects, generally independent of the type of estrogen, are thought to be mediated by specific estrogen receptors. The existence of specific vascular receptors for estrogen and progesterone in reproductive tissues (uterine and mammary arteries), as well as nonuterine vascular tissues such as myocardium, coronary vessels, aorta, and endothelial cells, is well documented (Colburn and Buonassesse 1978; Lin et al. 1982; McGill 1989; Karas et al. 1994; Losordo et al. 1994).

It is clear now that at least two distinct estrogen receptors exist, ERα and ERβ (Kuiper et al. 1996; 1997; Lewin 1997), but it is not currently known which receptor mediates the estrogen-induced increase in uterine blood flow. Recently a study by Tschugguel et al. (1998) has suggested that α and β receptors are located in endothelial cells, but that different receptors may be located at different levels of the vascular tree. These investigators showed that endothelial cells taken from larger vessels in humans (aorta and pulmonary artery) contain mainly α receptors, whereas endothelial cells from smaller vessels (basilar, uterine, and umbilical arteries) contain mainly β receptors. Thus, small arterioles may have ERβ receptors that could mediate the local uterine response. However, it is currently unclear whether this occurs, since Iafrati and co-workers (1997), using an ERα-deficient transgenic mouse, have shown that α knockout mice are not fertile nor do they appear to have a normal estrus-associated hyperemic response.

In recent years the role of NO in mediating estrogen-induced increases in uterine blood flow has been extensively investigated. In 1992 our laboratory was the first to demonstrate that estradiol-17β-induced increases in uterine blood flow could be antagonized by inhibitors of NOS (Van Buren et al. 1992). In these studies, administration of estradiol-17β led to uterine vasodilation, which was significantly attenuated by systemic as well as local uterine artery administration of L-NAME, an inhibitor of NOS.

At least three isoforms of NOS exist: eNOS (endothelial) and bNOS (brain), which are calcium-dependent, and iNOS (inducible), which is calcium-independent. Estradiol-17β has been shown to increase levels of eNOS in uterine and systemic vessels (Veille et al. 1996; Figueroa et al. 1998; Magness et al. 1998; Salhab et al. 1998). The increase in endothelial NO appears to occur because of increased NOS production as a result of estrogen receptor-mediated up-regulation of the eNOS enzyme. Although no studies to date have reported changes in iNOS in the uterine vasculature in response to estradiol-17β, changes have been observed in the rat kidney. Estradiol-17β appears to regulate both eNOS and iNOS in the rat kidney (Neugarten et al. 1997), and ovariectomy of female rats causes both eNOS and iNOS to fall significantly, whereas hormone supplementation with estrogen causes them both to rise. Attempts to consistently document up-regulation by estradiol-17β of iNOS in the uterine vasculature of sheep using Western blots have been hampered by the lack of antibodies and molecular probes specific for sheep (unpublished observations). Recent studies in our laboratory have also shown that not all estrogen-induced vasodilation is attenuated by the NOS inhibitor L-NAME. Ethinylestradiol, which produces vasodilation in the uterine circulation equivalent to that produced by estradiol-17β, with an identical time course, is not inhibited following systemic or local uterine administration of L-NAME (Clark et al. 1998), but it is completely blocked by aminoguanidine, a compound thought to be somewhat selective for iNOS (unpublished observations). The observation that the response to estrogen takes 35–45 minutes to begin to occur and does not peak until 120 minutes is also suggestive of an inducible form of NOS. Thus, although the response appears to be mediated by NO the isoform responsible for the production of the NO is currently unclear.

Recent studies from our laboratory have shown that administration of L-NAME either locally or systemically decreases basal uterine blood flow by 35%-50%. The decrease in basal uterine blood flow appears to be due solely to NOS inhibition, since uterine vasodilation produced by a second vasodilator, prostacyclin, is not altered by L-NAME. These data are interpreted to mean that L-NAME is not a directly acting vasoconstrictor but rather acts specifically by inhibiting NOS, thus resulting in increased vascular tone. This observation is further supported by the observation that the D-isomer of L-NAME has no effect on basal uterine blood flow. The basal concentrations of NO being produced in blood vessels appear to have significant effects on both vascular responses to vasoactive substances and baseline vascular resistance. Weiner and co-workers (1994) have shown in guinea pigs that chronic estrogen therapy up-regulates the eNOS enzyme. Similar observations have been made in sheep chronically treated with estrogen (Magness et al. 1998; Salhab et al. 1998). This up-regulation would be expected to decrease tone. Although estrogen can up-regulate NOS and nitrate levels, recently work by Rosselli and co-workers (1995) has shown that, at least in postmenopausal women, elevated nitrite and nitrate levels associated with estrogen therapy can be eliminated by the addition of progestins.

Menopause

It is becoming increasingly clear that ovarian steroids, which played an important role in the uterus, also play a critical role in the overall cardiovascular health of women as they complete their reproductive years and enter menopause. Studies by Stampfer et al. (1991) have shown that during menopause the risk of death from heart disease for women approaches that for men, once the ovaries stop producing estrogen, and that this protection can be reinstated by hormonal therapy. It seems reasonable to speculate that the loss of ovarian steroids is associated with reductions in eNOS, which in turn would lead to decreased NO in the local uterine and systemic vasculature. As estrogen levels decrease, basal uterine, vaginal, and endometrial blood flow are thought to decrease, leading to endometrial atrophy and increased vaginal dryness. Hormone replacement therapy with estrogens can reverse these changes and has been shown also to have dramatic effects on the systemic vasculature. Recent studies from our laboratory have shown that estrogen not only increases uterine blood flow but also increases coronary blood flow in conscious animals (Lang et al. 1997), and these effects are mediated at least in part by NO. As indicated above, hormone therapy has been shown to reduce heart disease in women by as much as 50%, and this reduction may be partially mediated by estrogen-induced increases in NO in the vessel walls, which may also inhibit cellular proliferation (Morey et al. 1997). Since estrogen therapy increases endothelial NO, it seems reasonable to suggest that the elevation of eNOS and the resultant NO concentrations may in part explain the therapeutic benefits of estrogens in multiple vasculatures in postmenopausal women.

Pregnancy

Pregnancy represents a major cardiovascular challenge to the organism, since in most species uterine blood flow increases by as much as 100-fold over the non-pregnant state. This dramatic increase in uteroplacental blood flow occurs in response to direct vasodilation and vessel growth (angiogenesis) as well as the development of the placental vascular bed. In addition to the uterine vasculature, the systemic vasculature dilates during pregnancy, with cardiac output increasing by as much as 30%–40% as gestation progresses. These changes are thought to be mediated in part by increased synthesis of NO and are most likely under the control of the increased levels of circulating sex steroids, that occur in pregnancy.

Several studies have shown that NO levels are significantly elevated during pregnancy in humans, sheep, and rats. In most cases, these studies evaluated both plasma and urinary nitrate and nitrite levels and showed significant elevation (Conrad et al. 1993; Jaekle et al. 1994; Yang et al. 1996). Studies using in vitro isolated blood vessels from nonpregnant and pregnant animals have also shown significant increases in NO synthesis in vessels from pregnant animals. More recently, eNOS activity has been shown to be significantly elevated in vessels from pregnant sheep (Magness et al. 1998).

As previously indicated, NO can be released in response to vasoactive agonists. There is some evidence so that the release is greater in vessels from pregnant animals than in vessels from nonpregnant animals, probably become of the elevation of basal levels of endothelial NO in pregnancy. Nitric oxide also appears to be elevated in the myometrium in pregnancy. This may be important, because, in contrast to many other vascular beds, uterine blood flow can be dramatically affected by uterine contractile activity, with increased myometrial tone leading to reduction in uterine blood flow. (The role of NO in the regulation of myometrial quiescence is discussed in greater detail below.)

Blood flow to the uterus is responsible for delivering all the nutrients and oxygen that are required by the fetus. If uterine blood flow is not adequate to meet the needs of the fetus, the fetus will fail to grow normally. Failure of these pregnancy-related vasodilator systems has been implicated in pregnancy-associated diseases such as preeclampsia (Rutherford et al. 1995; Davidge et al. 1996) and intrauterine growth restriction (IUGR), but as described below, there is clear evidence that in the fetal circulation these systems may be activated to allow compensation rather than depressed, as would be expected. Although it is possible that inadequate production of NO in the maternal vasculature leads to diseases such as hypertension of pregnancy or preeclampsia, the number of studies supporting these theories is approximately equivalent to the number of studies that do not support it. Thus, no clear-cut conclusion has been reached at this time.

Regulation of Umbilical Blood Flow

Since there is no autonomic innervation in the human placenta, blood flow to the placenta is determined by fetal cardiac output and umbilical–placental vascular resistance (Adamson et al. 1998). Humoral factors in blood or local autocrine on paracrine effectors must determine fetal placental vascular resistance. Many in vitro investigations have shown regional differences in the efficacy of various families of vasoactive autocoids in regulating resistance in the different regions of the fetal placental circulation (umbilical vasculature, chorionic plate vessels, and! villous tree) (Myatt 1992). In the isolated umbilical cord, prostacyclin appears to be a more potent vasodilator than NO, whereas in the terminal villous tree of the placenta, NO appears to be a more potent vasodilator than prostacyclin.

In vitro investigations have shown that NO appears to maintain low resistance in the fetal placental vasculature and to attenuate the action of vasoconstrictors such as endothelin, thromboxane, and angiotensin II (Myatt et al. 1991, 1992). Levels of the NO breakdown product nitrate are elevated in fetuses, although no venous–arterial difference has been shown across the placenta (Yang et al. 1996). None of the classical vasodilator agonists that were shown to cause release of endothelium-derived relaxing factor (EDRF) appear able to cause release of NO in the human placental vasculature. The major stimulus to NO release in the placental vasculature appears to be flow or shear stress across the surface of endothelial cells (Wieczorek et al. 1995; Learmont and Poston 1996). Therefore, alterations in flow or shear stress on endothelial cells, the viscosity or flow rate of blood, and diameter of blood vessels might be important determinants of resistance in human placental vasculature.

In pregnancy complicated by preeclampsia and/or IUGR, both of which are associated with increased fetal morbidity and mortality, increases in fetal–placental vascular impedance can be demonstrated by abnormal umbilical flow velocity waveforms (Erskine and Ritchie 1985; Giles et al. 1985). This increase in fetal–placental vascular impedance may arise from alterations in vasoactive mechanisms that regulate the fetal–placental vasculature, but also may be due to primary defects in placental angiogenesis. Indeed, in placentas of IUGR fetuses with absent end-diastolic flow velocity, there appears to be a failure of angiogenesis, since capillary loops, that are sites of low resistance in the fetal placental circulation, are not properly formed, leaving the fetal placental circulation as a high-resistance circuit (Krebs et al. 1996).

Paradoxically, recent investigations have shown that in pregnancies complicated by preeclampsia and/or IUGR, there is up-regulation of eNOS expression in the villous vasculature of the placenta (Myatt et al. 1997). This is also associated with increased concentrations of nitrate, the breakdown product of NO found in the umbilical cord in such pregnancies (Lyall et al. 1995, 1996). This suggests that increased expression of NOS may be a compensatory or adaptive response to the increased resistance to flow seen in the placenta in these pregnancies, with the stimulus to up-regulation being increased shear stress over endothelial cells in the vasculature (Wieczorek et al. 1995; Learmont and Poston 1996).

Alternatively, the up-regulation may be a response to the relative hypoxia that may occur in the placenta of preeclampsia or IUGR pregnancies. There are several reports that maternal administration of the NO donors nitroglycerin or isosorbide dinitrate (Giles et al. 1992; Grunewald et al. 1995; Makino et al. 1997) acutely improved the abnormal flow velocity waveforms in these pregnancies, suggesting that maternally administered NO donors may cross the placenta to vasodilate the fetal–placental vasculature. However, in a certain proportion of such pregnancies, administration of the NO donors was unable to normalize the umbilical cord flow velocity waveforms (Giles et al. 1992), suggesting that the primary defect in these pregnancies might be a failure of vascular development rather than a defect in vascular reactivity.

The bioactivity of the NO radical is curtailed by its interaction with superoxide radical; however, the interaction of NO and superoxide produces the powerful long-lived oxidant peroxynitrite. Recently, evidence has been provided that in the placenta in pregnancies complicated by preeclampsia and/or IUGR, formation of peroxynitrite is occurring, as evidenced by the formation of nitrotyrosine residues (Myatt et al. 1996). Preeclampsia is thought of as a state of oxidant-mediated vascular endothelial dysfunction (Roberts et al. 1991), which may lead to the altered maternal vascular reactivity seen in such pregnancies. The presence of nitrotyrosine residues in the vasculature of placentas from these pregnancies suggests that oxidant (peroxynitrite)-mediated endothelial dysfunction is also occurring in the placenta, which may contribute to the increased fetal–placental vascular resistance characteristics of such pregnancies. As indicated above, maintenance of adequate and increasing blood flows throughout pregnancy on both sides of the placenta is necessary to ensure optimal uptake of oxygen and nutrients and removal of carbon dioxide and waste products from the placenta, and thus to ensure adequate fetal growth. In conclusion, NO appears to be a major regulator of human fetal placental vascular resistance, and there is evidence that alteration in NO synthesis and action may be occurring in pathologic pregnancies.

Nitric Oxide and the Regulation of Myometrial Contractility

The discovery that NO had potent vascular smooth muscle relaxatory effects very soon led to interest in and investigation of its role in the regulation of myometrial smooth muscle contractility. A variety of indirect descriptive studies have suggested that NO may have a role in the regulation of myometrial contractility; however, there is a paucity of direct mechanistic studies. Complicating the investigations in this area is the fact that there are many potential cellular sites of synthesis of NO within the nonpregnant or pregnant uterus. This makes elucidation of the role of NO derived from these various cell types difficult. In vitro isolated strips of pregnant and labored rat myometrium can be relaxed by the NO donor sodium nitroprusside or by 8-bromo cyclic GMP (Izumi et al. 1995), and these effects can be antagonized by methylene blue, an inhibitor of guanylate cyclase. Similarly, the substrate for NO L-arginine and other donors of NO can relax pregnant rat myometrial strips, but at pharmacologic concentrations (Yallampalli et al. 1994a). This response to L-arginine appeared to be increased by progesterone treatment, whereas antiprogesterone treatment decreased cGMP-induced relaxation (Yallampalli et al. 1994a). Further, myometrial tissue has been reported to produce NO, and responses to NO and cGMP were decreased during delivery (Yallampalli et al. 1993a). Overall, these findings suggest that an L-arginine–NO–cGMP system is present in the uterus that may regulate myometrial contractility during pregnancy, but the effect is lost at the time of delivery. Studies using production of nitrite and cGMP released from rat uterine tissue as an index of NO synthesis reported that NO production is substantially increased during pregnancy but decreased during labor and postpartum, when the formation of both NO and GMP is similar to that in nonpregnant rats (Yallampalli et al. 1994b). These data suggest that the NO–GMP system is up-regulated during pregnancy, possibly to maintain uterine quiescence, and the influence of progesterone on the system suggests that during pregnancy myometrial NO synthesis may be hormonally regulated. Recent studies have shown differences between in vitro and in vivo effects of NO donors on the pregnant rat myometrium. Although in vitro there are decreased inhibitory responses to NO donors during labor, studies in vivo demonstrate that NO donors can decrease uterine contractility even more effectively during delivery at term than at mid-pregnancy (Buhimschi et al. 1997). This suggests that there may be some indirect effect of NO donors on the myometrium in vivo, as opposed to the direct effects seen in vitro.

When pregnant human myometrial strips were studied for responses to NO, they were also shown to release nitrites and cGMP (Buhimschi et al. 1995). Indeed, production of cGMP was increased by treatment with L-arginine, the substrate for NO, and inhibited by treatment with L-NAME, an inhibitor of NOS. This report also claimed that the spontaneous contractility of human myometrium in vitro was increased by treatment with L-NAME but decreased by NO donors in a dose-dependent manner, with the response of tissue from pregnant, nonlaboring women being substantially greater than that of tissue from laboring or nonpregnant women. Overall, these studies again provided indirect evidence that a NO–cGMP system present in the human uterus inhibits contractility, and that responses to NO change throughout gestation. The interpretation of these studies has been questioned because of the pharmacologically high concentrations L-arginine, that were necessary to cause relaxation of spontaneous and oxytocin-induced contractions of pregnant rat myometrium (Kato et al. 1995). Indeed, basic amino acids have been reported to give similar patterns of relaxation at these concentrations, due to increases in extracellular pH (Kato et al. 1995). Similarly, other work reported that there was no effect of L-arginine or of the NO synthesis inhibitor L-NAME on human myometrial strips obtained either at term or preterm prior to labor (Jones and Poston 1996). This has brought into question, therefore, the early observations that NO donors were able to relax the pregnant uterus. Chronic in vivo treatment of pregnant animals with inhibitors of NO synthesis has also failed to reveal a direct role for NO synthesis in the control of uterine contractility (Yallampalli et al. 1993b; Diket et al. 1994; Molnar et al. 1994). In all of these studies, the inhibitory effect on NO synthesis was manifest as increases in maternal blood pressure, IUGR, and hind limb defects in the offspring, but none of the studies resulted in preterm delivery, suggesting that NO synthesis does not mediate the maintenance of uterine quiescence and that reduction in NO synthesis is not required for the onset of labor.

Several groups have made direct measurement of NOS activity in uterine tissue homogenates from pregnant animals. In pregnant rabbit decidual tissue, NOS activity decreased by 80% on the day of parturition (Sladek et al. 1993). Characterization of this enzyme activity showed it to be the iNOS, isoform, which is insensitive to calcium and calmodulin. The activity measured in the myometrium in the same study, however, was much less than that of the decidua and did not appear to change throughout gestation. Similarly, in the pregnant rat, NOS activity in the uterus was reported to be higher at 16 days of gestation than at term (day 22 of gestation) (Natuzzi et al. 1993). Again, the finding is consistent with an association of high NOS activity at mid-gestation with myometrial quiescence and an association of a decrease in activity at term with the onset of contractility. A more recent study with a greater number of time points has shown a decrease in rat myometrial NOS activity between days 15 and 21 of gestation, but with no further decrease at day 22 either before or after the onset of labor (Sladek et al. 1996). This activity was described again as calcium-insensitive (i.e., the iNOS isoform). However, in the pregnant guinea pig, myometrial NOS activity appeared to decline slowly throughout gestation but was never different from that of the non-pregnant guinea pig myometrium (Weiner et al. 1994), suggesting there may be differences in uterine NOS activities between species.

Immunohistochemistry has allowed localization of the different NOS isoforms in the pregnant uterus. In the pregnant rat uterus, both the inducible and the endothelial NOS isoforms were localized to the granulated metrial gland (GMG) cells. These are cells of a natural killer cell lineage, and the number of NOS positive GMG cells corresponded to the activity of NOS measured in this tissue (Sladek et al. 1998). Since NOS activity decreased when the number of GMG cells containing NOS decreased, this suggests that uterine NOS activity is localized to a specific cell type whose appearance or disappearance may be hormonally regulated. Indeed, in the mouse uterus, expression of the iNOS isoform in specific cell types appears to be regulated by ovarian steroid hormones (Huang et al. 1995). Although changes in the activity of NOS in the uterus during gestation have been suggested to have a role in the regulation of myometrial contractility, the presence of the iNOS isoform in the metrial gland of the rat suggests that NO may play a role in the immune defenses of the fetus or in regulation of placental blood flow in the uterus.

In pregnant human myometrium, both calcium-dependent and calcium-independent NOS activities have been measured (Ramsey et al. 1996). These workers found no significant differences in calcium-dependent NOS activity be-

tween nonpregnant and pregnant women either before or after the onset of spontaneous labor at term. They reported that the calcium-independent myometrial
NOS activity was significantly lower in pregnant myometrium than in nonpregnant myometrium; however, individuals in labor had significantly higher activities than those who were not in labor. A more recent study has presented conflicting data (Bansal et al. 1977) showing expression of iNOS in smooth muscle cells
of pregnant myometrium, with the highest expression in myometrium of preterm
patients who were not in labor. At term expression fell by 75% and was barely detectable in preterm-in-labor or term-in-labor specimens. There was no staining for
iNOS in myocytes of nonpregnant myometrium. Interestingly, although activity
was highest in preterm patients who were not in labor, only approximately 8% of
the myocytes stained positive for iNOS. The authors suggest that this selective expression of NOS in scattered myocytes may act to prevent large organized contractions of the myometrium during gestation. Currently, therefore, data are conflicting regarding NOS activities in myometrium throughout gestation and its
relationship to myometrial contractility. Studies in rats and rabbits, in which
higher activities are found in myometrium during gestation, are consistent with a
role in the maintenance of myometrial quiescence; however, the absence of a clear
consensus that there is a fall in NOS activity at the time of the onset of labor does
not support a role for NO in the switch from myometrial quiescence to contractility. Similarly, the inability of NOS inhibitors to cause the onset of labor in animals, even though they exert profound effects on blood pressure, fails to support
a role for NO in mediation of uterine contractility. The studies on the distinct cellular localization of NOS isoforms in the uterus may give us a better understanding of the roles that NO plays, perhaps in regulation of the placental and uterine
blood flows and in immune defense mechanisms.

Nitric oxide synthase is also found in cervical tissue (Buhimschi et al. 1996),
and an increase in the amount of iNOS in the cervix occurred during labor of rats
at term, compared with cervices collected from animals at day 19 of pregnancy.
This was the opposite of changes in concentration of the isoform in the uterus of
the same animals. Induction of preterm labor by antiprogesterone treatment (on-
apristone) in these animals also significantly increased the iNOS concentration in
the cervix. Treatment of pregnant animals with L-NAME to prevent NO synthesis significantly prolonged the duration of delivery and decreased cervical exten-
sibility. Therefore, NO in the cervix may play a role in cervical dissolution at the
time of labor. Treatment of pregnant animals with lipopolysaccharide (LPS)
caused a significant increase in NO production in the cervix and also in the uterus
(Buhimschi et al. 1996, Nakaya et al. 1996)

The iNOS isoform is induced by a variety of cytokines and growth factors and
by LPS, particularly from macrophages of polymorphonucleocytes. Expression of
the iNOS isoform has been demonstrated in the amnion and chorion of pregnant
women. Expression is very prevalent in the invading host cells seen in fetal membranes of patients with infection-induced preterm labor (Eis et al. 1997). Women
with infection-induced preterm labor also had significantly increased concentrations of plasma and urine nitrates as compared with women not in labor (Jaekle et

al. 1994). This apparently paradoxical finding of an increase in NO, a supposed smooth muscle relaxant, in preterm labor may be the result of increased NO from host cells that are activated to fight the infection as a cytostatic or cytotoxic defense mechanism. The increase in NO in these women, however, may explain why some women with overwhelming intrauterine infection show a dysfunctional intrauterine activity pattern. Nitric oxide donors such as amyl nitrites (Hendricks et al. 1992) and nitroglycerin (Peng et al. 1989) have been used clinically to relax the contracted uterus. In an uncontrolled study, nitroglycerin administered by patch was also reported to stop preterm labor and prolong gestation in women (Lees et al. 1993). Similarly, in the pregnant sheep, nitroglycerin infusion stops active labor (Heymann et al. 1993), and NO donors inhibit preterm labor caused by hysterotomy in the pregnant rhesus monkey (Jennings et al. 1993). Although it appears that exogenous NO donors can relax myometrial smooth muscle, we are still uncertain whether endogenous NO production in the uterus of the pregnant rat plays a role in the maintenance of uterine quiescence during most of pregnancy. Similarly, large-scale, well-controlled clinical trials are needed to determine whether NO donors have a place in the clinical armamentarium to prevent preterm delivery.

In summary, NO appears to play an important role in regulating uterine blood flow during both the menstrual (estrous) cycle and pregnancy. It is clear that it plays an important role in regulating basal uterine vascular tone, and that this tone is modulated by increased NOS activity in the endothelium and underlying vascular smooth muscle and is under the control of circulating estrogens. The onset of menopause results in decreased NOS in the endothelium and vascular smooth muscle in both reproductive and nonreproductive tissues, and this is reversed by hormone replacement therapy, which has been shown to have significant beneficial affects on the cardiovascular system. Nitric oxide also plays a significant role in maintaining reduced vascular resistance in the umbilical circulation, and its basal production may also be regulated by estrogen in this vascular bed; however, it is clear that shear force is also important. Finally, NO produces relaxation of the myometrium, which is important in maintaining uterine quiescence and preventing preterm labor. Thus, it is clear that NO has an important role in the hemodynamic regulation of the uterine and umbilical circulations as well as reproduction in general.

References

Adamson SL, Myatt L, & Byrne BMP. 1998. Regulation of umbilical blood flow. In: Polin RA, Fox WW, eds. *Fetal and Neonatal Physiology*. 2nd ed. Philadelphia: WB Saunders, pp 977–988.

Bansal RK, Goldsmith PC, He Y, Zaloudek CT, Ecker JL, & Reimer RK. 1997. A decline in myometrial nitric oxide synthase expression is associated with labor and delivery. *J Clin Invest 99*:2502–2508.

Brody MJ, Clark KE, Edvinsson L, Owman CL, & Sjoberg NO. 1974. Determination of uterine blood volume and correlation with ovarian function. *Proc Soc Exp Biol Med 147*:91–96.

Buhimschi I, Yallampalli C, Dong YL, & Garfield RE. 1995. Involvement of a nitric oxide-cyclic guanosine monophosphate pathway in control of human uterine contractility during pregnancy. *Am J Obstet Gynecol 172:*1577–1584.

Buhimschi I, Ali M, Jain V, Chwalisz K, & Garfield RE. 1996. Differential regulation of nitric oxide in the rat uterus and cervix during pregnancy and labour. *Hum Reprod 11:*1755–1766.

Buhimschi C, Buhimschi I, Yallampalli C, Chwalisz K, & Garfield RE. 1997. Contrasting effects of diethyleneriamine-nitric oxide, a spontaneously releasing nitric oxide donor, on pregnant rat uterine contractility in vitro versus in vivo. *Am J Obstet Gynecol 177:*690–701.

Carpenter L, Baker RS, Greenberg SG, & Clark KE. 1997. Adenosine-induced uterine vasodilation is mediated in part by nitric oxide. *J Soc Gynecol Invest 4(1S):*251A.

Clark KE, Baker RS, & Kopernik G. 1998. Ethinyl-estradiol induced increases in uterine blood flow are not mediated by nitric oxide. *J Soc Gynecol Invest 5(1S):*145A.

Colburn P, & Buonassesse V. 1978. Estrogen-binding sites in endothelial cell cultures. *Science 201:*817.

Conrad KP, Joffe GM, Kruszyna H, Kruszyna R, Rochelle LG, Smith RP, Chavez JE, & Mosher MD. 1993. Identification of increased nitric oxide biosynthesis during pregnancy in rats. *FASEB J 7:*566–571.

Dainty IA, McGrath JC, Spedding M, & Templeton AGB. 1990. The influence of the initial stretch and the agonist-induced tone on the effect of basal and stimulated release of EDRF. *Br J Pharmacol 100:*767–773.

Davidge ST, Stranko CP, & Roberts JM. 1996. Urine but not plasma nitric oxide metabolites are decreased in women with preeclampsia. *Am J Obstet Gynecol 174:*1008–1013.

De Nucci G, Thomas R, D'Orleans-Juste P, Antures E, Walder C, Warner TD, & Vane JR. 1988. Pressor effects of circulating endothelin are limited by its removal in the pulmonary circulation and release of prostacyclin and endothelium-derived relaxing factor. *Proc Natl Acad Sci USA 85:*9797–9800.

Diket AL, Pierce MR, Munshi UK, et al. 1994. Nitric oxide inhibition causes intrauterine growth retardation and hind-limb disruptions in rats. *Am J Obstet Gynecol 171:*1243–1250.

Eis ALW, Brockman DE, & Myatt L. 1997. Immunolocalization of the inducible nitric oxide synthase isoform in human fetal membranes. *Am J Reprod Immunol 38:*289–294.

Erskine RLA, & Ritchie JWK. 1985. Umbilical artery blood flow characteristics in normal and growth-retarded fetuses. *Br J Obstet Gynaecol 92:*605–610.

Figueroa JP, Zhang J, Massmann GA, & Mirable CP. 1998. Differential regulation by estrogen of type I and type III nitric oxide synthase (NOS) mRNA expression in the uterus of non pregnant sheep. *J Soc Gynecol Invest 5 (1S):*182A.

Giles WB, Trudinger BJ, & Baird PJ. 1985. Fetal ambilical artery flow velocity waveforms and placental resistance: pathological correlation. *Br J Obstet Gynaecol 92:*31–38.

Giles W, O'Callaghan S, Boura A, & Walters W. 1992. Reduction in human fetal umbilical-placental vascular resistance by glyceryl trimirate. *Lancet 340:*856.

Grunewald C, Kublickas M, Carlstrom K, Lunell NO, & Nisell H. 1995. Effects of nitroglycerin on the uterine and umbilical circulation in severe preeclampsia. *Obstet Gynecol 86:*600–604.

Hendricks SK, Ross B, Colverd MA, Cahill A, Shy K, & Benedetti TJ. 1992. Amylnitrite: use as a smooth muscle relaxant in difficult preterm cesarean section. *Am J Perinatol 9:*289–292.

Heymann MA, Bookstaylor B, Roman C, et al. 1993. Glyceryl trinitrate stops active labor in sheep. In: Moncada S, Feelish M, Busse R, Higgs EA, eds. *The Biology of Nitric Oxide*. London and Chapel Hill, NC: Portland Press, pp 201–203.

Huang J, Roby KF, Pace JL, Russell SW, & Hunt JS. 1995. Cellular localization and hormonal regulation of inducible nitric oxide synthase in cycling mouse uterus. *J Leukocyte Biol 57*:27–35.

Iafrati MD, Karas RH, Aronovitz M, Kim S, Sullivan TR, Lubahn DB, O'Donnell TF, Korach KS, & Mendelsohn ME. 1997. Estrogen inhibits the vascular injury response in estrogen receptor α deficient mice. *Nature Med 3*:545–548.

Ignarro LJ. 1989. Endothelium-derived nitric oxide: actions and properties. *FASEB J 3*:31–36.

Izumi H, & Garfield RE. 1995. Relaxant effects of nitric oxide and cyclic GMP on pregnant rat uterine longitudinal smooth muscle. *Eur J Obstet Gynecol Reprod Biol 60*:171–180.

Jaekle RK, Lutz PD, Rosenn B, Siddiqi TA, & Myatt L. 1994. Nitric oxide metabolites and preterm pregnancy complications. *Am J Obstet Gynecol 171*:1115–1119.

Jennings RW, MacGillivray TE, & Harrison MR. 1993. Nitric oxide inhibits preterm labor in the rhesus monkey. *J Matern Fet Med 2*:170–174.

Jones GD, & Poston L. 1996. The influence of modulations of nitric oxide synthesis on spontaneous human myometrial contractility in vitro. *J Soc Gynecol Invest 3*:346A.

Karas RH, Patterson BL, & Mendelsohn ME. 1994. Human vascular smooth muscle cells contain functional estrogen receptor. *Circulation 89*:1943–1950.

Kato S, Fomin VF, Lau K, & Ward RA. 1995. Effect of basic amino acids and alkaline pH on uterine contractility in vitro. *J Soc Gynecol Invest 2*:179.

Krebs C, Macara LM, Leiser R, Bowman AW, Greer IA, & Kingdom JCP. 1996. Intrauterine growth restriction with absent end-diastolic flow velocity in the umbilical artery is associated with maldevelopment of the placental terminal villous tree. *Am J Obstet Gynecol 175*:1534–1542.

Kuiper GGJM, Enmark E, Pelto-Huikko M, Nilsson S, & Gustafsson JA. 1996. Cloning of a novel estrogen receptor expressed in rat prostate and ovary. *Proc Natl Acad Sci USA 93*:5925–5930.

Kuiper GGJM, Carlsson B, Grandien K, Enmark E, Haggblad J, Nilsson S, & Gustafsson JA. 1997. Comparison of the ligand binding specificity and transcript tissue distribution of estrogen receptors α and β. *Endocrinology 138*:863–870.

Lang U, Baker RS, & Clark KE. 1997. Estrogen-induced increases in coronary blood flow are antagonized by inhibitors of nitric oxide synthesis. *Eur J Obstet Gynecol 74*:229–235.

Learmont JG, & Poston L. 1996. Nitric oxide is involved in flow-induced dilation of isolated human small fetoplacental arteries. *Am J Obstet Gynecol 174*:583–588.

Lees C, Campbell S, Jauniaux E, et al. 1993. Arrest of preterm labor and prolongation of gestation with glyceryl trinitrate, a nitric oxide donor. *Lancet 343*:1325–1326.

Lewin DJ. 1997. Researchers get down to the alphas and betas of estrogen's effects. *J NIH Res 9*:23–25.

Lin AL, McGill HC, & Shain SA. 1982. Hormone receptors of the baboon cardiovascular system: biochemical characterization of aortic and myocardial cytoplasmic progesterone receptors. *Circ Res 50*:610–616.

Losordo DW, Kearney M, Kim EA, Jekanowski J, & Isner JM. 1994. Variable expression of the estrogen receptor in normal and atherosclerotic coronary arteries of premenopausal women. *Circulation 89*:1501–1510.

Lyall F, Young A, & Greer IA. 1995. Nitric oxide concentrations are increased in the feto-placental circulation in pre-enclampsia. *Am J Obstet Gynecol 173:*714–718.

Lyall F, Greer IA, Young A, & Myatt L. 1996. Nitric oxide concentrations are increased in the fetoplacental circulation in intrauterine growth restriction. *Placenta 17:*165–168.

Magness RR, Phernetton TM, Shaw CE, & Long RA. 1998. Effect of estrogen treatment and pregnancy on ovine coronary vascular resistance and coronary artery endothelial nitric oxide synthase expression. *J Soc Gynecol Invest 5 (1S):*61A.

Makino Y, Izumi H, Makino I, & Shirakawa K. 1997. The effect of nitric oxide on uterine and umbilical artery flow velocity waveform in preeclampsia. *Eur J Obstet Gynecol Reprod Biol 73:*139–143.

McGill H. 1989. Sex steroid hormone receptors in the cardiovascular system. *Postgrad Med 85:*64–68.

Molnar M, Suto T, Toth T, & Hertelendy F. 1994. Prolonged blockade of nitric oxide synthesis in gravid rats produces sustained hypertension, proteinuria, thrombocytopenia and intrauterine growth retardation. *Am J Obstet Gynecol 170:*1458–1466.

Moore PK, Al-Swayeh OA, Chong NWS, Evans RA, & Gibson A. 1990. l-$N^{\underline{G}}$-nitro arginine (l-NOARG) a novel, l-arginine-reversible inhibitor of endothelium-dependent vasodilatation in vitro. *Br J Pharmacol 99:*408–412.

Morey AK, Pedram A, Razandi M, Prins BA, Hu RM, Biesiada E, & Levin ER. 1997. Estrogen and progesterone inhibit vascular smooth muscle proliferation. *Endocrinology 138:*3330–3339.

Myatt L. 1992. Control of vascular resistance in the human placenta. *Placenta 13:* 329–341.

Myatt L, Brewer A, & Brockman DE. 1991. The action of nitric oxide in the perfused human fetal-placental circulation. *Am J Obstet Gynecol 164:*687–692.

Myatt L, Brewer AS, Langdon G, & Brockman DE. 1992. Attenuation of the vasoconstrictor effects of thromboxane and endothelin in the human fetal-placental circulation. *Am J Obstet Gynecol 166:*224–230.

Myatt L, Rosenfield RB, Eis AL, Brockman DE, Greer IA, & Lyall F. 1996. Nitrotyrosine residues in placenta. Evidence of peroxynitrite formation and action. *Hypertension 28:*488–493.

Myatt L, Eis Al, Brockman DE, Greer IA, & Lyall F. 1997. Endothelial nitric oxide synthase in placental villous tissue from normal, pre-eclamptic and intrauterine growth-restricted pregnancies. *Hum Reprod 12:*167–172.

Nakaya Y, Yamamoto S, Hamada Y, Kamada M, Aono T, & Niwa M. 1996. Inducible nitric oxide synthase in uterine smooth muscle. *Life Sci 58:*249–255.

Natuzzi ES, Ursell PC, Harrison M, Buscher C, & Riemer RK. 1993. Nitric oxide synthase activity in the pregnant uterus decreases at parturition. *Biochem Biophys Res Common 194:*1–8.

Neugarten J, Ding Q, Friedman A, Lei J, & Silbiger S. 1997. Sex hormones and renal nitric oxide synthetase. *J Am Soc Nephrol 8:*1240–1246.

Palmer RMJ, Ashton DS, & Moncada S. 1988. Vascular endothelial cells synthesize nitric oxide from L-arginine. *Nature 333:*664–666.

Peng ATC, Garman RS, Shulman SM, DeMarchis E, Nyunt K, & Blancato LS. 1989. Intravenous nitroglycerin for uterine relaxation in the postpartum patient with retained placenta. *Anesthesiology 71:*172–175.

Ramsay B, Sooranna SR, & Johnson MR. 1996. Nitric oxide synthase activities in human myometrium and villous trophoblast throughout pregnancy. *Obstet Gynecol 87:*249–253.

Rees DD, Palmer RMJ, Schulz R, Hodson HF, & Moncada S. 1990. Characterization of three inhibitors of endothelial nitric oxide synthase in vitro and in vivo. *Br J Pharmacol 101:*746–752.

Roberts JM, Taylor RN, & Goldfien A. 1991. Clinical and biochemical evidence of endothelial cell dysfunction in the pregnancy syndrome preeclampsia. *Am J Hypertens 4:*700–708.

Rosenfeld CR, Killam AP, Battaglia FC, Makowski G, & Meschia G. 1973. Effect of estradiol-17β on the magnitude and distribution of uterine blood flow in nonpregnant oophorectomized ewes. *Pediatr Res 7:*139–148.

Rosseli M, Imthurn B, Keller PJ, Jackson EK, & Dubey RK. 1995. Circulating nitric oxide (nitrite/nitrate) levels in postmenopausal women substituted with 17β-estradiol and norethisterone acetate: a two year follow-up study. *Hypertension 25 (Part 2):*848–853.

Rutherford RA, McCarthy A, Sullivan MH, Elder MG, Polak JM, & Wharton J. 1995. Nitric oxide synthase in human placenta and umbilical cord from normal, intrauterine growth-retarded and pre-eclamptic pregnancies. *Br J Pharmacol 116:*3099–3109.

Salhab W, Shaul P, Cox B, & Rosenfield CR. 1998. Daily estradiol-17β increases nitric oxide synthesis (NOS) in uterine arteries of nonpregnant ewes. *J Soc Gynecol Invest 5(1S):*52A.

Sladek SM, Regenstein AC, Lykins D, & Roberts JM. 1993. Nitric oxide synthase activity in pregnant rabbit uterus decreases on the last day of pregnancy. *Am J Obstet Gynecol 169:*1285–1291.

Sladek SM, Kanbour-Shakir A, Watkins S, Berghorn KA, Hoffman GE, & Roberts JM. 1998. Granulated metrial gland cells contain nitric oxide synthases during pregnancy in the rat placenta. *Placenta 19:*55–65.

Sladek SM, & Roberts JM. 1996. Nitric oxide synthase activity in the gravid rat uterus decreases a day before the onset of parturition. *Am J Obstet Gynecol 175:*1661–1667.

Stampfer MJ, Colditz GA, Willett WC, Manson JE, Rosner B, Speizer FE, & Hennekens CH. 1991. Postmenopausal estrogen therapy and cardiovascular disease; ten-year follow-up from the nurses' health study. *N Engl J Med 325:*756–762.

Tschugguel W, Schneeberger C, Zhegu Z, Wieser F, Waselmayr B, Sator MO, Wojita J, Binder BR, & Huber JC. 1998. Distinct expression pattern of estrogen receptor alpha and beta mRNA in cultured human endothelial and vascular smooth muscle cells. *J Soc Gynecol Invest 5(1S):*143A.

Weiner CP, Knowles RG, Nelson SE, & Stegink LD. 1994. Pregnancy increases oxide synthesis. *Endocrinology 135:*2473–2478.

Wieczorek KM, Brewer AS, & Myatt L. 1995. Shear stress may stimulate release and action of nitric oxide in the human fetal-placental vasculature. *Am J Obstet Gynecol 175:*708–713.

Van Buren GA, Yang DA, & Clark KE. 1992. Estrogen-induced uterine vasodilation is antagonized by L-nitroarginine methyl ester, an inhibitor of nitric oxide synthesis. *Am J Obstet Gynecol 167:*828–833.

Veille JC, Li P, Eisenach JC, Massmann AG, & Figueroa JP. 1996. Effects of estrogen on nitric oxide biosynthesis and vasorelaxant activity in sheep uterine and renal arteries in vitro. *Am J Obstet Gynecol 174:*1043–1049.

Yallampalli C, & Garfield RE. 1993. Inhibition of nitric oxide synthesis in rats produces signs similar to those of preeclampsia. *Am J Obstet Gynecol 169:*1316–1320.

Yallampalli C, Garfield RE, & Byam-Smith M. 1993a. Nitric oxide inhibits uterine contractility during pregnancy but not during delivery. *Endocrinology 133:*1899–1902.

Yallampalli C, Byam-Smith M, Nelson SO, & Garfield RE. 1994b. Steroid hormones modulate the production of nitric oxide and cGMP in the rat uterus. *Endocrinology*
*134:*1971–1974.

Yallampalli C, Izumi H, Byam-Smith M, & Garfield RE: 1994a. An L-arginine-nitric oxide
cyclic guanosine monophosphate system exists in the uterus and inhibits contractility
during pregnancy. *Am J Obstet Gynecol 170:*175–185.

Yang DA, Lang U, Greenberg SG, Myatt L, & Clark KE. 1996. Elevation of nitrate levels
in pregnant ewes and their fetuses. *Am J Obstet Gynecol 174:*573–577.

18
NO Effect on Penile Blood Flow and Lower Genitourinary Tract Function

TRINITY J. BIVALACQUA, HUNTER C. CHAMPION, PHILIP J. KADOWITZ, AND WAYNE J.G. HELLSTROM

Introduction

Nitric oxide (NO) is an unstable radical that exists primarily as a gas. For decades, NO was thought to be an environmental contaminant. Highly toxic, it was considered an unlikely candidate as a biological mediator. However, a landmark article by Furchgott and Zawadzki in 1980 described the release of a substance by the endothelial lining of blood vessels that was responsible for the vasorelaxation of smooth muscle in response to acetylcholine. They named this novel molecule endothelium-derived relaxing factor (EDRF). Since that time, most authorities have concluded that EDRF is NO (Ignarro et al. 1987; Palmer et al. 1987), however some researchers postulate that EDRF is actually an S-nitrosothiol that acts through NO transfer (Meyers et al. 1990).

Within the last decade, it has become abundantly clear that endogenously produced NO serves as a second messenger in the central and peripheral nervous systems. Nitric oxide has been shown to possess powerful vasodilatory properties, in addition to serving as an essential factor in the physiological and pathophysiological regulation of blood flow. Endogenous NO has been associated with the regulation of blood pressure and vasomotor tone, platelet adhesion and aggregation, control of cell proliferation, modulation of myocardial contractility, and neurotransmission (Marin and Angeles Rodriguez-Martinez 1997). Nitric oxide has manifested itself as a new paradigm for the mechanism by which a second-messenger molecule can govern the biological environment in which it is active.

Recent evidence supports the role of NO as the principal neurotransmitter in penile erection (Burnett et al. 1992). Soon after this identification, a number of investigators elucidated the mechanism of action by which NO induces penile erection and, more important, its larger role in lower genitourinary tract function (Andersson and Persson 1995, Burnett 1995a).

Erectile Dysfunction

Erectile dysfunction is defined as an inability to achieve an erection of sufficient rigidity for satisfactory sexual intercourse (NIH Consensus Conference 1993). Impotence is a major quality-of-life issue, with an estimated 20 to 30 million men suffering from this condition in the United States alone (Feldman et al. 1994). Normal erectile function is a hemodynamic process, involving three synergistic and simultaneous processes: increase of arterial inflow, relaxation of the cavernosal smooth muscle, and restriction of venous outflow from the penis (Lue 1992).

Erections result from a complex interplay of neuroregulatory control mechanisms. The observation that neither cholinergic (parasympathetic) nor adrenergic (sympathetic) mechanisms can fully influence erectile function prompted researchers to search for a nonadrenergic, noncholinergic (NANC) neurotransmitter (Burnett 1997). For this reason, the mechanism of neurotransmitter release from the erectile tissue and nerves has become a major research focus in the field of sexual dysfunction.

The NANC mechanism has a number of putative mediators that may be involved in erections. Some proposed mediators include vasoactive intestinal polypeptide (VIP), calcitonin gene-related peptide (CGRP), substance P, purines (e.g., adenosine and ATP), decarboxylated amino acids, and other factors, including adrenomedullin (ADM), nociceptin, serotonin, prostaglandins, bradykinin, and histamine (Andersson 1993). Adrenomedullin and nociceptin have demonstrated the ability to cause penile erection in animal models, but their true physiological relevance in man has not been fully determined (Champion et al. 1997a,b). Of note is the fact that there has been no convincing evidence to suggest that any one of the aforementioned neurotransmitters is the sole physiological mediator of penile erection, but there have been an overwhelming number of publications implicating NO as the molecule responsible for cavernosal smooth muscle relaxation and penile erection (Burnett et al. 1992; Burnett 1995 a,b; Rajfer et al. 1992). Therefore, most impotence researchers have focused much of their investigative effort on NO as the central mediator for penile erection.

Nitric Oxide as a Second-Messenger Neurotransmitter

The modulation of cell function and its complex physiologic responses to free radicals is an area of profound research interest. The primary role of NO is to bind to the heme moiety of guanylate cyclase, which subsequently increases intracellular cGMP. Increased levels of cGMP in the vascular smooth muscle lead to a reduction of intracellular Ca^{2+} levels, thus causing smooth muscle relaxation (Marin and Angeles Rodriguez-Martinez 1997).

Nitric oxide is synthesized as a by-product of the enzymatic conversion of L-arginine to L-citrulline by the enzyme nitric oxide synthase (NOS). The NO synthases are a group of enzymes that catalyze the NADPH-dependent oxidation of L-arginine to NO and L-citrulline (Knowles and Moncada 1994). To date, three distinct isoforms of NOS have been purified, sequenced, and partially characterized: type I, or neuronal NOS (nNOS); type II, or inducible macrophage NOS (iNOS); and type III, or endothelial cell NOS (eNOS) (Knowles and Moncada 1994). The isoforms nNOS and eNOS are both regulated by calcium and calmodulin and have traditionally been referred to as constitutive, whereas iNOS is not limited by calcium or calmodulin and has been traditionally referred to as inducible (Knowles and Moncada 1994).

Nitric oxide synthase has been identified in a variety of cell types, including vascular endothelial cells, neurons, macrophages, and smooth muscle cells (Knowles and Moncada 1994; Kerwin et al. 1995). The constitutive isoforms, which commonly exist in endothelial cells and neurons, are Ca^{2+}-calmodulin-dependent enzymes that are inactive until intracellular calcium levels increase. Once intracellular calcium levels increase, calmodulin can bind to calcium, and then the calcium–calmodulin complex can bind to NOS and activate NO expression (Bredt et al. 1992; Moncada 1992). In comparison, iNOS is associated with macrophages and other cells of immune function. The NO produced by iNOS is pathogenic to surrounding cells, bacteria, and parasites (Bredt et al. 1992).

As a group, the NOSs are ubiquitous, since they are found in virtually all tissues; however, the type of NOS appears to be tissue-specific. The NOSs have been implicated in a wide array of physiological processes, including neuronal transmission, immune targeting, control of vascular wall tone, cell differentiation, gene expression, and apoptosis (Bredt et al. 1992; Marin and Angeles Rodriguez-Martinez 1997). Nitric oxide synthase enzyme activity and its protein product have been identified in the testes, epididymides, seminal vesicles, and prostate gland of a number of species, including humans. (Andersson and Persson 1994; Andersson 1996).

Role of NO in the Penis

Normal erectile function depends on the ability of penile corporal smooth muscle to undergo complete relaxation. In the genitourinary tract, NO has been commonly associated with penile erection, and a number of studies have characterized its control over both neurogenically mediated and endothelium-dependent relaxation of the vascular and trabecular smooth muscle of the penis (Lugg et al. 1995a; Burnett 1997). Several in vitro studies have demonstrated that NO is responsible for relaxation of human and rabbit cavernous smooth muscle (Ignarro et al. 1990; Bush et al. 1992a). In vivo studies, the first of which were performed in our laboratories, have shown that NO is an important mediator of penile erection in several species (Domer et al. 1978; Trigo-Rocha et al. 1993a; Hellstrom et al. 1994; Wang et al. 1994).

Ignarro et al. (1990) first demonstrated that electrical field stimulation (EFS) of isolated strips of rabbit corpus cavernosum caused relaxation by endogenous formation of NO. This corporal smooth muscle relaxation could be eliminated by the use of tetrodotoxin, proving that relaxation was neuronally mediated (Kim et al. 1991). Because this neurogenic response was resistant to cholinergic receptor blockade and selective adrenergic nerve inhibition, it was termed an NANC response. Further studies involving isolated corporal tissue specimens from several animal species and from humans implicated NO as the mediator of corpora cavernosal smooth muscle relaxation (Ignarro et al. 1990; Holmquist et al. 1991a; Kim et al. 1991; Bush et al. 1992a). Corporal tissue relaxation can be abolished by N^G-substituted analogues of L-arginine, which are known to inhibit NO synthesis (Holmquist et al. 1991a). Similarly, NANC-mediated relaxation of isolated erectile tissue could be blocked by methylene blue, a guanylate cyclase inhibitor, whereas inhibitors of cGMP phosphodiesterase caused a potentiation of the relaxant response (Bush et al. 1992a; Holmquist et al. 1993). These results support the concept that cavernosal tissue relaxation is dependent on the NO signal transduction pathway.

Further confirmation of NO involvement in the erection process comes from animal models of penile erection. In the rat and rabbit, electrically induced erections can be blocked by the administration of NOS inhibitors or methylene blue, whereas this blockade can be reversed with L-arginine or NO substrates (Holmquist et al. 1991b; Finberg et al. 1993). Similarly, NO-donor-induced canine erections can be blocked by N^G-Nitro-L-Arginine and subsequently restored by L-arginine (Trigo-Rocha et al. 1993a).

Tumescence can be elicited in a variety of species, including man, after the administration of NO, NO-releasing drugs, cGMP, and specific cGMP phosphodiesterase inhibitors (Andersson and Persson 1994; Burnett 1995b; Lugg et al. 1995a). The destruction of the penile endothelium by 3-[(3-cholaminclopropyl)-dimethylammonia]-1 propane sulfonate (CHAPS) attenuates the erectile response to intracavernosal injection of acetylcholine, but only partially inhibits the response to EFS. This may suggest that NO release from the cavernosal nerve has a more important role in penile erection than NO release from the penile endothelium (Trigo-Rocha et al. 1993b). In feline experiments, the NOS inhibitor N^G-nitro-L-arginine methylester (L-NAME) blocked acetylcholine-induced penile erections, whereas the erectile effects of intracavernosally injected NO donors were not altered (Wang et al., 1994). Once again, the importance of the NO–cGMP signal transduction pathway lays the groundwork for understanding the NO mechanism in human penile erection.

Nitric Oxide Synthase in the Penis

The precise localization of NOS in the penile tissue, pelvic structures, and genital organs was derived from enzyme histochemical and immunohistochemical methods. The discovery that NOS protein and NOS mRNA colocalized with NADPH diaphorase activity in these genital tissues led to the precise localization of NOS in the penis (Trigo-Rocha et al. 1993b). Burnett et al. (1995b) used both immuno-histochemistry with rat cerebellar cNOS antibody and NADPH diaphorase histo-chemistry for measuring the catalytic activity of NOS to localize high concentra-tions of NOS in the penis, membranous urethra, and spinal nuclei innervating the spinal ganglia of the male rat. NOS has similarly been identified in the human penis by Brock and colleagues (Brock et al. 1993b). Further biochemical and his-tochemical evidence in the rat and rabbit penis suggests that nNOS is the major isozyme that functions in penile erection (Bush et al. 1992b; Brock et al. 1993b; Burnett et al. 1995b). In a variety of species, nNOS has been localized to the pelvic plexus, the terminal nerve endings of the cavernous nerves, branches of the dorsal penile nerves, and the nerve plexuses within the adventitia of the deep cav-ernous nerves (Bush et al. 1992b; Brock et al. 1993b; Trigo-Rocha et al. 1993b; Burnett 1995a; Burnett et al. 1995b). This distribution of NOS-containing nerves gives strong evidence that NO neuronally modulates the local vascular smooth musculature of the penis. On this basis, NO can be considered the neuronal medi-ator of penile erection in man. In addition to the human penis, NOS has been iden-tified in the human clitoris, suggesting a role for NO in its erectile physiology (Burnett et al. 1997).

There are a number of cotransmitters that are induced or modulated by NO in the NO-dependent pathway of erectile physiology. NOS is colocalized with acetylcholine esterase, VIP, and neuropeptide Y, suggesting that NO may have a role both as a direct transmitter and as a modulator of efferent neurotransmission. Acetylcholine and bradykinin stimulate the NO–cGMP pathways to produce smooth muscle relaxation in human and rabbit corporal tissue (Knispel et al. 1991). Substance P and acetylcholine diffuse from adventitial nerves of the en-dothelium in concentrations sufficient to stimulate NO release (Anderson 1993). Furthermore, substance P has been found in high concentrations in nerve fiber groups beneath the epithelium of the glans penis and elicits erections in the cat by releasing NO (Andersson 1993; Wang et al. 1994).

In several neuronal systems, the coexistence of NO and VIP has been deter-mined both functionally and immunohistochemically, suggesting a synergistic action. Vasoactive intestinal peptide has been colocalized with NOS in human and rat penile neurons, and inhibitors of NOS or guanylate cyclase can attenu-ate VIP-induced relaxation of rabbit corporal tissue (Kim et al. 1994; Tamura et al. 1995). In some vascular beds, VIP may be responsible for atropine-resistant vasodilation following parasympathetic stimulation and may also induce relax-ation by directly stimulating endothelial NOS. Vasoactive intestinal peptide in-directly contributes to NO-mediated erections because of its high concentra-tions in the pudendal arteries and the corpora cavernosa (Domoto and Tsumori 1994). Further evidence for VIP as a cotransmitter is its ability to induce an erection in the cat when injected intracavernosally (Wang et al. 1993). Vasoac-tive intestinal peptide and NO colocalize in other areas of the body, such as the adventitial innervation of the cerebral arteries and the myenteric plexus (Morris et al. 1995).

Role of NO in Androgen-Dependent Penile Erection and Centrally Evoked Penile Erection

Although androgens are essential for the expression of normal male libido, their exact role in penile erection has not been defined. Recent studies have helped elucidate the role NO plays in androgen-dependent penile erection. It has been shown by EFS of the cavernosal nerve that androgen loss with aging or by castration in the rat induces erectile dysfunction and a subsequent decrease in the penile levels of NOS (Mills et al. 1994; Lugg et al. 1995b). Androgen replacement in these affected rats restores the erectile response and content of NOS (Mills et al. 1994). Lugg et al. (1995b) have shown that dihydrotestosterone is the active androgen responsible for restoration of the erectile response in castrated rats and suggest that this effect may be mediated, at least partially, by changes in penile NOS levels.

Centrally evoked erections have been characterized in rats using a number of pharmacologic agents. One such drug, apomorphine, induces a dual response of yawning and penile erection (Melis et al. 1989). These effects induced by subcutaneous apomorphine or intracerebroventricular oxytocin can be inhibited by the NOS inhibitor L-NAME (Melis and Argiolas 1993). This suggests a likely role for NO in centrally induced penile erections. It is important to note that the apomorphine erectile response is not the same as an electrically stimulated erection in animals, but this methodology will undoubtedly be of value in the future study of psychogenic erectile reflexes. Further clinical applications of apomorphine to elicit erections in men with psychogenic and organic impotence are currently being conducted.

Nitric Oxide, a Cure for Erectile Dysfunction?

The clinical relevance of NO as a remedy for erectile dysfunction may revolutionize future management strategies. Since the pioneering work of Virag and Brindley, pharmacological erection programs have gained widespread acceptability and popularity among men suffering from erectile dysfunction (Virag 1982; Brindley 1986). Stief et al. (1992) demonstrated the therapeutic efficacy of NO in the management of erectile dysfunction by administering linsidomine chlorhydrate (SIN-1) intracavernously. This pharmacological agent releases NO nonenzymatically, and when injected intracavernosally in 63 patients, it induced a dose-related erectile response, with 46% of the patients experiencing full tumescence (Stief et al. 1992). Additionally, this NO donor caused no unexpected inflammatory reaction or patient discomfort following intracavernosal delivery. However, later reports comparing intracavernosal injections of SIN-1 with prostaglandin E_1 in impotent men demonstrated that SIN-1 produced inferior erectile responses (Porst 1993). However, in another study employing intracavernosal sodium nitroprusside (SNP), severe hypotension with only minimal tumescence was exhibited (Brock et al. 1993a). Intracavernosal administration of cGMP in 15 patients induced penile erection, but in 13 of these men the duration of the erection was much shorter than that obtained with the standard drug mixture (Burnett 1997). Nonetheless, the use of NO-releasing agents with minimal side effects suggests future pharmaceutical clinical applications for the treatment of erectile dysfunction.

The major drawback of NO is that it is a labile, short-lived gaseous molecule possessing a half-life of less than 5 seconds when in physiologic solution. Therefore, potential treatments of erectile dysfunction with NO will be based on the synthesis of stable compounds that efficiently release or generate NO to the cavernosal smooth muscle of the penis over time. New classes of nucleophile/NO complexes that spontaneously and predictably release NO over time are adducts 1-[N-(3-Ammoniopropyl)-N-(n-propyl)amino]diazen-1-ium-1,2-diolate (PAPA/NO), 1-{N-Methyl-N-[6-(N-methylammoniohexyl)amino]} diazen-1-ium-1,2-diolate (MAHMA/NO), and sodium 1-(N,N-diethylamino)diazen-1-ium-1,2-diolate (DEA/NO) (Maragos et al. 1991). These NO nucleophile adducts contain the structure $X[N(0)NO]^-$, where X represents the nucleophile residue that allows the spontaneous release of nonenzymatically generated NO (Maragos et al. 1991). In practice, the extent and role of NO release can be modified by changing the structure of the nucleophile residue (Maragos et al. 1991). It appears likely that the diversity of half-lives for NO release among members of the NO/nucleophile adducts will allow for formulation of compounds in combination with other vasoactive agents in the future treatment of male erectile dysfunction.

Nitric Oxide in the Urethra, Bladder, and Male Reproductive Tract

NADPH Diaphorase Histochemistry

Evidence has accumulated that NO is a neurotransmitter responsible for the inhibitory NANC responses in the lower urinary tract (Andersson 1993). In the pig detrusor, trigone, and urethra, NADPH diaphorase-positive fibers and thick nerve branches were identified in and around the muscular bundles (Persson et al. 1993). NADPH diaphorase-positive nerve fibers were frequently found around arteries, but not veins, and were more abundant in the urethral/trigonal area, whereas they were less common in the detrusor (Persson et al. 1993). Numerous NADPH diaphorase-positive nerve fibers were found around arteries and in smooth muscle fibers of the female rabbit lamina propria (Zygmunt et al. 1993). A plexus of NADPH diaphorase-containing nerve fibers was also isolated in the lateral wall and trigone regions of the human bladder (Smet et al. 1994). NADPH activity has also been detected in the urethra and in the intramural ganglia of the human (Smet et al. 1994).

Nitric Oxide Synthase Immunohistochemistry

The density of NOS immunoreactivity was found to be significantly higher in the trigone and urethral tissue than in the detrusor (Persson et al. 1993). Such a distribution corresponds with the ability of the bladder to empty when responding to NO-mediated relaxation by nerve stimulation that was distinct in the trigone and the urethra but not in the detrusor (Persson and Andersson 1992). Nitric oxide synthase immunoreactivity has also been demonstrated in the bladder neck and membranous urethra of rats and in the nerves of the mucosal stroma and those encircling the small arteries (Alm et al. 1993). In the pig, colocalization of NOS-immunoreactive nerves was similar to that of nerves stained for VIP, neuropeptide Y, and acetylcholine esterase (Andersson and Persson 1994). These data clearly indicate the existence of NOS in the urethra and bladder and a functional role for NO in this region.

Nitric Oxide in the Urethra

Inhibitory NANC-mediated responses were described in smooth muscle preparations from rabbit and human urethra by Andersson and colleagues (Andersson et al. 1983). In vitro data from a variety of animal species have shown that agents that affect the biosynthesis and mechanism of action of NO can alter the neurogenic relaxation of isolated urethral smooth muscle strips. In the isolated female rabbit urethra, maximal relaxation was increased after pretreatment with L-arginine, and an inhibitory effect was exhibited after treatment with N^G-nitro-L-arginine (L-NA) (Andersson and Persson 1995). In similar in vitro experiments, the responses of the urethral lamina propria and the external urethral sphincter were found to be also dependent on NO relaxation of the smooth musculature (Andersson and Persson 1994). Dokita et al. found that a selective cGMP phosphodiesterase inhibitor potentiated the relaxation in rabbit urethra (Dokita et al. 1991). Nerve-induced relaxation of the rabbit urethra increased the smooth muscle content of cGMP, but not cAMP (Persson and Andersson 1994). In the presence of zaprinast, a specific type V cGMP phosphodiesterase inhibitor, the increase in cGMP levels was significantly higher after nerve-induced relaxation of the urethra (Persson and Andersson 1994). Further evidence for the existence of NO in the human urethra was provided by Leone et al. (1994), who found the NO metabolites nitrite and nitrate in the urethra. These data suggest that the NO–cGMP signal transduction pathway has a primary role in urethral function in a variety of species, including humans.

Nitric Oxide in the Bladder

The bladder is under neuronal control; however, the interactive effects of smooth muscle, connective tissue, and epithelial elements of the bladder influence its contractility. The demonstration of NOS-containing nerves in the human bladder and smooth muscle bundles of the detrusor has led researchers to believe that NO is important in the regulatory physiology of this organ (Smet et al. 1994). In theory, increased activity of NO-releasing inhibitory nerves allows the bladder to relax during the filling phase. In support of such a view, mice lacking neuronal NOS demonstrated enlarged bladders with smooth muscle hypertrophy and a decreased ability to relax and empty (Burnett et al. 1995a). However, detrusor sensitivity to NO and agents acting via the cGMP system makes it less likely that NO has a role as a relaxant neurotransmitter in this tissue (Burnett 1995a). Therefore, it is unlikely that NO is of vital importance in bladder relaxation, but this does not exclude the possibility that NO can modulate other neurotransmitters, such as VIP and neuropeptide Y, since NOS has been colocalized with these peptides in this region.

Nitric Oxide in the Male Reproductive Tract

Multiple neuroregulatory mechanisms have control over the male reproductive tract, and without a doubt NO has some influence. Histochemical and immuno-histochemical studies have precisely located NOS in the reproductive structures of the male rat (Burnett et al. 1995a). The first studies detected NOS by biochemical assays that measured NOS activity (Burnett et al. 1995a). Nitric oxide synthase has been detected in neuronal fibers of the smooth musculature and subepithelial regions of the epididymis, ejaculatory duct, and vas deferens and in the epithelial cells of the epididymis and coagulating gland (Burnett et al. 1995a). Absence of androgens directly affects the influence of NOS activity in the epididymis, prostate, and seminal vesicles (Chamness et al. 1995). Based on the evidence present from the work in this anatomical region, one can propose that NO is involved in the regulation of the male reproductive tract.

Experimental Results

Our laboratory has used the feline erection model to study the effects of NO on penile erection. Adult male cats were sedated with ketamine hydrochloride (10–15 mg/kg intramuscularly) and anesthetized with sodium pentobarbital (30 mg/kg intravenously). Supplemental doses of pentobarbital were administered when needed to maintain a uniform level of anesthesia. The animals were maintained at 37° C with a heating blanket. A vertical circumcision-like incision was made to expose the two ventral corpora cavernosa and the dorsal corpus spongiosum. A 30-gauge needle was placed into the right corpus cavernosum to permit administration of drugs into the penis. A 25-gauge needle was placed midway into the left corpus cavernosum for the measurement of intracavernosal pressure. Systemic arterial and intracavernous pressures (mmHg) were measured with Statham P23 transducers connected to a Grass model 7 polygraph. Penile length (in millimeters) was measured with a ruler. These procedures have been approved by the Tulane University Animal Care and Use Committee.

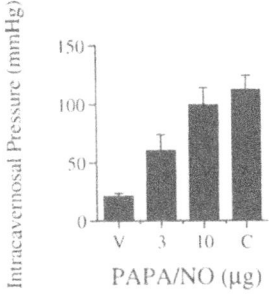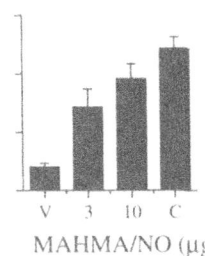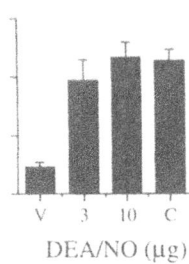

FIGURE 18.1. Bar graph showing the dose-dependent increase in penile pressure in response to intracavernosal injection of PAPA/NO, MAHMA/NO, and DEA/NO. C denotes response to the standard reference (papaverine, phentolamine, and PGE₁) administered at the end of the experiment.

PAPA/NO, MAHMA/NO, and DEA/NO were synthesized in the laboratory of Dr. L.K. Keefer (Laboratory of Comparative Carcinogenesis, National Cancer Institute, Frederick Cancer Research and Development Center) and were dissolved in 0.9% NaCl just before use. The standard drug combination, papaverine hydrochloride (Eli Lilly, Indianapolis, IN), prostaglandin E₁ (PGE₁) (Upjohn, Kalamazoo, MI), and phentolamine mesylate (CIBA-GEIGY, Summit, NJ), was prepared and used at the end of every experiment, as previously described (Champion et al. 1997a). The solvents of these agents had no significant effect on baseline corporal pressure or on responses to the vasoactive agents. All drug solutions were stored in a freezer in amber bottles. The NO donors were prepared before every experiment, and all other working solutions were prepared frequently and kept on crushed ice during an experiment.

The data were expressed as means ± standard error of the mean (SEM) and analyzed by Student's t-test for single-group comparison and by one-way analysis of variance (ANOVA) for multiple-group comparisons. A P value <0.05 was established as the criterion for statistical significance.

Responses to NO Donors

Intracavernosal injections of PAPA/NO (3 and 10 μg), MAHMA/NO (3 and 10 μg), and, DEA/NO (3 and 10 μg) caused dose-dependent increases in cavernosal pressure (Figure 18.1). Increases in cavernosal pressure in response to PAPA/NO, MAHMA/NO, and DEA/NO were compared to the standard control combination (1.65 mg papaverine, 25 μg phentolamine, and 0.5 μg PGE₁) commonly used in clinical practice for the treatment of impotence. Intracavernosal injections of PAPA/NO and MAHMA/NO at doses of 10 μg produced increases in cavernosal pressure that were comparable to those produced by the standard combination, whereas the 10-μg dose of DEA/NO elicited an erectile response greater than that elicited by the standard combination (Figure 18.1). The duration of the effect of

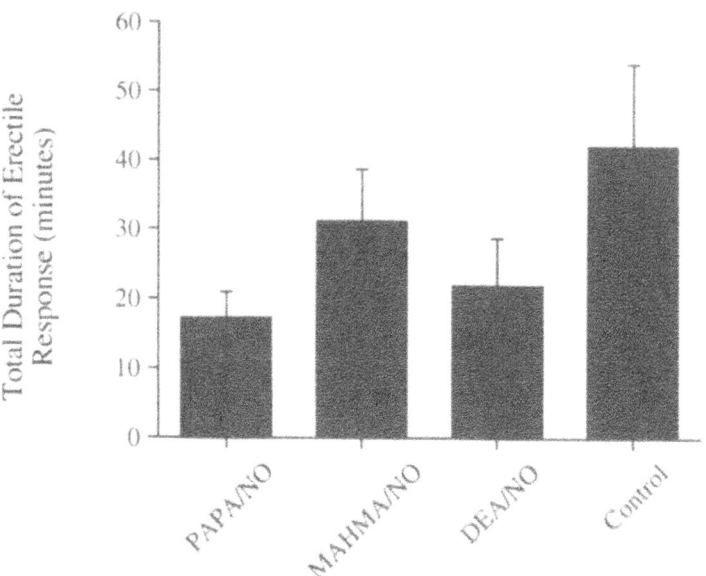

FIGURE 18.2. Bar graph showing the total duration of erectile response in response to intra-cavernosal injection of PAPA/NO, MAHMA/NO, and DEA/NO. Control denotes response to the standard reference (papaverine, phentolamine, and PGE$_1$) administered at the end of the experiment.

the drugs (in minutes) was shorter than the duration of the effect of the standard combination (Figure 18.2). Although the duration of erection was not as long with the NO donors as with the standard combination, it seems plausible to hypothesize that such compounds could be combined with other pharmacological agents to increase erectile response or to develop compounds that exhibit a greater diversity of half-lives for NO release.

Effects of NO Donors on Systemic Arterial Pressure

The effects of PAPA/NO, MAHMA/NO, and DEA/NO on systemic arterial pressure (SAP) were compared in the feline erection model. When PAPA/NO, MAHMA/NO, and DEA/NO were injected in doses of 10 μg, SAP was decreased significantly. The maximum decrease in SAP was obtained at a dose of 10 μg for PAPA/NO (-17.5 ± 3.1 mmHg), MAHMA/NO (-19.0 ± 2.2 mmHg), and DEA/NO (-15.0 ± 2.0 mmHg). The standard combination of papaverine, phentolamine, and PGE$_1$ decreased SAP by 42 ± 2 mmHg. The decreases in SAP produced by the NO donors are less than those produced by the standard drug, which induced systemic hypotension in these animals.

Conclusion

From a clinical perspective, our growing knowledge of NO and its function in the physiology of erection and control of the lower urinary tract will produce new pharmacologic management strategies for disorders of the genitourinary system. Many studies have characterized the role NO plays in the control over both neurogenically mediated and endothelium-dependent corporal smooth musculature relaxation. Penile erection, micturition, and peristalsis of the male excurrent duct system are all functions that seem to be dependent on the induction and control of NO. Impotence or urinary obstruction may in part be due to alterations in NO production. Therefore, pharmacologic control of NO or its mechanism of action may help treat or restore the desired urologic function in afflicted individuals. The future development of NO donors with minimal side effects will provide even better pharmacologic therapies for erectile dysfunction. Advancing knowledge and applications of NO in the genitourinary tract will hold promise for future clinical developments.

Acknowledgments. The authors would like to thank Melanie Cross for her critical review of the manuscript. This work was supported by the American Foundation for Urological Disease Summer Scholar Program (TJB).

References

Alm P, Larsson B, Ekblad E, Sundler F, & Andersson KE. 1993. Immunohistochemical localization of peripheral nitric oxide synthase containing nerves using antibodies raised against synthetized C- and N-terminal fragments of a cloned enzyme from rat brain. *Acta Physiol Scand 148:*421–429.

Andersson KE. 1993. Pharmacology of lower urinary tract smooth muscles and penile erectile tissues. *Pharmacol Rev 45:*253–308.

Andersson KE. 1996. Neurotransmitters and neuroreceptors in the lower urinary tract. *Curr Opin Obstet Gynecol 8:*361–365.

Andersson KE, & Persson K. 1994. Nitric oxide synthase and nitric oxide-mediated effects in lower urinary tract smooth muscles. *World J Urol 12:*274–280.

Andersson K, & Persson K. 1995. Nitric oxide synthase and the lower urinary tract: possible implications for physiology and pathophysiology. *Scand J Urol Nephrol 175:* 43–53.

Andersson KE, Mattiasson A, & Sjogren C. 1983. Electrically induced relaxation of the noradrenaline contracted isolated urethra from rabbit and man. *J Urol 129:*210–213.

Bredt DS, Ferris CD, & Snyder SH. 1992. Nitric oxide synthase regulatory sites. Phosphorylation by cyclic AMP-dependent protein kinase, protein kinase C, and calcium/camodulin protein kinase; identification of flavin and calmodulin binding sites. *J Biol Chem 267:*10976–10981.

Brindley GS. 1986. Pilot experiments on the actions of drugs injected into the human corpus cavernosum penis. *Br J Pharmacol 87:*495–500.

Brock G, Breza J, & Lue TF. 1993a. Intracavernous sodium nitroprusside: inappropriate impotence treatment. *J Urol 150:*864–867.

Brock G, Nunes L, Padma-Nathan H, Boyd S, & Lue TF. 1993b. Nitric oxide synthase: a new diagnostic tool for neurogenic impotence. *Urology 42:*412–417.

Burnett AL. 1995a. Nitric oxide control of lower genitourinary tract function a review. *Urology 42:*1071–1083.

Burnett AL. 1995b. Role of nitric oxide in the physiology of erection. *Biol Reprod 52:*485–489.

Burnett AL. 1997. Nitric oxide in the penis: physiology and pathology. *J Urol 157:*320–324.

Burnett AL, Lownstein CJ, Bredt DS, Chang TSK, & Snyder SH. 1992. Nitric oxide: a physiologic mediator of penile erection. *Science 257:*401–403.

Burnett AL, Calvin DC, Silver RI, Peppas DS, & Docimo SG. 1997. Immunohistochemical description of nitric oxide synthase isoforms in human clitoris. *J Urol 158:*75–78.

Burnett AL, Ricker DD, Chamness SL, Maguire MP, Crone JK, Bredt DS, Snyder SH, & Chang TS. 1995a. Localization of nitric oxide synthase in the reproductive organs of the male rat. *Biol Reprod 52:*1–7.

Burnett AL, Saito SS, Maquire MP, Yamaguchi H, Chang TS, & Hanley DF. 1995b. Localization of nitric oxide synthase in spinal nuclei innervating pelvic ganglia. *J Urol 153:*212–217.

Bush PA, Aronson WJ, Buga GM, Rajfer J, & Ignarro LJ. 1992a. Nitric oxide is a potent relaxant of human and rabbit corpus cavernosum. *J Urol 147:*1650–1655.

Bush PA, Gonzalez NE, & Ignarro LJ. 1992b. Biosynthesis of nitric oxide and citrulline from L-arginine by constitutive nitric oxide synthase present in rabbit corpus cavernosum. *Biochem Biophys Res Commun 186:*308–314.

Chamness SL, Ricker DD, Crone JK, Dembeck CL, Maguire MP, Burnett AL, & Chang TS. 1995. The effect of androgen on nitric oxide synthase in the male reproductive tract of the rat. *Fertil Steril 63:*1101–1107.

Champion HC, Wang R, Hellstrom WJ, & Kadowitz PJ. 1997a. Nociceptin, a novel endogenous ligand for the ORL1 receptor, has potent erectile activity in the cat. *Am J Physiol 273:*E214–E219.

Champion HC, Wang R, Santiago JA, Murphy WA, Coy DH, Kadowitz PJ, & Hellstrom WJG. 1997b. Comparison of responses to adrenomedullin and calcitonin gene-related peptide in the feline erection model. *J Androl 18:*513–521.

Dokita S, Morgan WR, Wheeler MA, Yoshida M, Latifpour J, & Weiss RM. 1991. N^G-nitro-L-arginine inhibits non-adrenergic, non-cholinergic relaxation in rabbit urethral smooth muscle. *Life Sci 4:*2429–2436.

Domer FR, Wessler G, Brown RL, & Charles HC. 1978. Involvement of the sympathetic nervous system in the urinary bladder internal sphincter and in penile erection in the anesthetized cat. *Invest Urol 15:*404–407.

Domoto T, & Tsumori T. 1994. Co-localization of nitric oxide synthase and vasoactive intestinal polypeptide immunoreactivity in neurons of the major pelvic ganglion projecting to the rat rectum and penis. *Cell Tissue Res 278:*273–278.

Feldman HA, Goldstein I, & Hatzichristou DG. 1994. Impotence and its medical and psychosocial correlates: results of the Massachusetts male aging study. *J Urol 151:*54.

Finberg JPM, Levy S, & Vardi Y. 1993. Inhibition of nerve stimulation-induced vasodilation in corpora cavernosa of pithed rat by blockade of nitric oxide synthase. *Br J Pharmacol 108:*1038–1042.

Furchgott RF, & Zawadzki J. 1980. The obligatory role of endothelial cells in the relaxation of arterial smooth muscle by acetycholine. *Nature 228*:373–376.

Hellstrom WJ, Monga M, Wang R, Domer FR, Kadowitz PJ, & Roberts JA. 1994. Penile erection in the primate: induction with nitric-oxide donors. *J Urol 151*:1723–1727.

Holmquist F, Hedlund H, & Andersson KE. 1991a. L-N^G-nitro arginine inhibits non-adrenergic, non-cholinergic relaxation of human isolated corpus cavernosum. *Acta Physiol Scand 141*:441–442.

Holmquist F, Stief CG, Joras U, & Andersson KE. 1991b. Effects of the nitric oxide synthase inhibitor N^G-nitro-L-arginine on the erectile response to cavernous nerve stimulation in the rabbit. *Acta Physiol Scand 143*:299–304.

Holmquist F, Fridstrand M, Hedlund H, & Andersson KE. 1993. Actions of 3-morpholinosydnonimin (SIN-1) on rabbit isolated penile erectile tissue, *J Urol 150*: 1310–1315.

Ignarro LJ, Buga GM, Wood KS, Byrns RE, & Chaudhuri G. 1987. Endothelium-derived relaxing factor produced and released from artery and vein is nitric oxide. *Proc Natl Acad Sci USA 84*:9265–9269.

Ignarro LJ, Bush PA, Buga GM, Wood KS, Fukuto JM, & Rajfer J. 1990. Nitric oxide and cyclic GMP formation upon electrical field stimulation cause relaxation of corpus cavernosum smooth muscle. *Biochem Biophys Res Commun 170*:843–850.

Kerwin JF, Lancaster JR Jr, & Feldman PL. 1995. Nitric oxide: a new paradigm for second messengers. *J Med Chem 38*:4343–4362.

Kim N, Azadzoi KM, Goldstein I, & Saenz de Tejada I. 1991. A nitric oxide-like factor mediates nonadrenergic, noncholinergic neurogenic relaxation of penile corpus cavernosum smooth muscle. *J Clin Invest 88*:112–118.

Kim YC, Choi HK, Ahn YS, Kim KH, Hagen PO, & Carson CC. 1994. The effect of vasoactive intestinal polypeptide (VIP) on rabbit cavernosal smooth muscle contractility. *J Androl 15*:392–397.

Knispel HH, Goessl C, & Beckmann R. 1991. Basal and acetycholine-stimulated nitric oxide formation mediates relaxation of rabbit cavernous smooth muscle. *J Urol 146*:1429–1433.

Knowles RG, & Moncada S. 1994. Nitric oxide synthases in mammals. *Biochem J 298*:249–258.

Leone AM, Wiklund NP, Hokfelt T, Brundin L, & Moncada S. 1994. Release of nitric oxide by nerve stimulation in the human urogenital tract. *Neuro Rep 5*:733–736.

Lue TF. 1992. Physiology of erection and pathophysiology of impotence. In: Walsh PC, Retik AB, Stamey TA, Vaughan ED Jr, editors. *Campbell's Urology*. 6th ed. Philadelphia: W.B. Saunders Co. p. 709–728.

Lugg JA, Gonzales-Cadavid NF, & Rajfer J. 1995a. The role of nitric oxide in erectile function. *J Androl 16*:2–5.

Lugg JA, Rajfer J, & Gonzalez-Cadavid NF. 1995b. Dihydrotestosterone is the active androgen in the maintenance of nitric oxide-mediated penile erection in the rat. *Endocrinology 136*:1495–1501.

Maragos CM, Morley D, Wink DA, Dunams TM, Saavedra JE, Hoffman A, Bove AA, Isaac L, Hrabie JA, & Keefer LK. 1991. Complexes of NO with nucleophiles as agents for the controlled biological release of nitric oxide. Vasorelaxant effects. *J Med Chem 34*:3242–3247.

Marin J, & Angeles Rodriguez-Martinez A. 1997. Role of vascular nitric oxide in physiological and pathological conditions. *Pharmacol Ther 75*:111–134.

Melis MR, & Argiolas A. 1993. Nitric oxide synthase inhibitors prevent apomorphine- and oxytocin-induced penile erection and yawning in male rats. *Brain Res Bull* *32*:71–74.

Melis MR, Argiolas A, & Gressa GL. 1989. Evidence that apomorphine induces penile erection and yawning by releasing oxytocin in the central nervous system. *Eur J Pharmacol* *164*:565–570.

Meyers PR, Minor PL Jr, Guerra R Jr, Bates JN, & Harrison DG. 1990. Vasorelaxant properties of the endothelium-derived relaxing factor more closely resemble *S*-nitrocysteine than nitric oxide. *Nature* *345*:161–163.

Mills TA, Stopper VS, & Wiedmeier VT. 1994. Effects of castration and androgen replacement on the hemodynamics of penile erection in the rat. *Biol Reprod* *51*:234–238.

Moncada S. 1992. The L-arginine:nitric oxide pathway. *Acta Physiol Scand* *145*:201–227.

Morris JL, Gibbins IL, Kadowitz PJ, Herzog H, Kreulen DL, Toda N, & Claing A. 1995. Roles of peptides and other substances in cotransmission from vascular autonomic and sensory neurons. *Can J Physiol Pharmacol* *75*:521–532.

NIH Consensus Conference. 1993. Impotence, NIH Consensus Development Panel of Impotence. *JAMA* *270*:83.

Palmer RMJ, Ferrige AG, & Moncada S. 1987. Nitric oxide release accounts for the biological release of endothelium-derived relaxing factor. *Nature* *327*:524–526.

Persson K, & Andersson KE. 1992. Nitric oxide and relaxation of the pig lower urinary tract. *Br J Pharmacol* *106*:416–422.

Persson K, & Andersson KE. 1994. Non-adrenergic, non-cholinergic relaxation and levels of cyclic nucleotides in rabbit lower urinary tract. *Eur J Pharmacol* *268*:159–167.

Persson K, Alm P, Johansson K, Larsson B, & Andersson KE. 1993. Nitric oxide synthase in pig lower urinary tract: immunohistochemistry, NADPH diaphorase histochemistry, and functional effects. *Br J Pharmacol* *110*:521–530.

Porst H. 1993. Prostaglandin E$_1$ and the nitric oxide donor linsidomine for erectile failure: a diagnostic comparative study of 40 patients. *J Urol* *149*:1280–1283.

Rajfer J, Bush PA, Dorey FJ, & Ignarro LJ. 1992. Nitric oxide as a mediator of relaxation of the corpus cavernosum in response to nonadrenergic, noncholinergic neurotransmission. *N Engl J Med* *326*:90–94.

Smet PJ, Edyvane KA, Jonavicius J, & Marshall VR. 1994. Distribution of NADPH-diaphorase-positive nerves supplying the human urinary bladder. *J Auton Nerv Syst* *47*:109–113.

Stief CG, Holmquist F, Djamilian M, Krah H, Andersson KE, & Jonas U. 1992. Preliminary results with the nitric oxide donor linsidomine chlorhydate in the treatment of human erectile function. *J Urol* *148*:1437–1440.

Tamura M, Kagawa S, Kimura K, Kawanishi Y, Tsuruo Y, & Ishimura K. 1995. Coexistence of nitric oxide synthase, tyrosine hydroxylase and vasoactive intestinal polypeptide in human penile tissue—a triple histochemistry and immunohistochemical study. *J Urol* *153*:530–534.

Trigo-Rocha F, Aronson WJ, Hohenfellner M, Ignarro LJ, Rajfer J, & Lue TF. 1993a. Nitric oxide and cGMP: mediators of pelvic nerve-stimulated erection in dogs. *Am J Physiol* *264*:H419–H422.

Trigo-Rocha F, Hsu GL, Donatucci CF, & Lue TF. 1993b. The role of cyclic adenosine monophosphate, cyclic guanosine monophosphate, endothelium and nonadrenergic, noncholinergic neurotransmission in canine penile erection. *J Urol* *149*:872–877.

Virag R. 1982. Intracavernous injection of papaverine for erectile failure. *Lancet* *2*:938.

Wang R, Higuera TR, Sikka SC, Minkes RK, Bellan JA, Kadowitz PJ, Domer FR, & Hellstrom WJG. 1993. Penile erections induced by vasoactive intestinal peptide and sodium nitroprusside. *Urol Res 21:*75–78.

Wang R, Domer FR, Sikka SC, Kadowitz PJ, & Hellstrom WJG. 1994. Nitric oxide mediates penile erection in cats. *J Urol 151:*234–237.

Zygmunt PKE, Persson K, Alm P, Larsson B, & Andersson KE. 1993. The L-arginine/nitric oxide pathway in the rabbit urethral lamina propria. *Acta Physiol Scand 148:*431–439.

E. Renal

19
Role of Nitric Oxide in the Regulation of Renal Blood Flow

Dewan S.A. Majid and L. Gabriel Navar

Introduction

The regulation of renal blood flow is of vital importance to the overall control of renal function and, thus, to the maintenance of body fluid and electrolyte home-ostasis. Among the many factors regulating blood flow to the kidney, the contributing roles of various intrarenal paracrine systems and their complex interactions are of paramount importance (Navar et al. 1996). Over the past decade, an enormous amount of research has established nitric oxide (NO) as an important regulator of renal hemodynamics and excretory function (Navar et al. 1996; Kone and Baylis 1997; Majid and Navar 1997; Mattson et al. 1997). In the kidney, NO interacts with vascular smooth muscle and with mesangial, juxtaglomerular, and tubular cells to profoundly affect vascular and tubular function (Navar et al. 1996; Kone and Baylis 1997). In this chapter, we have focused on the importance of intrarenal NO in the regulation of renal vascular resistance and its influence on the renal autoregulatory mechanism. In addition to the direct effects of NO, we also discuss the various interactions of NO with other renal paracrine factors.

Nitric Oxide Synthase in the Kidney

Nitric oxide is formed and released by endothelial and other cell types upon the oxidation of the amino acid L-arginine by the action of the enzyme nitric oxide synthase (NOS). Studies have provided substantial evidence that endothelial, epithelial, and other cells can generate NO in the kidney (Star 1997). Nitric oxide synthase exists in three isoforms named for the tissues in which they were first characterized. Endothelial NOS (eNOS) and neuronal NOS (nNOS), which are also NOS III and NOS I, respectively, are constitutively present and are dependent on calcium and calmodulin for activation (Lamas et al. 1992). An inducible NOS (iNOS), termed NOS II, is usually expressed primarily after transcriptional induction and seems less dependent on calcium for its activation because of its very high affinity for calmodulin, which thus forms a constitutive subunit (Cooke and Dzau 1997). Nitric oxide synthase II can form abundant quantities of NO

TABLE 19.1. Localization of NOS isoforms in the kidney

	Constitutive		Inducible
	NOS III	NOS I	NOS II
1) Endothelium			
a) large vessels	+ + +	−	−
b) afferent arterioles	+ + +	−	
c) efferent arterioles	+ +	+	−
d) glomerular capillaries	+ + +	−	−
2) Epithelium			
a) macula densa	−	+ + +	−
b) thick ascending limb	−	+	+
c) collecting duct	+	−	+
d) proximal tubule	−	−	+
3) Vascular Smooth Muscle	−	−	+
4) Mesangial Cells	−	+	+ +
5) Intrarenal Neuron	−	+	−
6) Medullary Interstitial Cells	−	−	+

+, Present; −, Absent; NOS, Nitric Oxide Synthase

once stimulated by cytokines (Xie et al. 1992; Shultz et al. 1994). All these iso-forms are heme-containing enzymes that catalyze the oxidation of L-arginine to NO and L-citrulline and are also expressed in the kidney. Table 19.1 provides the localization of NOS isoforms in the kidney. Generally, the endothelium of corti-cal and medullary vessels and also some tubular epithelial cells express NOS III (Ujiie et al. 1993; Bachmann et al. 1995) whereas macula densa cells and efferent arterioles express NOS I (Bachman et al. 1995). The inducible NOS II is ex-pressed in mesangial cells and several tubular epithelial cells (Bachman and Mundel 1994; Kone and Baylis 1997; Star 1997). Since NOS reacts directly with L-arginine, the formation and release of NO can be competitively inhibited by structural analogues of L-arginine such as N^G-monomethyl-L-arginine (L-NMMA) and nitro-L-arginine (NLA) (Moncada et al. 1991). Because of the short half-life and labile nature of NO, studies evaluating the in vivo effects of endogenous NO in the kidney have relied primarily on the use of the L-arginine analogues to in-hibit the synthesis of NO.

Nitric Oxide and Renal Vascular Resistance

Abundant evidence is now available that a tonic release of NO plays an important role in maintaining normal vascular tone in the kidney, which is known to have substantially lower vascular resistance than most other organs or vascular beds. Studies from our laboratory and many others have clearly demonstrated that con-stitutively released NO helps to maintain the normally low renal vascular resis-tance (RVR), being responsible for up to one-third of the normal renal blood flow (RBF) (Baumann et al. 1992; Beierwaltes et al. 1992; Majid and Navar 1992; Navar et al. 1996). The role of NO in the kidney has been assessed primarily with

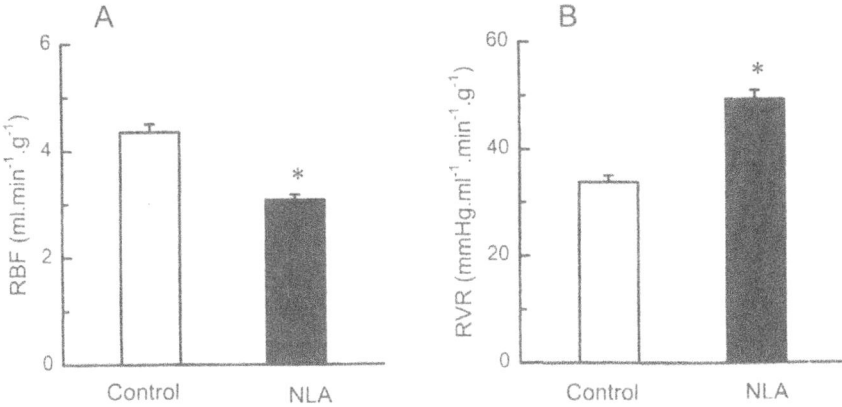

FIGURE 19.1. (A) Renal blood flow (RBF) and (B) renal vascular resistance (RVR) responses to intraarterial administration of nitro-L-arginine (NLA) (50 μg/kg/min) in anesthetized dogs ($N = 44$).

pharmacological intervations that enhance or inhibit endogenous production of NO. In addition, the administration of agents that release NO in the circulation, usually known as nitrovasodilators, has provided important information. Agonist agents that stimulate the endogenous release of NO, such as acetylcholine and bradykinin, have long been known to produce marked renal vasodilation (Baer et al. 1970; Lahera et al. 1990; Majid and Navar 1992).

Several critical studies have evaluated the contribution of endogenously released NO to the basal level of RVR. In response to either acute or chronic administration of L-arginine analogues to inhibit NO synthesis in anesthetized or conscious animals, there is an increase of 30%–60% in RVR and a decrease of 25%–40% in RBF (Baylis et al. 1990; Lahera et al. 1991; Baumann et al. 1992; Bierwaltes et al. 1992; Majid and Navar 1992; Salom et al. 1992; Denton and Anderson 1994; Bech et al. 1996). Figure 19.1 shows the responses of RBF and RVR to L-arginine analogues that block NO synthesis. In these experiments conducted in our laboratory (Majid and Navar 1992; Majid et al. 1993a,b, 1994, 1996), renal responses to intraarterial administration of NLA at a rate of 50 μg/kg/min for more than 30 minutes were evaluated in anesthetized dogs. This dose of NLA was found to be sufficient to achieve an effective blockade of intrarenal NO formation, as evidenced from the complete reversal of the renal vasodilatory effects of ATP infused intraarterially (Majid and Navar 1992). Administration of NLA in these dogs resulted in a 48% increase in RVR and a 28% decrease in RBF. The responses to NOS inhibition were also reversed by the administration of L-arginine (Elsner et al. 1992; Majid and Navar 1992; Yukimura et al. 1992), substantiating that the responses to L-arginine analogues are due specifically to inhibition of endogenous NO formation and release. Outer cortical blood flow measured by laser-Doppler flowmetry with a surface probe (Chen et al. 1992; Majid et al. 1993b), as

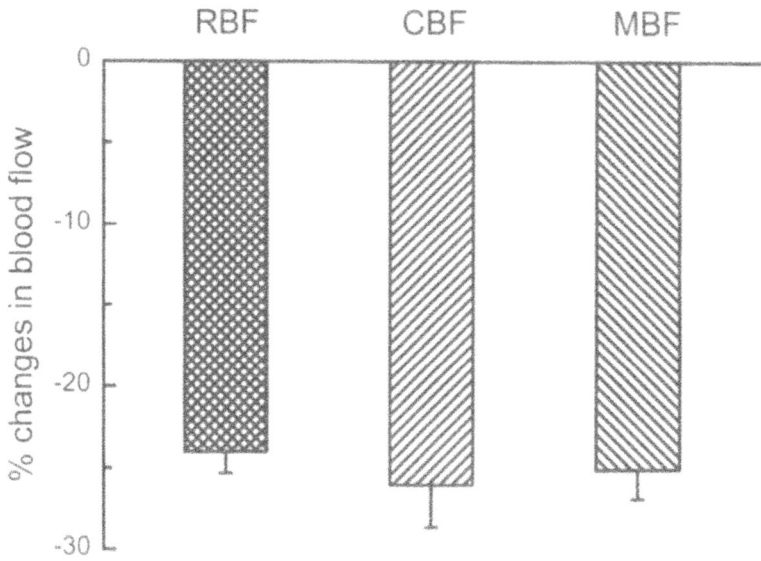

FIGURE 19.2. Comparison of the renal regional blood flow responses to intraarterial administration of nitro-L-arginine (NLA) (50 µg/kg/min) in anesthetized dogs ($N = 13$). RBF, Total renal blood flow; CBF, cortical blood flow; MBF, medullary blood flow.

well as blood flow in single cortical capillaries measured by fluorescent videomicroscopy (Lockhart et al. 1994), showed similar responses to L-arginine analogues. In further studies, laser-Doppler flowmetry with needle probes was used to assess renal medullary blood flow and critical blood flow responses to NOS inhibition (Majid et al. 1996). These studies demonstrated that blockade of NO synthesis in the kidney by NLA administration resulted in reductions in both cortical blood flow (26%) and medullary blood flow (25%), that were very similar to the reductions in total renal blood flow (24%) (Fig. 19.2). These findings indicate that the blood flows to the cortical and medullary regions of the kidney are equally dependent on endogenous NO. However, several other studies performed in rats have suggested that NO may exert a greater influence on the medullary circulation than on the cortical circulation (Brezis et al. 1991; Biondi et al. 1992; Mattson et al. 1992; Zou et al. 1997). Comparison of the responses to NOS inhibition of blood flow and vascular resistances in other vascular beds such as femoral, coronary, and brain has shown that the renal vasculature is much more dependent on endogenous NO than other organs to maintain normal perfusion of the organ (Sonntag et al. 1992; Sigmon et al. 1993). Collectively, these data generated in various species provide clear evidence that basal release of NO helps to maintain the relatively low vascular resistance that is characteristic of the kidney.

Several studies have evaluated the direct effects of elevated intrarenal NO levels using nitrovasodilators, also known as NO donor compounds. Nitrovasodila-

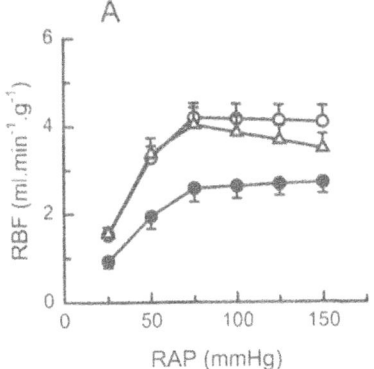
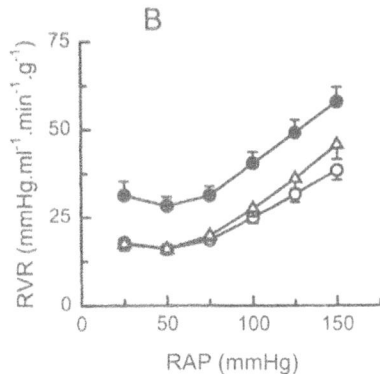

Figure 19.3. (A) Renal blood flow (RBF) and (B) renal vascular resistance (RVR) responses to acute reductions in renal arterial pressure (RAP) before (○) and during (●) intraarterial infusion of nitro-L-arginine (NLA) and during (△) infusion of S-nitroso-n-acetylpenicillamine (SNAP) in anesthetized dogs (N = 8). From Majid et al. (1993a). Copyright 1993 by Lippincott Williams & Wilkens. Reprinted with permission.

tors have been used clinically for about 100 years and are still widely used in conditions such as angina pectoris, hypertensive emergencies, and pulmonary hypertension in order to improve perfusion of blood and reduce vascular resistance (Abrams 1987). Among the commonly used nitrovasodilators, S-nitrosothiols are potent vasodilators and have been used in different studies to mimic the actions of endogenously formed NO (Ignarro et al. 1981; Moncada et al. 1991). To evaluate the unique and specific actions of NO on RBF and RVR, experiments were performed in which S-nitroso-N-acetylpenicillamine (SNAP), an S-nitrosothiol compound, was administered intraarterially in anesthetized dogs that were pretreated with NLA to inhibit endogenous NO release (Majid et al. 1993a). As depicted in Figure. 19.3, administration of SNAP at a dose of 2 µg/kg/min in these NLA-treated dogs elicited renal vasodilator responses and restored RBF and RVR to values existing before NOS inhibition, confirming the specificity of the vascular actions of NO in the kidney.

Renal Microvascular Actions of NO

Studies involving direct assessment of renal microvascular responses to NOS inhibition in rats and rabbits confirm that tonically released NO regulates both pre- and postglomerular vascular resistance (Zatz and DeNucci 1991; Imig and Roman 1992; Ohishi et al. 1992; Deng and Baylis 1993; Edwards and Trizina 1993; Ito and Ren 1993; Denton and Anderson 1994). The microcirculatory responses have been evaluated in both in vitro preparations and in vivo micropuncture experiments. Although all these studies consistently report decreases in vessel diameter and increases in vascular resistance in the afferent arterioles, the efferent arteriolar responses have shown a more variable response to NOS inhibition. In an isolated rabbit glomerulus preparation, Ito and Ren (1993) reported no change in efferent arteriolar diameter during arteriolar microperfusion of NOS inhibitors. Similar to this finding, Deng and Baylis (1993) reported that intraarterial infusion of NOS inhibitors did not cause detectable increases in efferent arteriolar resistance, glomerular pressure, or glomerular flow, although systemic infusion of NOS inhibitors did cause efferent arteriolar constriction. Using an in vitro blood-perfused juxtamedullary nephron preparation, Ohishi et al. (1992) demonstrated that both afferent and efferent arteriolar diameter in rat kidneys decreased to about 15% after superfusion of renal tissue with 1 mM NLA. Similar observations were also reported in several other studies (Imig and Roman 1992; Edwards and Trizina 1993; Hoffend et al. 1993; Imig et al. 1993). Decreases in glomerular flow and associated increases in resistances of afferent and efferent arterioles were also noted during systemic administration of L-arginine analogues in micropuncture studies in vivo (Zatz and DeNucci 1991; Baylis et al. 1992; Denton and Anderson 1994). Thus, on balance, the results support the conclusion that endogenous NO levels contribute to the vascular tone of both preglomerular and efferent arterioles.

Nitric Oxide and Tubuloglomerular Feedback Responses

Recent studies have demonstrated that NO may also be an important modulator of tubuloglomerular feedback (TGF) responsiveness (Wilcox et al. 1992; Thorup and Persson 1994; Braam and Koomans 1995; Vallon and Thomson 1995; Wilcox and Welch 1996). Studies in which NOS inhibitors were infused intravenously or microperfused into the interstitium or into the tubular segments encompassing the macula densa cells have shown that local NO production can inhibit TGF-mediated vasoconstriction of the afferent arterioles (Wilcox et al. 1992; Thorup and Persson 1994). Using the doubly perfused isolated glomerular segment with macula densa attached, Ito and Ren (1993) demonstrated that addition of an NOS inhibitor to the macula densa perfusate led to afferent arteriolar constriction when the tubules were perfused with isotonic Krebs-Ringer solution but not when they were perfused with a hypotonic solution. Furthermore, Wilcox and Welch (1996) reported that the TGF response was enhanced by NOS blockade during high-salt- but not in low-salt-intake rats. Blockade of macula densa reabsorption by furosemide was also shown to abolish this enhancement of TGF responsiveness to NOS inhibitors (Wilcox et al. 1992). Thus, it is generally thought that NO synthesis is initiated by enhanced NaCl and solute reabsorption by the macula densa cells, and the increased NO release partially counteracts the vasoconstrictor stimuli mediating TGF responses. Thus, NO is not a mediator of the TGF mechanism but apparently serves to modulate the responsiveness. Recently, Ichihara et al. (1998) demonstrated that superfusion of blood-perfused rat juxtaglomerular preparations with the specific neuronal NOS inhibitor S-methyl-L-thiocitrulline (L-SMTC) decreased afferent and efferent arteriolar diameters, and these decreases in arterial diameters were prevented during interruption of distal volume delivery by papillectomy. Afferent, but not efferent, arteriolar vasoconstrictor responses to L-SMTC were also enhanced during increases in volume delivery to the macula densa segment by the use of acetazolamide, and this effect was completely prevented after papillectomy. In contrast, the arteriolar diameter responses to the nonselective NOS inhibitor NLA were only partially attenuated by papillectomy. These findings indicate that neuronal NOS in the macula densa exerts a modulating influence on TGF-mediated adjustments in afferent arteriolar tone. It seems clear that NO is produced in the macula densa by the action of neuronal NOS (NOS I), and that macula densa-derived NO, not endothelial NO, is the apparent modulator of TGF responsiveness.

Nitric Oxide and RBF Autoregulation

The autoregulatory adjustments in the renal vasculature in response to changes in arterial pressure predict a change in blood flow velocity through the small arteries and arterioles. Since the entire preglomerular arteriolar vasculature changes dimensions in response to changes in arterial pressure (Carmines et al. 1990; Navar et al. 1996), there could be shear-stress-induced alterations in NO release that could modulate renal vascular tone (Marshall and Kontos 1990; King and Brenner 1991). Decreases in vessel diameter in the preglomerular arterioles in response to increases in arterial pressure following autoregulatory adjustment would indicate an increased velocity for the same absolute blood flow. Such changes in arteriolar blood flow velocity could exert shear stress on the vessel wall, thus causing alterations in NO synthesis during changes in perfusion pressures. Several studies in dogs and rats have been performed to determine how NO might contribute to renal autoregulatory behavior (Baumann et al. 1992; Beierwaltes et al. 1992; Majid and Navar 1992; Majid et al. 1993a,b). These experiments have shown that during NOS inhibition, there is essentially complete preservation of the ability of the kidney to autoregulate RBF in response to alterations in renal arterial pressure; however, there is suppression of the autoregulation plateau due to the decrease in basal RBF during NOS inhibition. In experiments conducted in anesthetized dogs in our laboratory (Majid et al. 1993a), RBF and RVR responses to stepwise reductions in renal arterial pressure were evaluated before and after NOS inhibition and during administration of a NO donor compound, SNAP. As illustrated in Figure 19.3, the autoregulatory efficiency of RBF remained intact during both NOS inhibition and NO replacement by SNAP. Most studies in both anesthetized and unanesthetized animals have reported similar results; NOS inhibition does not impair steady-state autoregulatory responses. However, an impairment of RBF autoregulatory efficiency during intrarenal infusion of an NOS inhibitor was reported in a study conducted in anesthetized dogs (Kiyomoto et al. 1992). In that particular study, the responses were evaluated within 30 minutes of the start of intrarenal infusion of NOS inhibitors. It is possible that pressure–flow relationships during graded reductions in renal perfusion pressure were evaluated before the full effects of NOS inhibition were achieved.

Studies using laser-Doppler flowmetry in anesthetized dogs have shown that the autoregulatory efficiency of cortical blood flow also remains intact during intrarenal infusion of NOS inhibitors (Majid et al. 1993b). There are mixed reports with regard to medullary blood flow. It has been reported that an impaired autoregulatory efficiency of papillary blood flow in volume-expanded rats, measured by laser-Doppler flowmetry, could be restored by administering NOS inhibitors (Fenoy et al. 1995). This finding indicates that an enhancement of NO production during volume expansion in the rat may be responsible for such impairment of autoregulatory efficiency in blood flow to the renal medulla, which was not observed in euvolumic rats (Roman et al. 1988; Mattson et al. 1993). However, it has been demonstrated that medullary blood flow maintains efficient autoregulatory responses in both euvolumic and volume-expanded dogs (Majid and Navar 1996; Majid et al. 1997a,b). Thus, it seems clear from the bulk of the data that intrarenal NO activity does not influence the normal autoregulatory efficiency of RBF. Although increases in renal arterial pressure increase NO release from the vascular endothelium due to shear stress, the amount of released NO is not sufficient to counteract the autoregulation mediated vasoconstriction.

Interactions of NO with Other Vasoactive Factors

The actions of NO in the kidney may be modulated through interactions with other paracrine factors. A number of studies have suggested that, in addition to its direct vasodilator effect via cGMP actions on vascular smooth muscle cells, NO may also counteract the vasoconstrictor effects of other paracrine factors (Navar et al. 1996). Although the effects of L-arginine analogues are mediated by elimination of the direct vasodilatory influences of NO on vascular smooth muscle cells, part of this effects may also be due to unmasking of the pressor influence of other endogenous vasoconstrictors. Several studies have indicated that part of the vasoconstrictor response to NOS inhibitors is mediated by enhanced formation and/or vascular responsiveness to angiotensin II (Ito et al. 1991; Pucci et al. 1992; Salazar et al. 1992; Sigmon et al. 1992a; Takenaka et al. 1993; Alberola et al. 1994; Baylis et al. 1994; Evans et al. 1994). In a study using anesthetized dogs, Alberola et al. (1994) demonstrated that treatment with NOS inhibitors enhanced the renal vasoconstricting effects of angiotensin II, indicating that NO and angiotensin II interact with each other in regulating the blood flow to the kidney. However, other studies in dogs have reported contrasting findings. It was shown that blockade of the renin–angiotensin system by angiotensin-converting enzyme (ACE) inhibitors or angiotensin II receptor antagonists had minimal effects on the RBF responses to NOS inhibitors (Perrella et al. 1991; Majid and Navar 1992; Majid et al. 1993b). In another study in anesthetized dogs, Baker et al. (1995) demonstrated that the magnitude of the reduction in RBF in response to angiotensin II infusion before and during treatment with NOS inhibitors was similar, indicating that an independent additive rather than a potentiating influence exists between the responses to angiotensin II and NOS inhibitors. It was also shown that the urinary excretion rate of the NO metabolites nitrate and nitrite did not change during angiotensin II infusion, indicating that angiotensin II, per se, does not stimulate NO production in vivo (Baker et al. 1995). Also, chronic treatment of conscious dogs with ACE inhibitors failed to modify the responses to NOS inhibition (Manning et al. 1993a,b), although plasma renin activity was substantially increased during NOS inhibitor treatment in conscious dogs (Salazar et al. 1992).

The degree of angiotensin II dependency in the RBF responses to NOS inhibition seems to be related to the species as well as to the experimental conditions. Studies in anesthetized rats demonstrated that the vasoconstrictor responses to NOS inhibitors were abolished or greatly attenuated by treatment with either an angiotensin II receptor antagonist or an ACE inhibitor (Sigmon et al. 1992a,b, 1993). In contrast, studies in conscious rats indicate that angiotensin II does not contribute greatly to the renal responses to NOS inhibition (Baylis et al. 1993; Sigmon and Bierwaltes 1993; Qiu et al. 1994). In vitro studies using the blood-perfused juxtamedullary nephron preparation (Ohishi et al. 1992) demonstrated that the vasoconstrictor responses in afferent and efferent arterioles evoked by NOS inhibitors were only attenuated by treatment with either the angiotensin II receptor antagonist losartan or the ACE inhibitor enalaprilat. The degree of attenuation was similar in both afferent and efferent arterioles examined in that study. However, Ito et al. (1993) reported that there was a greater contribution of angiotensin II in mediating afferent than efferent arteriolar responses to NOS inhibition in rabbits. Thus, the literature in this area remains somewhat inconsistent in terms of the magnitude of the responses, but there is general agreement that angiotensin II and NO interact to modulate the sensitivity of the renal vasculature.

The influences of other common endogenous vasoconstrictors, such as thromboxane, epinephrine, and endothelin, on the RBF responses to NOS inhibition have also been examined. Bank et al. (1994) reported that RBF responses to chronic administration of NOS inhibitors in conscious rats were not altered by pretreatment with inhibitors of thromboxane, epinephrine, endothelin, or with angiotensin II receptor antagonists. Similarly, the renal hemodynamic effects of NOS inhibitors in the rabbit were unaffected by α_1-adrenergic receptor blockade with prazosin (Hajj-Ali and Zimmerman 1991). However, Kumagai et al. (1994) reported that intravenous infusion of L-arginine in rabbits increased RBF and decreased renal nerve activity, which were greatly attenuated by prior treatment with NOS inhibitors, indicating that a neural factor may be involved in modulating the actions of NO in the kidney.

Thus, the interactions between NO and other vasoactive agents in regulating renal blood flow are not yet clearly defined. Many factors, such as species differences, state of anesthesia, and other experimental conditions, influence these interactions between NO and other paracrine systems. Further studies are needed to examine the exact nature of the interactions among these various mechanisms in the kidney.

Concluding Comments

In this short review, we have provided a summary of recent developments related to the role of NO in the regulation of RBF. Nitric oxide helps to maintain the normally low RVR necessary to ensure adequate perfusion of the kidney. Nearly one-third to one-fourth of total RBF is dependent on an intact NO influence. Nitric oxide regulates both pre- and postglomerular vascular resistance. However, the basic autoregulatory mechanism is not dependent on an intact NO system, although NOS blockade does change the absolute RBF. In addition to its direct effect, NO exerts indirect effects on RBF and renal vascular resistance through interactions with other intrarenal paracrine factors, in particular angiotensin II. However, the exact nature of these interactions remains to be evaluated more extensively in future investigations.

Acknowledgment. The experimental work conducted by the authors was supported by grants from the National Heart Lung and Blood Institute (HL 18426, HL 51306), Louisiana Education Quality Support Funds, and National Kidney Foundation. The authors are grateful to Agnes C. Buffone for excellent secretarial assistance.

References

Abrams J. 1987. A symposium: nitroglycerin therapy—a contemporary perspective. *Am J Cardiol 60:*1H–3H.

Alberola AM, Salazar FJ, Nakamura T, & Granger JP. 1994. Interaction between angiotensin II and nitric oxide in control of renal hemodynamics in conscious dogs. *Am J Physiol (Regul Integr Comp Physiol 36) 267:*R1472–R1478.

Bachmann S, & Mundel P. 1994. Nitric oxide in the kidney: synthesis, localization, and function. *Am J Kidney Dis 24:*112–129.

Bachmann S, Bosse HM, & Mundel P. 1995. Topography of nitric oxide synthesis by localizing constitutive NO synthases in mammalian kidney. *Am J Physiol (Renal Fluid Electrolyte Physiol 37) 268:*F885–F898.

Baer PG, Navar LG, & Guyton AC. 1970. Renal autoregulation, filtration rate, and electrolyte excretion during vasodilatation. *Am J Physiol 219:*619–625.

Baker R, Majid DSA, Godfrey M, & Navar LG. 1995. Modulation of renal responses to angiotensin II (ANG II) by nitric oxide (NO). *FASEB J. 9:*A843.

Bank N, Aynedjian HS, & Khan GA. 1994. Mechanism of vasoconstriction induced by chronic inhibition of nitric oxide in rats. *Hypertension 24:*322–328.

Baumann JE, Persson PB, Ehmke H, Nafz B, & Kirchheim HR. 1992. Role of endothelium-derived relaxing factor in renal autoregulation in conscious dogs. *Am J Physiol (Renal Fluid Electrolyte Physiol 32) 263:*F208–F213.

Baylis C, Harton P, & Engels K. 1990. Endothelial derived relaxing factor controls renal hemodynamics in the normal rat kidney. *J Am Soc Nephrol 1:*875–881.

Baylis C, Mitruka B, & Deng A. 1992. Chronic blockade of nitric oxide synthesis in the rat produces systemic hypertension and glomerular damage. *J Clin Invest 90:*278–281.

Baylis C, Engels K, Samsell L, & Harton P. 1993. Renal effects of acute endothelial-derived relaxing factor blockade are not mediated by angiotensin II. *Am J Physiol (Renal Fluid Electrolyte Physiol 33) 264:*F74–F78.

Baylis C, Harvey J, & Engels K. 1994. Acute nitric oxide blockade amplifies the renal vasoconstrictor actions of angiotensin II. *J Am Soc Nephrol 5:*211–214.

Bech JN, Nielsen CB, & Pedersen EB. 1996. Effects of systemic NO synthesis inhibition on RPF, GFR, U_{Na} and vasoactive hormones in healthy humans. *Am J Physiol (Renal Fluid Electrolyte Physiol 39) 270:*F845–F851.

Beierwaltes WH, Sigmon DH, & Carretero OA. 1992. Endothelium modulates renal blood flow but not autoregulation. *Am J Physiol (Renal Fluid Electrolyte Physiol 31) 262:*F943–F949.

Biondi ML, Bolterman RJ, & Romero JC. 1992. Zonal changes of guanidine 3',5'-cyclic monophosphate related to endothelium-derived relaxing factor in dog renal medulla. *Renal Physiol Biochem 15:*16–22.

Braam B, & Koomans HA. 1995. Reabsorption of nitro-L-arginine infused into the late proximal tubule participates in modulation of TGF responsiveness. *Kidney Int 47:*1252–1257.

Brezis M, Heyman SN, Dinour D, Epstein FH, & Rosen S. 1991. Role of nitric oxide in renal medullary oxygenation: studies in isolated and intact rat kidneys. *J Clin Invest 88:*390–395.

Carmines PK, Inscho EW, & Gensure RC. 1990. Arterial pressure effects on preglomerular micro-vasculature of juxtamedullary nephrons. *Am J Physiol 258:* F94–F102.

Chen C, Mitchell KD, & Navar LG. 1992. Role of endothelium-derived nitric oxide in the renal hemodynamic response to amino acid infusion. *Am J Physiol (Regul Integrative Comp Physiol 32) 263:*R510–R516.

Cooke JP, & Dzau VJ. 1997. Nitric oxide synthase: role in the genesis of vascular disease. *Annu Rev Med 48:*489–509.

Deng A, & Baylis C. 1993. Locally produced EDRF controls preglomerular resistance and ultrafiltration coefficient. *Am J Physiol (Renal Fluid Electrolyte Physiol 33) 264:*F212–F215.

Denton KM, & Anderson WP. 1994. Intrarenal haemodynamic and glomerular respnses to inhibition of nitric oxide formation in rabbits. *J Physiol 475:*159–167.

Edwards RM, & Trizna W. 1993. Modulation of glomerular arteriolar tone by nitric oxide synthase inhibitors. *J Am Soc Nephrol 4:*1127–1132.

Elsner D, Muntze A, Kromer EP, & Riegger GAJ. 1992. Inhibition of synthesis of endothelium-derived nitric oxide in conscious dogs: hemodynamic, renal, and hormonal effects. *Am J Hypertens 5:*288–291.

Evans RG, Rankin AJ, & Anderson WP. 1994. Interactions of blockade of nitric oxide synthase and angiotensin-converting enzyme on renal function in conscious rabbits. *J Cardiovasc Pharmacol 24:*542–551.

Fenoy FJ, Ferrer P, Carbonell L, & Garcia-Salom M. 1995. Role of nitric oxide on papillary blood flow and pressure natriuresis. *Hypertension 25:*408–414.

Hajj-ali AF, & Zimmerman BG. 1991. Kinin contribution to renal vasodilator effect of captopril in rabbit. *Hypertension 17:*504–509.

Hoffend J, Cavarape A, Endlich K, & Steinhausen M. 1993. Influence of endothelium-derived relaxing factor on renal microvessels and pressure-dependent vasodilation. *Am J Physiol (Renal Fluid Electrolyte Physiol 34) 265:*F285–F292.

Ichihara A, Inscho EW, Imig JD, & Navar LG. 1998. Neuronal nitric oxide synthase modulates rat renal microvascular function. *Am J Physiol (Renal Physiol 43) 274:* F516–24.

Ignarro LJ, Lippton H, Edwards JC, Baricos WH, Hyman AL, Kadowitz PJ, & Gruetter CA. 1981. Mechanism of vascular smooth muscle relaxation by organic nitrates, nitrites, nitroprusside and nitric oxide: evidence for the involvement of *S*-nitrosothiols as active intermediates. *J Pharmacol Exp Ther 218:*739–749.

Imig JD, & Roman RJ. 1992. Nitric oxide modulates vascular tone in preglomerular arterioles. *Hypertension 19:*770–774.

Imig JD, Gebremedhin D, Harder DR, & Roman RJ. 1993. Modulation of vascular tone in renal microcirculation by erythrocytes: role of EDRF. *Am J Physiol (Heart Circ Physiol 33) 264:*H190–H195.

Ito S, & Ren Y. 1993. Evidence for the role of nitric oxide in macula densa control of glomerular hemodynamics. *J Clin Invest 92:*1093–1098.

Ito S, Johnson CS, & Carretero OA. 1991. Modulation of angiotensin II-induced vasoconstriction by endothelium-derived relaxing factor in the isolated microperfused rabbit afferent arteriole. *J Clin Invest 87:*1656–1663.

Ito S, Arima S, Ren YL, Juncos LA, & Carretero OA. 1993. Endothelium-derived relaxing factor/nitric oxide modulates angiotensin II action in the isolated microperfused rabbit afferent but not efferent arteriole. *J Clin Invest 91:*2012–2019.

King AJ, & Brenner BM. 1991. Endothelium-derived vasoactive factors and the renal vasculature. *Am J Physiol 260:*R653–R662.

Kiyomoto H, Matsuo H, Tamaki T, Aki Y, Hong H, Iwao H, & Abe Y. 1992. Effect of L-N^G-nitro-arginine, inhibitor of nitric oxide synthesis, on autoregulation of renal blood flow in dogs. *Jpn J Pharmacol 58:*147–155.

Kone BC, & Baylis C. 1997. Biosynthesis and homeostatic roles of nitric oxide in the normal kidney. *Am J Physiol (Renal Fluid Electrolyte Physiol 41) 272:*F561–F578.

Kumagai K, Suzuki H, Ichikawa M, Jimbo M, Murakami M, Ryuzaki M, & Saruta T. 1994. Nitric oxide increases renal blood flow by interacting with the sympathetic nervous system. *Hypertension 24:*220–226.

Lahera V, Salom MG, Fiksen-Olsen MJ, Raij L, & Romero JC. 1990. Effects of N^G-monomethyl-*l*-arginine and *l*-arginine on acetylcholine renal response. *Hypertension 15:*659–663.

Lahera V, Salom MG, Fiksen-Olsen MJ, & Romero JC. 1991. Mediatory role of endothelium-derived nitric oxide in renal vasodilatory and excretory effects of bradykinin. *Am J Hypertens 4:*260–262.

Lamas S, Marsden PA, Li GK, Tempst P, & Michel T. 1992. Endothelial nitric oxide synthase: molecular cloning and characterization of a distinct constitutive enzyme isoform. *Proc Natl Acad Sci USA 89:*6348–6352.

Lockhart JC, Larson TS, & Knox FG. 1994. Perfusion pressure and volume status determine the microvascular response of the rat kidney to N^G-monomethyl-L-arginine. *Circ Res 75:*829–835.

Majid DSA, & Navar LG. 1992. Suppression of blood flow autoregulation plateau during nitric oxide blockade in canine kidney. *Am J Physiol (Renal Fluid Electrolyte Physiol 31) 262:*F40–F46.

Majid DSA, & Navar LG. 1994. Blockade of distal nephron sodium transport attenuates pressure natriuresis in dogs. *Hypertension 23:*1040 – 1045.

Majid DSA, & Navar LG. 1996. Medullary blood flow responses to changes in arterial pressure in canine kidney. *Am J Physiol (Renal Fluid Electrolyte Physiol 39) 270:*F833 – F838.

Majid DSA, & Navar LG. 1997. Nitric oxide in the mediation of pressure natriuresis. *Clin Exp Pharmacol Physiol 24:*595 – 599.

Majid DSA, Williams A, Kadowitz PJ, & Navar LG. 1993a. Renal responses to intra-arterial administration of nitric oxide donor in dogs. *Hypertension 22:*535 – 541.

Majid DSA, Williams A, & Navar LG. 1993b. Inhibition of nitric oxide synthesis attenuates pressure-induced natriuretic responses in anesthetized dogs. *Am J Physiol (Renal Fluid Electrolyte Physiol 33) 264:*F79 – F87.

Majid DSA, Godfrey M, Grisham MB, & Navar LG. 1995. Relation between pressure natriuresis and urinary excretion of nitrate/nitrite in anesthetized dogs. *Hypertension 25 (Part 2):*860 – 865.

Majid DSA, Omoro SA, & Godfrey M, et al. 1996. Assessment of differential intrarenal blood flow responses using single fiber laser-Doppler flowmetry. [Abstract] *J Am Soc Nephrol 7:*1584.

Majid DSA, Godfrey M, & Navar, LG. 1997a. Pressure natriuresis and renal medullary blood flow in dogs. *Hypertension 29:*1051 – 1057.

Majid DSA, Godfrey M, & Omoro, SA. 1997b. Pressure natriuresis and autoregulation of inner medullary blood flow in canine kidney. *Hypertension 29 (Part 2):*210 – 215.

Manning RD Jr, Hu L, Mizelle HL, & Granger JP. 1993a. Role of nitric oxide in long-term angiotensin II-induced renal vasoconstriction. *Hypertension 21:*949 – 955.

Manning RD Jr, Hu L, Mizelle HL, Montani J-P, & Norton MW. 1993b. Cardiovascular responses to long-term blockade of nitric oxide synthesis. *Hypertension 22:*40 – 48.

Marshall JJ, & Kontos HA. 1990. Endothelium-derived relaxing factors—a perspective from in vivo data. *Hypertension 16:*371 – 386.

Mattson DL, Roman RJ, & Cowley AW Jr. 1992. Role of nitric oxide in renal papillary blood flow and sodium excretion. *Hypertension 19:*766 – 769.

Mattson DL, Lu S, & Cowley AW Jr. 1997. Role of nitric oxide in the control of the renal medullary circulation. *Clin Exp Pharmacol Physiol 24:* 587 – 590.

Mattson DL, Lu S, Roman RJ, & Cowley AW Jr. 1993. Relationship between renal perfusion pressure and blood flow in different regions of the kidney. *Am J Physiol (Regul Integrative Comp Physiol 33) 264:*R578 – R583.

Moncada S, Palmer RMJ, & Higgs EA. 1991. Nitric oxide: physiology, pathophysiology, and pharmacology. *Pharmacol Rev 43:*109 – 142.

Navar LG, Inscho EW, Majid DSA, Imig JD, Harrison-Bernard LM, & Mitchell KD. 1996. Paracrine regulation of the renal microcirculation. *Physiol Rev 76:*425 – 536.

Ohishi K, Carmines PK, Inscho EW, & Navar LG. 1992. EDRF-angiotensin II interactions in rat juxtamedullary afferent and efferent arterioles. *Am J Physiol (Renal Fluid Electrolyte Physiol 32) 263:*F900 – F906.

Perrella MA, Hildebrand FL Jr, Margulies KB, & Burnett JC Jr. 1991. Endothelium-derived relaxing factor in regulation of basal cardiopulmonary and renal function. *Am J Physiol (Regul Integrative Comp Physiol 30) 261:*R323 – R328.

Pucci ML, Lin L, & Nasjletti A. 1992. Pressor and renal vasoconstrictor effects of N^G-nitro-L-arginine as affected by blockade of pressor mechanisms mediated by the sympathetic nervous system, angiotensin, prostanoids and vasopressin. *J Pharmacol Exp Ther 261:*240 – 245.

Qiu C, Engels K, & Baylis C. 1994. Angiotensin II and α-adrenergic tone in chronic nitric oxide blockade-induced hypertension. *Am J Physiol (Renal Fluid Electrolyte Physiol 35) 266:*R1470–R1476.

Roman RJ, Cowley AW Jr, Garcia-Estan J, & Lombard JH. 1988. Pressure-diuresis in volume-expanded rats: cortical and medullary hemodynamics. *Hypertension 12:* 168–176.

Salazar FJ, Pinilla JM, Lopez F, Romero JC, & Quesada T. 1992. Renal effects of prolonged synthesis inhibition of endothelium-derived nitric oxide. *Hypertension 20:*113–117.

Salom MG, Lahera V, Miranda-Guardiola F, & Romero JC. 1992. Blockade of pressure natriuresis induced by inhibition of renal synthesis of nitric oxide in dogs. *Am J Physiol (Renal Fluid Electrolyte Physiol 31) 262:*F718–F722.

Shultz PJ, Archer SL, & Rosenberg ME. 1994. Inducible nitric oxide synthase mRNA and activity in glomerular mesangial cells. *Kidney Int 46:*683–689.

Sigmon DH, & Beierwaltes WH. 1993. Angiotensin II: nitric oxide interaction and the distribution of blood flow. *Am J Physiol (Regul Integrative Comp Physiol 34) 265:*R1276–R1283.

Sigmon DH, Carretero OA, & Beierwaltes WH. 1992a. Angiotensin dependence of endothelium-mediated renal hemodynamics. *Hypertension 20:*643–650.

Sigmon DH, Carretero OA, & Beierwaltes WH. 1992b. Plasma renin activity and the renal response to nitric oxide synthesis inhibition. *J Am Soc Nephrol 3:*1288–1294.

Sigmon DH, Carretero OA, & Beierwaltes WH. 1993. Renal versus femoral hemodynamic response to endothelium-derived relaxing factor synthesis inhibition. *J Vasc Res 30:*218–223.

Sonntag M, Deussen A, & Schrader J. 1992. Role of nitric oxide in local blood flow control in the anaesthetized dog. *Pflügers Arch 420:*194–199.

Star RA. 1997. Intrarenal localization of nitric oxide synthase isoforms and soluble guanylyl cyclase. *Clin Exp Pharmacol Physiol 24:*607–610.

Takenaka T, Mitchell KD, & Navar LG. 1993. Contribution of angiotensin II to renal hemodynamic and excretory responses to nitric oxide synthesis inhibition in the rat. *J Am Soc Nephrol 4:*1046–1053.

Thorup C, & Persson AEG. 1994. Inhibition of locally produced nitric oxide resets tubuloglomerular feedback mechanism. *Am J Physiol (Renal Fluid Electrolyte Physiol 36) 267:*F606–F611.

Ujiie K, Drewett JG, Yuen PST, & Star RA. 1993. Differential expression of mRNA for guanylyl cyclase-linked endothelium-derived relaxing factor receptor subunits in rat kidney. *J Clin Invest 91:*730–734.

Vallon V, & Thomson S. 1995. Inhibition of local nitric oxide synthase increases homeostatic efficiency of tubuloglomerular feedback. *Am J Physiol (Renal Fluid Electrolyte Physiol 38) 269:*F892–F899.

Wilcox CS, & Welcb WJ. 1996. TGF and nitric oxide: effects of salt intake and salt-sensitive hypertension. *Kidney Int 49 (suppl 55):*S-9–S-13.

Wilcox CS, Welch WJ, Murad F, Gross SS, Taylor G, Levi R, & Schmidt HHHW. 1992. Nitric oxide synthase in macula densa regulates glomerular capillary pressure. *Proc Natl Acad Sci USA 89:*11993–11997.

Xie Q-W, Cho HJ, Calaycay J, Mumford RA, Swiderek KM, Lee TD, Ding A, Troso T, & Nathan C, 1992. Cloning and characterization of inducible nitric oxide synthase from mouse macrophages. *Science 256:*225–228.

Yukimura T, Yamashita Y, Miura K, Okumura M, Yamanaka S, & Yamamoto K. 1992. Renal effects of the nitric oxide synthase inhibitor, L-N^G-nitroarginine, in dogs. *Am J Hypertens 5:*484–487.

Zatz R, & De Nucci G. 1991. Effects of acute nitric oxide inhibition on rat glomerular microcirculation. *Am J Physiol (Renal Fluid Electrolyte Physiol 30) 261:*F360–F363.

Zou A-P, & Cowley AW Jr. 1997. Nitric oxide in renal cortex and medulla. An in vivo microdialysis study. *Hypertension 29 (part 2):*194–198.

20
Role of Nitric Oxide in the Regulation of Renal Function in Conscious Animals

R. Davis Manning, Jr., Lufei Hu, and Dunyong Y. Tan

The seminal discovery of endothelium-derived relaxing factor (EDRF) by Furchgott and Zawadzki in 1980 (11) has led to a myriad of studies on endothelial factors. Even though there may be several factors released from the endothelium that cause vasodilation, the EDRF discovered by Furchgott was later found to be nitric oxide (NO). Fifteen years ago, NO, an environmental pollutant found in cigarette smoke and smog, was suspected to be a carcinogen, to destroy the ozone layer in the atmosphere, and to produce acid rain (Star 1993). Over the last 15 years, research in diverse fields such as immunology, toxicology, and cardiovascular physiology has shown that this highly active biological molecule has mediated physiological actions that range from blood pressure regulation to antimicrobial defense to intracellular signal transduction.

Nitric oxide is a simple molecule made from the substrate, L-arginine, in the presence of several cofactors and the enzyme nitric oxide synthase (NOS). The reaction forms one molecule of NO and one molecule of citrulline for every molecule of L-arginine. Several isoforms of NOS exist in the body, including the constitutive forms endothelial NOS (eNOS) and neural NOS (nNOS), and three isoforms of inducible NOS (iNOS) found in vascular smooth muscle, macrophages and hepatocytes. Of particular significance to this chapter, NO has important effects on renal hemodynamics, sodium excretion, and the control of arterial pressure.

Numerous studies have been performed on the acute and chronic roles of NO in the regulation of renal function. Also, an increasing amount of evidence supports the theory that changes in renal function, through its pressure–natriuresis mechanism, are the main controller of the long-term level of arterial pressure (Guyton et al. 1972). Therefore, this chapter will address not only the direct effect of NO on renal function, but also the effects of NO on long-term arterial pressure control through the renal pressure–natriuresis mechanism.

Distribution of NOS in the Kidney

Nitric oxide can affect kidney function by local production of NO by the different NOS isoforms or by active NO entering the kidney through the arterial blood. The relative contributions of these sources of NO in the regulation of renal function are not known, but evidence that local production of NO in the kidney is highly important, continues to accumulate. In addition, the half-life of NO in the blood is probably around 6 seconds, which limits the usefulness of extrarenal NO in the control of renal function.

Studies using reverse transcriptase-polymerase chain reaction analyses have found eNOS in the arcuate and interlobular arteries, afferent arterioles, and glomerulus (Ujiie et al. 1994). In addition recent studies using immunohistochemistry have shown that eNOS is found in particularly high concentrations in the inner and outer medulla of Sprague-Dawley rats (Mattson and Higgins 1996). Therefore, nNOS, iNOS, and eNOS have been located throughout the kidney and may participate in the production of NO, which can significantly alter renal function.

Renal Hemodynamic Role of NO

Several investigators, in studies primarily in the rat, have shown that NO synthesis inhibition results in both short-term and long-term renal vasoconstriction. The first studies on the renal hemodynamic effects of blocking NO synthesis were performed on a short-term basis. Gardiner et al. (1990a,b) found that an intravenous bolus of the NO inhibitors N^G-monomethyl-L-arginine (L-NMMA) or N^G-nitro-L-arginine methylester (L-NAME) produced renal vasoconstriction in conscious Long-Evans rats in experiments that lasted between 5 and 60 minutes. In other short-term experiments in conscious (Baylis et al. 1990) or anesthetized (Lahera et al. 1991) rats, intravenous bolus administration of NO synthesis inhibitors caused hypertension and decreases in both renal plasma flow and glomerular filtration rate (GFR).

The first study on the effects of NO synthesis inhibition on renal hemodynamics over several hours showed that the arterial pressure increased and renal blood flow decreased over a 9 hour period in Brattleboro rats drinking water containing L-NAME (Gardiner et al. 1984). Also, rats on oral L-NAME for 2 weeks to 2 months had large decreases in GFR and renal blood flow (Baylis et al. 1992; Ribeiro et al. 1992; Recklelhoff and Manning 1993). In addition, intravenous infusion of a very low dose of L-NAME (0.05 μg/kg/min) for 3 days in dogs caused no change in arterial pressure but a decrease in GFR (Salazar et al. 1992); this dose of L-NAME probably only mildly decreased the systemic production of NO.

Studies in our laboratory confirmed that NO synthesis inhibition in dogs causes sustained renal vasoconstriction and hypertension (Manning and Hu 1994). Continuous intravenous infusion of 10 μg/kg/min of L-NAME for 11 days caused significant hypertension, which was partially reversed by L-arginine infusion and was associated with a decrease in urinary sodium excretion on the first day of NO synthesis inhibition (Manning et al. 1993), a transient decrease in GFR for several days, and a sustained decrease in renal plasma flow (Manning and Hu 1994; Manning et al. 1997). However, during normal sodium intake, urinary sodium excretion and fractional excretion of sodium returned to normal by the second day of L-NAME infusion, despite a significant increase in arterial pressure (Manning et al. 1993). The reason sodium excretion was unchanged at this time was probably the occurrence of a rightward shift along the arterial pressure axis in the relationship between urinary sodium excretion and arterial pressure (Manning and Hu 1994; Manning et al. 1997). Thus, a shift in this long-term pressure–natriuresis relationship could be responsible for the long-term increase in arterial pressure that occurs during NO synthesis inhibition in dogs. The exact intrarenal mechanism responsible for this change in sodium excretory ability cannot be discerned from our data, but other researchers have shown that NO causes a decrease in sodium reabsorption in several nephron segments, including the cortical collecting duct (Stoos et al. 1992).

We have also found that NO synthesis inhibition in the rat causes long-term hypertension. Studies in our laboratory showed that continuous intravenous infusion of L-NAME into Sprague-Dawley rats at 2 µg/kg/min caused sustained hypertension throughout a 9-day treatment (Hu et al. 1994). This hypertension was blocked by intravenous infusion of L-arginine at a dose that caused no change in arterial pressure. These effects of L-arginine on L-NAME hypertension are shown in Figure 20.1. As seen in the top panel, L-arginine was infused alone on day 1, and during days 2 to 4 L-NAME infusion was added, but no change in arterial pressure occurred. Then, when L-arginine infusion was stopped, L-NAME caused significant hypertension. The bottom panel of this figure shows that D-arginine did not prevent L-NAME hypertension. These data give evidence that L-NAME acts by inhibiting the L-arginine–NO system.

We have also studied the role of NO in the control of regional blood flow in the rat. Chronic L-NAME treatment in young Sprague-Dawley rats resulted in widespread vasoconstriction and a decrease in regional blood flow over a 4-week L-NAME treatment (Huang et al. 1997). Renal blood flow decreased from a control value of 5.5 ± 0.4 mm/min/g to 3.0 ± 0.3 ml/min/g ($P < 0.01$), which represents a 46% decrease in renal blood flow. Cardiac output decreased 31% and arterial pressure increased 47% in this study. These results indicate that renal blood flow is highly dependent on NO—more so than the cardiac output—and is a long-term modulator of arterial pressure.

In a study on aging in rats during oral intake of L-NAME for 2 weeks, the renal vascular resistance increased fivefold in older rats (21–24 months old) but only twofold in younger rats (Recklehoff and Manning 1993). Both afferent and efferent resistance increased in all rats, and the glomerular ultrafiltration coefficient increased in all rats, but more in the older rats. These data suggest that NO plays a progressively more important role in controlling renal function with advancing age.

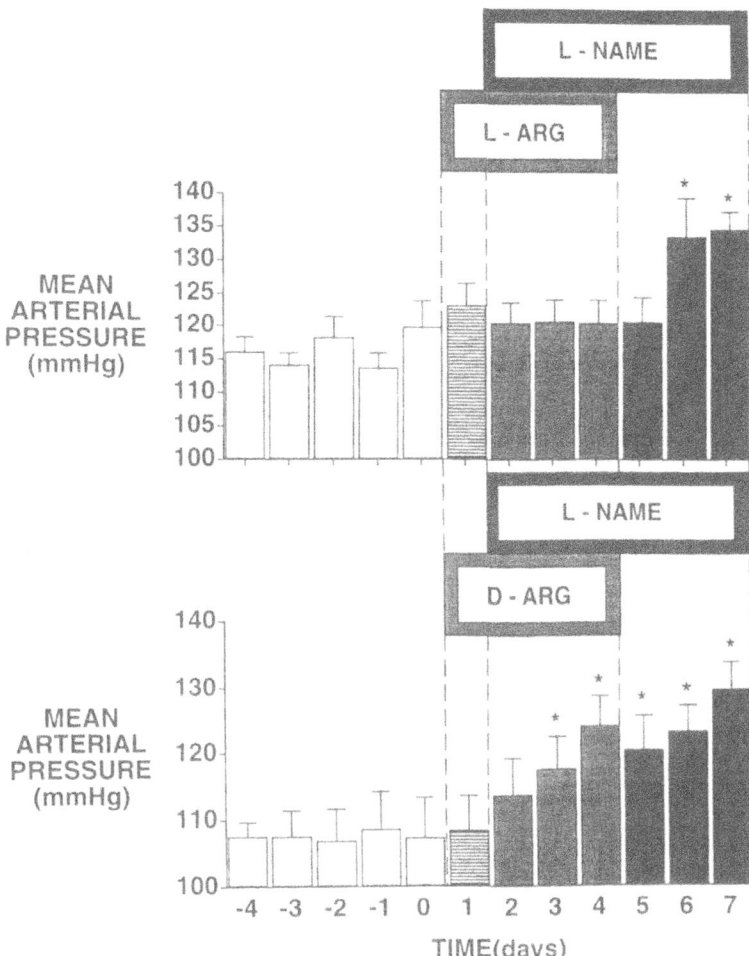

FIGURE 20.1. Bar graphs showing effects of continuous intravenous infusion of 10 μg/kg/min of N^G-nitro-L-arginine (L-NAME) and L-arginine HCl (L-ARG) or D-arginine HCl (D-ARG) on mean arterial pressure. *$P < 0.05$ compared with average control value. Graphs redrawn from Hu et al. (1994). Copyright 1994 by Lippincott Williams & Wilkens. Adapted with permission.

Interaction of NO and Vasoconstrictors

A number of acute studies have shown that local release of NO opposes the vaso-constrictor actions of several neurohumoral systems (Gardiner et al. 1991; Ito et al. 1991; Conrad and Whittemore 1992; Sakuma et al. 1992). Therefore, it is conceivable that some of these pressor and renal vasoconstrictor effects of NO synthesis inhibition could be mediated by other neurohumoral factors.

Short-term Interactions of NO and Vasoconstrictors

Application of L-NAME to isolated afferent arterioles has been shown to amplify the acute vasoconstrictor effects of angiotensin II (Ito et al. 1991). This suggests that the release of NO from the afferent arteriole opposes the vasoconstrictor actions of angiotensin II. Other studies on the interaction of NO and angiotensin II have indicated that the short-term interactions between angiotensin II and NO are significant. Angiotensin II was infused for 2 hours into the renal artery in conscious dogs and dogs with or without L-NAME infusion (Alberola et al. 1994). At a dose of 0.5 μg/kg/min, angiotensin II decreased renal plasma flow by 19%, while having no effect on GFR in control dogs. In contrast, angiotensin II decreased renal plasma flow by 54% and GFR by 40%, and increased renal vascular resistance by 125% in the presence of intrarenal NO synthesis blockade (Alberola et al. 1994). Therefore, short-term administration of NO synthesis inhibitors greatly magnified the vasoconstrictor effects of angiotensin II on renal hemodynamics.

We also studied the roles of the sympathetic nervous system, angiotensin II and arginine vasopressin, and the cardiovascular–renal responses to NO synthesis inhibition in conscious dogs (Manning et al. 1994). Dogs were subjected to NO synthesis inhibition with 10 μg/kg/min of L-NAME for 140 minutes in the conscious state, and the same dogs, after a 1-week recovery, were pretreated for 2 days with α_1 blockade, $\alpha_1 + \beta$ sympathetic blockade, angiotensin II-AT$_1$ receptor blockade, or AVP-V$_1$ blockade. L-NAME infusion was then repeated. L-NAME infusion in the control group increased mean arterial pressure 16% and renal vascular resistance 71%; renal blood flow, GFR, urine flow, and urinary sodium excretion were decreased by 33%, 16%, 61% and 64%, respectively. The decrement in renal blood flow and GFR during L-NAME administration was unaffected by any of the neurohumoral blockers. During either angiotensin II-AT$_1$ or V$_1$ blockade, L-NAME resulted in only a very small increase in arterial pressure and attenuation of the renal vascular resistance response. In addition, the decrease in urinary sodium excretion was attenuated in the $\alpha_1 + \beta$ blockade group. In conclusion, short-term renal and cardiovascular effects of NO synthesis inhibition in conscious dogs may be mediated, in order of importance, by arginine vasopressin, angiotensin II, and the sympathetic nervous system (Manning et al. 1994).

Another study has shown that the effects of intrarenal artery infusion of norepinephrine on GFR, renal blood flow, and renal vascular resistance are greatly enhanced during short-term administration of NO synthesis inhibitors (Granger et al. 1993). This suggests that NO normally opposes the acute renal vasoconstrictor effects of norepinephrine.

Other short-term studies showed that the renal nerves mediate the acute effects of NO synthesis inhibition on renal hemodynamics. In preliminary studies, L-NAME was infused intravenously for 3 hours, and renal hemodynamic effects were determined in either denervated or innervated kidneys in the same dog (Granger et al. 1992; Reinhart et al. 1997). L-NAME caused large decreases in renal plasma flow and GFR in the innervated kidney, but no changes in GFR and smaller changes in renal plasma flow in the denervated kidney (Granger et al. 1992). Also, as discussed above, we showed that the acute antinatriuretic effects of L-NAME in dogs are mediated by the sympathetic nervous system (Manning et al. 1994). However, in another study, the presence of renal nerves caused no acute changes in the renal hemodynamic or excretory responses to L-NAME infusion (Reinhart et al. 1997).

Long-term Interactions of NO and Vasoconstrictors

Even though short-term interactions between NO and several vasoconstrictors are important in the regulation of renal hemodynamics, only a few studies on the long-term interactions have been performed. We recently studied the role of long-term angiotensin II and NO interactions in dogs. Studies were conducted in 16 conscious dogs that received angiotensin-AT_1 receptor inhibition with the Merck compound L158809 or vehicle for 12 days. During the last 6 days of this infusion, NO synthesis was inhibited by infusing 10 µg/kg/min of L-NAME (Manning and Hu 1997). By itself, L-NAME infusion caused significant increases in arterial pressure and renal vascular resistance and significant decreases in GFR and renal plasma flow. Addition of AT_1 inhibition to the L-NAME infusion caused no significant difference in the responses of arterial pressure, renal vascular resistance, GFR, and renal plasma flow (Manning and Hu 1997). Therefore, the arterial pressure and renal hemodynamic responses to NO synthesis inhibition were not dependent on having intact angiotensin II-AT_1 receptors, suggesting that NO and angiotensin II have little long-term interaction in the dog.

In contrast, other studies have shown that the long-term vasopressor response to NO synthesis inhibition in the rat is dependent on angiotensin II. Oral administration of L-NAME for 4–6 weeks caused hypertension in rats that could be partially prevented with administration of the angiotensin II-AT_1 receptor antagonist losartan (Ribeiro et al. 1992). Plasma renin activity increased in these rats, possibly due to renal damage, and losartan reversed the decrease in GFR and partially reversed the decrease in renal blood flow due to L-NAME administration (Ribeiro et al. 1992). In another study, administration of losartan by gavage for 25 days totally prevented L-NAME-induced hypertension (Jover et al. 1993). In addition, L-NAME administration for 4 weeks in the presence of the AT_1 blockers losartan or Abbott A-81988 caused no significant increase in arterial pressure (Pollock et al. 1993). Therefore, in the rat but not the dog, angiotensin II-AT_1 receptor blockade causes attenuation of long-term L-NAME-induced hypertension.

The reason why long-term hypertension induced by L-NAME in the dog was not dependent on having intact AT_1 receptors is not clear, but several possibilities exist. There is a possibility that neither renin release nor the cardiovascular-renal sensitivity to angiotensin II increase in the dog during systemic administration of L-NAME, but the rat may respond differently. Neither acute (Manning et al. 1994) nor chronic (Manning et al. 1993) administration of L-NAME to conscious dogs in our laboratory caused an increase in renin activity, but another study showed that a nonpressor dose of L-NAME in dogs for 3 days caused a small increase in plasma renin activity (Salazar et al. 1992). On the other hand, plasma renin activity in the rat increased (Ribeiro et al. 1992; Hu et al. 1994), did not change (Jover et al. 1993), or decreased (Pollock et al. 1993) in different studies. Yet, in some studies in rats, the angiotensin II pressor sensitivity may have increased, because renin activity decreased, and angiotensin II blockers prevented the L-NAME-induced hypertension (Pollock et al. 1993). Therefore, the rat may show increased arterial pressure sensitivity to angiotensin II during L-NAME hypertension, whereas the dog does not.

Even though the short-term renal hemodynamic and sodium excretory effects of NO synthesis inhibition may be partially mediated by the sympathetic nervous system (Granger et al. 1992; Manning et al. 1994), the long-term effects of L-NAME have not been shown to be dependent on the renal nerves. During infusion of L-NAME at either 10 or 25 μg/kg/min for 5–14 days, the GFR, renal plasma flow, and renal excretory responses were not dependent on intact renal nerves (Granger et al. 1996; Reinhart et al. 1997). Therefore, the short-term but not the long-term renal hemodynamic excretory effects of NO synthesis inhibition may be mediated by the renal nerves.

Decreased NO Production May Cause Long-term Hypertension

Several studies have suggested that deficits in NO production may lead to the development of hypertension. Indeed, Luscher et al. (1987a,b) found that endothelium-dependent relaxations in response to acetylcholine, adenosine diphosphate, and thrombin were significantly decreased in the aortic rings from the Dahl salt-sensitive rat on a high-salt diet. They also found that pretreatment of prehypertensive salt-sensitive rats with antihypertensive therapy for 8 weeks prevented the blunted endothelium-dependent relaxations during a high-salt diet (Luscher et al. 1987b). Also, Panza et al. (1990) reported that endothelium-dependent vasodilatory responses to acetylcholine were decreased in hypertensive patients compared with those of normotensive subjects.

Role of NO and NOS Isoforms in Salt-Sensitive Hypertension

During increased sodium intake, the kidneys of some individuals retain significant amounts of sodium, thus increasing arterial pressure. These individuals have been classified as "salt-sensitive" as opposed to "salt-resistant" individuals, who retain very little sodium during high salt intake and whose arterial pressure therefore does not increase. One mechanism that could significantly increase the sodium sensitivity of arterial pressure is a decrease in NO production. Indeed, a recent study in humans showed that agonist-induced release of NO was lower in salt-sensitive essential hypertensives than in salt-resistant essential hypertensives

(Ghiadoni et al. 1997). Since administration of nonpressor doses of L-NAME caused decreases in urinary sodium excretion in dogs (Salazar et al. 1993), NO must aid the kidney in the excretion of sodium, and a decrease in NO should cause sodium retention.

Chen and Sanders (1991, 1993a) have shown that hypertension in the salt-sensitive Dahl rat can be completely prevented by parenteral or oral administration of L-arginine but not D-arginine. Their data also suggested that an increase in dietary sodium chloride increased NO production in salt-resistant rats but not in salt-sensitive rats by using L-NMMA as a probe to estimate NO production. Patel et al. (1993) reported that long-term administration of L-arginine prevented the changes in the pressure–natriuresis relationship as measured in anesthetized Dahl salt-sensitive rats. These studies support the concept that salt-sensitive hypertension may be caused by decreased NO production. However, NO production in the above studies was estimated using L-NMMA as a probe (Chen and Sanders 1991) or from the effects on the pressure-natriuresis relationship in anesthetized rats over a short period of time (Patel et al. 1993).

Studies in our laboratory and in others have shown that increases in sodium intake normally increase urinary nitrate and nitrite excretion (Manning et al. 1997; Tolins and Shultz 1994), an index of NO production, and enhance the renal hemodynamic response to NO synthesis inhibition (Deng et al. 1994; Tolins and Shultz 1994). However, this increase in NO production during increased sodium intake is blunted in the salt-sensitive rat (Chen and Sanders 1991; Hu and Manning 1995), and therefore, decrements in NO in these rats could lead to sodium retention and thus salt-sensitive hypertension.

In our laboratory, we studied the effects of intravenous L-arginine administration on the development of hypertension in Dahl salt-sensitive and salt-resistant rats (Hu and Manning 1995). Nitric oxide production, as measured by urinary nitrate and nitrite oxidation, was lower in salt-sensitive than in salt-resistant rats (Hu and Manning 1995). The top panel of Figure 20.2 shows the responses of mean arterial pressure of salt-resistant and -sensitive rats to changes in sodium intake without L-arginine infusion; the salt-sensitive rats but not the salt-resistant rats became hypertensive on high sodium intake. The bottom panel shows that intravenous infusion of L-arginine prevented salt-induced hypertension in the salt-sensitive rat but had little effect on the arterial pressure of the salt-resistant rat. In the same study, intravenous L-arginine infusion increased the urinary nitrate and nitrite excretion of salt-sensitive rats during the high-sodium-intake period to values very close to that of the salt-resistant rat (Hu and Manning 1995). The relationship between arterial pressure and the sodium excretory responses of salt-resistant and -sensitive rats with or without L-arginine infusion are shown in Figure 20.3. Sodium sensitivity is considered to be the slope of this relationship between normal and high sodium excretion. L-Arginine prevented the increase in sodium sensitivity of arterial pressure in salt-sensitive rats, and this sensitivity, during L-arginine infusion, was not different from that of the salt-resistant rats, as seen in Figure 20.3. These data suggest that the increase in NO production played an important role in preventing salt-loading hypertension in the salt-sensitive rat.

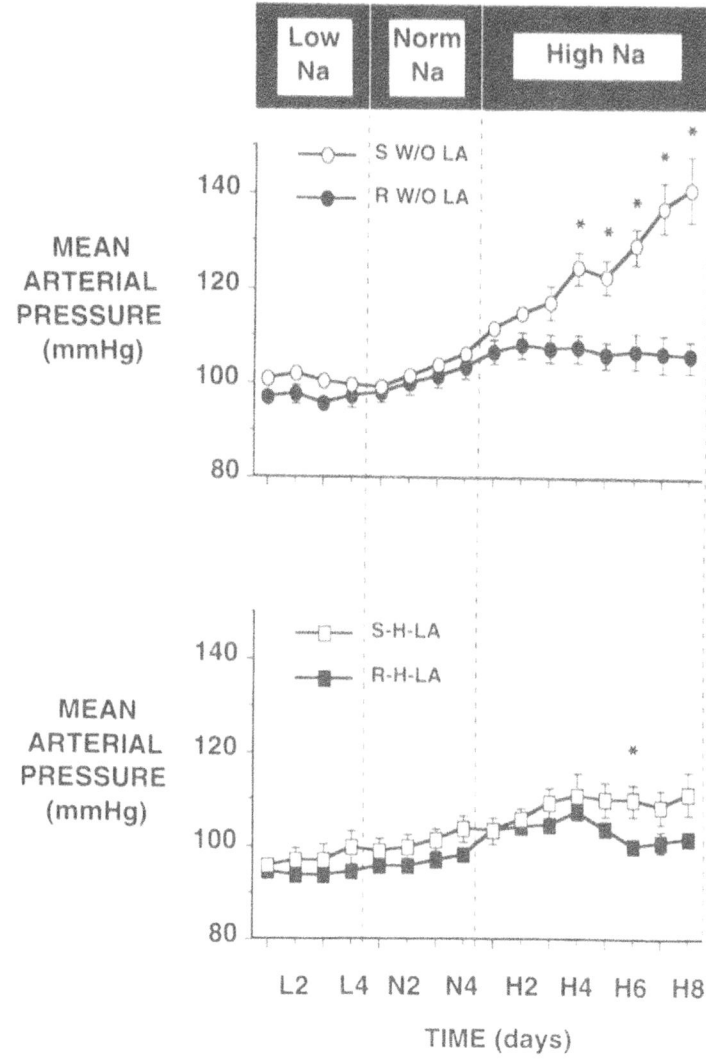

FIGURE 20.2. Line graphs showing responses of mean arterial pressure to changes in sodium intake and L-arginine (LA) in Dahl salt-resistant (R) and salt-sensitive (S) rats. W/O LA refers to the control group without L-arginine infusion, and H-LA refers to groups that received intravenous infusion of 4 mg/kg/min L-arginine beginning 2 days before the experiment began. L2–L4 refers to low-sodium days 2–4; N2–N4 refers to normal sodium days 2–4; and H2-H8 refers to high sodium days 2–8. *$P < 0.01$ when salt-sensitive rats are compared with salt-resistant rats during the same time period. Graphs redrawn from Hu and Manning (1995). Copyright 1995 by the American Physiological Society. Adapted with permission.

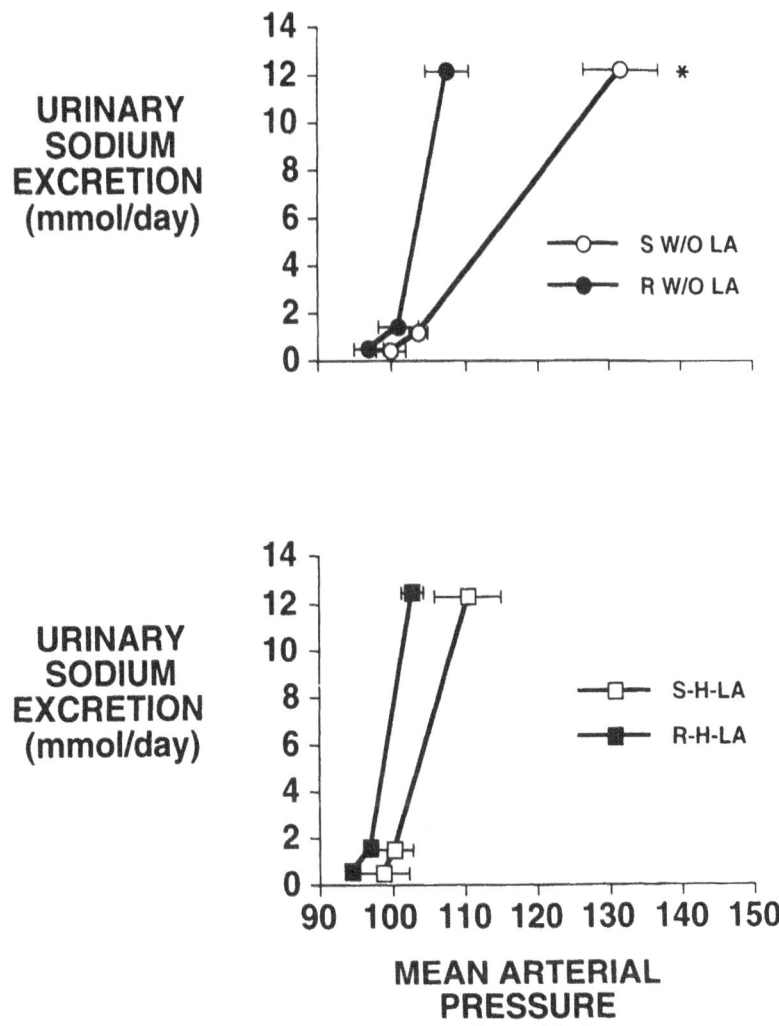

FIGURE 20.3. Line graphs showing long-term responses in the relationship between urinary sodium excretion and arterial pressure in Dahl rats subjected to changes in sodium intake and L-arginine. Refer to Figure 20.2 for abbreviations. *$P < 0.01$ when salt-sensitive rats are compared with salt-resistant rats during the same time period. Graphs redrawn from Hu and Manning (1995). Copyright 1995 by the American Physiological Society. Adapted with permission.

However, urinary nitrate and nitrite operation is an index of whole-body production of NO and does not indicate how much NO was produced in the kidney or how much NO is produced by the different isoforms of NOS.

The mechanisms by which NO production is stimulated with increased sodium intake are not clear, since few studies have been performed on the changes in NOS isoforms in the kidney during changes in dietary sodium intake. Increasing sodium intake in Sprague-Dowley rats caused large increases in eNOS, iNOS, and nNOS protein, particularly in the renal inner medulla (Mattson and Higgins 1996). During increased sodium intake in Sprague-Dawley rats, eNOS and iNOS mRNA were unchanged in the cortex, and nNOS mRNA decreased in the cortex but was unchanged in the inner medulla (Singh et al. 1996). The eNOS and iNOS mRNA levels in the inner medulla were not reported in this study. During high sodium intake, renal nNOS activity decreased, and eNOS and iNOS activities did not change in the Dahl Iwaí salt-sensitive rat compared with activities in the salt-resistant rat (Ikeda et al. 1995). However, this study analyzed NOS activity in the whole kidney, which is very similar to measuring activity in the cortex. Therefore, changes in the medulla may have been overlooked within this approach.

Role of Renal Medullary NO in Salt-Sensitive Hypertension

Several in vitro studies that have described the anatomic distribution of NOS iso-forms have shown that the renal medullary tissue has a greater capacity to synthesize NO than the renal cortex (Terada et al. 1992; Wilcox et al. 1992; Tojo et al. 1994; Bachmann et al. 1995; Zou and Cowley 1997). Western blotting has shown that nNOS, iNOS, and eNOS protein are all present in the rat renal medulla in greater amounts than in the cortex (Mattson and Higgins 1996). Also, in vivo microdialysis techniques showed that the renal medulla produces more NO than the cortex (Zou and Cowley 1997). During infusion of L-NAME into the renal medullary interstitium, renal medullary blood flow selectively decreased without affecting cortical flow, and acetylcholine infusion into the medulla prior to the L-NAME infusion increased medullary flow without affecting cortical flow (Mattson et al. 1994). During L-NAME infusion into the medulla, acetylcholine infusion did not significantly change either medullary or cortical flow (Mattson et al. 1994).

Since the blood pressure response to nonselective NOS inhibition is sodium-dependent (Deng et al. 1994; Tolins and Shultz 1994), the up-regulation of NOS in the renal medulla may be an important adaptation to elevations in dietary sodium intake. In fact, in the Sprague-Dawley rat, increased sodium intake markedly increased the concentrations of both outer and inner medullary eNOS, nNOS, and iNOS protein. Since the renal medulla has an abundance of all iso-forms of NOS that modulate during changes in sodium intake, these data suggest an important role of renal medullary NO in the long-term control of sodium homeostasis and arterial pressure.

Role of iNOS in Salt-Sensitive Hypertension

Both biochemical and functional studies suggest that renal iNOS may play an im-portant role in salt-sensitive hypertension. Messenger RNA for iNOS has been found in renal tubular and vascular segments. The highest level of iNOS mRNA has been found in the medullary thick ascending limb and the inner medullary collecting duct (Ahn et al. 1994; Mohaupt et al. 1994; Stoos et al. 1995).

The physiological function of iNOS in the kidney has been studied by several investigators. Renal proximal tubules reabsorb ~ 65% of the filtered sodium load through a process driven by activity of basolateral Na^+-K^+-ATPase, and hormonal factors may alter this ATPase activity (Rodriquez-Sargent et al. 1981). In fact, induction of iNOS in proximal tubular cells in culture decreased Na^+-K^+-ATPase activity, and this response was blocked by NO synthesis inhibition or removal of L-arginine from the medium (Guzman et al. 1995). Decreases in iNOS activity in the kidney, such as is found in the Dahl salt-sensitive rat, may increase proximal tubular Na^+-K^+-ATPase activity and thus play a major role in initiating salt-sensitive hypertension. In addition, as discussed below changes in renal medullary, NOS actively may be important in the control of urinary sodium excretion.

Chen and Sanders (1991) and our laboratory (Hu and Manning 1995) showed that L-arginine administration in salt-sensitive rats prevented the hypertensive effects of high sodium chloride intake. Our laboratory also showed that during high sodium intake, the salt-sensitive rat has a lower NO production than the salt-resistant rat, as determined by decreased urinary nitrate and nitrite excretion (Hu and Manning 1995). However, nitrate and nitrite in the urine originate from both renal and extrarenal sources. Therefore, urinary nitrate and nitrite excretion cannot be used to measure NO produced in the kidney, and other techniques must be used to determine renal NO production, including intravenous and intramedullary infusion of selective NOS inhibitors and measurement of the arterial pressure and renal hemodynamic responses.

The blood pressure-lowering effect of L-arginine in the salt-sensitive rat on high sodium chloride intake is prevented by infusion of dexamethasone, which putatively inhibits iNOS activity (Chen and Sanders 1993b). Mattson and Higgins (1996) recently showed that medullary iNOS level increased markedly during high sodium intake in Sprague-Dawley rats. In another study, unilaterally nephrectomized Sprague-Dawley rats maintained on a high-sodium diet had aminoguanidine, a selective inhibitor of iNOS (Corbett and McDaniel 1996, intravenously) infused for 6 days, and mean arterial pressure increased 11 mmHg; NOS activity in the renal medulla decreased 49%, but cerebellar NOS (presumably nNOS) was not affected (Mattson and Higgins 1996). Therefore, decreases in iNOS activity in the kidney may play a significant role in salt-sensitive hypertension.

Role of nNOS in Salt-Sensitive Hypertension

Several techniques, including immunohistochemistry, reverse transcriptase coupled to polymerase chain reaction of microdissected renal vessels and tubules, and in situ hybridization, have demonstrated the presence of nNOS protein and mRNA in the inner and outer medullary collecting ducts, glomerulus, macula densa, vasa recta, arcuate artery, and renal nerves (Terada et al. 1992; Tojo et al.

1994; Bachmann et al. 1995; Fukuto and Chaudhuri 1995). Functionally, nNOS blunts the tubuloglemerular feedback response of the afferent arteriole (Beierwaltes 1995, 1997; Wilcox and Welch 1996) and mediates the macula densa control of renin secretion (Singh et al. 1996; Beierwaltes, 1997). However, the effect of sodium intake on nNOS synthesis and expression in different renal parenchymal zones is controversial. Messenger RNA for nNOS may increase in the renal cortex during low sodium intake (Singh et al. 1996). However, other investigators showed that nNOS protein increased markedly in the inner medulla during increased sodium intake (Mattson and Higgins 1996). In support of the latter finding, renal medullary infusion of 7-nitroindazole, a specific inhibitor of nNOS, decreased medullary nNOS activity 37% and increased arterial pressure over a 6-day period in Sprague-Dawley rats on high sodium intake (Mattson and Bellehuneur 1996). This suggests that nNOS may enhance renal sodium excretion and thus help to prevent salt-loading hypertension.

The Role of NO in the Regulation of Renal Segmental Resistances, K_f, and Macula Densa Feedback

Oral administration of L-NAME for 2–8 weeks in Sprague Dawley rats caused increases in mean arterial pressure, decreases in renal plasma flow, increases in afferent and efferent resistance, and a decrease in the glomerular capillary ultrafiltration coefficient (K_f) (Baylis et al. 1992; Recklehoff and Manning 1993). Also, glomerular capillary hydraulic pressure increased due to the increase in mean arterial pressure and the increase in renal efferent resistance (Baylis et al. 1992; Recklehoff and Manning 1993). Since NO relaxes the mesangial cells (Raij and Baylis 1995), NO synthesis inhibition may have decreased K_f by causing mesangial constriction. Because of the sustained glomerular blood pressure in this model of hypertension, these rats exhibit moderate proteinuria and histological evidence of structural damage in the kidney, with mild increases in focal and segmental glomerular sclerosis (Baylis et al. 1992). Ribeiro et al. (1992) used a higher dose of L-NAME over a period of 4–6 weeks. This produced severe and sometimes malignant hypertension, with widespread renal damage, large decreases in GFR, and large increases in plasma renin activity. In this model, withdrawal of the growth-inhibitory actions of NO (Garg and Hassid 1989) may have contributed to the development of glomerular injury.

Nitric oxide also modulates the tubuloglomerular feedback mechanism. Excess delivery of sodium and chloride to the macula densa is accompanied by a feedback constriction of the afferent arteriole, and NO opposes this constriction (Wilcox and Welch 1996). In addition, selective inhibition of nNOS with 7-nitroindazole inhibits both furosemide-induced renin release (Beierwaltes 1995) and long-term release of renin during low dietary sodium intake (Beierwaltes 1997).

Acknowledgments. Thanks to Ivadelle Heidke for typing the manuscript. This research was supported by NIH-HL 51971.

References

Ahn KY, Mohaup MG, Madsen KM, Kone BC. 1994. In situ hybridization localization of mRNA encoding inducible nitric oxide synthase in rat kidney. *Am J Physiol* 267:F748–F757.

Alberola AM, Salazar FJ, Nakamura T, & Granger JP. 1994. Interaction between angiotensin II and nitric oxide in control of renal hemodynamics in conscious dogs. *Am J Physiol* 267:R1472–R1478.

Bachmann S, Bosse HM, & Mundel P. 1995. Topography of nitric oxide synthesis by localizing constitutive NO synthases in mammalian kidney. *Am J Physiol Renal Fluid Electrolyte Physiol* 268:F885–F898.

Baylis C, Harton P, & Engels K. 1990. Endothelial derived relaxing factor controls renal hemodynamics in the normal rat kidney. *J Am Soc Nephrol* 1:875–881.

Baylis C, Mitruka B, and Deng A. 1992. Chronic blockade of nitric oxide synthesis in the rat produces systemic hypertension and glomerular damage. *J Clin Invest* 90:278–281.

Beierwaltes WH, 1995. Selective neuronal nitric oxide synthase inhibition blocks furosemide-stimulated renin secretion in vivo. *Am J Physiol* 269:F134–F139.

Beierwaltes WH, 1997. Macula densa stimulation of renin is reversed by selective inhibition of neuronal nitric oxide synthase. *Am J Physiol Regul Integrative Comp Physiol* 272:R1359–R1364.

Chen PY, & Sanders PW. 1991. L-arginine abrogates salt-sensitive hypertension in Dahl/Rapp rats. *J Clin Invest* 88:1559–1567.

Chen PY, & Sanders PW. 1993a. Role of nitric oxide synthesis in salt-sensitive hypertension in Dahl/Rapp rats. *Hypertension* 22:812–818.

Chen PY, & Sanders PW. 1993b. Role of nitric oxide synthesis in salt-sensitive hypertension in Dahl/Rapp rats. *Hypertension* 22:812–818.

Conrad KP, & Whittemore SL. 1992. N^G-monomethyl-L-arginine and nitroarginine potentiate pressor responsiveness of vasoconstrictors in conscious rats. *Am J Physiol Regul Integrative Comp Physiol* 262:R1137–R1144.

Corbett JA, & McDaniel ML. 1996. Selective inhibition of inducible nitric oxide synthase by aminoguanidine. *Methods Enzymol.* 268:398–408.

Deng X, Welch WJ, & Wilcox CS. Wilcox. 1994. Renal vasoconstriction during inhibition of NO synthase: effects of dietary salt. *Kidney Int.* 46:639–646.

Fukuto, JM, & Chaudhuri G. 1995. Inhibition of constitutive and inducible nitric oxide synthase: potential selective inhibition. *Annu Rev Pharmacol Toxicol* 35:165–194.

Furchgott RF, & Zawadzki JV. 1980. The obligatory role of endothelial cells in the relaxation of arterial smooth muscle by acetylcholine. *Nature* 288:373–376.

Gardiner SM, Compton AM, Bennett T, Palmer RMJ, & Moncada S. 1984. Regional haemodynamic changes during oral ingestion of N^G-monomethyl-L-arginine or N^G-nitro-L-arginine methyl ester in conscious Brattleboro rats. *Br J Pharmacol 101:* 10–12.

Gardiner SM, Compton AM, Bennett T, Palmer RMJ, & Moncada S. 1990a. Control of regional blood flow by endothelium-derived nitric oxide. *Hypertension* 15:486–492.

Gardiner SM, Compton AM, Kemp PA, & Bennett T. 1990b. Regional and cardiac haemo-dynamic effects of N^G-nitro-L-arginine methyl ester in conscious, Long Evans rats. *Br J Pharmacol 101:*625–631.

Gardiner SM, Compton AM, Kemp PA, & Bennett T. 1991. Effects of N^G-nitro-L-arginine methyl ester of indomethacin on differential regional and cardiac haemodynamic actions of arginine vasopressin and lysine vasopressin in conscious rats. *Br J Pharmacol 102:*65–72.

Garg UC, & Hassid A. 1989. Inhibition of rat mesangial cell mitogenesis by nitric oxide-generating vasodilators. *Am J Physiol 257:*F60–F66.

Ghiadoni L, Virdis A, Taddei S, Gonzales J, & Salazar J. 1997. Defective nitric oxide pathway in salt-sensitive essential hypertensive patients. *Am J Hypertens 10:*20A.

Granger JP, Salazar FJ, Mizelle HL, Alberola A, & Nakamura T 1992. Role of the renal nerves in mediating the renal hemodynamic effects of systemic edno synthesis blockade. *FASEB J 6:*A1811.

Granger JP, Alberola A, Salazar FJ, & Nakamura Y. 1993. Nitric oxide protects the vasculature against norepinephrine-induced vasoconstriction in conscious dogs. *FASEB J 7:*A187.

Granger J, Novak J, Schnackenberg C, Williams W, and Reinhart GA. 1996. Role of renal nerves in mediating the hypertensive effects of nitric oxide synthesis inhibition. *Hypertension 27:*613–618.

Guyton AC, Coleman TG, Cowley AW Jr, Scheel KW, Manning RD Jr, & Norman RA. 1972. Arterial regulation: overriding dominance of the kidneys in long-term regulation and in hypertension. *Am J Med 52:*584–594, 1972.

Guzman NJ, Fang MZ, Tang SS, Inglefinger JR, & Garg LC. 1995. Autocrine inhibition of Na^+/K^+ ATPase by nitric oxide in mouse proximal tubule cells. *J Clin Invest 95:*2083–2088.

Hu L, & Manning RD Jr. 1995. Role of nitric oxide in regulation of long-term pressure-natriuresis relationship in Dahl rats. *Am J Physiol 268:*H2375–2383.

Hu L, Manning RD Jr, & Brands MW. 1994. Long-term cardiovascular role of nitric oxide in conscious rats. *Hypertension 23:*185–194.

Huang M, Manning RD Jr, LeBlanc MH, & Hester RL. 1997. Overall hemodynamic studies after the chronic inhibition of endothelial-derived nitric oxide in rats. *Am J Hyperten 8:*358–364.

Ikeda, Y, Saito K, Kim J-I, & Yokoyama M. 1995. Nitric oxide synthase isoform activities in kidney of Dahl salt-sensitive rats. *Hypertension 26:*1030–1034.

Ito S, Johnson CS, & Carretero OA. 1991. Modulation of angiotensin II-induced vasoconstriction by endothelium-derived relaxing factor in the isolated microperfused rabbit afferent arteriole. *J Clin Invest 87:*1656–1663.

Jover B, Herizi A, Ventre F, Dupont M, & Mimran A. 1993. Sodium and angiotensin in hypertension induced by long-term nitric oxide blockade. *Hypertension 21:*944–948.

Lahera V, Salom MG, Miranda-Guardiola F, Moncada S, & Romero JC. 1991. Effects of N^G-nitro-L-arginine methyl ester on renal function and blood pressure. *Am J Physiol Renal Fluid Electrolyte Physiol 261:*F1033–F1037.

Luscher TF, Raij L, & Vanhoutte PM. 1987a. Endothelium-dependent vascular responses in normotensive and hypertensive Dahl rats. *Hypertension 9:*157–163.

Luscher TF, Raij L, & Vanhoutte PM. 1987b. Antihypertensive treatment normalizes decreased endothelium-dependent relaxations in salt-induced hypertension of the rat. *Hypertension 9:*III193–III197.

Manning RD Jr, & Hu L. 1994. Nitric oxide regulates renal hemodynamics and urinary sodium excretion in dogs. *Hypertension 23:*619–625.

Manning RD Jr, & Hu L. 1997. Cardiovascular-renal responses to long-term nitric oxide inhibition during angiotensin II-AT$_1$ receptor inhibition. *Am J Hypertens 11:* 328–339, 1998.

Manning RD Jr., Hu L, Mizelle HL, Montani JP, & Norton MW. 1993. Cardiovascular responses to long-term blockade of nitric oxide synthesis. *Hypertension 22:*40–48.

Manning RD Jr, Hu L, & Williamson TW. 1994. Mechanisms involved in the cardiovascular-renal actions of nitric oxide inhibition. *Hypertension 23:*951–956.

Manning RD Jr, Hu L, & Recklehoff JF. 1997. Role of nitric oxide in arterial pressure and renal adaptations to long-term changes in sodium intake. *Am J Physiol 272:* R1162–R1169.

Mattson DL, & Bellehumeur TG. 1996. Neural nitric oxide synthase in the renal medulla and blood pressure regulation. *Hypertension 28:*297–303.

Mattson DL, & Higgins D. 1996. Influence of dietary sodium intake on renal medullary nitric oxide synthase. *Hypertension 27:*688–692.

Mattson DL, Lu S, Nakanishi K, Papanek PE, & Cowley AW Jr. 1994. Effect of chronic renal medullary nitric oxide inhibition on blood pressure. *Am J Physiol Heart Circ Physiol 266:*H1918–H1926.

Mattson DL, Maeda CY, & Cowley AW Jr. 1996. Inducible nitric oxide synthase and blood pressure. *Hypertension 28:*551.

Mohaupt MG, Elzie JL, Ahn KY, Clapp WL, Wilcox CS, & Kone BC. 1994. Differential expression and induction of mRNAs encoding two inducible nitric oxide synthases in rat kidney. *Kidney Int 46:*653–665.

Panza JA, Quyyumi AA, Brush JE Jr, & Epstein SE. 1990. Abnormal endothelium-dependent vascular relaxation in patients with essential hypertension. *N Engl J Med 323:*22–27.

Patel A, Layne S, Watts D, & Kirchner KA. 1993. L-Arginine administration normalizes pressure natriuresis in hypertensive Dahl rats. *Hypertension 22:*863–869.

Pollock DM, Polakowski JS, Divish BJ, & Opgenorth TJ. 1993. Angiotensin blockade reverses hypertension during long-term nitric oxide synthase inhibition. *Hypertension 21:*660–666.

Raij L, & Baylis C. 1995. Nitric oxide and the glomerulus. *Kidney Int 48:*20–32.

Recklehoff JF, & Manning RD Jr. 1993. Role of endothelial derived nitric oxide in the control of the renal microvasculature in the aging male rat. *Am J Physiol 265:* R11126–R1131.

Reinhart GA, Lohmeier TE, & Mizelle HL. 1997. Temporal influence of the renal nerves on renal excretory function during chronic inhibition of nitric oxide synthesis. *Hypertension 29:*199–204.

Ribeiro MO, Antunes E, de Nucci G, Lovisolo SM, & Zatz R. 1992. Chronic inhibition of nitric oxide synthesis: a new model of arterial hypertension. *Hypertension 20:*298–303.

Rodriquez-Sargent C, Cangiano U, Opava-Stitzer S, & Martinez-Maldonado M. 1981. Renal Na$^+$/K$^+$-ATPase in Okamoto and Dahl hypertensive rats. *Hypertension 3:*II86–1191.

Sakuma I, Togashi H, Yoshioka M, Saito H, Yanagida M, Tamura M, Kobayashi T, Yasuda H, Gross SS, & Levi R. 1992. N^G-methyl-L-arginine, an inhibitor of L-arginine-derived nitric oxide synthesis, stimulates renal sympathetic nerve activity in vivo. *Circ Res 70:*607–611.

Salazar FJ, Pinilla JM, Lopez F, Romero JC, & Quesada T. 1992. Renal effects of prolonged synthesis inhibition of endothelium-derived nitric oxide. *Hypertension*
*20:*113–117.

Salazar FJ, Alberola A, Pinilla JM, Romero JC, & Quesada T. 1993. Salt-induced increase
in arterial pressure during nitric oxide synthesis inhibition. *Hypertension 22:*49–55.

Singh I, Grams M, Wang WH, Yang TX, Killen P, Smart A, Schnermann J, & Briggs JP.
1996. Coordinate regulation of renal expression of nitric oxide synthase, renin, and an-
giotensinogen mRNA by dietary salt. *Am J Physiol Renal, Fluid Electrolyte Physiol*
*270:*F1027–F1037.

Star RA. 1993. Southwestern Internal Medicine Conference: nitric oxide. *Am J Med Sci.*
*306:*348–358.

Stoos BA, Carretero OA, Farhy RD, Scicli G, & Garvin JL. 1992. Endothelium-derived relaxing factor inhibits transport and increases cGMP content in cultured mouse cortical
collecting duct cells. *J Clin Invest 89:*761–765.

Stoos BA, Garcia NH, & Garvin JL. 1995. Nitric oxide inhibits sodium reabsorption in the
isolated perfused cortical collecting duct. *J Am Soc Nephrol 6:*89–94.

Terada Y, Tomita K, Nonoguchi H, & Marumo F. 1992. Polymerase chain reaction localization of constitutive nitric oxide synthase and soluble guanylate cyclase messenger
RNAs in microdissected rat nephron segments. *J Clin Invest 90:*659–665.

Tojo A, Gross SS, Zhang L, Tisher CC, Schmidt HHHW, Wilcox CS, & Madsen K. 1994.
Immunocytochemical localization of distinct isoforms of nitric oxide synthase in the
juxtaglomerular apparatus of normal rat kidney. *J Am Soc Nephrol 4:*1438–1447.

Tolins JP, & Shultz PJ. 1994. Endogenous nitric oxide synthesis determines sensitivity to
the pressor effect of salt. *Kidney Int 46:*230–236.

Ujiie K, Yuen J, Hogarth L, Danziger R, & Star RA. 1994. Localization and regulation of
endothelial NO synthase mRNA expression in rat kidney. *Am J Physiol*
*267:*F296–F302.

Wilcox CS, & Welch WJ. 1996. TGF and nitric oxide: effects of salt intake and salt-
sensitive hypertension. *Kidney Int 49(Suppl 55):*S9–S13.

Wilcox CS, Welch WJ, Murad F, Gross SS, Taylor G, Levi R, & Schmidt HHHW. 1992. Nitric oxide synthase in macula densa regulates glomerular capillary pressure. *Proc Natl*
*Acad Sci USA 89:*11993–11997.

Zou AP, & Cowley AW Jr. 1997. Nitric oxide in renal cortex and medulla. An in vitro microdialysis study. *Hypertension 29:*194–198.

21
Modulation of Renal Microvascular Responsiveness by Nitric Oxide

Edward W. Inscho

Introduction

Our understanding of the role of nitric oxide (NO) as an important paracrine regulator of vascular and microvascular responsiveness to vasoactive agonists is steadily growing. Since its discovery, numerous reports and reviews have been published describing the role of NO in virtually every organ system, including the kidney. Indeed, vascular resistance in the kidney may be more heavily influenced by endogenous NO production than in other organ systems. Studies have shown that renal vascular resistance is more sensitive to manipulation of NO synthesis than resistance in other vascular beds, such as the femoral, coronary and cerebral circulations (Sonntag et al. 1992; Sigmon et al. 1993). Several reviews have been written detailing the effect of NO on renal function and whole-kidney renal hemodynamics; however, less has been reported on the direct effects of NO on renal microvascular function. For a current assessment of the role of NO in the regulation of renal blood flow, the interested reader is referred to Chapter 19 by Majid and Navar as well as to several excellent reviews that have been published recently (Navar et al. 1996; Kone and Baylis 1997; Majid and Navar 1997; Mattson et al. 1997). This chapter will summarize what is currently known about the influence of endothelium-derived NO in regulating the renal microvasculature. Emphasis will be placed on detailing the direct effects of NO on renal microvascular responsiveness to vasoactive physiological agonists, tubuloglomerular feedback, and local renal microvascular control.

Renal hemodynamic control is exquisitely accomplished through alterations in preglomerular and postglomerular resistance. These resistance changes are accomplished primarily by altering the diameter of the afferent arteriole and, to a lesser extent, by altering the diameter of the postglomerular efferent arteriole. Studies have shown that both the afferent and efferent arterioles are responsive to the vasodilatory influence of NO and thus NO may be importantly involved in directly regulating renal microvascular responsiveness to vasoactive agonists and stimuli (Imig and Roman 1992; Imig et al. 1993; Ito and Ren 1993; Ito et al. 1993; Ito 1995; Navar et al. 1996). More recent work has shown that renal microvascular regulation by NO may involve NO generated by endothelial nitric oxide synthase (eNOS), as well as NO generated by the neuronal form of NOS (nNOS). Through the remainder of this chapter, we will endeavor to highlight the important observations regarding the role of intrarenal NO in the control of renal microvascular function.

Direct Effects of NO on the Renal Microvasculature

Evaluation of the direct effects of NO or NOS inhibition on renal microvascular function has largely been accomplished using in vitro techniques developed to provide direct access to the renal microvasculature. These techniques include isolated afferent and efferent arterioles from rat and rabbit, videomicroscopic observations from hydronephrotic kidneys that do not possess functional renal tubules, and using the blood-perfused juxtamedullary nephron technique where vascular and tubular structures remain intact and functional. We will not elaborate on the details of the particular methods used but rather will focus on the findings and their interpretation.

Studies have shown that the endothelium of afferent and efferent arterioles, glomerular capillaries, descending vasa recta, and intrarenal arteries expresses the eNOS isoform of NOS (Ujiie et al. 1994; Bachmann et al. 1995; Kone and Baylis 1997). In addition, the nNOS isoform is expressed by the efferent arteriolar endothelium and the macula densa, which is an important regulatory site of renal microvascular function (Mundel et al. 1992; Wilcox et al. 1992; Ujiie et al. 1994; Bachmann et al. 1995; Kone and Baylis 1997). Therefore, these endothelial and nonendothelial tissues appear readily able to generate NO, which can participate in the local regulation of microvascular function. Consistent with this postulate, administration of NO donors or inhibitors of NOS has been shown to have a significant impact on preglomerular microvascular diameter and responsiveness to vasoactive agonists.

Nitric Oxide Donors

Some investigators have examined renal vascular responses to NO by elevating the endogenous NO environment using excess quantities of the amino acid substrate for NOS, L-arginine, or by administering nitrovasodilators as NO donors. These agents generate NO as the final common effector molecule and include such substances as sodium nitroprusside, glyceryl trinitrite, and S-nitrosothiols. Infusion of the S-nitrosothiol compound S-nitroso-N-acetylpenicillamine (SNAP) into the canine renal artery stimulates a pronounced increase in renal blood flow reflective of a marked reduction in renal vascular resistance (Majid et al. 1993). Consistent with this hemodynamic response, Ichihara et al. (1998a) recently reported that SNAP administration caused a concentration-dependent vasodilation of rat juxtamedullary afferent and efferent arterioles. Sodium nitroprusside also directly stimulates NO-dependent, but endothelium-independent, vasorelaxation by functioning as an NO donor. Topical administration of sodium nitroprusside to the adventitial surface of the renal microvasculature rapidly increases rat arcuate and interlobular artery diameter as well as afferent and efferent arteriolar diameter (Carmines et al. 1986; Inscho et al. 1990; Ortenberg et al. 1992; Ikenaga et al. 1996). Treatment with exogenous L-arginine evoked a significant vasodilation of arcuate and interlobular arteries and afferent and efferent arterioles in the hydronephrotic kidney (Hoffend et al. 1993). The vasodilation was dependent on NOS activity, since the vascular response to L-arginine administration was attenuated by treatment with N^G-nitro-L-arginine methylester (L-NAME) (Hoffend et al. 1993). These data demonstrate that all preglomerular arteries and arterioles as well as the postglomerular arterioles are responsive to local concentrations of NO leading to renal microvascular vasodilation.

Nitric Oxide Inhibition

The alternative approach of determining the effect of NO synthesis inhibition has also been evaluated. If NO is an important regulator of ambient renal microvascular resistance, then it should be constitutively released from the renal microvascular endothelium. Inhibition of that constitutive NO release should eliminate the vasodilatory influence of endogenously produced NO on renal vascular tone and lead to vasoconstriction. Investigators have used several approaches to test this hypothesis, including inhibition of NOS activity, scavenging of the endogenously released NO, and removal of the endothelium from renal microvascular segments, thus eliminating the endothelial source of NO. Inhibition of NOS has consistently resulted in a significant decrease in renal microvascular diameter, despite the fact that the experiments were performed in markedly different settings. In studies using the isolated, blood-perfused juxtamedullary nephron preparation, Ohishi et al. (1992) reported that inhibition of NOS with N^G-nitro-L-arginine (L-NA) decreased both afferent and efferent arteriolar diameter in a concentration-dependent manner. Similarly, Imig and co-workers demonstrated that L-NA administration significantly reduced rat juxtamedullary afferent and efferent arteriolar diameter and showed that the upstream arcuate and interlobular arteries were also vasoconstricted by NOS inhibition (Imig et al. 1993). In studies using the hydronephrotic kidney, inhibition of NO synthesis with L-NAME elicited varying degrees of vasoconstriction in different microvascular segments (Gulbins et al. 1993; Hoffend et al. 1993). The greatest response was observed in the arcuate arteries and efferent arterioles. Slightly smaller responses were observed from interlobular arteries, whereas afferent arterioles exhibited the smallest overall reduction in luminal diameter. Finally, in studies performed using isolated rabbit afferent and efferent arterioles, NO synthesis inhibition with either L-NA, L-NMMA, L-NAME, or N^G-monomethyl-L-arginine (L-NMMA) consistently caused vasoconstriction of both afferent and efferent arterioles (Ito et al. 1991; Edwards and Trizna 1993; Ito and Ren 1993; Ito et al. 1993). L-NA, L-NAME, and L-NMMA are structural analogues of the principal amino acid substrate for NOS, L-arginine, and function as competitive inhibitors of the enzyme. Since they are competitive antagonists, their inhibitory actions should be overcome if excess substrate is made available to the enzyme. In fact, in many of the studies reported above where these L-arginine analogues were used as NOS inhibitors, subsequent administration of L-arginine, but not D-arginine, overcame the competitive inhibition of NOS activity, restored microvascular diameter to normal, and restored microvascular responsiveness to vasoactive stimuli (Imig and Roman 1992; Edwards and Trizna 1993; Ito and Ren 1993; Ito et al. 1993). Others have determined the effect of elimination of endogenously generated NO by administering NO scavengers such as red blood cells and hemoglobin. In studies using the rat juxtamedullary nephron preparation, the diameter of the preglomerular microvascular segments was assessed while the kidney was perfused with a cell-free perfusate and again during sequential addition of washed red blood cells to progressively increase the hematocrit (Imig et al. 1993). Elevating the hematocrit to as little as 0.1% resulted in a significant reduction in afferent arteriolar diameter, with the maximum response observed at a hematocrit of 1%. Interlobular artery diameter was also similarly affected by the presence of red blood cells in the perfusate. Subsequent addition of L-NA, in the presence of red blood cells, had no additional vasoconstrictor influence. These data demonstrate that the presence of red blood cells and the associated increase in the perfusate concentration of hemoglobin (an NO scavenger) sequestered the endogenously produced NO and attenuated the vasodilatory influence of this constitutive NO production on preglomerular microvascular caliber.

Endothelium Removal

The underlying hypothesis regarding the role of NO in the regulation of renal microvascular function is that NO is produced by the endothelium and diffuses into the adjacent vascular smooth muscle cells, where it initiates a cGMP-mediated vasodilation. The endothelial contribution to the regulation of large-caliber artery function has been examined by measuring stimulus-induced vascular smooth muscle tension before and after removal of the endothelium. Denudation of the endothelium is typically accomplished by mechanical stripping, enzymatic digestion, or chemical destruction. Although technically feasible for large arteries, these techniques are poorly suited for use in very small-caliber resistance arterioles with diameters in the range of 10–30 microns. Therefore, Juncos et al. (1994) developed a novel method for immune denudation of renal microvessels in order to test the hypothesis that NO generated by the microvascular endothelium modulated renal microvascular responsiveness to vasoactive stimuli. Selective endothelial cell lysis was accomplished using an antibody/antigen complex combined with complement in isolated, perfused rabbit afferent arterioles. Acetylcholine elicited an afferent arteriolar vasodilation when the endothelium was intact; however, the vasodilatory response was abolished following destruction of the endothelium. Importantly, afferent arteriolar responses to nonendothelium-dependent vasoactive stimuli were retained. These data demonstrate that the endothelium lining the afferent arteriole does generate a vasodilatory substance that modulates afferent arteriolar smooth muscle function. These data are also consistent with earlier studies indicating that the vasodilatory substance released from the endothelium is NO.

Role of NO in Renal Microvascular Responses to Vasoactive Agonists

Acetylcholine and Platelet Activating Factor

The initial description of endothelium-dependent vasodilator influence came from the work of Furchgott and Zawadzki (1980). Their observation revolutionized the way we look at vascular regulation in all organ systems. Clearly, endothelium-derived NO plays a significant role in the regulation of basal vascular tone and also the vascular response to vasoactive agonists. Consistent with the observations of Furchgott and co-workers (Furchgott and Zawadzki 1980), acetylcholine vasodilates the afferent and efferent arterioles of rat and rabbit kidneys (Ortenberg et al. 1992; Edwards and Trizna. 1993; Hayashi et al. 1994). These vasodilatory responses were blocked by NOS inhibition with L-NA or by destruction of the endothelium (Juncos et al. 1994). However, interaction between extracellular vasoactive agonists and NO synthesis is not unique to acetylcholine. Numerous agonist-induced renal microvascular vasoconstrictor responses are also influenced by endogenous NO or NOS activity. For example, experiments using isolated perfused rabbit afferent arterioles preconstricted with norepinephrine demonstrated that picomolar concentrations of platelet activating factor (PAF) elicited a dose-dependent afferent arteriolar vasodilation (Juncos et al. 1993). This vasodilatory response was unaffected by cyclooxygenase inhibition but was abolished by NOS inhibition with L-NA. Similarly, picomolar concentrations of PAF also vasodilated isolated, perfused rabbit efferent arterioles through an NOS-dependent mechanism (Juncos et al. 1993; Arima et al. 1996). Interestingly, this vasodilatory response was only observed when the efferent arteriole was perfused in the orthograde direction, with the perfusate passing through the glomerulus before entering the efferent arteriolar lumen. This observation suggests that PAF stimulates the release of NO from the glomerulus directly. Alternatively, PAF may stimulate release of a paracrine factor from the glomerulus, which, in turn, stimulates NO synthesis and release in the downstream efferent arteriole.

Angiotensin II

Angiotensin has also been reported to interact with NO in the renal circulation. In vitro studies using isolated perfused rabbit arterioles have shown that the afferent, but not the efferent, vasoconstriction to angiotensin II is significantly enhanced following NOS inhibition with L-NA (Ito et al. 1991, 1993). These data suggest that NO production by the renal microvessels modulates angiotensin II-mediated afferent arteriolar vasoconstriction; however, they do not indicate whether this modulatory influence stems from endogenous levels of NO or from angiotensin II-dependent stimulation of NO synthesis. This question was recently addressed by Ikenaga et al. (1996) using the blood-perfused juxtamedullary nephron preparation. As with isolated afferent arterioles, NOS inhibition augmented angiotensin II-mediated vasoconstriction of rat afferent arterioles; however, in contrast to what is observed with isolated rabbit efferent vessels, NOS inhibition also enhanced the rat efferent arteriolar response to angiotensin II. Similar results were observed using vasopressin, demonstrating that this augmentation is not unique to angiotensin II-dependent events. As previously discussed, NOS inhibition with L-NA consistently results in a renal microvascular vasoconstriction that could alter microvascular responsiveness. To control for this possibility, Ikenaga and co-workers reversed the L-NA-induced microvascular vasoconstriction by administering a carefully titrated solution of the NO donor sodium nitroprusside. This presumably restored the level of endogenous NO to the pre-L-NA level and eliminated the acute effect of L-NA on baseline diameter and responsiveness. Under conditions of NO supplementation during NOS inhibition, the magnitudes of the angiotensin II-mediated afferent and efferent arteriolar vasoconstrictions were similar to the responses observed before NOS inhibition and were not enhanced, as would have been predicted if angiotensin II stimulated increased NOS activity. Similar results were obtained in another study using the same experimental approach but with kidneys from hypertensive rats (Ichihara et al. 1998a). These data suggest that the modulatory influence exerted by NO on renal microvascular responsiveness to angiotensin II results from the level of endogenous NO production rather than from the direct simulation of enhanced NOS activity by angiotensin II. Nevertheless, Thorup et al. (1998) recently reported that angiotensin II-stimulated NO production by isolated, perfused rat intrarenal resistance arteries could be detected using an NO-sensitive microelectrode. Furthermore, angiotensin II-stimulated NO production was markedly attenuated by blockade of angiotensin II-AT$_1$ receptors with losartan or candesartan. Thus, it appears that angiotensin II can stimulate NO production from isolated rat renal resistance arteries perfused with a cell-free perfusate. It is difficult to reconcile these observations with the work of Ikenaga et al. (1996) and Ichihara et al. (1998a); however, some potentially important differences exist between the two sets of data. Ikenaga et al. (1996) and Ichihara et al. (1998a) focused their observations on afferent and efferent arterioles, which average approximately 20 microns in diameter. These arterioles are substantially smaller than the resistance arteries used by Thorup et al. (1998). It is possible that angiotensin II-stimulated generation of NO varies in a vascular segment-specific manner, and that angiotensin II does not stimulate release from afferent or efferent arterioles. Alternatively, arteriolar perfusion with blood could scavenge NO released in response to angiotensin II, masking its physiological effect. Finally, the NO electrode may be sensitive enough to detect NO concentrations that are too low to produce a significant physiological effect. Nevertheless, the potential for vasoconstrictor agonists to stimulate simultaneous generation of a diffusible vasodilator suggests an intriguing interaction between vasodilatory and vasoconstrictor substances in the regulation of renal microvascular function.

Nitric Oxide and the Tubuloglomerular Feedback Mechanism

Renal hemodynamics are under the control of multiple regulatory mechanisms. One such mechanism, referred to as the tubuloglomerular feedback mechanism, postulates that macula densa cells of the distal nephron respond to changes in distal tubular fluid composition or flow rate by adjusting preglomerular resistance in an effort to maintain a stable glomerular capillary pressure and a constant rate of glomerular filtration (Navar et al. 1996). With the discovery that macula densa cells express the neuronal isoform of NOS and with the recognition that the macula densa plays a major role in regulating preglomerular resistance through the tubuloglomerular feedback mechanism, the postulate has been put forth that macula densa control of preglomerular microvascular function might be influenced by NO generated in the macula densa by nNOS (Wilcox et al. 1992; Schnermann 1998). Evidence obtained at the whole kidney level as well as data derived specifically from the microvasculature support this hypothesis. In whole-kidney experiments, inhibition of neuronal NO by chronic administration of the selective nNOS inhibitor 7-nitro indazole (7-NI) enhanced tubuloglomerular feedback responsiveness in rats (Ollerstam et al. 1997). Microperfusion of the nephron with 7-NI also enhanced tubuloglomerular feedback-mediated preglomerular responsiveness, as indicated by an augmented perfusion-mediated decrease in glomerular capillary pressure (Thorup and Persson 1996). The direct effects of nNOS inhibition on microvascular function have also been specifically addressed. Distal tubular perfusion of isolated juxtaglomerular apparatus with a solution containing the nonselective NOS inhibitor L-NAME and high concentrations of NaCl resulted in an enhanced afferent arteriolar vasoconstriction compared with the response obtained with low NaCl concentrations in the perfusate (Ito and Ren 1993). These findings suggest that NO generated as a result of distal tubular fluid composition plays a significant role in modulating afferent arteriolar diameter in a manner consistent with tubuloglomerular feedback-mediated regulation of glomerular filtration.

Recent studies have taken advantage of the more selective nNOS inhibitor, S-methyl-L-thiocitrulline, to evaluate the direct effects of nNOS inhibition on renal microvascular function. Ichihara et al. (1998b) demonstrated that exposure of rat juxtamedullary afferent and efferent arterioles to the selective nNOS inhibitor S-methyl-L-thiocitrulline (L-SMTC) resulted in a concentration-dependent vasoconstriction consistent with a tonic modulatory influence of endogenous macula densa-derived NO on afferent and efferent arteriolar diameter (Ichihara et al. 1998b). This response was observed at a concentration of L-SMTC that did not significantly alter the NO-dependent microvascular response to acetylcholine, suggesting that the vasoconstriction arose from selective inhibition of nNOS in the macula densa. The L-SMTC-mediated vasoconstrictor response was eliminated when the flow of tubular fluid past the macula densa region of the distal nephron was interrupted by papillectomy. In addition, the afferent arteriolar vasoconstriction was enhanced when volume delivery to the macula densa segment of the nephron was increased by acetozolamide treatment (Ichihara et al. 1998b). These data demonstrate that preglomerular and postglomerular arteriolar diameter is under the tonic influence of NO generated from both endothelium-derived eNOS and macula densa-derived nNOS and suggest that nNOS activity may play a unique role in regulating microvascular resistance and glomerular hemodynamics.

Closing Remarks

The renal microcirculation is substantially influenced by the vasodilatory influence of NO through endogenous production as well as agonist-stimulated production. This diffusible paracrine regulator of vascular tone is just one of a number of vasodilatory and vasoconstrictor endothelial factors that are under investigation. The discovery of these paracrine substances has led to the realization that the endothelium is more than just a vascular lining, but rather is an organ system actively involved in the local regulation of vascular resistance. This awareness has opened entire new areas of research into the physiological regulation of organ perfusion, generated new explanations for pathophysiological conditions, and provided important new targets for therapeutic intervention for cardiovascular diseases. As can be seen from this discussion and the other chapters in this volume, endothelium-derived NO is a very important participant in the regulation of renal and nonrenal vascular function. To the extent that cardiovascular health is tightly coupled to appropriate kidney function and regulation of renal hemodynamics, the significance of NO in the overall regulation of renal physiology and in disease will continue to grow.

References

Arima S, Ren Y, Juncos LA, & Ito S. 1996. Platelet-activating factor dilates efferent arterioles through glomerulus-derived nitric oxide. *J Am Soc Nephrol 7*:90–96.

Bachmann S, Bosse HM, & Mundel P. 1995. Topography of nitric oxide synthesis by localizing constitutive NO synthase in mammalian kidney. *Am J Physiol 268*:F885–F898.

Carmines PK, Morrison TK, & Navar LG. 1986. Angiotensin II effects on microvascular diameters of in vitro blood-perfused juxtamedullary nephrons. *Am J Physiol 251*:F610–F618.

Edwards RM, & Trizna W. 1993. Modulation of glomerular arteriolar tone by nitric oxide synthase inhibitors. *J Am Soc Nephrol 4*:1127–1132.

Furchgott RF, & Zawadzki, JV. 1980. The obligatory role of endothelial cells in the relaxation of arterial smooth muscle by acetylcholine. *Nature 288*:373–376.

Gulbins E, Hoffend J, Zou AP, Dietrich MS, Schlottmann K, Cavarape A, & Steinhausen M. 1993. Endothelin and endothelium-derived relaxing factor control of basal renovascular tone in hydronephrotic rat kidneys. *J Physiol (Lond) 469*:571–582.

Hayashi K, Loutzenhiser R, Epstein M, Suzuki H, & Saruta T. 1994. Multiple factors contribute to acetylcholine-induced renal afferent arteriolar vasodilation during myogenic and norepinephrine- and KCl-induced vasoconstriction. *Circ Res 75:*821–828.

Hoffend J, Cavarape A, Endlich K, & Steinhausen M. 1993. Influence of endothelium-derived relaxing factor on renal microvessels and pressure-dependent vasodilation. *Am J Physiol 265:*F285–F292.

Ichihara A, Imig JD, Inscho EW, & Navar LG. 1998a. Interactive nitric oxide-angiotensin II influences on renal microcirculation in angiotensin II-induced hypertension. *Hypertension 31:*1255–1260, 1998.

Ichihara A, Inscho EW, Imig JD, & Navar LG. 1998b. Neuronal nitric oxide synthase modulates rat renal microvascular function. *Am J Physiol 274:*F516–F524.

Ikenaga HR, Fallet W, & Carmines PK. 1996. Basal nitric oxide production curtails arteriolar vasoconstrictor responses to ANG II in rat kidney. *Am J Physiol 271:*F365–F373.

Imig JD, & Roman RJ. 1993. Nitric oxide modulates vascular tone in preglomerular arterioles. *Hypertension 19:*770–774.

Imig JD, Gebremedhin D, Harder DR, & Roman RJ. 1993. Modulation of vascular tone in renal microcirculation by erythrocytes: role of EDRF. *Am J Physiol 264:*H190–H195.

Inscho EW, Carmines PK, Cook AK, & Navar LG. 1990. Afferent arteriolar responsiveness to altered perfusion pressure in renal hypertension. *Hypertension 15:*748–752.

Ito S. 1995. Nitric oxide in the kidney. *Curr Opin Nephro Hypertens 4:*23–30.

Ito S, & Ren Y. 1993. Evidence for the role of nitric oxide in macula densa control of glomerular hemodynamics. *J Clin Invest 92:*1093–1098.

Ito S, Johnson CS, & Carretaro OA. 1991. Modulation of angiotensin II-induced vasoconstriction by endothelium-derived relaxing factor in the isolated microperfused rabbit afferent arteriole. *J Clin Invest 87:*1656–1663.

Ito S, Arima S, Ren Y, Juncos LA, & Carretero OA. 1993. Endothelium-derived relaxing factor/nitric oxide modulates angiotensin II action in the isolated microperfused rabbit afferent but not efferent arteriole. *J Clin Invest 91:*2012–2019.

Juncos LA, Ren Y, Arima S, & Ito S. 1993. Vasodilator and constrictor actions of platelet-activating factor in the isolated microperfused afferent arteriole of the rabbit kidney: role of endothelium-derived relaxing factor/nitric oxide and cyclooxygenase products. *J Clin Invest 91:*1374–1379.

Juncos LA, Ito S, Carretero OA, & Garvin JL. 1994. Removal of endothelium-dependent relaxation by antibody and complement in afferent arterioles. *Hypertension 23 (Suppl 1):*I-54–I-59.

Kone BC, & Baylis C. 1997. Biosynthesis and homeostatic roles of nitric oxide in the normal kidney. *Am J Physiol 272:*F561–F578.

Majid DSA, & Navar LG. 1997. Nitric oxide in the mediation of pressure natriuresis. *Clin Exp Pharmacol Physiol 24:*595–599.

Majid DSA, Williams A, Kadowitz PJ, & Navar LG. 1993. Renal responses to administration of nitric oxide donors in dogs. *Hypertension 22:*535–541.

Mattson DL, Lu SH, & Cowley AW Jr. 1997. Role of nitric oxide in the control of the renal medullary circulation. *Clin Exp Pharmacol Physiol 24:*587–590.

Mundel P, Bachmann S, Bader M, Fischer A, Kummer W, Mayer B, & Kriz W. 1992. Expression of nitric oxide synthase in kidney macula densa cells. *Kidney Int 42:*1017–1019.

Navar LG, Inscho EW, Majid DSA, Imig JD, Harrison-Bernard LM, & Mitchell KD. 1996. Paracrine regulation of the renal microcirculation. *Physiol Rev 76:*425–536.

Ohishi K, Carmines PK, Inscho EW, & Navar LG. 1992. EDRF-angiotensin II interactions in rat juxtamedullary afferent and efferent arterioles. *Am J Physiol 263:*F900–F906.

Ollerstam A, Pittner J, Persson AEG, & Thorup C. 1997. Increased blood pressure in rats after long-term inhibition of the neuronal isoform of nitric oxide synthase. *J Clin Invest 99:*2212–2218.

Ortenberg JM, Cook AK, Inscho EW, & Carmines PK. 1992. Attenuated afferent arteriolar response to acetylcholine in Goldblatt hypertension. *Hypertension 19:*785–789.

Schnermann J. 1998. Juxtaglomerular cell complex in the regulation of renal salt excretion. *Am J Physiol 274:*R263–R279.

Sigmon DH, Carretaro OA, & Beierwaltes WH. 1993. Renal versus femoral hemodynamic response to endothelium-derived relaxing factor synthesis inhibition. *J Vasc Res 30:*218–223.

Sonntag M, Deussen A, & Schrader J. 1992. Role of nitric oxide in local blood flow control in the anaesthetized dog. */Pflugers Arch 420:*194–199.

Thorup C, & Persson AEG. 1996. Macula densa derived nitric oxide in regulation of glomerular capillary pressure. *Kidney Int 49:*430–436.

Thorup C, Kornfeld M, Winaver JM, Goligorsky MS, & Moore LC. 1998. Angiotensin-II stimulates nitric oxide release in isolated perfused renal resistance arteries. *Pflugers Arch 435:*432–434.

Ujiie K, Yuen J, Hogarth L, Danziger R, & Star RA. 1994. Localization and regulation of endothelial NO synthase mRNA expression in the rat kidney. *Am J Physiol 267:*F296–F302.

Wilcox CS, Welch WJ, Murad F, Gross SS, Taylor G, Levi R, & Schmidt HHHW. 1992. Nitric oxide synthase in macula densa regulates glomerular capillary pressure. *Proc Natl Acad Sci USA 89:*11993–11997.

Index

Acetylcholine
   effects on blood vessel tone, 51
      dependence on vascular endothelium,
         197, 229
      regional differences in, 87–88
      response to L-NAME and glyben-
         clamide, 207–9
      in the kidneys, 341
      relaxant response to, variation with age,
         186–87
      renal vasodilation produced by, 307
Acetylcholine esterase, colocalization with
      nitric oxide synthase, 290
Actin
   ADP ribosylation of, caused by nitric
      oxide, 53
   effect of nitric oxide on expression in
      stellate cells, 250
Acute chest syndrome, effect of nitric
      oxide therapy in, 173
Acute respiratory distress syndrome. *See*
      Adult respiratory distress syndrome
Adenosine
   P1 purinergic receptors for, 65
   role in vasodilation during hypoxemia, 119
   vasodilation by, 66
Adenosine, A_1 receptors for, on perivascu-
      lar nerves, 74
Adenosine diphosphate (ADP), role in in-
      hibition of platelet aggregation, 77
Adenosine triphosphate (ATP)
   as cotransmitter in vasoconstriction, 67,
      72–73
   release from sympathetic perivascular
      nerves, 69

   vasodilation by, 66
   vasodilation and vasoconstriction by, de-
      pendence on receptor location, 75
Adenylate cyclase (adenylylcyclase)
   P_{2T} receptor mediation of inhibition of,
      74–75
   $P2Y_1$ receptor mediation of inhibition
      of, 71
Adhesion molecules, role in leukocyte-
      endothelial cell interactions, 262
Adrenergic system
   1-adrenergic blockade, effect on hyper-
      tension, 101–2
   mediation hypertensive response to
      L-NAME and L-NA in, 228–29
Adrenomedullin (ADM)
   effect on penile erection, 287
   as a naturetic peptide, 232–34
Adult respiratory distress syndrome
      (ARDS), 152–57, 199
   animal model of, 149–50
   complications in care of patients with, 148
   treatment with inhaled nitric oxide,
      167–69
Adults, lung injury in, 167–71
Age
   and effect of L-NAME on renal vascular
      resistance, 322
   and response to acetylcholine, 186–87
Albuterol, vasodilator responses to, effect
      of charybdotoxin on, 213–15
Almitrine, effect on response to inhaled ni-
      tric oxide, 156
Alveolar hypoxia, in severe lung injury
      and disease, 199–200

Alveolocapillary membrane, diffusion capacity of, determination with inhaled nitric oxide, 173–74
Anesthetics, intravenous, in the pulmonary circulation, 131–47
Angiogenesis, in intrauterine growth restriction, 275
Angioplasty, effect on normal nitric oxide function in, 7
Angiotensin II
  effects of L-NAME on responses to, 324
    pulmonary circulation, 205–7
    renal function, 325
  interaction with nitric oxide in the renal circulation, 341–43
  interaction with nitric oxide synthase inhibitors, 312–13
  mediation of short-term effect of nitric oxide synthase inhibition by, 324
6-Anilino-5,8-quinolinedione (LY83583), inhibition of guanylate cyclase by, 200
Apomorphine, induction of yawning and penile erection by, 291
Arginine/arginine analogues
  mediation of short-term effect of nitric oxide synthase inhibition by, 324
  pressor response to, 228–29
L-Arginine
  conversion to nitric oxide and citrulline, 15–16
  effect on hypertension in salt-sensitive rats, 327–30
  effect on superoxide production by nitric oxide synthase, 33, 41
  treatment with, effect on salt-induced hypertension, 104, 331
Arterial autoregulation, hepatic, 245–46
Arteries
  cerebral
    colocalization of vasoactive intestinal peptide and nitric oxide in, 290
  hepatic, physiology of, 244–45
  peripheral, relaxation of, mediation by purinergic A_{2A} receptors, 90–93
  pressure regulation though the pressure-natriuresis mechanism, 320, 322
  pudendal, concentration of vasoactive intestinal peptide in, 290

regulation of sodium ion-potassium ion-ATPase activity in, 50–51
relaxation of, 89–90
  mediation by purinergic A_{2A} receptors, 69–70
systemic pressure
  effect of ketamine on, 132–33
  effects of nitric oxide donors on, 296
Asthma, treating with inhaled nitric oxide, 170
Astrocytes, expression of nitric oxide synthases in, 114
Atherosclerosis
  effect on endothelium-dependent relaxation, 43
  effect on nitric oxide functioning, 9, 42
Atropine, effect on response to ventilatory hypoxia, 203–5
Autocrine effects, of nitric oxide in liver, 249–50
Autoregulation
  of ATP neurotransmission, 74
  of cerebrovascular adjustments to changing perfusion pressure, 119–20
  hepatic
    arterial, 245–46
    responses to inflammation, 251
  of platelet aggregation, 74–75
  pressure-flow, in the gastrointestinal circulation, 260
  of renal blood flow, 311–12

Beraprost, effect on pulmonary hypertension, 172
Bladder, nitric oxide in, 293–94
Blood flow
  cerebral, roles of nitric oxide in regulation of, 115
  coronary
    effect of estrogen on, 191, 273
    restoring in aged humans, 89
  penile, 286–301
  regional, 65–84
  renal
    autoregulation of, 311–12
    effect of SNAP on, 338
    in L-NAME treated rats, 322

roles of nitric oxide in regulation of,
   1–12, 66, 259–60, 306–9
organ specificity of, 245
steal of, in pulmonary arteries induced
   by nitric oxide vasodilation,
   148–49
umbilical, regulation of, 275–76
uterine
   response to estradiol-17β, 272
   roles of nitric oxide synthase in, 87
uteroplacental, in pregnancy, 274–75
See also Hemodynamics
Blood pressure
   effect of endothelin on, 55
   effect of ketamine on, 132–33
   effect of propofol on, 131
   glomerular, after L-NAME administra-
      tion, 332
   See also Arteries, pressure; Hyperten-
      sion
Bradykinin
   renal vasodilation produced by, 307
   role in inflammatory response and car-
      diovascular function, 229
Bridge to transplantation, inhaled nitric
   oxide, in primary pulmonary hypertension, 169

Calcium
   dependence on, of endothelial nitric
      oxide synthase activation, 6
   involvement in ketamine responses in
      the lung, 144
Calcium ion (Ca²⁺)
   activation of Ca²⁺-ATPase, 18–19
   intracellular, regulation by cyclic GMP-
      dependent protein kinase, 15–32
   reduction of concentration, mechanisms
      of cGMP mediation, 18–23
   release by the sarcoplasmic reticulum,
      inhibition of, 20–21
Calcium ion channels
   blockade of, with verapamil, 102
   inhibition of influx through, 22–23
Captopril, effect of L-noradrenalin pressor
   response, 228
Carbon dioxide, cerebral vasodilation in
   response to, 117–18

Cardiac myocytes, effect of cGMP on calcium ion in, 22–23
Cardiac output, in L-NAME treated rats, 322
Cardiac tissue, inotropic action of digoxin
   in, 58
Cardiovascular hemodynamics, effects of
   acute nitric oxide synthase inhibition on, 100
Catalase
   effect on inflammatory responses to
      L-NAME, 263
   vascular, inhibition by nitric oxide,
      38–39
Central nervous system, effect of nitric
   oxide deficiency on, 104
Cerebral arteries, relaxation of, factors affecting, 89–90
Cerebral circulation, nitric oxide in control
   of, 113–27
Cerebral microcirculation, effect of oxidants in, 42
Cervical tissue, nitric oxide synthase in, 279
Charybdotoxin
   effect of
      on relaxation in muscle prepara-
         tions, 22
      on responses to nitroglycerin, nitro-
         prusside, albuterol, and isopro-
         terenol, 213–17
   inhibition of calcium ion-activated
      potassium channels, 186
Chronic hypoxic pulmonary hypertension,
   effect of ketamine on relaxation
   mediated by nitric oxide, 133
Chronic nitric oxide synthase inhibition
   (CNOSI) model
   mechanisms of hypertension in,
      100–103
   organ damage secondary to nitric oxide
      deficiency, 103–4
Chronic obstructive pulmonary disease, effect of inhaled nitric oxide in,
   169–70, 190–93, 199
Cimetidine, effect on histamine-induced
   vasodilation, 238
Clinical applications, of inhaled nitric
   oxide, 166–84
Contractile protein, regulation of the sensitivity of, 23

Contraction
  of aortic rings, effect of L-arginine ana-
    logues on, 86
  uterine, evaluation of the role of nitric
    oxide, 277–78
Copper, binding to guanylate cyclase, 4
Copper, zinc-superoxide dismutase, inhi-
    bition of, by diethyldithiocarba-
    mate, 35
Coronary blood flow, effect of estrogen on,
    191, 273
Coronary circulation, response to estro-
    gens, 271
Cross activation, of cGMP kinase, and in-
    hibition of calcium ion channels,
    22–23
Cyclic adenosine monophosphate kinase,
    phosphorylation of phospholamban
    by, 19
Cyclic guanosine monophosphate
  binding to vimentin, 20
  effect of inhaled nitric oxide on, in lung
    injury, 150
  mechanism of action on cerebral vascu-
    lar tone, 116–17
  protein phosphorylation facilitated by, 198
  role in cerebral artery relaxation, 90
  role in inhibition of platelet aggrega-
    tion, 77
Cyclic guanosine monophosphate-
    dependent protein kinase, regula-
    tion of intracellular calcium ion by,
    15–32
Cyclic guanosine monophosphate kinase,
    mediation of relaxation of vascular
    smooth muscle by, 24
Cysteine, role in activation of guanylate
    cyclase by nitroglycerin, 2
Cytochrome oxidase, mitochondrial, re-
    duction of nitric oxide by, 35
Cytokines, nitric oxide synthase induction
    triggered by in liver, 250
Cytotoxicity, of nitric oxide from Kupffer
    cells, 251

Depolarization, mediation by purinergic
    P2X receptors, 72–73
Desmin, expression by stellate cells, 249

Diabetes
  complications of, role of the nitric
    oxide-sodium ion-potassium ion-
    ATPase system, 59–60
  endothelium-dependent relaxation in,
    impaired, 43
Diacylglycerol (DAG), formation following
    activation of phospholipase C, 71
Diaphragmatic hernia, congenital, inhaled
    nitric oxide for treating, 172
Diffusion capacity, measuring with inhaled
    nitric oxide, 173–74
Digoxin, effect on penile erection, 58
Dihydrotestosterone, restoration of erectile
    response with, 291
Dilator mechanisms, for control of the
    cerebral circulation, 115–17
5,7-Dimethyl-2-ethyl-3[[2'-((1H-tetrazol-
    5yl)-[1,1]β-biphenyl-4yl]-
    ]methyl]-3H-imidazo[4,5-b]pyri-
    dine. See L158809
Donors, nitric oxide, responses to in impo-
    tence, 295–96
Dosage, of inhaled nitric oxide, 176–77
Double-blind studies, of nitric oxide in-
    halation therapy in ARDS, 157

Ectonucleotidases, effect on ATP and
    UTP, 67
Electrical field stimulation (EFS)
  of the corpus cavernosum, relaxation
    caused by, 288–89
  of perivascular nerves, effect of nitric
    oxide blockade on, 75
Endothelial cell-derived nitric oxide, pro-
    tective effects of, 264–66
Endothelial cells, interactions with inflam-
    matory cells, 260–63
Endothelin-1
  effect on stellate cells, 249
  levels of, in newborns, 188–89
  modulation of nitric oxide release by, 55
Endothelin, effect on inhibition on hyper-
    tension, 102
Endothelium
  barrier function, regulation by nitric
    oxide, 263–64
  basal release of nitric oxide from, 75

nitric oxide derived from, 85–98
role in regulation of sodium ion pump activity, blood vessels, 50–51
sinusoidal cells of, in liver, 252
stripping from small vessels, 340–41
Endothelium-derived hyperpolarizing factor (EDHF), role in relaxation, 88, 186, 229
Endothelium-derived relaxing factor (EDRF), 15–16, 186
adrenomedulin as, 232
nitric oxide as, 4–5
release of, and vasodilation by adenosine and ATP, 66
response to agonist activation of receptors, 75
Endotoxins
action on hepatocytes, 252–53
pulmonary hypertension induced by, effects of nitric oxide on, 149
role of nitric oxide synthase in pathology of, 86–87
Environmental safety, risks of inhaled nitric oxide, 177–78
Epileptic seizures, focal, nitric oxide as a mediator of vasodilation during, 121
Erectile function
androgen-dependent, 291
in diabetes, 60
role of nitric oxide in, 7, 57, 287
Estradiol-17β, uterine blood flow increase caused by, 272
Estrogens
effect on coronary blood flow, 191
effect on endothelial nitric oxide synthase, 273
effect on vascular beds, 191
response of uterine blood flow to, 271
Ethics, of inhaled nitric oxide therapy, 177
Ethinylestradiol, effects of blockade of inducible nitric oxide synthase on activity of, 272
Excitatory junctional potentials (EJPs), mediation by adenosine triphosphate, 67
Exhaled nitric oxide, evaluating abnormalities with, 174

Experimental procedure
effect of anesthetics in pulmonary circulation, 133–42
mechanism of hypoxic pulmonary vasoconstriction, 200–203
mechanism of penile erection, 294–95
Extracorporeal membrane oxygenation (ECMO), for persistent pulmonary hypertension of the newborn, 157–58

Felodipine, attenuation of hypertension in rats receiving L-NAME, 102
Fetal lung liquid production, 189–93
Five-coordinate complex, nitric oxide–heme, 4
Forskolin, activation of cGMP kinase by, 20–21

Gastrointestinal circulation, 259–69
Genitourinary tract function, lower, effect of nitric oxide on, 286–301
G kinase, effect on cerebral vascular tone, 116–17
Glibenclamide, effect on reactive hyperemia, 236
Glomerular blood pressure, after L-NAME administration, 332
Glomerular flow, effect of endogenous nitric oxide on, 310
Glucocorticoids, effect on nitric oxide synthase induction, 87
Glucose, effect of nitric oxide on metabolism of, 236–37
Glutamate
effect on neuronal nitric oxide synthase, 114
NMDA receptors for, and cerebral vasodilation, 121
Glybenclamide, 200
effect on baseline tone, pulmonary circulation, 203
effect on pulmonary perfusion pressure, 211–12
effect on response to ventilatory hypoxia, 203–5
Glycogenolysis, effect of nitric oxide on, 247

Glycolysis, mechanism of regulation by nitric oxide, 236–37

G proteins
P1 and P2 receptors coupled to, 65
P_{2T} receptors coupled to, effects on adenylate cylase activity, 74–75

Granulated metrial gland (GMG) cells, nitric oxide synthase isoforms in, 278

Guanine nucleotide regulatory (G) protein, role in inositol triphosphate formation, 19–21

Guanylate cyclase
activation in cerebral arteries, 90
activation by nitric oxide, 1–2
mechanism, 3–4
and activation of vascular smooth muscle cells, 198
expression in skeletal muscle, 237
and platelet aggregation, 77

Guyton's hypothesis, defining hypertension by, 100

Heart
congenital disease of, inhaled nitric oxide treatment in, 172
pulmonary hypertension after surgery, effect of inhaled nitric oxide on, 170
See also Cardio entries

Heme/hemoglobin
binding of nitric oxide by, 114, 198
of nitric oxide synthase, 306
role in activation of guanylate cyclase, 3–4, 16

Hemodynamics
cardiovascular, effects of acute nitric oxide synthase inhibition on, 100, 154
cerebral, effect of nitric oxide in, 113–27
regulation of, roles of neuronal nitric oxide, 94
renal
role of nitric oxide in, 321–22
role of nitric oxide synthase inhibition in, 100, 324–25
role of the vascular endothelium in, 50–51
See also Blood flow

Hemostasis, role of nitric oxide in, 6–7

Hepatic circulation, 243–58

Hepatocytes, 252–53

Histamine, vasoconstriction and vasodilation in response to, 230–31

Homeostasis, dependence on constant hepatic blood flow, 243

Hydroperoxyeicosatetraenoic acids, effect on activating protein 1 in smooth muscle cells, 35

13-Hydroperoxyoctadecadienoic acid, effect on expression of nitric oxide synthase, 34–35

Hypercapnia, cerebral vasodilation in, 117–18

Hyperemia
gastrointestinal, role of nitric oxide in, 260
reactive, role of nitric oxide in, 235–36

Hyperglycemia, effect on endothelium-dependent responses in blood vessels, 59–60

Hyperpolarization
effect on smooth muscle, 49, 234
maintenance of calcium ion influx by, 234–38
nitric oxide mediation of, in vascular smooth muscle cells, 21–22
due to sodium ion-potassium ion ATPase activity, 51

Hypertension
angiotensin II model, 39
effect of nitric oxide deficit, 326
increase in, effect of nitric oxide synthase selective inhibitors, 228
induction by nitric oxide synthase inhibitors, 93, 321–22
and nitric oxide, 99–109
pulmonary
in adult respiratory distress syndrome, 152–57
reducing, 148
salt-sensitive, 326–30
role of renal medullary nitric oxide, 329
superoxide ion production in, 43
treatment of, effect on endothelium-dependent relaxation, 89
See also Blood pressure

Hypotension
effect on cerebral prostanoid production, 119–20

sepsis-induced, treating with nitric oxide synthase inhibitors, 149–50

Hypoxemia, rebound, after inhaled nitric oxide treatment, 169

Hypoxia
ATP release from erythrocytes during, 67
cerebrovasodilation during, 118–19
effect of charybdotoxin on ventilatory response to, 215–17
effect of L158809 on response to, 209–13
effect of propofol on ventilatory response to, 131

Hypoxic pulmonary vasoconstriction, 199–200
mechanism of, 200–217

Hypoxic respiratory failure, in neonates, treatment with inhaled nitric oxide, 171

Idiopathic pulmonary hypertension, inhaled nitric oxide as treatment for, 199

Immune defenses
fetal, role of nitric oxide in, 278
role of inducible nitric oxide synthase in, 99

Immunohistochemistry, nitric oxide synthase, 293

Impotence
etiology of, 7
as a quality-of-life issue, 287
See also Erectile function

Indomethacin, effect of L-noradrenalin pressor response, 228

Inflammation, nitric oxide as an autoregulator of, in liver, 251

Inflammatory cells, interactions with endothelial cells, 260–63

Inhalation, lung injury caused by, effect of nitric oxide on, 150

Inhaled nitric oxide, 191–93
for selective reduction of pulmonary vasoconstriction, 198–99

Inositol phosphate (IP₃)
following activation of phospholipase C, 71
in gastric smooth muscle cells, 20
regulation of production of, 19–21
tissue-specific expression of receptors for, 21

Insulin, stimulation of sodium ion-potassium ion-ATPase activity by, 56

Interferon-1 (IFN-Δδ)
role in inducible nitric oxide synthase expression, 252
role in nitric oxide release in liver, 250

Intrauterine growth restriction (IUGR)
failure of vasodilator systems in, 274–75
increase in fetal-placental vascular impedance in, 275–76

Intravascular adhesion, role of nitric oxide in preventing, 260

Intravital studies
of FITC-albumin leakage from mesenteric venules, 264–65
of nitric oxide synthase inhibitors and nitric oxide donors, 261

Ischemia, reactive hyperemia following, 235–36

Isobutylmethylxanthine (IBMX), effect on sodium ion-potassium ion-ATPase activity, 52

Isoforms, of nitric oxide synthase, 227

Isoproterenol, vasodilator responses to, effect of charybdotoxin on, 213–15

Kallidin, vasodilator responses to, 229

Ketamine
cerebrovasodilation by, 132–33
experimental determination of pulmonary effects of, 133–43

Kidney, damage to, secondary to nitric oxide deficiency, 103. See also Renal entries

Kinases, cGMP
mechanism of reduction of calcium ion concentration by, 18
in smooth muscle cells, 17–18

Kinins, vasoactive responses to, 229

Kupffer cells, 253–54
inducible nitric oxide synthase produced by, 250–51

L158809
angiotensin II type I receptor antagonist, 200–201
effects on response to hypoxia, 209–13

Labor, inducible nitric oxide synthase activity during, 279
Laboratory models, evaluation of inhaled nitric oxide, 148–52
Leukocytes
    aggregation of
        inhibition by nitric oxide, 148–49
        role of nitric oxide, 260
    in vitro studies of, 261
    effects on, by nitric oxide, 261
Leukotriene B_4, role in gastrointestinal circulation, 262–63
Levcromakalim
    effect on blood vessel tone, response to L-NAME and glybenclamide, 207–9
    effect on pulmonary arterial pressure, 137
Ligand-gated ion channels, P2 purinergic receptors as, 65
Linsidomine chlorhydrate (SIN-1), effect on erectile function, 291
Lipid mediators, role in cell adhesion, 262–63
    leukocyte-endothelial cells, 262–63
Lipopolysaccharide (LPS)
    effects on hepatocytes, 252–53
    effects on vascular tissue, 39
        in pregnant animals, 279
Liver
    cell types in, 249–53
    chronic disease of, capillarization in, 244–45
    hepatic circulation, 243–58
    regeneration of, role of vascular nitric oxide in, 248–49
    See also Hepatic entries
L-NAME
    effect on baseline tone, pulmonary circulation, 203
    effect in cat intestine, 264
    effect in hypoxic pulmonary vasoconstriction, 200–217
    effect on responses to angiotensin II and serotonin, pulmonary circulation, 205–7
    effect on response to nerve stimulation, 69–70
    effect on response to ventilatory hypoxia, 203–5

effect on uterine blood flow, 272–73
effect in ventilatory hypoxia, 209–11
inhibition of nitric oxide synthase by, 76
Lung
    inflation of, and effects of inhaled nitric oxide in lung injury, 150
    inhaled nitric oxide therapy after surgery on, 170–71
    oxidant induced injury to, effects of nitric oxide on, 150–51
    See also Pulmonary entries
LY83583, effect on response to ventilatory hypoxia, 203–5

Macrophages, Kupffer cell, 251
Macula densa
    feedback in, 332
    neuronal nitric oxide synthase of, effect on afferent artery, 310–11, 343–44
Mast cells, role in cell adhesion, 262
Mechanism
    of hypoxic vasoconstriction, 200–217
    of nitric oxide synthase action, 37–39
        influence of oxidant processes on, 37–39
    of nitroglycerin action, 2–3
    permissive, in control of cerebral circulation, role of nitric oxide, 118
    pressure-natriuresis, for arterial pressure regulation, 320
    of pulmonary vasodilation by inhaled nitric oxide, 167
    of tolerance to vasorelaxants, 3
Meclofenamate, effect on vasodilation by propofol, ketamine and arachidonic acid, 141
Menopause, cardiovascular health during, 273–74
Methemoglobin, oxidation of nitrosylhemoglobin to, 198
Methemoglobinemia, from inhaled nitric oxide therapy, 175
Methemoglobin reductase, 198
S-Methyl-L-citrulline, evaluating renal microvascular function using, 343–44
Michaelis constant, for GTP, effects of nitric oxide on, 4

Microvascular responses, renal, 337–46
    to nitric oxide synthase, 309–10
Modulation, of tubuloglomerular feedback,
    310–11
Monoclonal antibodies
    effect on L-NAME-induced albumin leak-
        age from mesenteric venules, 265
    prevention of L-NAME-induced oxidant
        stress, 263
Muscle
    contractile protein force generation in
        fibers, 23
    role of nitric oxide in contractile func-
        tion of, 237
    role of nitric oxide in metabolism in,
        236–37
Myenteric plexus, cerebral, colocalization
        of vasoactive intestinal peptide and
        nitric oxide in, 290
Myocardium, damage to, secondary to ni-
        tric oxide deficiency, 103–4
Myocytes
    cardiac, effect of cGMP on calcium ion
        in, 22–23
    effect of ketamine on, 133
Myometrium
    nitric oxide levels in, in pregnancy,
        274–75
    regulation of contractility of, 276–80
Myosin light chains, dephosphorylation of,
        and relaxation of vascular smooth
        muscle cells, 198

NADH oxidase
    effect on vascular nitric oxide signal-
        ing, 39
    results of prolonged hypertension
        on, 43
NADPH diaphorase histochemistry, 292
Natriuresis
    role of adrenomedulin in, 232
    role of nitric oxide in, 102–3
    See also Renal entries
Nerves
    conduction of, in diabetes, 60
    renal, effect on hemodynamics follow-
        ing nitric oxide synthase inhibition,
        324–26

source of nitric oxide in control of cere-
        bral hemodynamics, 120–21
    vasodilator, 85–98
    See also Perivascular nerves/neurons
Neuromodulation, by nitric oxide, 246–47
Neuropeptide Y, nitric oxide synthase colo-
        calization with, 290
Neurotransmitter
    purinergic release of, modulation by ni-
        tric oxide, 66
Neurotransmitter, nitric oxide as, 120–21
    in penile erection, 286–301
Newborns, risks of inhaled nitric oxide
        therapy in, 175–76
Nicotine, effect on cerebral artery relax-
        ation, 89–90
Nifedipine, inhibition of calcium ion chan-
        nels by, 22–23
    and attenuation of hypertension, 102
Nitrate, excretion of, and salt intakes,
        327–30
Nitric oxide synthase (NOS), 305–6
    activation by calcium ion-calmodulin
        complex, 15–16
    constitutive, 227
        calcium ion-dependent, 86, 259–60
        endothelial, 114
        endothelial and brain, 99–109
        renal vasculature, 338
        role in relaxation of vascular smooth
            muscle, 197–98
    inducible, 227, 305–6
        association with sinusoidal endothe-
            lial cells, 252
        damage from nitric oxide produced
            by, 266
        response to cytokines, 86
        response to cytokines in hepato-
            cytes, 252
        role in immune defense, 99–109
        role in salt-sensitive hypertension,
            330–31
        in stellate cells, 250
        synthesis by Kupffer cells, 251
    in the kidney, 305–6
        distribution of, 321
        mechanisms of action, 187
        influence of oxidant processes on,
            37–39

Nitric oxide synthase (*cont.*)
  neuronal, 305–6
    as a source of cerebral nitric oxide, 114
    role in salt-sensitive hypertension,
      331–32
    in the macula densa, 338
  in platelets, 77
  role in liver regeneration, 248–49
Nitric oxide synthase (NOS) inhibitors
  effect on systemic blood pressure, 6–7
  effects in the kidneys, 339–40
  interaction with angiotensin II, 312, 324
  L-NAME, 272
    effect on blood vessel constriction, 76
Nitrogen dioxide (NO$_2$), from oxidation of
      nitric oxide, toxicity of, 175
Nitroglycerin
  effect on preterm and active labor, 280
  effects on blood vessel tone, response to
      L-NAME and glybenclamide,
      207–9
  as a vascular smooth muscle relaxant,
      1–3
  vasodilator responses to, effect of
      charybdotoxin on, 213–15
S-Nitrosohemoglobin, release of nitric
      oxide from, in hypercapnia, 118
S-Nitrosothiols
  as nitric oxide donor agents, 2
  vasodilation by, 309
Nitrotyrosine residues, in placentas, in
      complicated pregnancies, 276
Nitrovasodilators
  historic use of, 309
  nitroprusside
    effect of charybdotoxin on, 213–15
    renal responses to, 338–39
  See also Nitroglycerin; Vasodilation
Nitroxidergic nerve, 90–91
  regulation of vascular tone in peripheral
      arteries involving, 92–93
N-Methyl-D-aspartate (NMDA) receptors
  cerebral vasodilation induced by, 121
  regulation of activity by nitric oxide, 53
N^G-monomethyl-L-arginine (L-NMMA)
  contractile response enhancement by, 76
  effect of
    on nitric oxide release, enantiomer-
      specific, 86

  on sodium ion-potassium ion-ATPase
      activity, 55, 86
  mediation hypertensive response to
      L-NAME and L-NA in, 228–29
N^G-nitro-L-arginine (L-NA)
  inhibition of nitric oxide synthase by, 76
  pressor response to, 228–29
N^G-nitro-L-arginine benzylester (L-NABE),
      effect of vasodilator responses, 139
N^G-nitro-L-arginine methyl ester (L-NAME).
      See L-NAME
Nociceptin, effect on penile erection, 287
Nociception, role for P2X receptors in, 74
Nonadrenergic, noncholinergic (NANC)
      nerves, in the cerebral arteries,
      89–90
Nonadrenergic, noncholinergic (NANC)
      neurotransmitter, role in erectile
      function, 287
Nonadrenergic, noncholinergic (NANC)
      relaxation, neurogenic, mediation
      by adenosine triphosphate, 68–69
Noradrenergic vasoconstrictor, regulation
      of vascular tone in peripheral arter-
      ies involving, 92–93
Nucleophile/nitric oxide complexes, re-
      lease of nitric oxide from, 292

Okadaic acid, blocking of hyperpolariza-
      tion by, 22
Oleic acid, lung injury induced by,
      149–50
Opioids, role in vasodilation during hypox-
      emia, 119
Organ specificity
  of nitric oxide regulation of vasocon-
      striction, 247
  of response to shear stress, 245
Ouabain
  effect on penile erection, 57
  inhibition of nitric oxide relaxation by, 51
Oxidant processes
  effect on mechanisms of action of nitric
      oxide, 37–39
  and the stability of nitric oxide, 35–37
Oxidants, vascular nitric oxide signaling
  effects of, 33–48
  sources of, 39–43

Oxygen free radicals, scavenging of, 148–49

Oxyhemoglobin, reaction of nitric oxide with, 35

P1 receptors
  types of, 65
P2 receptors, types of, 65
Paracrine system
  effects of nitric oxide
    in the kidney, 312
    in liver, 249–50
  role in control of cerebral vascular tone, 116–17
Pathophysiology, vascular
  effects of oxidants and nitric oxide in, 43
  in release of ATP, ADP and UTP from platelets, 67
Pediatric patients, treating with inhaled nitric oxide, 171–73
Penis
  nitric oxide synthase in, 288–90
  role of nitric oxide in, 288–89
  See also Erectile function
Pericytes, stellate cells as, 249
Perinatal pulmonary circulation, 185–96
Perivascular nerves/neurons
  effects on cerebral arteries, 89–90
  nitric oxide derived from, 85–98, 114
  purinergic receptors on, 74
Permissive mechanism, in control of cerebral circulation, 116–18
Peroxides, effects in endothelium-dependent relaxation, 33–48
Peroxynitrite anion
  effects on enzymes involved in energy metabolism, 39
  exacerbation of cellular injury by, 151
  formation of, 5
    by high concentrations of vasodilators, 16
    in the placenta, 276
    role in nitric oxide synthase signaling, 34
Persistent pulmonary hypertension of the newborn (PPHN), 157–59
  nitric oxide treatment for, 171, 199

Phenylephrine, effect on sodium ion-potassium ion-ATPase activity, 55
Phorbol 12-13, dibutyrate (PDBu), 55
Phosphodiesterases
  cGMP regulation by, 16
  cGMP-specific, 151–52
  inhibitors of
    effects on pulmonary smooth muscle relaxation, 188–89
    effect on responses to histamine, 232
    regulation of cGMP levels by, 17
Phospholamban, phosphorylation of, 19
Phospholipase C (PLC)
  activation by purinergic receptors P2Y$_1$ and P2Y$_2$, 71
  interaction with G protein, 19–20
Placenta, blood flow to, factors affecting, 275–76
Platelet activating factor (PAF)
  effect on afferent arteriolar vasodilation, kidneys, 341
  role in gastrointestinal circulation, 262–63
Platelet aggregation
  inhibition by nitric oxide, 1–2, 5, 76–77, 148–49, 260
  modulation by nitric oxide, 66
Platelets
  purinergic receptors on, 74–75
  storage of ATP, ADP and UTP in, 67
Portal vein, physiology of, 244–45
Portal vein, physiology of, 244–45
Potassium ion-ATP channel, inhibition of activity by ketamine, 133
Potassium ion channels
  calcium ion activation of, 21–22
  independence of cGMP, 53–54
  effect of histamine in opening, 231
  inhibition of, in hypoxia, 213–14, 216
  nitric oxide regulation of, 38–39, 49
  role in agonist-induced nitric oxide release, 234–38
  role in mediating propofol responses, 132
Prazosin, effect of L-noradrenalin pressor response, 228
Preeclampsia
  failure of vasodilator systems in, 274–75
  increase in fetal-placental vascular impedance in, 275–76

Pregnancy, calcium ion-dependent nitric
    oxide synthase activity in, cardio-
    vascular challenge of, 274–75
Pregnancy, calcium ion-dependent nitric
    oxide synthase activity in, 191
Pressor response, to hypoxia, ketamine ef-
    fect on, 136
Pressure-flow autoregulation, in the gas-
    trointestinal circulation, 260
Priapism, idiopathic recurrent venocclu-
    sive, treating with digoxin, 58–59
Primary pulmonary hypertension
    (PPHT), effect of nitric oxide
    treatment on, 169
Proadrenomedullin NH_2-terminal 20 pep-
    tide (PAMP), 233–34
Progesterone, effect on cGMP-induced re-
    laxation, myometrium, 277
Proliferating factors, expression of, after
    partial hepatectomy, 248–49
Propofol, pulmonary effects of, 131
    experimental determination, 133–43
Prostacyclin (PGI_2)
    contribution to control of cerebral vas-
        cular tone, 116–17
    production of
        in the arachidonic acid metabolic
            pathway, 186–87
        in the nonpregnant uterus, 270
    role in relaxation of blood vessels,
        88–89
    role in vasodilation, 66, 88
        in the lung, 186–87
        in the umbilical cord, 275–76
        uterine, 273
Prostaglandin E_1 (PGE_1), effect on erectile
    response, 291
Prostaglandin E_2 (PGE_2)
    effect on stellate cells, 249
Prostaglandin $F_{2\alpha}$ ($PGF_{2\alpha}$), effect on stel-
    late cells, 249
Prostaglandins (PGs)
    production by vascular tissue, 38
    role in platelet aggregation, 75
Prostanoids
    contribution to control of cerebral vas-
        cular tone, 116–17
    vasodilating
        effect of acetylcholine on, 88

production caused by hypotension,
    119–20
Protein kinase, 49–64
    cGMP-dependent, 17–18, 49
    regulation by calcium ion, 15–32
Protein kinase C (PKC), activation by dia-
    cylglycerol, 71
P-selectin, effect of nitric oxide on expres-
    sion of, 262
Pulmonary arterial pressure
    in adult respiratory distress syndrome,
        effect of nitric oxide on, 154
    effect of ketamine and propofol
        on, 136
    effect of levcromakalim on, 137
Pulmonary circulation
    adult, and nitric oxide, 197–223
    details of, 199–217
    effects of nitric oxide on, 187–89
    intravenous anesthetics in, 131–47
    perinatal, and nitric oxide, 185–96
Pulmonary edema
    after endotoxin administration, effect of
        nitric oxide on, 149
    after inhaled nitric oxide evaluation of
        pulmonary hypertension, 173
Pulmonary fibrosis, treating with inhaled
    nitric oxide, 169–70
Pulmonary hypertension
    evaluating reversibility of, with inhaled
        nitric oxide, 173
    reducing, 148
    in respiratory distress syndrome
        adult, 152–57
        severe, 171
    after surgery
        heart, 170
        in pediatric patients, 172
    See also Hypertension
Pulmonary vascular bed, rat, experimental
    determination of the effects of
    propofol and ketamine on, 133–43
Pulmonary vascular resistance, effects of
    nitric oxide on, 166–67, 169–70
Pulmonary vascular tone
    hypoxic vasoconstriction
        defined, 199–200
        mechanism of, 200–217
        mediation by nitric oxide, 190–93

vasodilation, effect on lung liquid production, 189–93
Purinergic receptors
    effect on blood flow, 65–84
    P2, vasoconstriction or vasospasm associated with, 67
Purines
    in blood vessels, 66–69
    nitric oxide modulation of vasoconstrictor responses to, 75
Pyrilamine, effect on histamine-induced vasodilation, 238
Pyrimidines, in blood vessels, 66–69

Reactive oxygen metabolites (ROMs), role in leukocyte-endothelial cell adhesion, 263
Receptors
    adenosine, on sympathetic and sensorimotor nerves, 74
    for estrogen, 272
    heteromeric P2X$_{2/3}$, 74
    purinergic P2Y and P2Y$_1$, 73
Recombinant smooth muscle receptors, purinergic P2X$_1$, 72–73
Relaxation
    attenuation in the presence of superoxide, 37
    digoxin inhibition of, in vascular smooth muscle cells, 58
    mediation by nitric oxide
        in penile corporal smooth muscle, 288
        in vascular smooth muscle cells, 21–22
    mediation by smooth muscle P2Y receptors, 73
    neurogenic
        of canine cerebral arteries, 85–98
        nonadrenergic, noncholinergic, mediated by adenosine triphosphate, 68–69
    nitric oxide-induced, 49–64
    responses to ketamine, 133
    See also Vasodilation; Vasorelaxation
Renal function
    protection of, with felodipine, 102
    regulation in conscious animals, 320–36

Renal hemodynamics
    effects of acute nitric oxide synthase inhibition on, 100
    regulation of blood flow, 305–19
Renal segmental resistances, regulation of, 332
Renal vascular resistance, 306–9
    effect of noradrenalin infusion on, 228
Renin, effect of L-NAME on activity, 325–26
Renin-angiotensin system
    activation of, in nitric oxide deficiency, 104
    effect on, of angiotensin blockade in CNOSI, 101
    mediation of hypertensive response to L-NAME and L-NA in, 228–29
Reperfusion, of ischemic tissues, 260
Reproductive tract, male, nitric oxide in, 294
Respiration, in tissues, regulation by nitric oxide, 38. See also Pulmonary entries
Respiratory distress syndrome (hyaline membrane disease), effect of nitric oxide treatment in, 159, 192
Respiratory failure
    acute, nitric oxide therapy for, 148–65
    neonatal, nitric oxide therapy for, 157–59

Salt retention, effect of nitric oxide in, 102–3
Sarcoplasmic reticulum
    calcium ion release by, inhibition of, 20–21
    localization of cGMP kinase in, 20
    roles of nitric oxide in, 237
    uptake of calcium ion by, 18–19
Second messenger, endogenous nitric oxide as, 286–88
Selectivity, of inhaled nitric oxide, 151–52
Sepsis, lung, signal transduction downregulation in, 154, 156
Serine threonine kinases, cGMP activation of, 17
Serotonin, effects of L-NAME on responses to, pulmonary circulation, 205–7

Severe respiratory distress syndrome
(SRDS), in pediatric patients, 171
Shear stress
  ATP release from erythrocytes
    during, 67
  correlation with basal release of nitric
    oxide, 228, 253, 260
  at the endothelial cell surface, 6–7, 115
    in the placental vasculature,
    275–76
  and renal vascular tone, 311–12
  from vasoconstriction, release of nitric
    oxide by, 246–47
  wall, rate in postcapillary venules, 261
Shock, endotoxin, role of inducible nitric
  oxide synthase in, 86–87
Sickle cell disease, acute chest syndrome
  associated with, 173
Side effects, of inhaled nitric oxide therapy, 192–93
Signaling systems, vascular nitric oxide,
  effects of oxidants on, 33–48
Sinusoidal endothelial cells, 252
Skeletal muscle vascular bed, role of nitric
  oxide in, 227–40
Smooth muscle, penile corporal, relaxation
  of, 288
Smooth muscle tone
  myometrial, regulation of, 276–80
  regulation by the vascular endothelium,
    50–51, 259–60
S-Nitroso-N-acetylpenicillamine
  (SNAP), 309
  effect on renal blood flow, 338
  effect of superoxide dismutase inhibitors
    on, 35–36
Sodium intake, and nitric oxide synthase in
  the renal inner medulla, 329
Sodium ion-potassium ion-ATPase
  role in nitric oxide-induced relaxation,
    49–64
  stimulation of activity by nitric oxide,
    51–53
Sodium ion pump
  effect in diabetes and hyperglycemia, 60
  inhibitor of, for treating idiopathic re-
    current venocclusive priapism,
    58–59

Specificity
  of nitric oxide synthase, 288
  organ
    of nitric oxide regulation of vasocon-
      striction, 247
    of response to shear stress, 245
Stellate cells, 249–50, 253–54
Stress fibers, formation from stellate cells
  in liver injury, 250
Sulfhydryl groups, and stimulation of
  sodium ion-potassium ion-ATPase
  activity, 53–54
Superoxide
  cytotoxicity from interaction with nitric
    oxide, 251
  nitric oxide as a scavenger for, 151
  production catalyzed by nitric oxide
    synthase, 33–48
  and toxicity of inhaled nitric oxide, 175
Superoxide dismutase (SOD)
  effect on inflammatory responses to
    L-NAME, 263
  effect on nitric oxide in vitro, 261
  prolonging half-life of nitric oxide
    with, 5
  rate of reaction with superoxide, 35
Sympathetic nervous system
  activity of, and hypertension, 101–2
  mediation of short-term effect of nitric
    oxide synthase inhibition by, 324
Systemic arterial pressure
  effect of ketamine on, 132–33
  effects of nitric oxide donors
    on, 296

Tetraethylammonium (TEA), inhibition
  of potassium ion channels by,
  234–35
Tetrahydrobiopterin, effect on superoxide
  production, 33, 42
Thioperamide, effect on histamine-induced
  vasodilation, 238
Thromboxane receptor (TR), contribution
  to control of cerebral vascular tone,
  116–17
Tolerance, to vasorelaxants, mechanisms, 3
Toxicity, of inhaled nitric oxide, 174–78

Transmural electrical stimulation (TES)
  effect in cerebral arteries, 90
  effect in peripheral arteries, 90–91
  effect in temporal arteries, 92
Transmural pressure, effect on rate of pro-
       duction of nitric oxide, 260
Treatment
  nitric oxide therapy for acute respiratory
       failure, 148–65
  for sepsis-induced hypotension, 149–50
Tubuloglomerular feedback (TGF),
       310–11, 332, 343–44
Tumor necrosis factor α, role in nitric
       oxide production in hepatocytes,
       252–53

U-37883A, effect on reactive hyperemia, 236
Umbilical blood flow, regulation of,
       275–76
Urethra, nitric oxide in, 293
Uridine nucleotide-specific receptors, 73
Uterine blood flow, effect of constitutive
       nitric oxide synthase on, 87
Uterus, effects of nitric oxide on, 270–85

Vascular endothelium
  nitric oxide derived from, 86–89
  purinergic receptors on, 69–72
Vascular permeability, effect of L-NAME
       on, 264
Vascular physiology, hepatic, 244–45
Vascular resistance
  pulmonary, effects of nitric oxide on,
       166–67, 169–70
  renal, 306–9
Vascular smooth muscle
  cyclic guanosine monophosphate re-
       sponse to nitric oxide inhalation
       in, 150
  effect of nitric oxide on sodium ion
       pump activity in, 51
  effects of insulin on, 56
  modulation of contractile responses by
       nitric oxide, 66
  nitric oxide and cGMP signaling in,
       15–16, 197–98

purinergic receptors on, 72–73
  regulation of tone by nitric oxide, 227–29
  role of cGMP in, 16–17
  relaxation of, 5–7
  uterine, 271
  type of guanylate cyclases in cells of, 16
Vasculature, uterine
  nonpregnant, 270–73
  pregnant, 274–75
Vasoactive agonists
  microvascular responses to, 341–43
  stimulation of sodium ion-potassium
       ion-ATPase activity by, 54–56
Vasoactive intestinal peptide (VIP), colocal-
       ization with nitric oxide synthase, 290
Vasoactive substances, effects on nitric
       oxide and prostacyclin produc-
       tion, 271
Vasoconstriction
  interaction of nitric oxide with com-
       pound causing, 323–26
  interaction with nitric oxide, 325–26
  in the liver, shear stress-dependent mod-
       ulation of, 247
  unopposed, in the chronic nitric oxide
       inhibition model, 101
Vasodilation
  agonists to, mediation of responses by
       nitric oxide, 229–34
  cortical, mediation by nitric oxide, 121
  by ketamine, 133
  by propofol, 131–32
  pulmonary
       by inhaled nitric oxide, 148–49
       by inhaled nitric oxide prolonged by
         zaprinast, 152
  See also Nitrovasodilators
Vasodilator nerves, nitric oxide-mediated,
       85–98
Vasopressin, mediation of short-term effect
       of nitric oxide synthase inhibition
       by, 324, 342–43
Vasorelaxation
  cyclic guanine monophosphate induc-
       tion of, 51
  property of nitric oxide, 1–2
Ventilation, partial liquid, combination with
       inhaled nitric oxide therapy, 152

Ventilatory hypoxia, effect of charybdo-
    toxin on, 215–17
Verapamil
    effect on ketamine-induced vasodila-
        tion, 144
    inhibition of calcium ion channels by,
        22–23
    and hypertension reversal, 102
Vimentin, reversible binding to cGMP ki-
    nase, 20

Xanthine oxidase inhibitor, effect on en-
    dothelial function in hypercholes-
    terolemia, 43

Zaprinast, effect of
    on cyclic guanosine monophosphate,
        151–52, 293
    on pulmonary arterial pressure and vas-
        cular resistance, 153

The manufacturer's authorised representative in the EU is Springer
Nature Customer Service Centre GmbH, Europaplatz 3, 69115 Heidelberg,
Germany. If you have any concerns regarding our products, please
contact ProductSafety@springernature.com

Printed and bound by CPI Group (UK) Ltd, Croydon, CR0 4YY
23/04/2026
02095593-0005